Graphics Technology

Graphics Technology

Second Edition
AutoCAD® 2005

James H. Earle

Professor Emeritus
Texas A&M University

PEARSON
Prentice
Hall

Upper Saddle River, New Jersey
Columbus, Ohio

Library of Congress Cataloging-in-Publication Data

Earle, James H.
  Graphics technology : AutoCAD 2005 / James H. Earle.--2nd ed.
    p. cm.
  Includes index.
  ISBN 0-13-147643-2
    1. Engineering graphics. 2. AutoCAD. I. Title.

T353.E35 2005
620′.0042′0285536--dc22

2004050279

Executive Editor: Debbie Yarnell
Managing Editor: Judith Casillo
Editorial Assistant: ReeAnne Davies
Production Editor: Louise N. Sette
Production Coordination: Lisa Garboski, *bookworks*

Design Coordinator: Diane Ernsberger
Cover Designer: Kristina Holmes
Cover art: Conoco Phillips
Production Manager: Deidra M. Schwartz
Marketing Manager: Jimmy Stephens

This book was set in Utopia by *The GTS Companies*/ York, PA Campus. It was printed and bound by Courier Kendallville, Inc. The cover was printed by Phoenix Color Corp.

Certain images and materials contained in this publication were reproduced with the permission of Autodesk, Inc., © 2004. All rights reserved.

AutoCAD® is a registered trademark of Autodesk, Inc., 111 McInnis Parkway, San Rafael, CA 94903.
Windows is a registered trademark of the Microsoft Corporation, One Microsoft Way, Redmond, WA 98052-6399.

Pearson Education Ltd.
Pearson Education Singapore Pte. Ltd.
Pearson Education Canada, Ltd.
Pearson Education—Japan

Pearson Education Australia Pty. Limited
Pearson Education North Asia Ltd.
Pearson Educación de Mexico, S.A. de C.V.
Pearson Education Malaysia Pte. Ltd.

10 9 8 7 6 5
ISBN: 0-13-147643-2

Dedicated to
Theresa Earle

Preface

New and Better

This second edition of *Graphics Technology* is an improved version of the 1995 edition in all categories: content, format, readability, clarity, quality of illustrations, and economy. The challenge with each revision has always been, "How can the book be written and illustrated to make it easier for the student to learn and the teacher to teach?" Also asked is, "What should be the content for today's course that will fit tomorrow's needs?"

Meeting this goal is difficult for an author in any discipline, but it is especially awesome in the area of engineering design graphics where 1500 illustrations, 620 problems, and a multitude of topics must be merged into a cohesive textbook as compactly as possible. Content must include fundamentals, design, computer graphics, industrial applications, and meaningful problems. We believe that we have met these goals.

Classic Content

Every paragraph and illustration has been revisited for evaluation, revision, improvement, or elimination so that all content makes a worthwhile contribution. No space has been squandered to make room for exotic illustrations of examples that are beyond the scope of a beginning freshman course. Instead, that valuable space has been used to better illustrate and present fundamental concepts in an understandable format to reduce the amount of classroom tutoring needed by the student.

Major content areas covered in this text are:

- design and creativity,

- computer graphics,

- engineering drawing,

- descriptive geometry, and

- problem solving.

Design and Creativity

Chapter 2, which is devoted to the introduction of design and creativity, has been revised and improved. New design examples from industry, along with examples of worksheets applying the steps of design, guide the student through the process of design.

Care has been taken to offer realistic design problems that are within the grasp of beginning students rather than overwhelming them with projects beyond their capabilities. Since the primary objective of design instruction is to introduce the process of design, meaningful design assignments are given to make the process fun and to encourage the application of creativity and intuition.

A variety of design problems are given in Chapter 2 that can be used as quickie problems, short assignments, or semester-long design projects. Additional design exercises are included at the ends of the chapters throughout the book.

Computer Graphics

An introduction to AutoCAD® 2005 is presented in a step-by-step format to aid the student in learning how to use this popular software, not just read about it in the abstract. Steps of each illustration show the reader what will be seen on the screen as the example is followed. Chapter 25 gives an introduction to two-dimensional computer graphics by AutoCAD.

The main purpose of *Graphics Technology* is to help students learn the fundamental principles of graphics, whether done on the drawing board or on the computer. This book can be used in courses where the entire course is taken on the computer, where part of the course is taken on the computer, or where none of the course is taken on the computer.

Format: Make It Easy

Much effort has been devoted to the creation of illustrations separated into multiple steps to present the concepts as clearly and simply as possible. A second color is applied as a functional means of emphasizing sequential steps, key points, and explanations, not simply as decoration. Explanatory information and text are closely associated with the steps of each example.

Many three-dimensional pictorials have been drawn, modified, and refined that will aid the student in visualizing the example at hand. Photographs of actual industrial parts and products have been merged with explanatory examples of principles being covered.

The illustrations in this book have been developed and drawn by the author, not by illustrators who have never had the experience of trying to explain a concept to a student. Only after years of classroom practice and trial-and-error testing is it possible to cover principles of graphics in a format that enables the student and teacher to cover the most content with fewest learning obstacles.

The two-color, step-by-step format of presentation with conveniently located text has been classroom tested to validate its effectiveness over a number of years. Although substantially better than the conventional format used in other texts, this format required about twice as much effort by the author and twice as much expense by the publisher to produce.

We hope you agree that the results justify the added effort and expense.

Streamlined

The content in all chapters has been compressed, but no material essential to the adequate coverage of a topic has been eliminated. Chapters on gears and cams, nomography, empirical equations, pipe drafting, and electronic graphics have been eliminated to save over 100 pages, thus providing this book at a lower cost for programs not covering this content.

Many new problems and illustrations have been added and most of the existing figures have been edited and improved to make them more effective. No space in this book has been wasted.

A Book to Keep

Some material in this book may not be formally covered in the course for which it was adopted because of time limitations or variations in emphasis by different instructors. These briefly covered topics may be the ones that will be needed in later courses or in practice. Therefore, this book should be retained as permanent reference by the engineer, technologist, or technician.

A Teaching System

Graphics Technology used in combination with the supplements listed below comprises a complete teaching system.

Textbook problems: Approximately 620 problems are offered to aid the student in mastering the principles of graphics and design.

Instructor's solution manual: A manual containing the solutions to most of the problems in this book is available from Prentice Hall to assist the teacher with grading.

Problem manuals: Nineteen problem books and teachers' guides (with outlines, problem solutions, tests, and test solutions) that are keyed to this book are available from Creative Publishing. Fifteen of the problem books are designed to allow problem solution by computer, by sketching, or on the drawing board. A list of these manuals is given inside the back cover of this book.

Acknowledgments

We are grateful for the assistance of many who have influenced the development of this volume. Many industries have furnished photographs, drawings, and applications that have been acknowledged in the corresponding legends. The Engineering Design Graphics staff of Texas A&M University has been helpful in making suggestions for the revision of this book. Professor Tom Pollock provided valuable information on metallurgy for Chapter 11. Professor Leendert Kersten of the University of Nebraska, Lincoln, kindly provided his descriptive geometry computer programs for inclusion; his cooperation is appreciated.

We are indebted to Jimm Meloy and Denis Cadu of Autodesk, Inc. for their assistance and cooperation. We are appreciative of the assistance of David Ratner of the Biomechanics Corporation, Inc., for providing HUMANCAD® software. Also giving support were Michael Flanagan of Bresslergroup, and Henry Keck of Keck-Craig, Inc.

We are appreciative of the fine editorial and production team assembled by Eric Svendsen and Debbie Yarnell at Prentice Hall. Lisa Garboski completed a complex project in a very skillful manner. Jill Horton scanned and translated the illustrations into final form for printing. Pat Wilson sorted through every word, comma, semicolon, and character to make the text as error-free as possible. I am appreciative to all of you.

Thank You

Above all we appreciate the many institutions who have thought enough of our publications to adopt them for classroom use. This is the highest honor that can be paid an author. We are hopeful that this textbook will fill the needs of engineering and technology programs. As always, comments and suggestions for improvement and revision will be appreciated.

Jim Earle
College Station, Texas

Contents

Graphics Technology

1

Engineering and Technology

1.1 Introduction

This book deals with the field of engineering design graphics and its application to the design process. Engineering graphics is the primary medium of design. Essentially all designs begin with graphics and end with graphical documents from which products and projects become realities. The solution of most engineering problems requires a combination of organization, analysis, problem-solving principles, graphics, skill, and communication (**Figure 1.1**).

This book is intended to help you use your creativity and develop your imagination because innovation is essential to a successful career in engineering and technology. Albert Einstein said, "Imagination is more important than knowledge, for knowledge is limited, whereas imagination embraces the entire world . . . stimulating progress, or, giving birth to evolution."

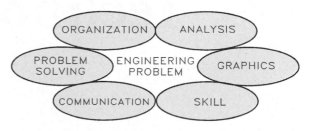

1.1 This text illustrates the total approach to engineering, with the engineering problem as the focal point.

1.2 Engineering Graphics

Engineering graphics covers the total field of graphical problem solving within two major areas of specialization: **descriptive geometry** and **documentation drawings.** Other areas of application are nomography, graphical mathematics, empirical equations, technical illustration, vector graphics, data analysis, and computer graphics. Graphics is one of the designer's most effective tools for developing design concepts and solving three-dimensional problems. It also is the designer's best means of communicating ideas to others.

MILESTONES OF THE 20TH CENTURY			
1900	Vacuum cleaner	1950	A—bomb tests
	Airplane		Optical fibers
	Dial telephone		Soviet satelite
	Light bulb		Microchip
	Model T Ford	1960	Commun. satelite
1910	Washing machine		Indus. robot
	Refrigerator		Nuclear reactor
	Wireless phone		Heart transplant
1920	Radio broadcasts		Man on moon
	Telephone service	1970	Silicon chip
	35mm camera		Personal computer
	Cartoons & sound		Videocassette record.
1930	Tape recorder		Supersonic jet
	Atom split		Neutron bomb
	Jet engine	1980	Stealth bomber
	Television		Space shuttle
1940	Elect. computer		Artificial heart
	Missile		Soviet space station
	Transistor	1990	Computer voice
	Microwave		recoginition
	Polaroid camera		Artificial intelligence
			E—Mail and the
			internet

1.2 This chronology lists some of the significant technological advances of the twentieth century.

Descriptive Geometry

Gaspard Monge (1746–1818), the "father of descriptive geometry," used graphical methods to solve design problems related to fortifications and battlements while a military student in France. His headmaster scolded him for not using the usual long, tedious mathematical process. Only after lengthy explanations and demonstrations of his technique was he able to convince his faculty that graphical methods (now called descriptive geometry) produced solutions in less time.

Descriptive geometry was such an improvement over mathematical methods that it was kept as a military secret for fifteen years before the authorities allowed it to be taught as part of the civilian curriculum.

Descriptive geometry is the projection of three-dimensional figures onto the two-dimensional plane of paper in a manner that allows geometric manipulations to determine lengths, angles, shapes, and other geometric information about the figures.

1.3 Technological Milestones

Many of the technological advancements of the twentieth century are engineering achieve-

1.3 Technological and design team members, with their varying experiences and areas of expertise, must communicate and interact with each other. *(Courtesy of Toshiba America Information Systems, Inc.)*

ments. Since 1900, technology has taken us from the horse-drawn carriage to the moon and back and more advancements are certain in the future.

Figure 1.2 shows a few of the many technological mileposts since 1900. It identifies products and processes that have provided millions of jobs and a better way of life for all. Other significant achievements were building a railroad from Nebraska to California that met at Promontory Point, Utah, in 1869 in less than four years; constructing the Empire State Building with 102 floors in thirteen and a half months in 1931; and retooling industry in 1942 for World War II to produce each day 4.5 naval vessels, 3.7 cargo ships, 203 airplanes, and 6 tanks, while supporting 15 million Americans in the armed forces.

One "miracle project" of the 1990s was the construction of the thirty-one mile "Chunnel," which connects England and France under the English Channel for high-speed shuttle trains. It consists of three tunnels that were drilled 131 feet under the channel floor; two of the tunnels are 24 feet in diameter. The trip from London to Paris can be made in 3-1/2 hours.

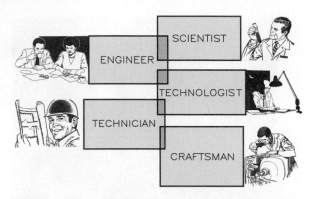

1.4 This ranking of the typical technological team is from the most theoretical level (scientists) to the least technical level (craftspeople).

1.5 These scientists are conducting research utilizing genomics for discovery of new insecticides. *(Courtesy of FMC Corporation.)*

1.4 The Technological Team

Technology and design have become so broad and complex that teams of specialists rather than individuals undertake most projects **(Figure 1.3).** Such teams usually consist of one or more scientists, engineers, technologists, technicians, and craftspeople, and may include designers and stylists **(Figure 1.4).**

Scientists

Scientists are researchers who seek to discover new laws and principles of nature

1.6 Engineers and technologist combine their knowledge to work on an avionics modification. *(Courtesy of Cessna Aircraft Company.)*

through experimentation and scientific testing **(Figure 1.5).** They are more concerned with the discovery of scientific principles than with the application of those principles to products and systems. Their discoveries may not find applications until years later.

Engineers

Engineers receive training in science, mathematics, and industrial processes to prepare them to apply the findings of the scientists **(Figure 1.6).** Thus engineers are concerned with converting raw materials and power sources into needed products and services. Creatively applying scientific principles to develop new products and systems is the design process, the engineer's primary function. In general, engineers use known principles and available resources to achieve a practical end at a reasonable cost.

Technologists

Technologists obtain backgrounds in science, mathematics, and industrial processes. Whereas engineers are responsible for analysis, overall design, and research, technologists are concerned with the application of engineering principles to planning, detail design, and

1.7 An engineering technician works on a phased-array radar antenna. *(Courtesy of Northrop Grumman Corporation.)*

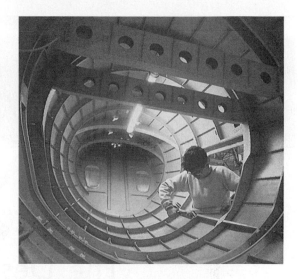

1.8 A craftsman assembles a portion of an experimental aircraft. *(Courtesy of Cessna Aircraft Company.)*

production **(Figure 1.6).** Technologists apply their knowledge of engineering principles, manufacturing, and testing to assist in the implementation of projects and production. They also provide support and act as liaison between engineers and technicians.

Technicians

Technicians assist engineers and technologists at a less theoretical level than technologists and provide liaison between technologists and craftspeople **(Figure 1.7).** They have backgrounds in mathematics, drafting, computer programming, and materials testing. Their work varies from conducting routine laboratory experiments to supervising craftspeople in manufacturing or construction.

Craftspeople

Craftspeople are responsible for implementing designs by fabricating them according to engineers' specifications. They may be machinists who make product parts or electricians who assemble electrical components. Their ability to produce a part according to design specifications is as necessary to the success of a project as an engineers' ability to design it. Craftspeople include electricians, welders, machinists,

fabricators, drafters, and members of many other occupational groups **(Figure 1.8).**

Designers

Designers may be engineers, technologists, inventors, or industrial designers who have special talents for devising creative solutions. Designers do not necessarily have engineering backgrounds, especially in newer technologies where there is little design precedent. Thomas A. Edison, for example, had little formal education, but he created some of the world's most significant inventions.

Stylists

Stylists are concerned with the appearance and market appeal of a product rather than its fundamental design. They may design an automobile body or the exterior of an electric iron. Automobile stylists, for example, consider the car's appearance, driver's vision, passengers' enclosure, power unit's space requirement, and so on. However, they are not involved with the design of the car's internal mechanical functions, such as the engine, steering linkage, and brakes. Stylists must have a high degree of aesthetic awareness and an instinct for styling to attract consumers.

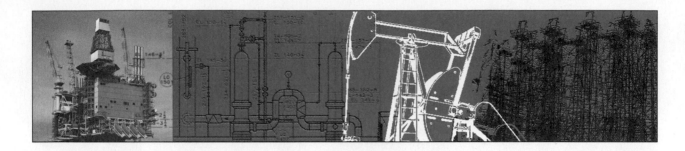

2

The Design Process

2.1 Introduction

The design process is the method of devising innovative solutions to problems that will result in new products or systems. Engineering graphics is the primary medium of design that is used for developing designs from initial concepts to final working drawings. Initially, a design consists of sketches that are refined, analyzed, and developed into precise detail drawings and specifications. They, in turn, become part of the contract documents for the parties involved in funding and implementing a project.

At first glance, the solution of a design problem may appear to involve merely the recognition of a need and the application of effort toward its solution, but most engineering designs are more complex than that. The engineering and design efforts may be the easiest parts of a project.

For example, engineers who develop roadway systems must deal with constraints such as ordinances, historical data, human

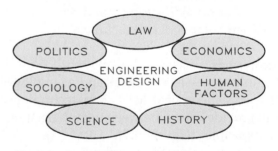

2.1 An engineering project may involve the interaction of people representing many professions and interests, with engineering design as the central function.

factors, social considerations, scientific principles, budgeting, and politics **(Figure 2.1)**. Engineers can readily design driving surfaces, drainage systems, overpasses, and other components of the system. However, adherence to budgetary limitations is essential, and funding is closely related to politics on public projects.

Traffic laws, zoning ordinances, environmental impact statements, right-of-way acquisition, and liability clearances are legal aspects of roadway design that engineers must deal with. Past trends, historical data, human factors (including driver characteristics), and safety features affecting the function of the traffic system must be analyzed. Social problems may arise if proposed roadways will be heavily traveled and will attract commercial development such as shopping centers, fast-food outlets, and service stations. Finally, designers must apply engineering principles developed through research and experience to obtain durable roads, economical bridges, and fully functional systems.

2.2 Types of Design Problems

Most design problems fall into one of two categories: **product design** and **systems design.**

Product Design

Product design is the creation, testing, and manufacture of an item that usually will be mass produced, such as an appliance, a tool, a toy, or larger products such as automobiles **(Figure 2.2).** In general, a product must have sufficiently broad appeal for meeting a specific need and performing an independent function to warrant its production in quantity. Designers of products, whether an automobile or a bicycle, must consider current market needs, production costs, function, sales, distribution methods, and profit predictions **(Figure 2.3).**

Products can perform one or many functions. For instance, the primary function of an automobile is to provide transportation, but it also contains products that provide communications, illumination, comfort, entertainment, and safety. Because it is mass produced for a large consumer market and can be purchased as a unit, the automobile is regarded as a product. However, because it consists of

2.2 Product design seeks to develop a product that meets a specific need, that can function independently, and that can be mass produced.

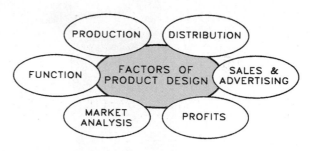

2.3 These are some of the major factors that must be considered when developing a product design.

many products that perform various functions, the automobile also is a system.

Systems Design

Systems design combines products and their components into a unique arrangement and provides a method for their operation. A residential building is a system of products consisting of heating and cooling, plumbing, natural gas, electrical power, sewage, appliances, entertainment, and others that form the overall system as shown in **Figure 2.4.**

Systems Design Example

Suppose that you were carrying luggage to a faraway gate in an airport terminal. It would be easy for you to recognize the need for a luggage cart that you could use and then leave behind for others to use. If you have this need, then others do too. The identification of this need could prompt you to design a cart like

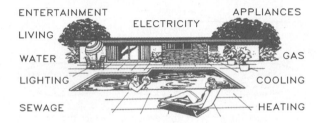

2.4 The typical residence is a system composed of many components and products.

2.6 Luggage carts are dispensed from cart management units at the major entrances of airport terminals. *(Courtesy of Smarte Carte, Inc.)*

2.5 A luggage cart in an airport terminal is a product and it is part of a system. This cart is equipped with a hand brake for added safety and ease of use. *(Courtesy of Smarte Carte, Inc.)*

the one shown in **Figure 2.5** to hold luggage, and even a child, and to make it available to travelers. The cart is a product.

How could you profit from providing such a cart? First you would need a method of holding the carts and dispensing them to customers, such as the one shown in **Figure 2.6.** You would also need a method for users to pay for cart rental, so you could design a cart management unit (**Figure 2.7**).

For an added customer convenience you could provide a vending machine that will take bills as well as coins and make change and

2.7 The management unit releases a cart to the customer after a credit card or bills are inserted at the unit. All major credit cards are accepted and bills (ones and fives) or coins can be used and change will be made. *(Courtesy of Smarte Carte, Inc.)*

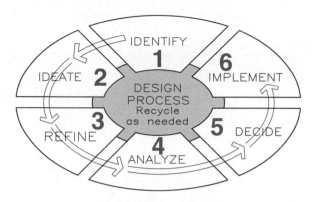

2.8 The design process consists of six steps, each of which can be repeated as necessary.

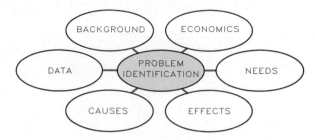

2.9 Problem identification requires that the designer accumulate as much information about a problem as possible before attempting a solution. The designer also should keep product marketing in mind at all times.

accept major credit cards **(Figure 2.7).** Can you think of an incentive for encouraging carts to be returned to a cart management unit?

The combination of these products and the method of using them is a system design. Such a system of products is more valuable than the sum of the products alone.

2.3 The Design Process

Design is the process of creating a product or system to satisfy a set of requirements that has multiple solutions by using any available resources. In essentially all cases, the final design must be completed at a profit or within a budget.

The steps of the design process shown in **Figure 2.8** are:

1. problem identification,
2. preliminary ideas (ideation),
3. refinement,
4. analysis,
5. decision, and
6. implementation.

Designers should work sequentially from step to step but they should review previous steps periodically and rework them if a new approach comes to mind during the process.

Problem Identification
Most engineering problems are not clearly defined at the outset and require identification before an attempt is made to solve them **(Figure 2.9).** For example, air pollution is a concern, but we must identify its causes before we can solve the problem. Is it caused by automobiles, factories, atmospheric conditions that harbor impurities, or geographic features that trap impure atmospheres?

Another example is traffic congestion. When you enter a street where traffic is unusually congested, can you identify the reasons for the congestion? Are there too many cars? Are the signals poorly synchronized? Are there visual obstructions? Has an accident blocked traffic?

Problem identification involves much more than simply stating, "We need to eliminate air pollution." We need data of several types: opinion surveys, historical records, personal observations, experimental data, physical measurements from the field, and more. It is important that the designer resist the temptation to begin developing a solution before the identification step has been completed.

Preliminary Ideas
The second step of the design process is the development of as many ideas for problem solution as possible **(Figure 2.10).** A brainstorming session is a good way to collect ideas at the outset that are highly creative, revolutionary, and even wild. Rough sketches, notes,

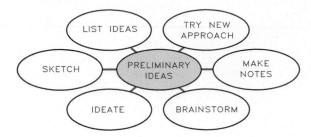

2.10 The designer gathers ideas from a brainstroming session and develops preliminary ideas for problem solution. Ideas should be listed, sketched, and noted to have a broad range of ideas to work with.

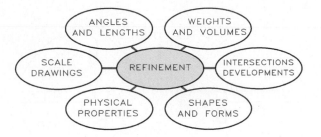

2.11 Refinement begins with the construction of scale drawings of the best preliminary ideas. Descriptive geometry and graphics are used to describe geometric characteristics.

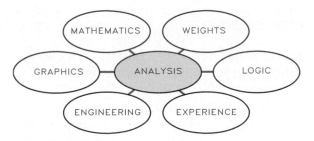

2.12 All available methods, from science to technology to graphics to experience, should be used to analyze a design.

and comments can capture and preserve preliminary ideas for further refinement. The more ideas, the better at this stage.

Refinement

Several of the better preliminary ideas are selected for refinement to determine their merits. The rough preliminary sketches are converted into scale drawings for spatial analysis, determination of critical measurements, and the calculation of areas and volumes affecting the design **(Figure 2.11).** Descriptive geometry aids in determining spatial relationships, angles between planes, lengths of structural members, intersections of surfaces and planes, and other geometric relationships.

Analysis

Analysis is the step during which engineering and scientific principles are used most intensively to evaluate the best designs and compare their merits with respect to function, strength, safety, cost, and optimization **(Figure 2.12).** Graphical methods play an important role in analysis, also. Data can be analyzed graphically, forces analyzed as graphical vectors, and empirical data can be analyzed, integrated, and differentiated by other graphical methods. Analysis is less creative than the previous steps.

Decision

After analysis, a single design, which may be a compromise among several designs, is decided on as the solution to the problem **(Figure 2.13).** The designer alone, or a team, may make the decision. The outstanding aspects of each design usually lend themselves to graphical comparisons of manufacturing costs, weights, operational characteristics, and other data essential in decision making.

Implementation

The final design must be described and detailed in working drawings and specifications from which the project will be built, whether it is a computer chip or a suspension bridge **(Figure 2.14).** Workers must have precise instructions for the manufacture of each component, often measured within thousandths of an inch to ensure proper fabrication assembly. Working drawings must be sufficiently explicit to serve as part of the legal contract with the successful bidder on the job.

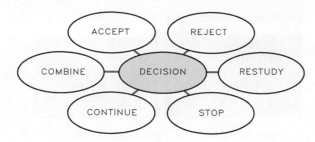

2.13 Decision involves the selection of the best design or design features to implement. This step may require an acceptance, rejection, or a compromise of the proposed solution.

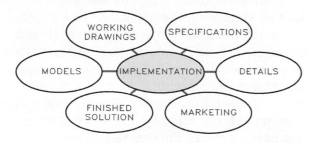

2.14 Implementation is the preparation of drawings, specifications, and documentation from which the product can be made. The product is produced and marketing is begun.

2.4 Graphics and Design

Whether by freehand sketching, with instruments, or on a computer, graphics is the medium of design. Engineering, scientific, and analytical principles must be applied throughout the process, and graphics is the medium that is employed in each step from problem identification to implementation.

An example of how graphics is used by a design firm to develop a product is shown in **Figure 2.15** through **Figure 2.19.** A freehand sketch of the basic concept with notes explaining how it is to be used to rinse out the eye of the user by spraying a solution is shown in **Figure 2.15.**

The designer makes a series of four sketches in **Figure 2.16** to study how the tank housing is snapped in position. These sketches represent his thinking process as well as a means of

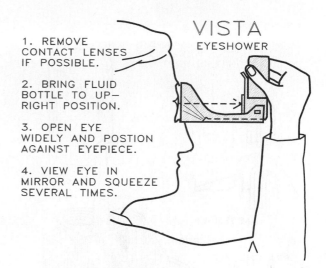

1. REMOVE CONTACT LENSES IF POSSIBLE.

2. BRING FLUID BOTTLE TO UP-RIGHT POSITION.

3. OPEN EYE WIDELY AND POSTION AGAINST EYEPIECE.

4. VIEW EYE IN MIRROR AND SQUEEZE SEVERAL TIMES.

2.15 This freehand sketch with its accompanying notes describes the concept for an eye shower device. (*Courtesy Keck-Craig Inc.*)

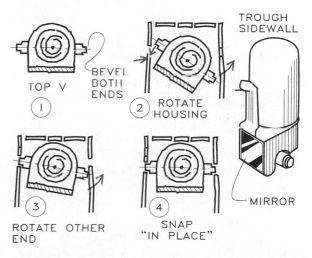

2.16 These freehand sketches show how the tank housing is snapped into position. (*Courtesy Keck-Craig Inc.*)

recording his ideas and for communicating with others.

A three-dimensional pictorial (**Figure 2.17**) illustrates additional design features as the product evolves. The preliminary designs are studied and evaluated by several designers as a means of "troubleshooting" the design and applying all available experience and expertise to the eye shower.

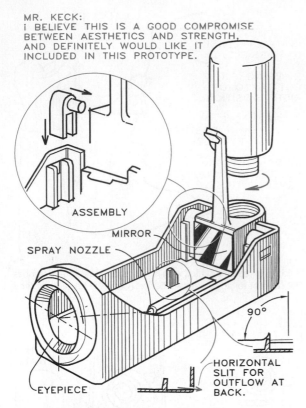

MR. KECK:
i BELIEVE THIS IS A GOOD COMPROMISE BETWEEN AESTHETICS AND STRENGTH, AND DEFINITELY WOULD LIKE IT INCLUDED IN THIS PROTOTYPE.

ASSEMBLY

MIRROR

SPRAY NOZZLE

90°

EYEPIECE

HORIZONTAL SLIT FOR OUTFLOW AT BACK.

2.17 This three-dimensional pictorial clarifies additional production details that must be provided for the eye shower device. (*Courtesy Keck-Craig Inc.*)

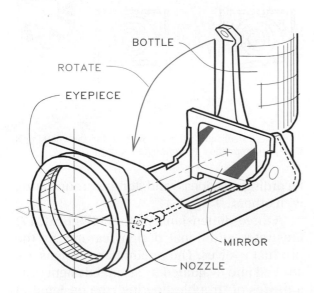

BOTTLE

ROTATE

EYEPIECE

MIRROR

NOZZLE

2.18 This pictorial shows the motion of the tank and the relationship of the mirror and nozzle to eye shower. (*Courtesy Keck-Craig Inc.*)

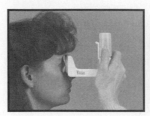

VISTA EYESHOWER

A. OPEN FOR USE   B. CLOSED FOR STORAGE

2.19 Photographs show how the eye shower is used and closed for carrying in a pocket or purse. (*Courtesy Keck-Craig Inc.*)

Additional design details are explained in the sketch in **Figure 2.18.** The movements of the adjacent parts can be shown better in a three-dimensional drawing than in multiview drawings. These concepts must be refined and modified to attain the final design ready for production as a prototype as shown in **Figure 2.19.** A more detailed coverage of the steps of the design process will be given in the following chapters.

2.5 Application of the Design Process

The following example illustrates the application of the design process to a simple problem.

Hanger Bracket Problem

A bracket is needed to support a 2-inch diameter hot-water pipe from a beam or column 9 inches from the mounting surface. This design project is typical of an in-house assignment by an engineer of a company specializing in pipe hangers and supports.

Problem Identification

First, write a statement of the problem and a statement of need **(Figure 2.20).** List limitations and desirable features and make descriptive sketches to better identify the requirements **(Figure 2.21).** Even if much of this information may be obvious, writing

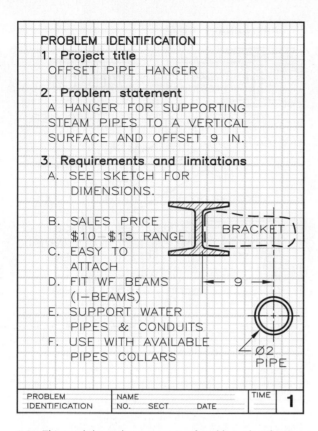

PROBLEM IDENTIFICATION
1. Project title
   OFFSET PIPE HANGER

2. Problem statement
   A HANGER FOR SUPPORTING
   STEAM PIPES TO A VERTICAL
   SURFACE AND OFFSET 9 IN.

3. Requirements and limitations
   A. SEE SKETCH FOR
      DIMENSIONS.
   B. SALES PRICE
      $10 $15 RANGE
   C. EASY TO
      ATTACH
   D. FIT WF BEAMS
      (I-BEAMS)
   E. SUPPORT WATER
      PIPES & CONDUITS
   F. USE WITH AVAILABLE
      PIPES COLLARS

BRACKET

9

Ø2 PIPE

| PROBLEM IDENTIFICATION | NAME NO.   SECT   DATE | TIME | 1 |

2.20 This worksheet shows aspects of problem identification, the first step of the design process.

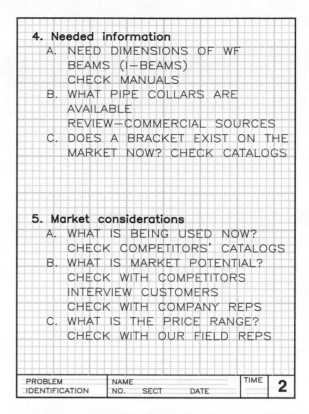

4. Needed information
   A. NEED DIMENSIONS OF WF
      BEAMS (I-BEAMS)
      CHECK MANUALS
   B. WHAT PIPE COLLARS ARE
      AVAILABLE
      REVIEW—COMMERCIAL SOURCES
   C. DOES A BRACKET EXIST ON THE
      MARKET NOW? CHECK CATALOGS

5. Market considerations
   A. WHAT IS BEING USED NOW?
      CHECK COMPETITORS' CATALOGS
   B. WHAT IS MARKET POTENTIAL?
      CHECK WITH COMPETITORS
      INTERVIEW CUSTOMERS
      CHECK WITH COMPANY REPS
   C. WHAT IS THE PRICE RANGE?
      CHECK WITH OUR FIELD REPS

| PROBLEM IDENTIFICATION | NAME NO.   SECT   DATE | TIME | 2 |

2.21 (Sheet 2) This worksheet shows information needed before the designer can proceed, including sources from which it can be obtained.

statements and making sketches about the problem will help you "warm up" to the problem and begin the creative process. Also, you must begin thinking about sales outlets and marketing methods.

Preliminary Ideas

Brainstorm the problem for possible solutions with others, or alone if necessary. List the ideas obtained on a worksheet (**Figure 2.22**), and summarize the best ideas and design features on a separate worksheet (**Figure 2.23**). Then translate and expand these verbal ideas into rapidly drawn freehand sketches (**Figure 2.24** and **Figure 2.25**). You should develop as many ideas as possible during this step because a large number of ideas represents a high level of creativity. This is the most creative step of the design process.

Problem Refinement

Describe the design features of one or more preliminary ideas on a worksheet for comparison (**Figure 2.26**). Draw the better designs to scale in preparation for analysis; you need to show only a few dimensions at this stage (**Figure 2.27**).

Use instrument-drawn orthographic projections, computer drawings, and descriptive geometry to refine the designs and ensure precision. Let's say that you select ideas 3 and 5 for analysis. **Figure 2.27** depicts orthographic and auxiliary views of the two designs.

Analysis

Use an analysis worksheet to analyze the cast-iron bracket design as shown in **Figures 2.28–2.31.** If you are considering more than one solution, analyze each design separately.

2.5 APPLICATION OF THE DESIGN PROCESS • 13

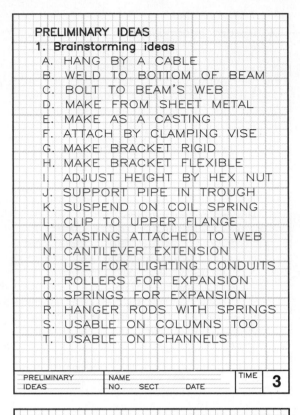

PRELIMINARY IDEAS

1. Brainstorming ideas
   A. HANG BY A CABLE
   B. WELD TO BOTTOM OF BEAM
   C. BOLT TO BEAM'S WEB
   D. MAKE FROM SHEET METAL
   E. MAKE AS A CASTING
   F. ATTACH BY CLAMPING VISE
   G. MAKE BRACKET RIGID
   H. MAKE BRACKET FLEXIBLE
   I. ADJUST HEIGHT BY HEX NUT
   J. SUPPORT PIPE IN TROUGH
   K. SUSPEND ON COIL SPRING
   L. CLIP TO UPPER FLANGE
   M. CASTING ATTACHED TO WEB
   N. CANTILEVER EXTENSION
   O. USE FOR LIGHTING CONDUITS
   P. ROLLERS FOR EXPANSION
   Q. SPRINGS FOR EXPANSION
   R. HANGER RODS WITH SPRINGS
   S. USABLE ON COLUMNS TOO
   T. USABLE ON CHANNELS

| PRELIMINARY IDEAS | NAME | | TIME | 3 |
| | NO. SECT DATE | | | |

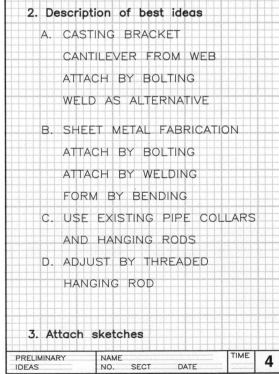

2. Description of best ideas

   A. CASTING BRACKET
      CANTILEVER FROM WEB
      ATTACH BY BOLTING
      WELD AS ALTERNATIVE

   B. SHEET METAL FABRICATION
      ATTACH BY BOLTING
      ATTACH BY WELDING
      FORM BY BENDING

   C. USE EXISTING PIPE COLLARS
      AND HANGING RODS

   D. ADJUST BY THREADED
      HANGING ROD

3. Attach sketches

| PRELIMINARY IDEAS | NAME | | TIME | 4 |
| | NO. SECT DATE | | | |

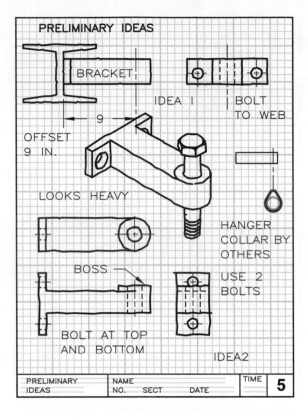

PRELIMINARY IDEAS

BRACKET
IDEA I   BOLT TO WEB
OFFSET 9 IN.
9
LOOKS HEAVY
HANGER COLLAR BY OTHERS
BOSS
USE 2 BOLTS
BOLT AT TOP AND BOTTOM
IDEA2

| PRELIMINARY IDEAS | NAME | | TIME | 5 |
| | NO. SECT DATE | | | |

2.22 (Sheet 3) A member of the brainstorming team records the ideas from the session.

2.23 (Sheet 4) The best ideas are selected from the brainstorming session to be developed as preliminary ideas.

2.24 (Sheet 5) Preliminary ideas are sketched and noted for further development. This is the most creative step of the design process.

2.25 (Sheet 6) Additional preliminary design solutions are sketched here.

2.26 (Sheet 7) Refinement of preliminary ideas begins with written descriptions of the better ideas.

2.27 (Sheet 8) Scale drawings of two designs are made to describe to designs. Almost no dimensions are needed.

2.28 (Sheet 9) A continuation of the analysis step of the design process.

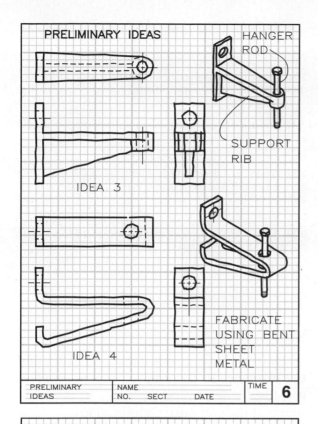

PRELIMINARY IDEAS

HANGER ROD

IDEA 3

SUPPORT RIB

IDEA 4

FABRICATE USING BENT SHEET METAL

PRELIMINARY IDEAS	NAME		TIME	6
	NO.	SECT DATE		

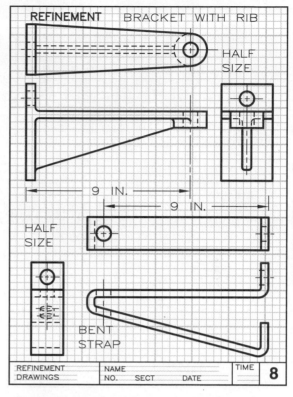

REFINEMENT   BRACKET WITH RIB

HALF SIZE

9 IN.

9 IN.

HALF SIZE

BENT STRAP

REFINEMENT DRAWINGS	NAME		TIME	8
	NO.	SECT DATE		

REFINEMENT
1. **Description of best ideas**
   A. BRACKET WITH RIB
      1. BRACKET MADE AS CASTING 10–12 EXTENSION FROM BEAM
      2. ATTACHED WITH A SINGLE BOLT ON UPPER SIDE
      3. SUPPORTS PIPE WITH HANGER ROD—STANDARD PART
      4. STURDY AND STABLE

   B. BENT STRAP BRACKET
      1. MADE OF BENT SHEET METAL
      2. ATTACHED BY A SINGLE BOLT OR BY WELDING
      3. SUPPORTS HANGER ROD SAME AS CAST BRACKET
      4. ABOUT 2 LBS IN WEIGHT

2. **Attach scale drawings**

DESIGN REFINEMENT	NAME		TIME	7
	NO.	SECT DATE		

ANALYSIS
1. **Function**
   A. EXTENDS HANGER ROD OUT 9 IN. FROM VERTICAL
   B. HOLDS 3/8 IN. ROD
   C. MOUNTS ON 1/2 BOLT

2. **Human engineering**
   A. EASY & QUICK TO INSTALL
   B. REQUIRES NO SPECIAL TOOLS
   C. PROVIDES THE REQUIRED SAFETY

3. **Market and consumer acceptance**
   A. KEEP PRICE UNDER $20
   B. CAN BE SOLD AS AN ADDITION TO OUR LINE
   C. SHOULD BE ADDED TO OUR PRODUCT CATALOG
   D. NEED TO FIND ESTIMATE OF OFFSET BRACKETS SOLD PER YR.

DESIGN ANALYSIS	NAME		TIME	9
	NO.	SECT DATE		

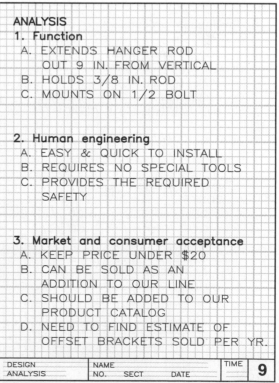

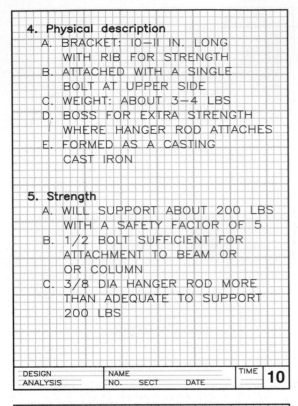

```
    4. Physical description
       A. BRACKET: 10-11 IN. LONG
          WITH RIB FOR STRENGTH
       B. ATTACHED WITH A SINGLE
          BOLT AT UPPER SIDE
       C. WEIGHT: ABOUT 3-4 LBS
       D. BOSS FOR EXTRA STRENGTH
          WHERE HANGER ROD ATTACHES
       E. FORMED AS A CASTING
          CAST IRON

    5. Strength
       A. WILL SUPPORT ABOUT 200 LBS
          WITH A SAFETY FACTOR OF 5
       B. 1/2 BOLT SUFFICIENT FOR
          ATTACHMENT TO BEAM OR
          OR COLUMN
       C. 3/8 DIA HANGER ROD MORE
          THAN ADEQUATE TO SUPPORT
          200 LBS




  DESIGN        NAME                  TIME
  ANALYSIS      NO.   SECT   DATE          10
```

```
    6. Production procedures
       A. FABRICATE AS A CASTING
          USING CAST IRON
       B. DRILL .50 DIA HOLE FOR
          .375 DIA HANGER ROD
       C. DRILL .625 DIA HOLE FOR
          .50 DIA BOLT TO ATTACH
          BRACKET
       D. PAINT TO GIVE RUST-PROOF
          COATING

    7. Economic analysis
       A. COSTS
          1. DEVELOPMENT  $0.20
          2. CAST IRON       .50
          3. CASTING COST   1.40  2.50
          4. DRILL 2 HOLES   .20
          5. PAINTING        .20
       B. LABOR                    .70
       C. PACKAGING               .30
       D. PROFIT                 2.00
       E. WHOLESALE PRICE        6.50
       F. RETAIL PRICE        $12.00

  DESIGN        NAME                  TIME
  ANALYSIS      NO.   SECT   DATE          11
```

2.29 (Sheet 10) A continuation of the analysis step of the design process.

2.30 (Sheet 11) This portion of the analysis focuses on the production and economic considerations.

2.31 (Sheet 12) Graphics and descriptive geometry are used in analysis to determine the load carried by the support bolt.

2.32 (Sheet 13) A decision table is used to evaluate alternative designs in arriving at the final selection of the design that will be implemented.

2.33 (Sheet 14) The decision step is summarized to conclude this step of the design process.

By using the maximum load of 200 lbs. and the geometry of the bracket, it is possible to graphically determine the angle of the reaction that the bolt must support (456 lbs) when the bracket carries its maximum design load of 200 pounds (**Figure 2.31**). Again, graphics is used as an important design tool.

Decision

The decision table (**Figure 2.32**) compares two designs: the bracket with a rib and the bent strap. Apportion weighting factors to features to be analyzed by assigning points to them so that the sum of all factors is 10. You can then rank the designs on a 10-point scale from highest (10) to lowest (0).

Draw conclusions and summarize the features of the recommended design, along with a projection of its marketability (**Figure 2.33**). In this case, we decide to implement the bracket with a rib.

Implementation

Detail the design of the bracket with a rib in a drawing that graphically describes and dimensions each individual part (**Figure 2.34**).

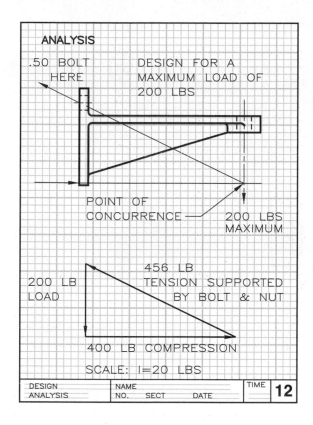

ANALYSIS

.50 BOLT
HERE

DESIGN FOR A
MAXIMUM LOAD OF
200 LBS

POINT OF
CONCURRENCE

200 LBS
MAXIMUM

200 LB
LOAD

456 LB
TENSION SUPPORTED
BY BOLT & NUT

400 LB COMPRESSION

SCALE: 1=20 LBS

| DESIGN ANALYSIS | NAME | NO. SECT DATE | TIME | 12 |

DECISION

1. Decision table for evaluation

DESIGN 1: BRACKET WITH RIB
DESIGN 2: BENT STRAP
DESIGN 3:
DESIGN 4:
DESIGN 5:

MAX	FACTORS	1	2	3	4	5
2	FUNCTION	2	1.5			
2	HUMAN FACT.	1.5	1.5			
1	MARKET ANAL	.5	.2			
1	STRENGTH	1	.5			
1	PRODUCTION	.5	.5			
1	COST	.5	.2			
2	PROFITABILITY	1.0	1.0			
0	APPEARANCE	0	0			
10	TOTALS	7	5.4			

| DESIGN DECISION | NAME | NO. SECT DATE | TIME | 13 |

CONCLUSIONS

THE BRACKET WITH A RIB DESIGN
APPEARS TO BE THE BEST AND
MOST MARKETABLE SOLUTION

IT WOULD BE MOST SUITED TO
THE MARKET NEEDS

GOOD POSSIBILITY OF SELLING AS
ADDITIONAL PRODUCT ACCESSORY

RETAIL	$12.00
SHIPPING EXPENSES	2.00
NUMBER TO SELL TO BREAK EVEN	1000
MANUFACTURED IN HOUSE	$3.50
PROFIT PER UNIT	$2.00

RECOMMEND IMPLEMENTATION
AND PRODUCTION OF THE
PRODUCT AND ADD TO LINE.

| DESIGN DECISION | NAME | NO. SECT DATE | TIME | 14 |

A drawing of this type from which a design can be fabricated is called a working drawing.

Use notes to specify standard parts—the nuts, bolts, and hanger rod, but it will be unnecessary to draw them because they will be bought as standard parts. This working drawing shows the details and specifications for making the bracket. Now, you need to build a prototype or model of the product and test it for function.

Figure 2.35 shows the final product, the cast-iron bracket with a rib. After determining how it will be packaged and distributed, your next task will be to add it to your company's product line, list it in your catalog, and introduce it to the marketplace.

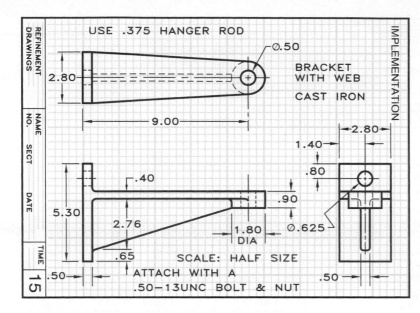

USE .375 HANGER ROD

Ø.50

2.80

BRACKET WITH WEB

CAST IRON

9.00

IMPLEMENTATION

2.80

1.40

.80

.40

.90

5.30

2.76

1.80 DIA

Ø.625

.65

SCALE: HALF SIZE

.50

ATTACH WITH A
.50—13UNC BOLT & NUT

.50

2.34 The completed bracket design can be produced from this working drawing and brought to market.

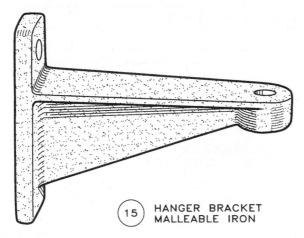

15 HANGER BRACKET
MALLEABLE IRON

2.35 The final hanger bracket is shown here ready to be marketed.

2.6 Design Problems

This section offers problems that are suitable for both individual assignments and team projects to provide experience in applying the methods of creative problem solving introduced in this textbook. All new products begin with sketches at the preliminary idea step

and end with working and assembly drawings at the implementation step.

2.7 Short Design Problems

The following short design problems can be completed in less than two hours.

1. Lamp bracket. Design a bracket to attach a desk lamp to a vertical wall for reading in bed. It should be removable for use as a conventional desk lamp.

2. Towel bar. Design a towel bar for a kitchen or bathroom. Determine optimum size and consider styling, ease of use, and method of attachment.

3. Bicycle rack. Design a rack (or basket) for a bicycle for carrying books and class materials.

4. Boot puller. Design a device for helping you remove cowboy boots from your feet.

2.36 Problem 6. *(Courtesy of Rockwell Automation.)*

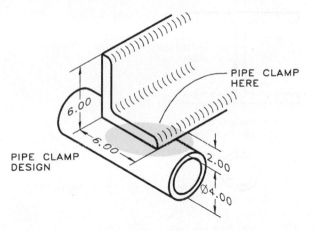

2.37 Problem 7.

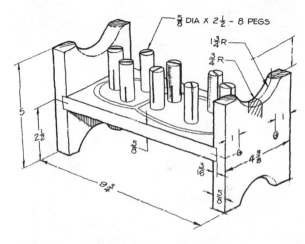

2.38 Problem 8.

5. Side-mounted mirror. Design an improved side-mounted rear-view mirror for an automobile. Consider aerodynamics, protection weather, visibility, and other factors.

6. Part modification (Figure 2.36). Modify the base that supports the 2-inch diameter shaft by changing the square base to a circular base with six holes instead of 4. Modify the ribs accordingly.

7. Pipe clamp (Figure 2.37). A 4-inch diameter pipe must be supported by angles spaced 8 foot apart. Design a clamp that will support the pipe without drilling holes in the angles.

8. Child's toy (Figure 2.38). Design an alternative design to this toy that is intended to develop a child's manual dexterity. Or, make detail drawings of this design.

9. Pipe column support. Design a base that can be attached to a concrete slab with bolts that would provide a base for 3-inch OD diameter pipe columns.

10. Motor bracket (Figure 2.39). Design a bracket to support a motor. The plate should be the upper part of the finished bracket.

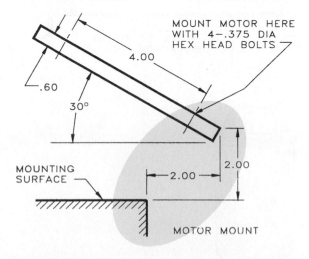

2.39 Problem 10.

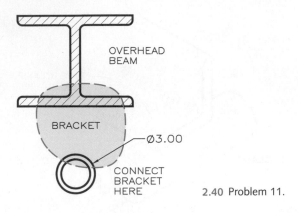

OVERHEAD
BEAM

BRACKET

Ø3.00

CONNECT
BRACKET
HERE

2.40 Problem 11.

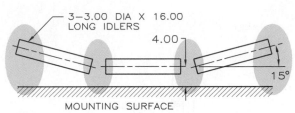

3–3.00 DIA X 16.00
LONG IDLERS

4.00

15°

MOUNTING SURFACE

2.41 Problem 13.

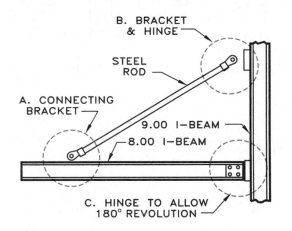

B. BRACKET
& HINGE

STEEL
ROD

A. CONNECTING
BRACKET

9.00 I-BEAM

8.00 I-BEAM

C. HINGE TO ALLOW
180° REVOLUTION

2.42 Problem 16.

PUSH

2.43 Problem 19.

11. Pipe bracket clamp (Figure 2.40). Design a clamp that can be attached to an overhead I-beam to support a pipe without welding or drilling holes in the beam.

12. Foot scraper. Design a device that can be attached to the sidewalk for scraping mud from your shoes.

13. Roller brackets (Figure 2.41). The rollers are to be positioned as shown to support a conveyor belt that carries bulk material. Design the brackets to support the rollers.

14. Audio cassette storage unit. Design a storage unit for an automobile that will hold several audio cassettes, making them accessible to the driver but not to a thief.

15. Slide projector elevator. Design a device for raising a slide projector to the proper angle for projection on a screen. It may be part of the original projector or an accessory to be attached to existing projectors.

16. Jib crane brackets (Figure 2.42). Design the brackets at the joints indicated at A, B, and C to form a jib crane made of an $8'' \times 4'' \times 8'$ long I-beam and a steel connecting rod. The crane should have at least a 180° swing.

17. Book holder 2. Design a holder to support a textbook or reference book at a workstation for ease of reading and accessibility.

18. Table leg design. Do-it-yourselfers build a variety of tables using hollow doors or plywood for the tops and commercially available legs. Design a family of legs that can be attached to table tops with screws.

19. Trashcan cover (Figure 2.43). Design a functional lid with an appropriate opening through which to put garbage.

20. Toothbrush holder. Design a toothbrush holder for a cup and two toothbrushes that can be attached to a bathroom wall.

21. Clothes hook. Design a clothes hook that can be attached to a closet door for hanging clothes on.

22. Conduit hanger (Figure 2.44). Design an attachment for a 3/4-inch conduit to support a channel used as a raceway for electrical wiring.

23. Hammock support. Design a hammock support that will fold up and that will require minimal storage space.

24. Door stop. Design a door stop that can be attached to a wall or floor to prevent a door knob from hitting the wall.

25. Basketball goal. Design a basketball goal that is easy to install for the 8–10 age range.

26. Channel bracket (Figure 2.45). The bracket shown is designed to fit on the flat side of the channel. Design a method of attaching the bracket to the slotted side so the inside nut will not drop down inside the channel during assembly.

27. Drawer handle. Design a handle for a standard file cabinet drawer.

28. Handrail bracket. Design a bracket that will support a tubular handrail to be used on a staircase.

29. Flagpole socket. Design a flagpole socket that is to be attached to a vertical wall.

30. Sit-up bench. Design a sit-up bench for exercising. Can you make it serve multiple purposes?

31. Hose spool. Design a spool/rack on which a garden hose can be wound and left neatly near the outside faucet.

32. Gate hinge. Design a hinge that can be attached to a 3-inch-diameter tubular post to support a 3-foot-wide wooden gate.

33. Cup holder. Design a holder that will support a soft-drink can or bottle in an automobile.

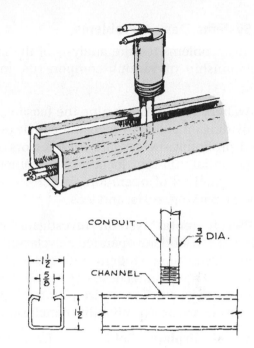

2.44 Problem 22.

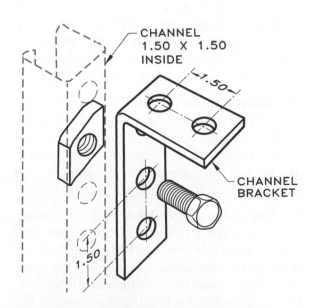

2.45 Problem 26. *(Courtesy of Tyco International Inc.)*

2.8 Systems Design Problems

Systems problems require analysis of the interrelationship of various components and products.

34. Archery range. Determine the feasibility of providing an archery range that can be operated profitably. Investigate such factors as the potential market, location, equipment needed, method of operation, utilities, concessions, parking, costs, and fees.

35. Bicycle rental system. Investigate the feasibility of a student-operated bicycle rental system. Determine student interest, costs, bikes needed, personnel needs, storage, maintenance, and so on. Summarize your findings, operating costs, and profitability conclusions.

36. Model airplane field. Investigate the need for a model-airplane field, including space requirements, types of surface needed, sound control, safety factors, and method of operation. Select a site on or near your campus that is adequate for this facility, and evaluate the equipment, utilities, and site preparation required.

37. Overnight campsite. Analyze the feasibility of converting a vacant tract of land near a major highway into sites for overnight campers. Determine the facilities required by the campers and the venture's profitability.

38. Skateboard facility study. Determine the cost of building and operating a skateboard facility on your campus. Consider where it could be located, how many students would use it, and the amount of equipment and labor necessary to operate it.

39. Hot water supply. Your weekend cottage does not have a hot-water supply, but cold water is available from a private well. Design a system that uses the sun's energy in the summer to heat water for bathing and kitchen use. Devise a system for heating the water in the winter by some other source. Determine whether one or both of these systems could be made portable for showers on camping trips.

40. Information center. Design a drive-by information center to help campus visitors find their way around. Determine the best location for it and the informational material needed, such as slides, photographs, maps, sound, and other audiovisual aids.

41. Golf driving range ball-return system. Balls at golf driving ranges are usually retrieved by hand or with a specially designed vehicle. Design a system capable of automatically returning balls to the tee area.

42. Car wash. Design a car wash facility for your campus that would be self-supporting and would provide the basic needs for washing cars. Think simple and economical.

43. Instant motel. Many communities need temporary housing for celebrations and sporting events. Investigate methods of providing an "instant motel" involving the use of tents, vans, trailers, train cars, or other temporary accommodations. Estimate profitability.

44. Computer wiring system. Computer installations become cluttered with wiring as accessories are attached and access to connections becomes difficult. Design a system whereby the electrical wiring can be more organized, convenient, and accessible.

45. Injury-proof playground. Design a playground that permits the greatest degree of participation by children with the least risk of injury.

46. Patio table production. Determine the procedures for production of a plywood patio table: materials, equipment, methods, storage, and sales. Determine the number needed to be manufactured per month to break even and the selling price. Establish quantity breaks for selling prices for quantities that exceed the break-even level.

2.9 Product Design Problems

Product design involves developing a device that will perform a specific function, be mass produced, and be sold to a large number of consumers.

47. Assembly jig (Figure 2.46). A concept for an assembly jig for holding together parts for bonding thin pieces to thick pieces by furnace brazing:
 a. Make the necesary drawings to explain the concept as shown.
 b. Design an alternative jig of your own.

48. Handrail tee fitting (Figure 2.47). Make the necessary drawings to complete the total design for the handrail fitting that connects three 1.07 inch O.D. pipes.

49. Rescue litter (Figure 2.48). Design a litter that can be folded in a number of positions in order to care for the injured. It should be lightweight, collapsible, and portable.

50. Hunting blind. Design a portable hunting blind adequate for hunting geese or ducks that can be easily taken to its site. Consider making the blind of degradable materials so it can be left at the site.

51. Mailbox. Design a residential mailbox that either attaches to the house or is supported on a pole at the street.

52. Writing table arm. Design a writing table arm that can be attached to a folding chair.

53. Drum truck (Figure 2.49). Design a truck for handling 55-gallon drums of turpentine (7.28 lb per gallon) one at a time. Drums are stored in a vertical position and used in a horizontal position. The truck should be useful in tipping a drum into a

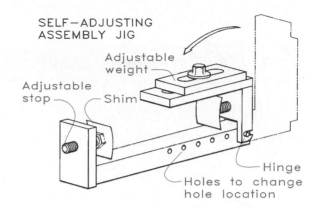

2.46 Problem 47. *(Courtesy of NASA.)*

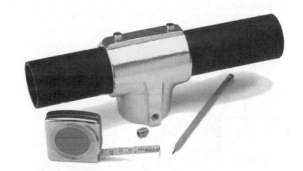

2.47 Problem 48. *(Courtesy of Hollaender Manufacturing Company.)*

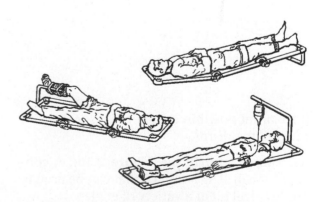

2.48 Problem 49. *(Courtesy of NASA.)*

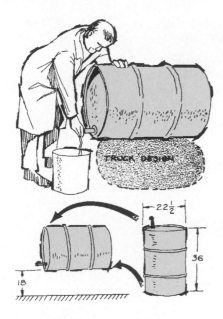

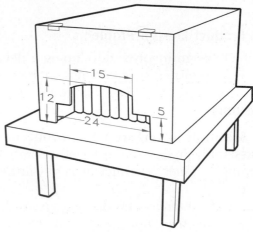

2.51 Problem 55.

2.49 Problem 53.

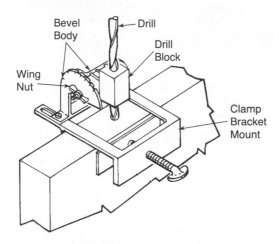

2.50 Problem 54. *(Courtesy of NASA.)*

horizontal position (as shown) as well as for moving the drum.

54. Universal drill jig (Figure 2.50). A concept for a drill jig is shown that can be used to guide a drill bit at a variety of angles.

    a. Make the necessary drawings to depict the design as it is given.

    b. Design your own version of a drill jig of this type. Consider how drills of various diameters could be accomodated.

55. Luggage template (Figure 2.51). To limit the size of carry-on luggage, airlines are testing templates that can be used to screen luggage at check-in time. Using the specifications shown, design a suitable template to solve this problem.

56. Utility rack. Design a rack that can be attached to a pickup truck to aid in hauling equipment and tools.

57. Yard helper. Design a movable container that can be used for gardening and yard work.

58. Computer mount. Design a device that can be clamped to a desktop for holding a computer, permitting it to be adjusted to various positions while leaving the desktop free to work on.

59. Workers' stilts. Design stilts to give workers access to an 8-foot high ceiling, permitting them to install 4 × 8 foot ceiling panels.

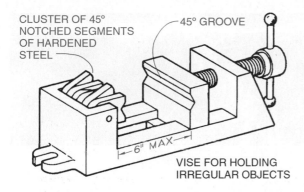

CLUSTER OF 45°
NOTCHED SEGMENTS
OF HARDENED
STEEL

45° GROOVE

6" MAX

VISE FOR HOLDING
IRREGULAR OBJECTS

2.52 Problem 66. *(Courtesy of NASA.)*

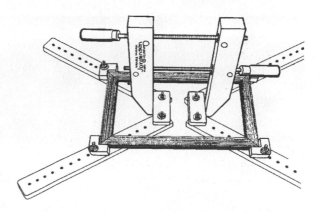

2.53 Problem 67. *(Courtesy of Adjustable Clamp Company.)*

60. Panel applicator. A worker applying a 4 × 8 foot plasterboard to a ceiling needs a helper to hold the panel in place while she nails it. Design a device to hold the panel and eliminate the need for an assistant.

61. Sportsman's chair. Design a sportsman's chair that can be used for camping, for fishing from a bank or boat, at sporting events, and for other purposes.

62. Can crusher. Design a kitchen device for flattening aluminum cans.

63. Monitor support. Design a computer monitor arm to support and position the screen for ease of use.

64. Projector cabinet. Design a cabinet to serve as an end table or some other function while housing a slide projector and slide trays ready for use.

65. Heavy appliance mover. Design a device for moving large appliances—stoves, refrigerators, and washers—about the house for the purposes of rearranging, cleaning, and servicing them.

66. Specialty vise (Figure 2.52). A concept for a vise for holding irregular objects while they are being machined is shown:

a. Make the necessary drawings for explaining the design concept as given.
b. Design a solution of your own for this problem.

67. Framing fixture (Figure 2.53). Design a device that can be used for the assembly of picture planes.

68. Map holder. Design a map holder to give the driver a view of the map in a convenient location in the car while driving. Provide a method of lighting the map that will not distract the driver.

69. Chimney cover. Design a chimney cover that can be closed from inside the house for repelling rain and reducing temperature loss.

70. Gate opener. An annoyance to farmers and ranchers is the necessity of opening and closing gates. Design a manually operated gate that could be opened and closed by the driver from his vehicle.

71. Automobile coffee maker. Design a device that will provide hot coffee from the dashboard of an automobile. Consider the method of changing and adding water, the spigot system, and similar details.

3

Drawing Instruments

3.1 Introduction

The preparation of technical drawings requires the ability to use a variey of drawing instruments and the computer. Even people with little artistic ability can produce professional technical drawings when they learn to use drawing instruments properly.

The drawing instruments covered in this chapter are traditional ones that are used by hand as opposed to computer instruments. Traditional instruments will always have an application in the development of drawings, but to a lesser degree than in the past.

Computer Instruments

Electronic drawing instruments—computers, plotters, scanners, and similar equipment—are covered in Chapter 25. The ability to use computer graphics and its associated hardware is a necessary skill for members of the engineering team because of its efficiency and uniformity in producing graphics for industrial applications.

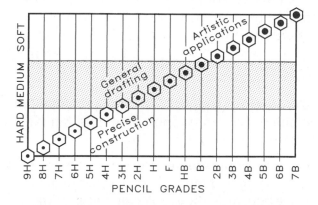

3.1 The hardest pencil lead is 9H and the softest is 7B. The diameters of hard leads are progressively smaller than those of the softer leads.

3.2 Drawing Media

Pencils

A good drawing begins with the correct pencil grade and its proper use. Pencil grades range from the hardest of 9H to the softest, 7B **(Figure 3.1).** The pencils in the medium-grade range of 4H to B are used most often

A. LEAD HOLDER

Holds any size lead; point must be sharpened.

B. FINE-LINE HOLDER

Must use different size holder for different lead sizes; does not need to be sharpened.

C. WOOD PENCIL

Wood must be trimmed and lead must be pointed.

D. THE PENCIL POINT

Sharpen point to a conical point with a lead pointer or a sandpaper pad.

3.2 Sharpen the drafting pencil to a tapered conical point (not a needle point) with a sandpaper pad or other type of sharpener.

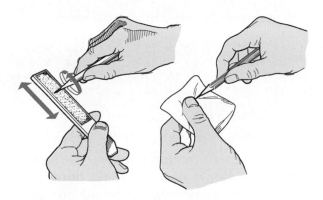

3.3 Revolve the drafting pencil about its axis while stroking the sandpaper pad to form a conical point. Wipe away the graphite from the point with a tissue.

for drafting work of the type covered in this textbook.

Figure 3.2 shows three standard pencils used for most drawings. The leads used in the lead holder shown in **Figure 3.2A** (the best all-around pencil of the three) are marked in white at the ends to indicate their grade.

The fine-line leads used in the lead holder in **Figure 3.2B** are more difficult to identify because their sizes are smaller and they are unmarked. A different fine-line holder must be used for each size of lead. The common sizes are 0.3 mm, 0.5 mm, and 0.7 mm, which are the diameters of the leads. A disadvantage of the fine-line pencil is the tendency of the lead to snap off when you apply pressure to it.

Although you have to sharpen it and point its lead, the wood pencil shown in **Figure 3.2C** is a very satisfactory pencil. The grade of the lead is marked on one end of the pencil; therefore, the opposite end should be sharpened so the identity of the grade of lead will be retained. The wood can be sharpened with a knife or a drafters pencil sharpener to leave about 3/8 inch of lead exposed.

You must sharpen a pencil's lead properly to obtain a point as shown in **Figure 3.2D.** To obtain a conical point, stroke the pencil lead against a sandpaper board and revolve the pencil about its axis in the process **(Figure 3.3).** Wipe excess graphite from the point with a cloth or tissue.

Although sharpening a pencil point with a sandpaper board may seem outdated, it is a very practical way to sharpen pencil points and about the only way to sharpen compass points. However, there are a multitude of pencil pointers available, some of which work well while others do not.

Papers and Films

Sizes   Sheet sizes are specified by the letters A through E. These sizes are multiples of either the standard 8-1/2 × 11-inch sheet (used by engineers) or the 9 × 12-inch sheet (used by architects) as shown in **Figure 3.4.** The metric sizes (A4 through A0) are equivalent to the 8-1/2 × 11-inch modular sizes.

Detail Paper   When drawings are not to be reproduced by the diazo (blue-line process), an opaque paper, called **detail paper,** can be used as the drawing surface. The higher

	ENGINEERS'	ARCHITECTS'		METRIC		
A	11" X 8.5"	12" X 9"	A4	297	X	210
B	17" X 11"	18" X 12"	A3	420	X	297
C	22" X 17"	24" X 18"	A2	594	X	420
D	34" X 22"	36" X 24"	A1	841	X	594
E	44" X 34"	48" X 36"	A0	1189	X	841

3.4 Standard sheet sizes vary by purpose.

the rag content (cotton additive) of the paper, the better is its quality and durability. You may draw preliminary layouts on detail paper and then trace them onto the final surface.

Tracing Paper Tracing paper, or tracing vellum, is a thin, translucent paper that permits light to pass through it, allowing reproduction by the blue-line process. Tracing papers that yield the best reproductions are the most translucent ones. Vellum is a tracing paper that has been chemically treated to improve its translucency, but vellum does not retain its original quality as long as high-quality, untreated tracing paper does.

Tracing Cloth Tracing cloth is a permanent drafting medium used for both ink and pencil drawings. It is made of cotton fabric and is coated with a starch compound to provide a tough, erasable drafting surface that yields excellent blue-line reproductions. Tracing cloth does not change shape as much as tracing paper with variations in temperature and humidity; it can withstand erasures to a higher degree than tracing paper.

Polyester Film An excellent drafting surface is polyester film, which is available under several trade names such as *Mylar*®. It is more transparent, stable, and tougher than paper or cloth and is waterproof. Mylar film is used for both pencil and ink drawings. A plastic-lead pencil must be used with some films, whereas standard lead pencils may be used with others.

3.5 The drafting machine is used for drawings made by hand.

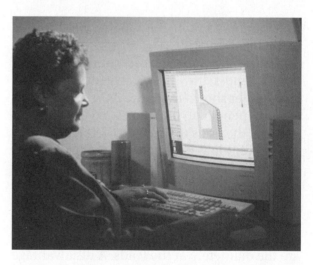

3.6 Many professional drafters and engineers work in environment of this type with the most sophisticated CAD software. *(Courtesy of Anvil International, Inc.)*

3.3 Drawing Equipment

Drafting Machine

Most professional drafters prefer the mechanical drafting machine **(Figure 3.5),** which is attached to the drawing table top and has fingertip controls for drawing lines at any angle. A modern, fully equipped drafting station is shown in **Figure 3.6.** Today, most offices are equipped with computer graphics stations,

3.7 A typical engineering workstation used in industry. *(Courtesy Jervis B. Webb Co.)*

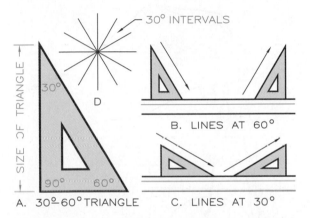

3.8 The 30°–60° triangle is used to draw lines at 30° intervals throughout 360°.

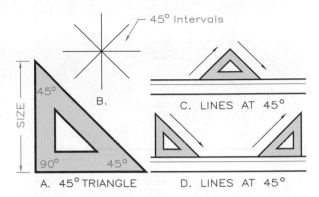

3.9 Use the 45° triangle to draw lines at 45° angles throughout 360°.

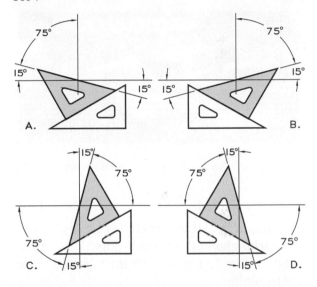

3.10 By using a 30°–60° triangle in combination with a 45° triangle, angles can be drawn at 15° intervals.

which have replaced much of the manual equipment **(Figure 3.7).**

Triangles

The two types of triangles used most often are the 45° triangle and the 30°–60° triangle. The size of a 30°–60° triangle is specified by the longer of the two sides adjacent to the 90° angle **(Figure 3.8).** Standard sizes of 30°–60° triangles range in 2-inch intervals from 4 to 24 inches.

The size of a 45° triangle is specified by the length of the sides adjacent to the 90° angle.

These range in 2-inch intervals from 4 to 24 inches, but the 6-inch and 10-inch sizes are adequate for most classroom applications. **Figure 3.9** shows the various angles that you may draw with this triangle. By using the 45° and 30°–60° triangles in combination, you may draw angles at 15° intervals throughout 360° **(Figure 3.10).**

Protractor

When drawing or measuring lines at angles other than multiples of 15°, a protractor is used **(Figure 3.11).** Protractors are available

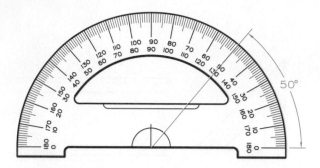

3.11 This semicircular protractor is used to measure angles.

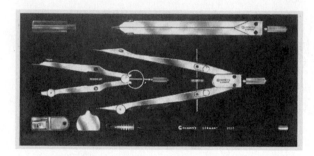

3.12 This is a typical cased set of drawing instruments. *(Courtesy of Gramercy Guild.)*

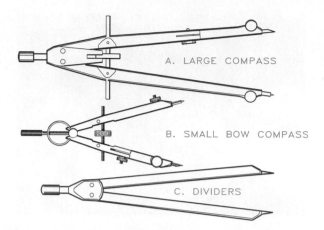

A. LARGE COMPASS

B. SMALL BOW COMPASS

C. DIVIDERS

3.13 Drawing instruments can be purchased individually or as a set.

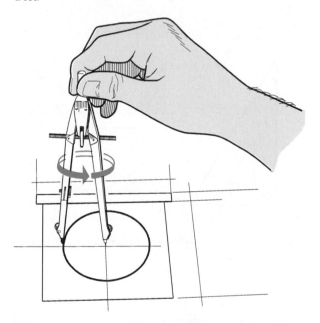

3.14 Use a compass for drawing circles.

as semicircles (180°) or circles (360°). Adjustable triangles with movable edges that can be set at different angles with thumbscrews also are available.

Instrument Set

Figure 3.12 shows a cased set of some of the basic drawing instruments, which are available individually as well. The three most important instruments for hand drawing are the large compass, small compass, and divider shown in **Figure 3.13**.

Compass Use the compass to draw circles and arcs in pencil or ink **(Figure 3.14)**. To draw circles well with a pencil compass, sharpen the lead on its outside with a sandpaper board to bevel cut as shown in **Figure 3.15.** You cannot draw a thick arc in pencil with a single sweep. You must draw a series of

thin concentric circles by adjusting the radius of the compass slightly.

When setting the compass pivot point in the drawing surface, insert it just enough for a firm set, not to the shoulder of the point. If your table top has a hard surface, you must place several sheets of paper under the drawing to provide a seat for the compass point.

Use a small bow compass **(Figure 3.16)** to draw circles of up to 2 inches in radius. For

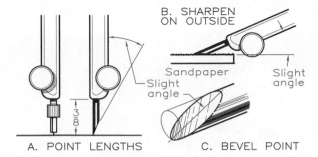

B. SHARPEN
ON OUTSIDE

Sandpaper
Slight
angle

Slight
angle

A. POINT LENGTHS

C. BEVEL POINT

3.15 Compass lead
A. Adjust the pencil point to be the same length as the compass point.
B. & C. Sharpen the lead from the outside with a sandpaper pad.

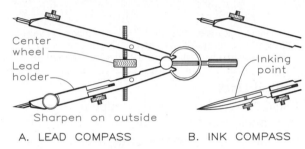

Center wheel

Lead holder

Inking point

Sharpen on outside

A. LEAD COMPASS

B. INK COMPASS

3.16 Use a small bow compass for drawing circles or up to a 2-inch radius in pencil or in ink.

larger circles, use an extension bar included in most sets to extend the range of the large bow compass. You may draw small circles conveniently with a circle template aligned with the centerlines of the circles **(Figure 3.17).**

Divider The divider looks like a compass without a drawing point. It is used for laying off and transferring dimensions onto a drawing. For example, you can step off equal divisions rapidly and accurately along a line **(Figure 3.18).** As you make each measurement, the divider's points make a slight impression mark in the drawing surface.

Also, use dividers to transfer dimensions from a scale to a drawing **(Figure 3.19)** or to divide a line into a number of equal parts. Bow dividers **(Figure 3.20)** are useful for transferring smaller dimensions, such as the spacing between lettering guidelines.

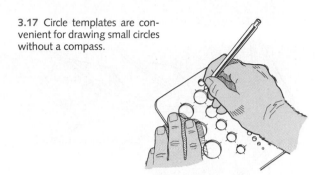

3.17 Circle templates are convenient for drawing small circles without a compass.

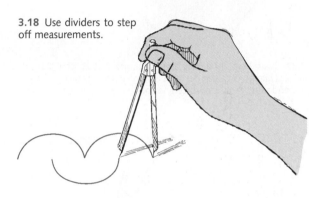

3.18 Use dividers to step off measurements.

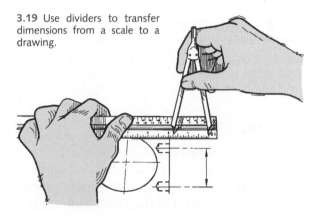

3.19 Use dividers to transfer dimensions from a scale to a drawing.

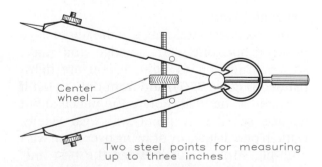

Center wheel

Two steel points for measuring up to three inches

3.20 Bow dividers are used to transfer small dimensions such as spacing for guidelines for lettering.

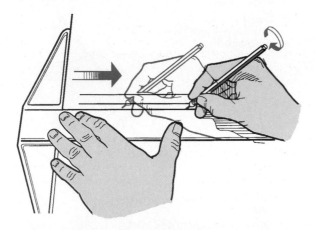

3.21 Draw horizontal lines along the upper edge of a straightedge while holding the pencil in a plane perpendicular to the paper and at 60° to the surface and rotate the pencil about its axis.

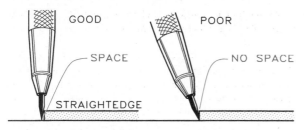

3.22 While drawing, hold the pencil or pen point in a plane perpendicular to the paper, leaving a space between the point and the straightedge.

3.4 Lines

The type of line produced by a pencil depends on the hardness of its lead, drawing surface, and your drawing technique. You must experiment in order to achieve the ideal combination.

Horizontal Lines

To draw a horizontal line, use the upper edge of your horizontal straightedge and make strokes from left to right, if you are right-handed **(Figure 3.21),** and from right to left if you are left-handed. Rotate the pencil about its axis so that its point will wear evenly. Darken pencil lines by drawing over them with multiple strokes. For drawing the best line, leave a small space between the straightedge and the pencil or pen point **(Figure 3.22).**

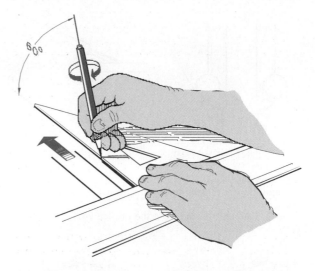

3.23 Draw vertical lines along the left side of a triangle (if you are right-handed) in an upward direction, holding the pencil or pen in a plane perpendicular to the paper and at 60° to the surface.

Vertical Lines

Use a triangle and a straightedge to draw vertical lines. Hold the straightedge firmly with one hand and position the triangle where needed and draw the vertical lines with the other hand **(Figure 3.23).** Draw vertical lines upward along the left side of the triangle if you are right-handed and upward along the right side of the triangle if you are left-handed.

Irregular Curves

Curves that are not arcs must be drawn with an irregular curve (sometimes called French curves). These plastic curves come in a variety of sizes and shapes, but the one shown in **Figure 3.24** is typical. Here, we use the irregular curve to connect a series of points to form a smooth curve.

Erasing Lines

Always use the softest eraser that will do a particular job. For example, do not use ink erasers to erase pencil lines because ink erasers are coarse and may damage the surface of the paper. When working in small areas, you should use an erasing shield to avoid accidentally erasing adjacent lines **(Figure 3.25).** Follow

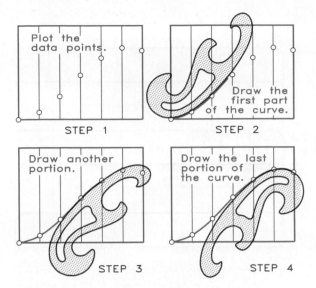

STEP 1

STEP 2

Plot the data points.

Draw the first part of the curve.

Draw another portion.

Draw the last portion of the curve.

STEP 3

STEP 4

3.24 Using the irregular curve
Step 1 Plot data points with a circle template.
Step 2 Position the curve to pass through as many points as possible and draw that portion of the curve.
Step 3 Reposition the irregular curve and draw another portion of the curve.
Step 4 Draw the last portion to complete the curve.

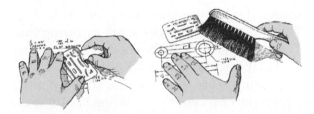

3.25 Use an erasing shield for erasing in tight spots. Use a brush, not your hand, to brush away the erasure crumbs.

erasing by brushing away the "crumbs" with a dusting brush. Wiping the crumbs away with your hands will smudge the drawing. A typical cordless electric eraser that can be used with several grades of erasers to meet your needs is shown in **Figure 3.26.**

3.5 Measurements

Scales

All engineering drawings require the use of scales for measuring lengths and sizes. Scales may be flat or triangular in cross section and

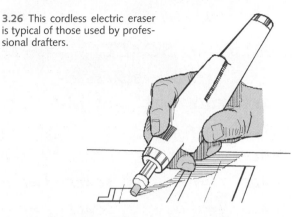

3.26 This cordless electric eraser is typical of those used by professional drafters.

A. ARCHITECTS' SCALE

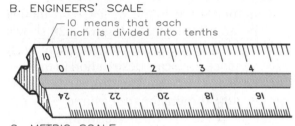

Each inch is divided into sixteenths — 1 inch

B. ENGINEERS' SCALE

10 means that each inch is divided into tenths

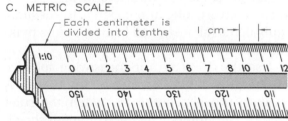

C. METRIC SCALE

Each centimeter is divided into tenths — 1 cm

3.27 The architects' scale (A) measures in feet and inches. The engineers' scale (B) and the metric scale (C) are calibrated in decimal units.

are made of wood, plastic, or metal. **Figure 3.27** shows triangular architects', engineers', and metric scales. Most scales are either 6 or 12 inches long.

Architects' Scale

Drafters use architects' scales to dimension and scale features such as room-size, cabinets, plumbing, and electrical layouts. Most indoor

FROM END OF SCALE

BASIC FORM *SCALE:* $\dfrac{X}{X}$ *=1'–0*

TYPICAL SCALES

SCALE: FULL SIZE (USE 16–SCALE)

SCALE: HALF SIZE (USE 16–SCALE)

SCALE: 3=1'–0   *SCALE:* $\dfrac{1}{4}$ *=1'–0*

SCALE: 1$\frac{1}{2}$=1'–0   *SCALE:* $\dfrac{3}{4}$ *=1'–0*

SCALE: $\dfrac{1}{2}$ *=1'–0*   *SCALE:* $\dfrac{3}{8}$ *=1'–0*

SCALE: $\dfrac{3}{16}$ *=1'–0*   *SCALE:* $\dfrac{1}{8}$ *=1'–0*

SCALE: $\dfrac{3}{32}$ *=1'–0*   *SCALE: 1=1'–0*

3.28 Use this basic form to indicate the scale on a drawing made with an architects' scale.

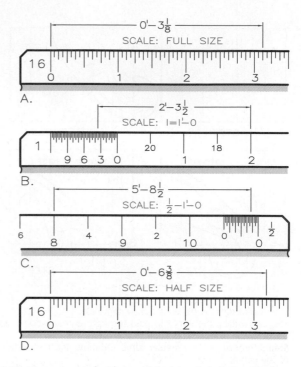

3.29 Lines measured with an architects' scale in feet and inches.

measurements are made in feet and inches with the architects' scale. **Figure 3.28** shows how to indicate the scale you are using on a drawing. Place this scale designation in the title block or in a prominent location on the drawing. Because dimensions measured with the architects' scale are in feet and inches, you must convert all dimensions to decimal equivalents (all feet or all inches) before making calculations.

Use the 16 scale for measuring full-size lines (**Figure 3.29**A). An inch on the 16 scale is divided into sixteenths to match the ruler used by carpenters. The measurement shown is 3-1/8″. When the measurement is less than 1 foot, a zero may precede the inch measurements, with inch marks omitted, or 0′-3-1/2. (Inch marks are omitted since inches are understood to the units of measurement in the English system.)

Figure 3.29B shows the use of the 1 = 1′-0 scale to measure a line. Read the nearest whole foot (2 ft in this case) and the remainder in inches from the end of the scale (3-1/2 in.) for a total of 2′-3-1/2. Note that at the end of each scale, a foot is divided into inches for

measuring fractional parts of feet in inches. The scale 1″ = 1′-0 is the same as saying that 1 inch is equal to 12 inches or that the drawing is 1/12 the actual size of the object.

When you use the 1/2 = 1′-0 scale, 1/2 inch represents 12 inches on a drawing (l/24th size). The line in **Figure 3.29C** measures 5′-8-1/2.

To obtain a half-size measurement, divide the full-size dimension by 2 and measure it with the 16 scale. Half size is sometimes specified as SCALE: 6 = 12 (inch marks omitted). The line in **Figure 3.29D** measures 0′-6-3/8.

Letter dimensions in feet and inches are shown in **Figure 3.30,** with fractions twice as tall as whole numerals.

Engineers' Scale

On the engineers' scale, each inch is divided into multiples of 10. Because it is used for making drawings of outdoor projects—streets, structures, tracts of land, and other topographical features—it is sometimes called the civil engineers' scale.

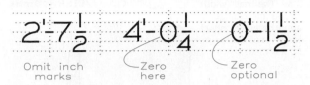

2'-7½   4'-0¼   0'-1½

Omit inch
marks

Zero
here

Zero
optional

3.30 Omit inch marks but show foot marks (according to current standards). When the inch measurement is less than a whole inch, use a leading zero.

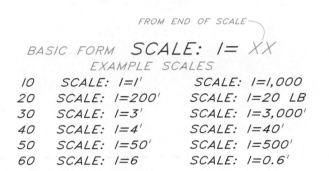

FROM END OF SCALE

BASIC FORM **SCALE: 1= XX**

EXAMPLE SCALES

10	SCALE: 1=1'	SCALE: 1=1,000
20	SCALE: 1=200'	SCALE: 1=20 LB
30	SCALE: 1=3'	SCALE: 1=3,000'
40	SCALE: 1=4'	SCALE: 1=40'
50	SCALE: 1=50'	SCALE: 1=500'
60	SCALE: 1=6	SCALE: 1=0.6'

3.31 Use this basic form to indicate the scale on a drawing made with an engineers' scale.

Figure 3.31 shows the form for specifying scales when using the engineers' scale. For example, scale: 1 = 10'. With measurements already in decimal form, calculations are easier because there is no need to convert common fractions as when using the architects' scale.

Each end of the scale is labeled 10, 20, 30, and so on, which indicates the number of units per inch on the scale **(Figure 3.32).** You may obtain many combinations simply by mentally moving the decimal places of a scale.

Figure 3.32A shows the use of the 10 scale to measure a line 32.0-feet-long drawn at the scale of 1 = 10'. **Figure 3.32B** shows use of the 20 scale to measure a line 540.0-feet-long drawn at a scale of 1 = 200'. **Figure 3.32C** shows use of the 30 scale to measure a line of 9.6 inches long at a scale of 1 = 3. **Figure 3.33** shows the proper format for indicating measurements in feet and inches.

English System of Units

The English (Imperial) system of units has been used in the United States, Great Britain

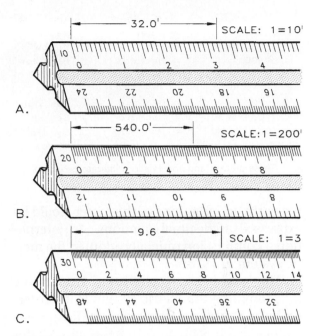

A.

B.

C.

3.32 These lines are measured with the engineers' scales.

Omit zero;
omit inch
marks

.13   2.13   0.15'   0.13"

GOOD
No zero
in front
of decimal

GOOD
Space for
decimal

GOOD
Zero for
fractional
feet

POOR
Decimal
point
crowded

3.33 For decimal fractions in inches, omit leading zeros and inch marks. For feet, leave adequate space for decimal points between numbers and show foot marks.

(until recently), and Canada since it was established. This system is based on arbitrary units (of length) of the inch, foot, cubit, yard, and mile. Because there is no common relationship among these units, calculations are cumbersome. For example, finding the area of a rectangle that measures 25 inches × 6-3/4 yards first requires conversion into common units.

Metric System (SI) of Units

France proposed the metric system in the fifteenth century. In 1793 the French National Assembly agreed that the meter (m) would be one ten-millionth of the meridian quadrant of

Value			Prefix	Symbol	Pronunciation
1 000 000	=	10^6	= Mega	M	"Megah"
1 000	=	10^3	= Kilo	k	"Keylow"
100	=	10^2	= Hecto	h	"Heck tow"
10	=	10^1	= Deka	da	"Dekah"
1	=				
0.1	=	10^{-1}	= Deci	d	"Des sigh"
0.01	=	10^{-2}	= Centi	c	"Cen'–ti"
0.001	=	10^{-3}	= Milli	m	"Mill lee"
0.000 001	=	10^{-6}	= Micro	μ	"Microw"

3.34 These prefixes and abbreviations indicate decimal placement for SI measurements.

the earth and fractions of the meter would be expressed as decimal fractions. An international commission officially adopted the metric system in 1875.

The worldwide organization responsible for promoting the metric system is the **International Standards Organization (ISO).** It has endorsed the Syst'me International d'Unites (International System of Units), abbreviated **SI.** Prefixes to SI units indicate placement of the decimal, as **Figure 3.34** shows.

Metric Scales

The meter is 39.37 inches. The basic metric unit of measurement for an engineering drawing is the millimeter (mm), which is one-thousandth of a meter, or one-tenth of a centimeter. Dimensions on a metric drawing are understood to be in millimeters unless otherwise specified.

The width of the fingernail of your index finger is a convenient way to approximate the dimension of one centimeter, or ten millimeters **(Figure 3.35).** Depicted in **Figure 3.36** is the format for specifying metric scales on a drawing.

Decimal fractions are unnecessary on most drawings dimensioned in millimeters. Thus dimensions are rounded off to whole numbers except for dimensions with specified tolerances. For measurements of less than 1, a zero goes in front of the decimal. In the English system, the zero is omitted from measurements of less than an inch **(Figure 3.37).**

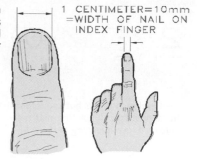

3.35 The nail width of your index finger is approximately equal to 1 centimeter, or 10 millimeters.

1 CENTIMETER=10mm =WIDTH OF NAIL ON INDEX FINGER

FROM END OF SCALE

BASIC FORM SCALE: 1= XX

EXAMPLE SCALES

SCALE: 1:1 (1mm=1mm; 1cm=1cm)

SCALE: 1:2 (1mm=2mm; 1mm=20mm)

SCALE: 1:3 (1mm=30mm; 1mm=0.3mm)

SCALE: 1:4 (1mm=4mm; 1mm=40mm)

SCALE: 1:5 (1mm=5mm; 1mm=500mm)

SCALE: 1:6 (1mm=6mm; 1mm=60mm)

3.36 Use this basic form to indicate the scale of a drawing made with a metric scale.

22.0 0.15 .15
GOOD GOOD POOR
Missing zero

Too little room for decimal

146 14.6 14.6
GOOD GOOD POOR

3.37 In the metric system, a zero precedes the decimal. Allow adequate space for it.

Metric scales are expressed as ratios: 1:20, 1:40, 1:100, 1:500, and so on. These ratios mean that one unit represents the number of units to the right of the colon. For example, 1:10 means that 1 mm equals 10 mm, or 1 cm equals 10 cm, or 1 m equals 10 m. The full-size metric scale **(Figure 3.38)** shows the relationship between the metric units of the decimeter, centimeter, millimeter, and micrometer. The line shown in **Figure 3.39A** measures

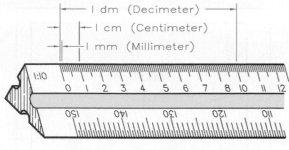

$$1 \text{ dm} = \frac{m}{10} \quad 1 \text{ cm} = \frac{m}{100} \quad 1 \text{ mm} = \frac{m}{1000} \quad 1 \text{ μm} = \frac{m}{1\,000\,000}$$

3.38 The decimeter is one-tenth of the meter, the centimeter is one-hundredth of a meter, a millimeter is one-thousandth of a meter, and a micrometer is one-millionth of a meter.

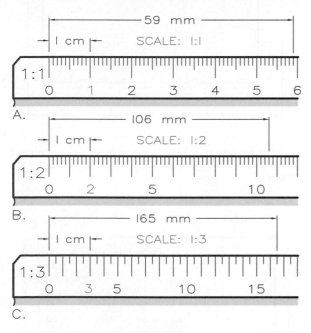

3.39 Measurements with metric scales.

59 mm. Use the 1:2 scale when 1 mm represents 2 mm, 20 mm, 200 mm, and so on. The line shown in **Figure 3.39B** measures 106 mm. **Figure 3.39C** shows a line measuring 165 mm, where 1 mm represents 3 mm.

Metric Symbols To indicate that drawings are in metric units, insert SI in or near the title block **(Figure 3.40).** The two views of the partial cone denote whether the orthographic views were drawn in accordance with the U.S. system

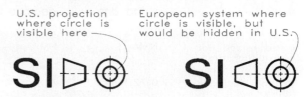

A. METRIC UNITS AND THIRD–ANGLE PROJECTION

B. METRIC UNITS AND FIRST–ANGLE PROJECTION

3.40 The large SI indicates that measurements are in metric units. The partial cones indicate whether the views are drawn in (A) the third-angle (U.S. system), or (B) the first-angle of projection (European system).

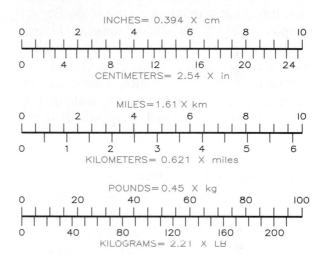

3.41 These scales show a comparison of the English system units with the metric system units.

(third-angle projection) or the European system (first-angle projection). **Figure 3.41** shows several comparisons of English and SI units.

Expression of Metric Units The general rules for expressing SI units are given in **Figure 3.42.** Do not use commas to separate digits in large numbers; instead, leave a space between them, as shown.

Scale Conversion

The Appendix gives factors for converting English to metric lengths and vice versa. For example, multiply decimal inches by 25.4 to obtain millimeters, and divide millimeters by 0.394 to obtain inches.

Omit commas and group into threes	I 000 000 GOOD	1,000,000 POOR
Use a raised dot for multiplication	N•M GOOD	NM POOR
Precede decimals with zeros	0.72 mm GOOD	.72 mm POOR
Methods of division	kg/m or kg•m⁻¹ GOOD GOOD	

3.42 Follow these general rules for showing SI units.

Multiply an architects' scale by 12 to convert it to an approximate metric scale. For example, Scale: 1/8 = 1'-0 is the same as 1/8 inch = 12 inches or 1 inch = 96 inches, which closely approximates the metric scale of 1:100. You cannot convert most metric scales exactly to English scales, but the metric scale of 1:60 does convert exactly to 1 = 5', which is the same as 1 in. = 60 in.

3.6 Presentation of Drawings

The following formats are suggested for the presentation of drawings. Most problems can be drawn and solved on size A (8-1/2 × 11-inch) sheets in a vertical format with a title strip as **Figure 3.43** shows. Size A sheets can also be laid out in a horizontal format as illustrated in **Figure 3.44.**

The standard sizes of sheets, from size A through size E and an alternative title strip for sizes A through E, are shown in **Figure 3.44.** Always use guidelines for lettering title strips.

Problems 1–9 (Figures 3.45–3.47): Draw these problems on size A sheets, with or without a printed grid, using the format shown in **Figure 3.45.** You may be assigned to solve two half-size drawings per sheet on size A sheets. Each grid represents 0.25 inch or 6 mm.

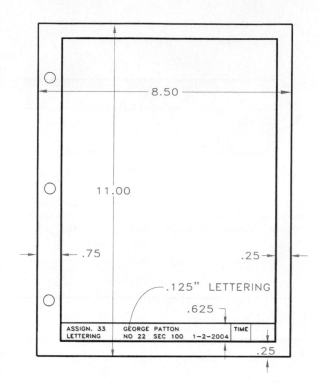

3.43 Use the format and title strip shown for a size A (vertical format) to present the solutions to problems at the end of each chapter.

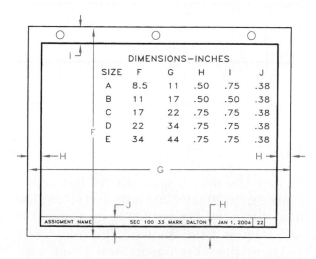

	DIMENSIONS—INCHES				
SIZE	F	G	H	I	J
A	8.5	11	.50	.75	.38
B	11	17	.50	.50	.38
C	17	22	.75	.75	.38
D	22	34	.75	.75	.38
E	34	44	.75	.75	.38

3.44 Format sizes for A through E sheets are shown here in columns F thru J.

Problems 10–20 (Figures 3.48–3.59): Draw full-size views on size A sheets (horizontal format) and omit the dimensions.

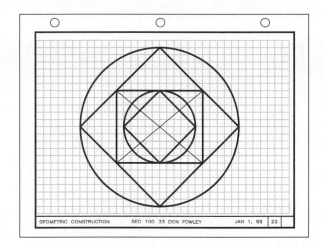

3.45 Problem 1 and sheet format.

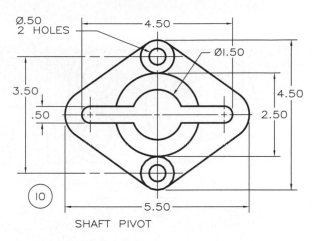

SHAFT PIVOT

3.48 Problem 10.

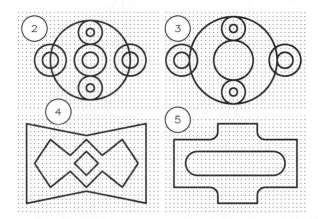

3.46 Problems 2–5.

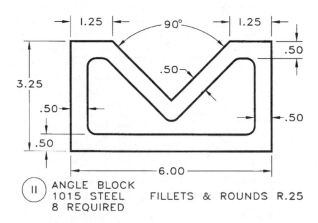

ANGLE BLOCK
1015 STEEL   FILLETS & ROUNDS R.25
8 REQUIRED

3.49 Problem 11.

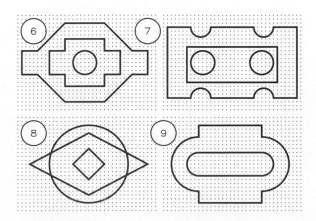

3.47 Problems 6–9.

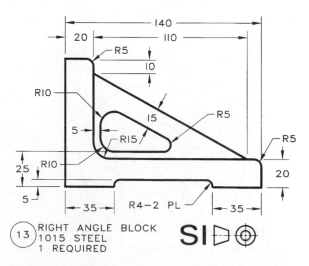

RIGHT ANGLE BLOCK
1015 STEEL
1 REQUIRED

3.50 Problem 12.

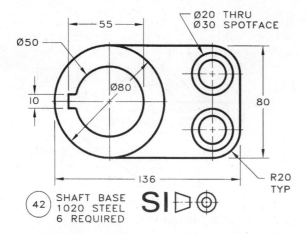

42 SHAFT BASE
1020 STEEL
6 REQUIRED SI ⊳⊕

3.51 Problem 13.

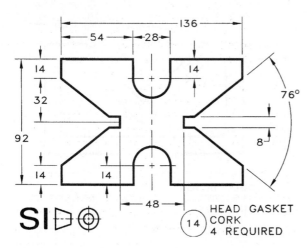

SI ⊳⊕

14 HEAD GASKET
CORK
4 REQUIRED

3.52 Problem 14.

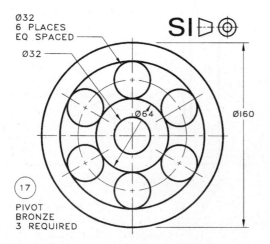

SI ⊳⊕

Ø32
6 PLACES
EQ SPACED

Ø32

Ø64

Ø160

17 PIVOT
BRONZE
3 REQUIRED

3.53 Problem 15.

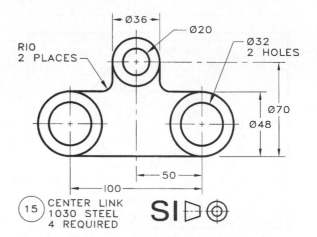

15 CENTER LINK
1030 STEEL
4 REQUIRED SI ⊳⊕

3.54 Problem 16.

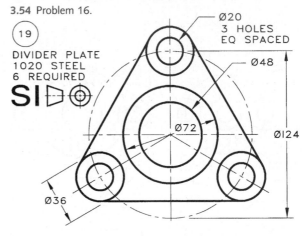

19 DIVIDER PLATE
1020 STEEL
6 REQUIRED

SI ⊳⊕

3.55 Problem 17.

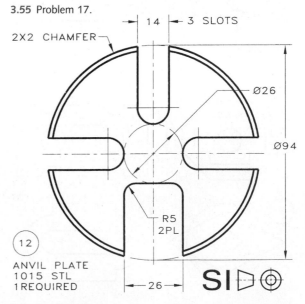

12 ANVIL PLATE
1015 STL
1 REQUIRED SI ⊳⊕

3.56 Problem 18.

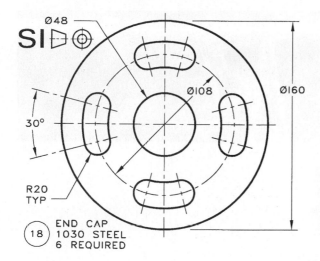

SI ⬤

Ø48

Ø108

Ø160

30°

R20
TYP

(18) END CAP
1030 STEEL
6 REQUIRED

3.57 Problem 19.

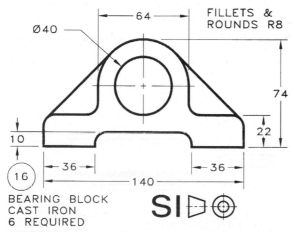

Ø40

64

FILLETS &
ROUNDS R8

74

10

22

36

36

140

(16)

BEARING BLOCK
CAST IRON
6 REQUIRED

SI ⬤

3.58 Problem 20.

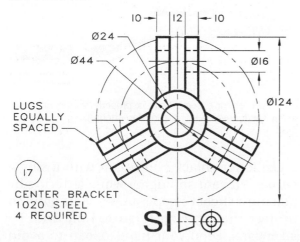

10 ⟶ 12 ⟵ 10

Ø24

Ø44

Ø16

Ø124

LUGS
EQUALLY
SPACED

(17)

CENTER BRACKET
1020 STEEL
4 REQUIRED

SI ⬤

3.59 Problem 21.

Design Application: Design 1.

The knob in **Figure 3.60** has a diameter of 3 inches (flat-to-flat). Make an instrument drawing of the descriptive end of this part; estimate the dimensions as if you were its designer.

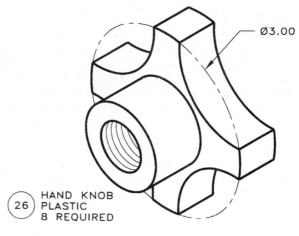

Ø3.00

(26) HAND KNOB
PLASTIC
8 REQUIRED

3.60 Design 1.

Design 2.

Make an instrument drawing of the film reel in **Figure 3.61** as it is shown, or design your own configuration of it using your own geometry while adherring to the overall dimensions.

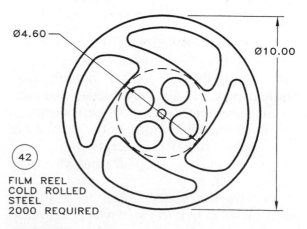

Ø4.60

Ø10.00

(42)

FILM REEL
COLD ROLLED
STEEL
2000 REQUIRED

3.61 Design 2.

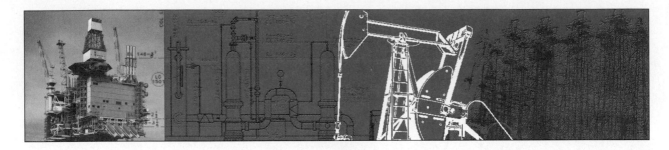

4

Lettering

4.1 Introduction

Notes, dimensions, and specifications, which must be lettered, supplement all drawings. The ability to letter freehand is an important skill to develop because it affects the use and interpretation of drawings. It also displays an engineer's skill with graphics, and may be taken as an indication of professional competence.

4.2 Lettering Tools

The best pencils for lettering on most surfaces are the H, F, and HB grades, with an F grade pencil being the one most commonly used. Some papers and films are coarser than others and may require a harder pencil lead. To give the desired line width, round the point of the pencil slightly **(Figure 4.1)** because a needle point is likely to snap off when you apply pressure to it.

Revolve the pencil about its axis with each stroke so that the lead will wear evenly. For good reproduction, bear down firmly to

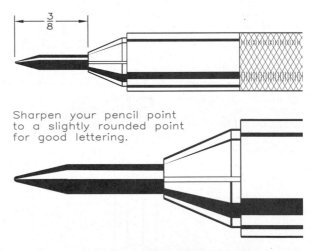

Sharpen your pencil point to a slightly rounded point for good lettering.

4.1 Good lettering begins with a properly sharpened pencil point. The F grade pencil is good for lettering.

make letters black and bright with a single stroke. Prevent smudging while lettering by placing a sheet of paper under your hand to protect the drawing **(Figure 4.2).** And, by all means, keep your hands clean to avoid smudges.

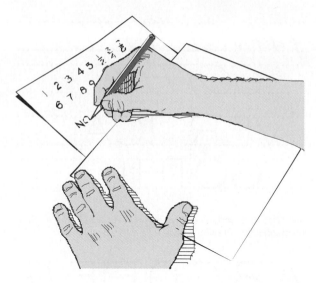

4.2 Place a protective sheet under your hand to prevent smudges and work from a comfortable position for natural strokes.

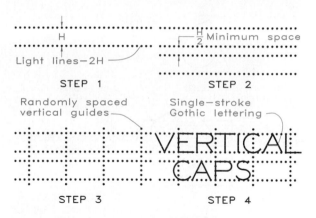

4.3 **Method of using lettering guidelines**
Step 1 Lay off letter heights, H, and draw light guidelines with a 2H pencil.
Step 2 Space lines no closer than H/2 apart.
Step 3 Draw vertical guidelines as light, thin, randomly spaced lines.
Step 4 Draw letters with single strokes using a medium-grade pencil: H, F, or HB. Leave the guidelines.

4.3 Guidelines

The most important rule of lettering is **use guidelines at all times,** whether you are lettering a paragraph or a single letter. **Figure 4.3** shows how to draw and use guidelines. Use a sharp pencil in the 2H, 3H, or 4H grade range and draw light guidelines, just dark enough to be seen.

Most lettering on an engineering drawing is done with capital letters that are 1/8 inch (3 mm) high. The spacing between lines of lettering should be no closer than half the height of the capital letters, or 1/16 inch in this case.

Lettering Guides

Two instruments for drawing guidelines are the **Braddock-Rowe** lettering triangle and the **Ames** lettering instrument.

The **Braddock-Rowe triangle** contains sets of holes for spacing guidelines (**Figure 4.4).** The numbers under each set of holes represent thirty-seconds of an inch. For example, the numeral 4 represents 4/32 inch or 1/8 inch for making uppercase (capital) letters. Some triangles have millimeter markings. Intermediate holes provide guidelines

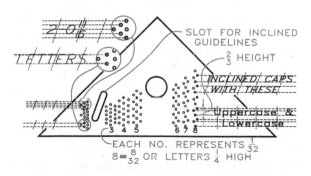

4.4 The Braddock-Rowe triangle is used for drawing guidelines for lettering. The numbers near the guideline holes represent thirty-seconds of an inch.

for lowercase letters, which are not as tall as capital letters.

While holding a horizontal straightedge firmly in position, place the Braddock-Rowe triangle against its upper edge. Insert a sharpened 2H pencil point in the desired guideline hole to contact the drawing surface and guide the pencil point across the paper, drawing the guideline while the triangle slides along the straightedge. Repeat this procedure by moving the pencil point to each successive

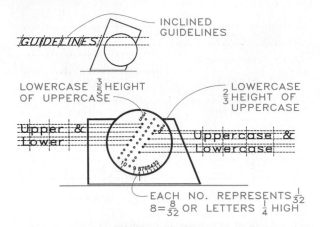

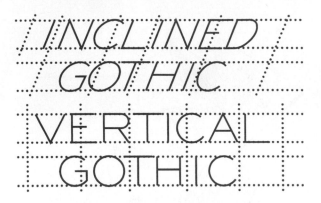

4.6 Use either vertical or inclined single-stroke Gothic lettering on engineering drawings.

4.5 The Ames guide is used for drawing guidelines for lettering. Set the dial to the desired number of thirty-seconds of an inch for the height of uppercase letters.

hole to draw other guidelines. Use the slanted slot in the triangle to draw guidelines, spaced randomly, for inclined lettering.

The **Ames lettering guide (Figure 4.5)** is a similar device but has a circular dial for selecting guideline spacing. The numbers around the dial represent thirty-seconds of an inch. For example, the number 8 represents 8/32 inch, or guidelines for drawing capital letters that are 1/4 inch tall.

4.4 Gothic Lettering

The lettering recommended for engineering drawings is single-stroke Gothic lettering, so called because the letters are a variation of the Gothic style made with a series of single strokes. Gothic lettering may be vertical or inclined **(Figure 4.6),** but only one style or the other should be used on a single drawing.

Vertical Letters
Uppercase The alphabet in the single-stroke Gothic uppercase (capital) letters is shown in **Figure 4.7.** Each letter is drawn inside a square box of guidelines to show their correct proportions. Draw straight lines with a single stroke; for example, draw the letter A with three single strokes. Letters composed of curves can best

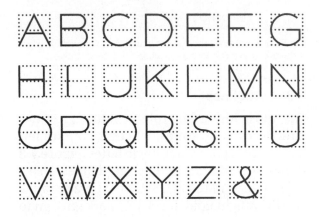

4.7 This alphabet shows the form of single-stroke Gothic vertical uppercase letters. Letters are drawn inside squares to show their proportions.

be drawn in segments; for example, draw the letter O by joining two semicircles.

The shape (form) of each letter is important. Small wiggles in strokes will not detract from your lettering if the letter forms are correct. **Figure 4.8** shows common errors in lettering that you should avoid.

Lowercase The alphabet of lowercase letters is shown in **Figure 4.9,** which should be either two-thirds or three-fifths as tall as uppercase letters. Both lowercase ratios are labeled on the Ames guide, but only the two-thirds ratio is available on the Braddock-Rowe triangle.

Some lowercase letters, such as the letter b, have ascenders that extend above the body

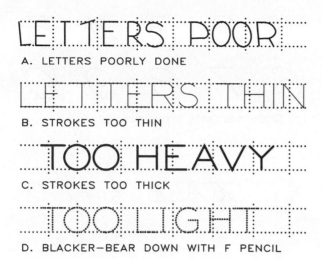

A. LETTERS POORLY DONE

B. STROKES TOO THIN

C. STROKES TOO THICK

D. BLACKER—BEAR DOWN WITH F PENCIL

4.8 Avoid these common errors when lettering.

Some lowercase letters, such as the letter b, have ascenders that extend above the body of the letter; some, such as the letter p, have descenders that extend below the body. Ascenders and descenders are equal in length.

The guidelines in **Figure 4.9** that form squares about the body of each letter are used to illustrate their proportions. Letters that have circular bodies may extend slightly beyond the sides of the guideline squares. **Figure 4.10** shows examples of capital and lowercase letters used together.

Numerals Vertical numerals for use with single-stroke Gothic lettering are shown in **Figure 4.11;** each numeral is enclosed in a square of guidelines. Numbers should be the same height as the capital letters being used, usually 1/8 inch. The numeral 0 (zero) is an oval, whereas the letter O is a circle in vertical lettering.

Inclined Letters

Uppercase Inclined uppercase (capital) letters have the same heights and proportions as vertical letters; the only difference is their 68° inclination **(Figure 4.12).** Guidelines for inclined lettering can be drawn with both the Braddock-Rowe triangle and the Ames guide.

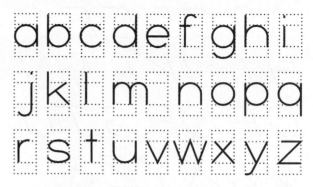

4.9 The alphabet is drawn here in single-stroke Gothic vertical lowercase letters.

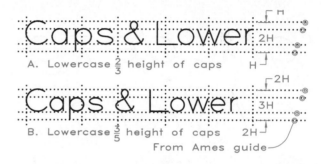

A. Lowercase $\frac{2}{3}$ height of caps

B. Lowercase $\frac{3}{5}$ height of caps

From Ames guide

4.10 The ratio of lowercase letters to uppercase letters should be either two-thirds (A) or three-fifths (B). The Ames guide has both ratios, but the Braddock-Rowe triangle has only the two-thirds ratio.

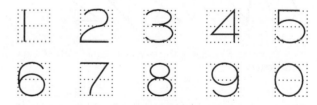

4.11 These numerals are used with single-stroke Gothic, vertical letters.

Lowercase Inclined lowercase letters are drawn in the same manner as vertical lowercase letters **(Figure 4.13),** but circular features are drawn as ovals (ellipses). Their angle of inclination is 68°, the same as for uppercase letters.

Numerals Examples of inclined numerals and letters used in combination are shown in

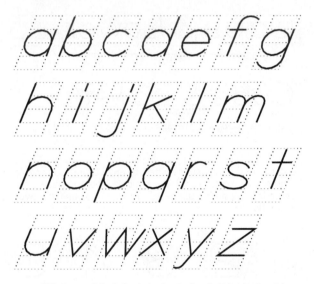

4.12 This is an alphabet of single-stroke Gothic, inclined upper case letters.

4.13 This is an alphabet of single-stroke Gothic, inclined lowercase letters.

Figure 4.14 and **Figure 4.15.** The ratio of lowercase to the uppercase letters should be either two-thirds (A) or three-fifths (B). The Ames guide has both ratios, but the Braddock-Rowe triangle has only the two-thirds ratio. Inclined letters and numbers are used in combination in **Figure 4.15.** Guidelines are drawn using the Braddock-Rowe triangle or the Ames lettering guide shown in **Figure 4.16.**

Spacing Numerals and Letters
Allow adequate space between numerals for the decimal point and fractions **(Figure 4.17).**

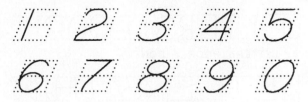

4.14 Single-stroke Gothic, inclined numerals.

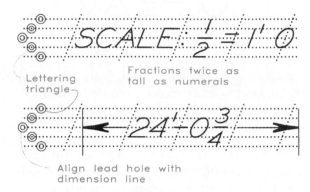

4.15 Inclined common fractions are twice as tall as single numerals. Omit inch marks in dimensions.

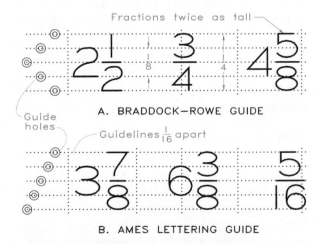

4.16 Guidelines are drawn for fractions so they can be twice as tall as single numerals.

Common fractions are twice as tall as single numerals **(Figure 4.17).** Both the Braddock-Rowe triangle and the Ames guide have separate sets of holes spaced 1/16 inch apart for common fractions. The center guideline locates the fraction's crossbar.

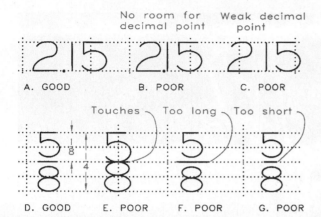

No room for decimal point Weak decimal point

A. GOOD   B. POOR   C. POOR

Touches   Too long   Too short

D. GOOD   E. POOR   F. POOR   G. POOR

4.17 Avoid making these errors in lettering fractions.

A. GOOD—EQUAL AREAS BETWEEN LETTERS

B. POOR—EQUAL LINEAR SPACING

4.18 Letters should be spaced so that the areas between them are about equal.

When grouping letters to spell words, make the areas between the letters approximately equal for the most visually pleasing result **(Figure 4.18)**. **Figure 4.19** shows the incorrect use of guidelines and other violations of good lettering practice that you should avoid.

Problems

Complete these lettering exercises on size A (11 × 8.5-inch, horizontal format) sheets, with or without printed grids, using the format shown in **Figure 4.20.**

1. Draw the alphabet in vertical (or inclined), uppercase letters as shown in **Figure 4.20.** Draw each letter three times.

2. Draw vertical (or inclined) numerals and the alphabet in lowercase letters as shown in **Figure 4.21.**

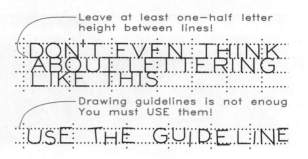

Leave at least one-half letter height between lines!

Drawing guidelines is not enoug You must USE them!

4.19 Always use guidelines (vertical and horizontal) whether you are lettering a paragraph or a single letter.

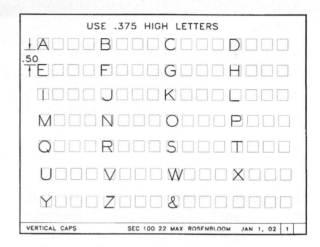

4.20 The sheet layout for lettering assignments.

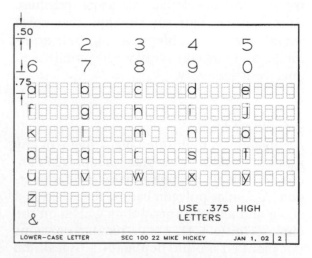

4.21 A typical sheet layout for lettering assignments.

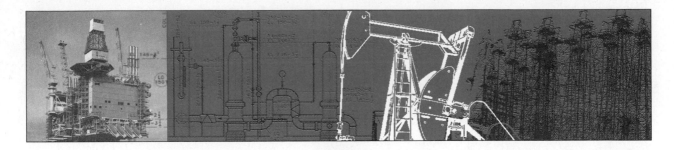

5

Geometric Construction

5.1 Introduction

The solution of many graphical problems requires the use of geometry and geometric construction. Because mathematics was an outgrowth of graphical construction, the two areas are closely related. The proofs of many principles of plane geometry and trigonometry can be developed by using graphics. Graphical methods can be applied to solve some types of problems in algebra, arithmetic, and virtually all types of problems in analytical geometry.

5.2 Polygons

A **polygon** is a multisided plane figure of any number of sides. If the sides of a polygon are equal in length, the polygon is a **regular polygon.** A regular polygon can be inscribed in a circle and all its corner points will lie on the circle **(Figure 5.1).** Other regular polygons not pictured are the **heptagon** (7 sides), the

| SQUARE | PENTAGON | HEXAGON | OCTAGON |
| 4 sides | 5 sides | 6 sides | 8 sides |

5.1 Regular polygons can be inscribed in circles.

nonagon (9 sides), the **decagon** (10 sides), and the **dodecagon** (12 sides).

The sum of the angles inside any polygon (interior angles) is $S = (n - 2) \times 180°$, where n is the number of sides of the polygon.

Triangles

A **triangle** is a three-sided polygon. The four types of triangles are **scalene, isosceles, equilateral,** and **right** triangles **(Figure 5.2).** The sum of the interior angles of a triangle is 180° $[(3 - 2) \times 180°]$.

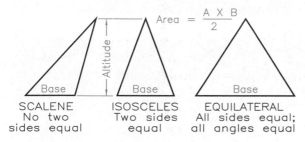

$$\text{Area} = \frac{A \times B}{2}$$

SCALENE	ISOSCELES	EQUILATERAL
No two sides equal	Two sides equal	All sides equal; all angles equal

5.2 Types of triangles and their definitions.

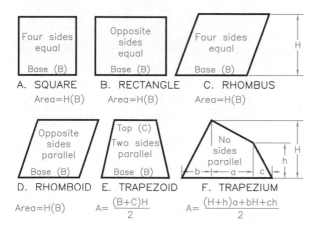

A. SQUARE	B. RECTANGLE	C. RHOMBUS
Four sides equal	Opposite sides equal	Four sides equal
Base (B)	Base (B)	Base (B)
Area=H(B)	Area=H(B)	Area=H(B)

D. RHOMBOID	E. TRAPEZOID	F. TRAPEZIUM
Opposite sides parallel	Top (C) Two sides parallel	No sides parallel
Base (B)	Base (B)	
Area=H(B)	$A=\dfrac{(B+C)H}{2}$	$A=\dfrac{(H+h)a+bH+ch}{2}$

5.3 Types of quadrilaterals, their definitions, and how to calculated their areas.

Quadrilaterals

A **quadrilateral** is a four-sided polygon of any shape. The sum of the interior angles of a quadrilateral is $360°$ $[(4 - 2) \times 180°]$. **Figure 5.3** shows the various types of quadrilaterals and the equations for their areas.

5.3 Circles

Figure 5.4 gives the names of the elements of a circle that are used throughout this textbook. A circle is constructed by swinging a radius from a fixed point through $360°$. The area of a circle equals πR^2.

5.4 Geometric Solids

Figure 5.5 illustrates the various types of solid geometric shapes and their characteristics.

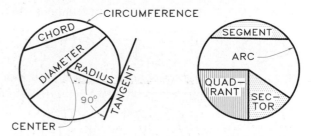

5.4 Elements of a circle and their definitions.

Polyhedra

A **polyhedron** is a multisided solid formed by intersecting planes. If its faces are regular polygons (having sides of equal shape), it is a regular polyhedron. Five regular polyhedra are the **tetrahedron** (4 sides), the **hexahedron** (6 sides), the **octahedron** (8 sides), the **dodecahedron** (12 sides), and the **icosahedron** (20 sides).

Prisms

A **prism** has two parallel bases of equal shape connected by sides that are parallelograms. The line from the center of one base to the center of the other is the axis. If its axis is perpendicular to the bases, the prism is a **right prism.** If its axis is not perpendicular to the bases, the prism is an **oblique prism.** A prism that has been cut off to form a base not parallel to the other is a **truncated prism.** A **parallelepiped** is a prism with bases that are either rectangles or parallelograms.

Pyramids

A **pyramid** is a solid with a polygon as a base and triangular faces that converge at a vertex. The line from the vertex to the center of the base is the **axis.** If its axis is perpendicular to the base, the pyramid is a **right pyramid.** If its axis is not perpendicular to the base, the pyramid is an **oblique pyramid.** A truncated pyramid is called a **frustum** of a pyramid.

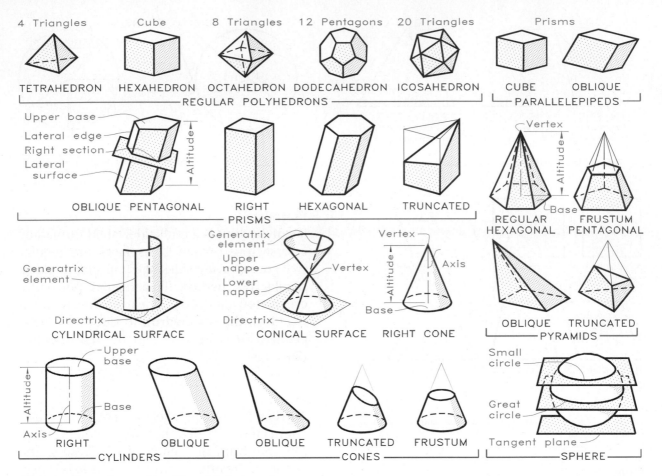

5.5 Various types of geometric solids and their elements.

Cylinders

A **cylinder** is formed by a line or an element (called a *generatrix*) that moves about the circle while remaining parallel to its axis. The axis of a cylinder connects the centers of each end of a cylinder. If the axis is perpendicular to the bases, it is the altitude of a **right cylinder.** If the axis does not make a 90° angle with the base, the cylinder is an **oblique cylinder.**

Cones

A **cone** is formed by a generatrix, one end of which moves about the circular base while the other end remains at a fixed vertex. The line from the center of the base to the vertex is the axis. If its axis is perpendicular to the base, the cone is a **right cone.** A truncated cone is called a **frustum** of a cone.

Spheres

A **sphere** is generated by revolving a circle about one of its diameters to form a solid. The ends of the axis of revolution of the sphere are **poles.**

5.5 Polygons

A regular polygon (having equal sides) can be inscribed in or circumscribed about a circle.

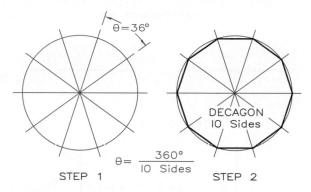

5.6 A regular polygon (sides of equal length):
Step 1 Divide circle into the correct number of sectors.
Step 2 Connect the division points with straight lines where they intersect the circle.

When it is inscribed, all corner points will lie on the circle **(Figure 5.6).** For example, constructing a 10-sided polygon involves dividing the circle into 10 sectors and connecting the points to form the polygon.

Triangles
When you know the lengths of all three sides of a triangle, you may construct it with a compass by triangulation, as **Figure 5.7** shows.

Hexagons
The **hexagon,** a six-sided regular polygon, can be inscribed in or circumscribed about a circle

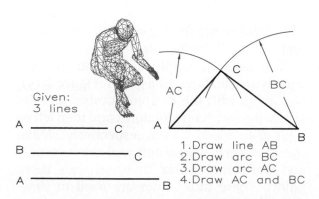

5.7 A triangle can be constructed by triangulation with a compass when three sides are given.

(Figure 5.8). Use a 30°–60° triangle to draw the hexagon. The circle represents the distance from corner to corner for an inscribed hexagon and from flat to flat when the hexagon is circumscribed about a circle.

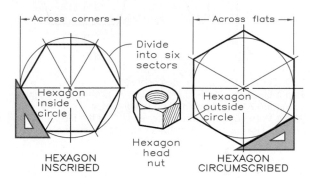

5.8 A hexagon can be inscribed in or circumscribed about a circle with a 30°–60° triangle.

Octagons
The **octagon,** an eight-sided regular polygon, can be inscribed in or circumscribed about a circle **(Figure 5.9).** Use a 45° triangle in both circumscribing and inscribing an octagon about a circle.

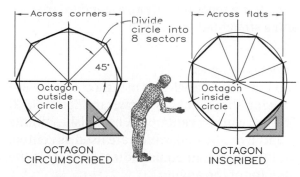

5.9 An octagon can be inscribed in or circumscribed about a circle with a 45° triangle.

Pentagons
The **pentagon,** a five-sided regular polygon, can be inscribed in or circumscribed about a circle. **Figure 5.10** shows the steps of constructing a

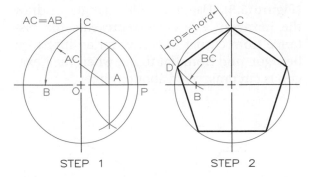

STEP 1 STEP 2

5.10 Constructing an inscribed pentagon:
Step 1 Bisect radius OP to locate point A. With A as the center and radius AC, locate point B on the diameter.
Step 2 With point C as the center and BC as the radius, locate point D. Use line CD as the chord to locate the other corners of the pentagon.

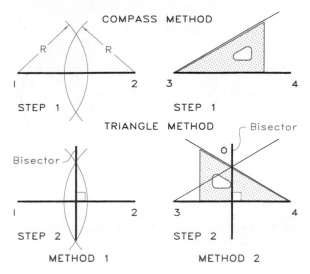

COMPASS METHOD

STEP 1 STEP 1

TRIANGLE METHOD

STEP 2 STEP 2

METHOD 1 METHOD 2

5.11 Bisecting a line:
Method 1 Use a compass and any radius.
Method 2 Use a standard triangle and a straightedge.

pentagon with a compass and straightedge where the vertices lie on the circle.

5.6 Bisecting Lines and Angles

Bisecting Lines

Two methods of finding the midpoint of a line with a perpendicular bisector are shown in **Figure 5.11.** In the first method, a compass is used to construct the perpendicular bisector to the line. In the second method, a standard triangle and a straightedge are used.

Bisecting Angles

You may bisect angles by using a compass and drawing three arcs, as shown in **Figure 5.12.** The arc with its center at A establishes points D and E on the line, which are used as centers for two arcs of an equal radius. Line AO is the bisector of the angle.

5.7 Division of Lines

Dividing a line into several equal parts is often necessary. **Figure 5.13** shows the method used to solve this type of problem where line

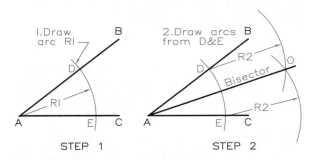

STEP 1 STEP 2

5.12 Bisecting an angle:
Step 1 Swing an arc R1 to locate points D and E.
Step 2 Draw equal arcs from D and E to locate point O. Line AO is the bisector of the angle.

AB is divided into five equal lengths by using a scale to lay off the five divisions.

The same principle applies to locating equally spaced lines on a graph **(Figure 5.14).** Lay scales with the desired number of units across the graph up and down and then left to right (0 to 3 and 0 to 5, respectively). Make marks at each whole unit and draw vertical and horizontal index lines through these points. These index lines are used to show data in a graph.

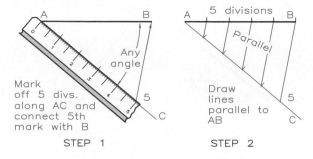

5.13 Dividing a line:
Step 1 To divide line AB into five equal lengths, lay off five equal divisions along line AC, and connect point 5 to end B with a construction line.
Step 2 Draw a series of five construction lines parallel to 5B to divide line AB.

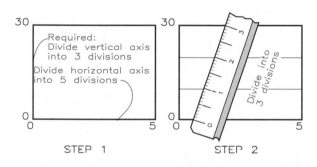

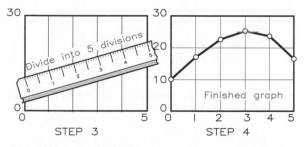

5.14 Dividing axes on a graph:
Step 1 Draw the outline of the graph.
Step 2 Divide the *y* axis into three equal segments by laying a scale across the graph with the 0 and 3 located on the top and bottom lines. Make marks at points 1 and 2 and draw horizontal lines through them.
Step 3 Divide the *x* axis into five equal segments by laying a scale across the graph with the 0 and 5 on the left- and right-hand vertical lines. Make marks at points 1, 2, 3, and 4 and draw vertical lines through them.
Step 4 Plot the data points and draw the curve to complete the graph.

5.8 An Arc Through Three Points

An arc can be drawn through three points by connecting the points with two lines and drawing perpendicular bisectors through each line to locate the center of the circle at C **(Figure 5.15).** Draw the arc, and lines AB and BD become chords of the arc.

To find the center of a circle or an arc, reverse this process by drawing two chords that intersect at a point on the circumference and bisecting them. The perpendicular bisectors intersect at the center of the circle.

5.9 Parallel Lines

You may draw one line parallel to another by using either method shown in **Figure 5.16.** In the first method, use a compass and draw two arcs having radius R to locate a parallel line at the desired distance (R) from the first line. In the second method, measure the desired perpendicular distance R from the first line, mark it, and draw the parallel line through it with your drafting machine.

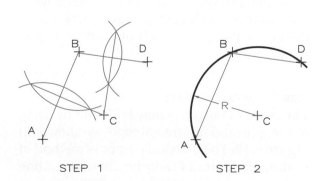

5.15 An arc through three points:
Step 1 Connect points A, B, and D with two lines and construct their perpendicular bisectors, which intersect at the center, C.
Step 2 Using the center C, and the distance to the points as the radius, R, draw the arc through the points.

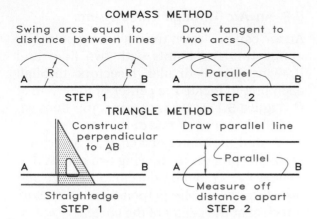

COMPASS METHOD

Swing arcs equal to distance between lines

Draw tangent to two arcs

A   R   R   B
STEP 1

A   Parallel   B
STEP 2

TRIANGLE METHOD

Construct perpendicular to AB

Draw parallel line

A   B
Straightedge
STEP 1

A   Parallel   B
Measure off distance apart
STEP 2

5.16 Drawing parallel lines:
Compass Method
Step 1 Swing two arcs from line AB.
Step 2 Draw the parallel line tangent to the arcs.
Triangle Method
Step 1 Draw a line perpendicular to AB.
Step 2 Measure the desired distance, R, along the perpendicular and draw the parallel line through it.

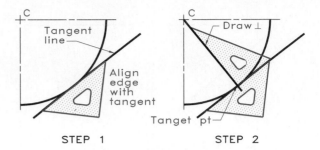

C
Tangent line

Align edge with tangent

STEP 1

C
Draw ⊥

Tanget pt

STEP 2

5.17 Locating a tangent point:
Step 1 Align a triangle with the tangent line while holding it against a firmly held straightedge.
Step 2 Hold the triangle in position, locate a second triangle perpendicular to it, and draw a line from the center to locate the tangent point.

5.10 Tangents

Marking Points of Tangency

A point of tangency is the theoretical point at which a line joins an arc or two arcs join without crossing. **Figure 5.17** shows how to find the point of tangency with triangles by constructing a perpendicular line to the tangent line from the arc's center. **Figure 5.18** shows the conventional methods of marking points of tangency.

Line Tangent to an Arc

The point of tangency may be found by using a triangle and a straightedge as shown in **Figure 5.19.** The classical compass method of finding the point of tangency between a line and a point is shown in **Figure 5.20.** Connect point A to the arc's center and bisect line AC (step 1); swing an arc from point M through point C locating tangent point T (step 2); draw the tangent to T (step 3); and mark the tangent point (step 4).

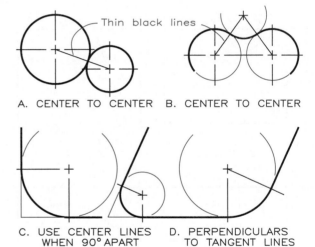

Thin black lines

A. CENTER TO CENTER   B. CENTER TO CENTER

C. USE CENTER LINES WHEN 90° APART   D. PERPENDICULARS TO TANGENT LINES

5.18 Use thin, black lines that extend from the centers slightly beyond the arcs to mark tangency points.

Arc Tangent to a Line from a Point

To construct an arc that is tangent to line DE at T and that passes through point P **(Figure 5.21),** draw the perpendicular bisector of line TP. Draw a perpendicular to line DE at point T to locate the center at point C and swing an arc with radius CT.

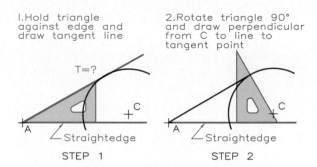

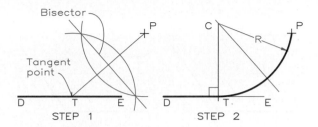

5.19 A tangent to an arc from a point:
Step 1 Hold a triangle against a straightedge and draw a line from point A that is tangent to the arc.
Step 2 Rotate your triangle 90° and locate the point of tangency by drawing a line through center C.

5.21 An arc through two points:
Step 1 To draw an arc through point P tangent to line DE at point T, draw the perpendicular bisector of TP.
Step 2 Construct a perpendicular to line DE at point T to intersect the bisector at point C and draw the arc from C with radius CT.

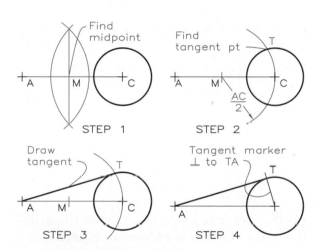

5.20 A line tangent to an arc from a point:
Step 1 Connect point A with center C and locate point M by bisecting AC.
Step 2 Using point M as the center and MC as the radius, locate point T on the arc.
Step 3 Draw the line from A to T tangent to the arc at T.
Step 4 Draw the tangent marker perpendicular to TA from the center past the arc as a thin, dark line.

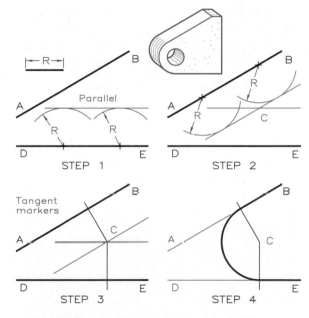

5.22 An arc tangent to lines making an acute angle:
Step 1 Construct a light construction line parallel to line DE with radius R.
Step 2 Draw a second light line parallel to and R distance from line AB to locate center C.
Step 3 Draw thin, dark lines from center C perpendicular to lines AB and DE to locate the tangency points.
Step 4 Draw the tangent arc and darken lines.

Arc Tangent to Two Lines

Figure 5.22 shows how to construct an arc of a given radius tangent to two nonparallel lines that form an acute angle. The same steps apply to constructing an arc tangent to two lines that form an obtuse angle **(Figure 5.23).** In both cases, the points of tangency are located with thin, dark lines drawn from their

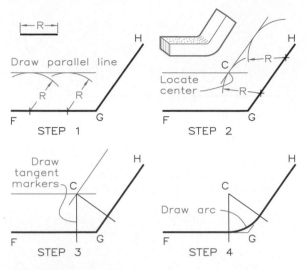

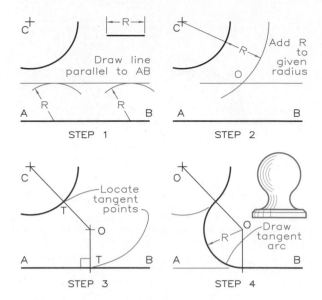

5.23 An arc tangent to lines making an obtuse angle:
Step 1 Using radius R, draw a light line parallel to FG.
Step 2 Construct a light line parallel to line GH that is R distance from it to locate center C.
Step 3 Draw thin lines from center C perpendicular to lines FG and GH to locate the tangency points.
Step 4 Draw the tangent arc and darken your lines.

5.25 An arc tangent to an arc and a line:
Step 1 Draw a line parallel to AB, R distance from it.
Step 2 Add radius R to the radius from center C. Swing the extended radius to find the center O.
Step 3 Lines OC and OT locate the tangency points.
Step 4 Draw the tangent arc between the points of tangency with radius R and center O.

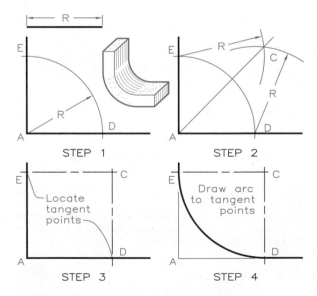

5.24 An arc tangent to perpendicular lines:
Step 1 Using radius R and center A, locate D and E.
Step 2 Find C by swinging two arcs with radius R.
Step 3 Perpendiculars CE and CD locate tangent pts.
Step 4 Draw the tangent arc and darken your lines.

centers perpendicular to and past the original lines. **Figure 5.24** shows a technique for finding an arc tangent to lines that are perpendicular.

Arc Tangent to an Arc and a Line
Figure 5.25 shows the steps for constructing an arc tangent to an arc and a line. **Figure 5.26** shows a variation of this technique for an arc drawn tangent to a given arc and line with the arc reversed.

Arc Tangent to Two Arcs
Figure 5.27 shows how to draw an arc tangent to two arcs. Lines drawn between the centers locate the points of tangency. The resulting tangent arc is concave from the top.

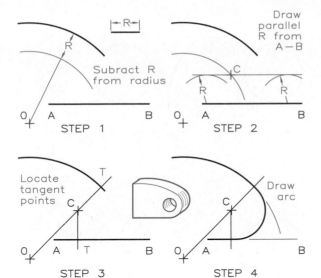

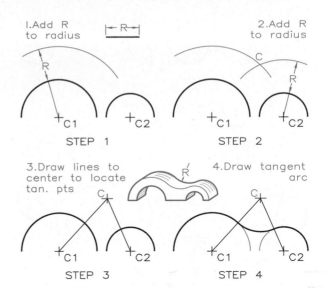

5.26 An arc tangent to an arc and a line:
Step 1 Subtract radius R from the radius through center O. Draw a concentric arc.
Step 2 Draw a line parallel to line 1-2 and R distance from it to locate the center at C.
Step 3 Locate the tangency points with lines OC and from C perpendicular to 1-2.
Step 4 Draw the tangent arc between the tangent points with radius R and center C.

5.27 A concave arc tangent to two arcs:
Step 1 Extend the radius by adding the radius R to it. Draw a concentric arc with the extended radius.
Step 2 Extend the radius of the other circle by adding radius R to it. Use this extended radius to construct a concentric arc to locate center C.
Step 3 Connect center C with centers C1 and C2 with thin, dark lines to locate the tangency points.
Step 4 Draw the tangent arc between the points of tangency using radius R and center C.

Drawing a convex arc tangent to the given arcs requires that its radius be greater than the radius of either of the given arcs, as shown in **Figure 5.28.**

One variation of this problem **(Figure 5.29)** is to draw an arc of a given radius tangent to the top of one arc and the bottom of the other. Another **(Figure 5.30)** is to draw an arc tangent to a circle and a larger arc.

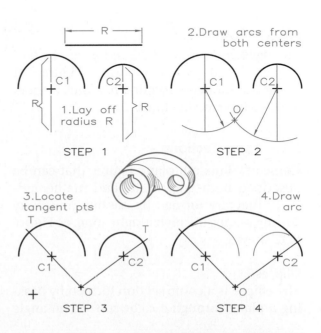

5.28 A convex arc tangent to two arcs:
Step 1 Extend each radius from the arc past its center by a distance of R, the radius, along these lines.
Step 2 Use the distance from each center to the ends of the extended radii to swing arcs to locate center 0.
Step 3 Draw thin, dark lines from center 0 through centers C1 and C2 to locate the points of tangency.
Step 4 Draw the tangent arc between the tangent points using radius R and center 0.

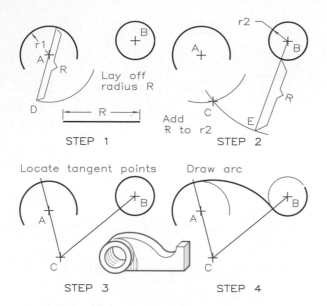

STEP 1 · Lay off radius R

STEP 2 · Add R to r2

STEP 3 · Locate tangent points

STEP 4 · Draw arc

5.29 An arc tangent to two circles:
Step 1 Lay off radius R from the arc along an extended radius to locate point D. Draw arc AD.
Step 2 Extend the radius from center B and add radius R to it to point E. Use radius BE to locate center C.
Step 3 Draw thin lines from center C through centers A and B to locate the points of tangency.
Step 4 Draw the tangent arc between the tangent points using radius R and center C.

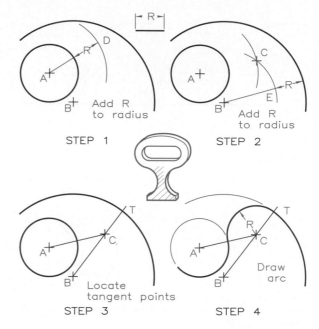

STEP 1 · Add R to radius

STEP 2 · Add R to radius

STEP 3 · Locate tangent points

STEP 4 · Draw arc

5.30 An arc tangent to two arcs:
Step 1 Add radius R to the radius from A. Use radius AD to draw a concentric arc from center A.
Step 2 Subtract radius R from the radius through B. Use radius BE to draw an arc to locate center C.
Step 3 Draw thin lines to connect the centers and mark the points of tangency.
Step 4 Draw the tangent arc between the tangency points using radius R and center C.

Ogee Curves

The **ogee curve** is an S curve formed by tangent arcs. The ogee curve shown in **Figure 5.31** is the result of constructing two arcs tangent to three intersecting lines.

5.11 Conic Sections

Conic sections are plane figures that can be described both graphically and mathematically; they are formed by passing imaginary cutting planes through a right cone, as **Figure 5.32** shows.

Ellipses

The ellipse is a conic section formed by passing a plane through a right cone at an angle

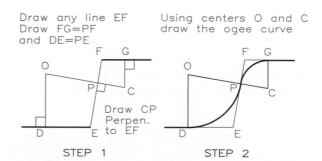

Draw any line EF · Draw FG=PF and DE=PE

Using centers O and C draw the ogee curve

Draw CP Perpen. to EF

STEP 1

STEP 2

5.31 An ogee curve:
Step 1 To draw an ogee curve between two parallel lines, draw light line EF at any angle. Locate P anywhere on EF. Find the tangent points by making FG equal to FP and DE equal to EP. Draw perpendiculars at G and D to intersect the perpendicular at O and C.
Step 2 Use radii CP and OP at centers O and C to draw two tangent arcs to complete the ogee curve.

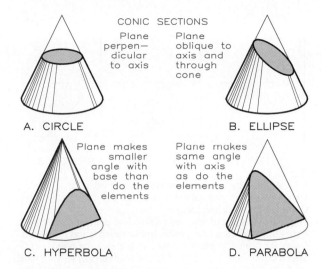

CONIC SECTIONS

A. CIRCLE
Plane perpen-dicular to axis

B. ELLIPSE
Plane oblique to axis and through cone

C. HYPERBOLA
Plane makes smaller angle with base than do the elements

D. PARABOLA
Plane makes same angle with axis as do the elements

5.32 The conic sections are the (A) circle, (B) ellipse, (C) hyperbola, and (D) parabola. They are formed by passing imaginary cutting planes through a right cone.

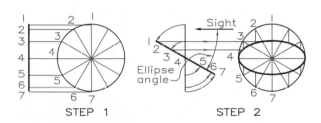

STEP 1   STEP 2

5.33 An ellipse by revolution:
Step 1 When the edge of a circle is perpendicular to the projectors between its adjacent view, it appears as a circle. Mark equally spaced points around the circle's circumference and project them to the edge.
Step 2 Revolve the edge of the circle and project the points to the circular view. Project the points vertically downward to obtain the elliptical view.

(Figure 5.32B). Mathematically, the ellipse is the path of a point that moves in such a way that the sum of the distances from two focal points is a constant. The largest diameter of an ellipse—the major diameter—is always the true length. The shortest diameter—the minor diameter—is perpendicular to the major diameter.

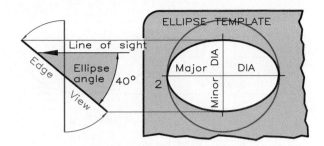

5.34 When the line of sight is not perpendicular to a circle's edge, it appears as an ellipse. The angle between the line of sight and the edge of the circle is the ellipse template angle.

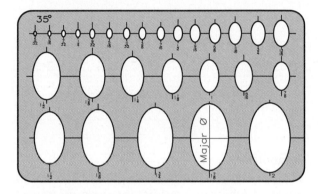

5.35 Ellipse templates are calibrated at 5° intervals from 15° to 60° with size increments of 1/8 inch.

Revolving the edge view of a circle yields an ellipse **(Figure 5.33).** The ellipse template shown in **Figure 5.34** is used to draw the same ellipse. The angle between the line of sight and the edge of the circle is the angle of the ellipse template (or the one closest to this size) that should be used. Ellipse templates are available in 5° intervals and in major diameter sizes that vary in increments of about 1/8 inch **(Figure 5.35).**

You can construct an ellipse inside a rectangle or parallelogram by plotting a series of points to form the ellipse **(Figure 5.36).**

An ellipse can be drawn on x and y axes by plotting x and y coordinates from the equation

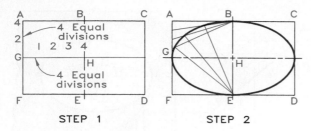

STEP 1 STEP 2

5.36 An ellipse by the parallelogram method:
Step 1 Draw an ellipse inside a rectangle or parallelogram by dividing the horizontal centerline AG and GH into the same number of equal divisions.
Step 2 The curve construction is shown for one quadrant. Rays from E and B cross at points on the curve.

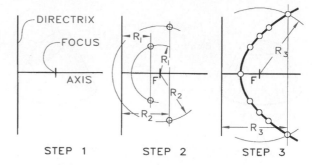

STEP 1 STEP 2 STEP 3

5.37 A parabola by the mathematical method:
Step 1 Draw an axis perpendicular to the directrix (a line). Choose a point for the focus, F.
Step 2 Use a series of selected radii to find points on the curve. For example, draw a line parallel to the directrix and R2 from it. Swing R2 from F to intersect the line and plot the point.
Step 3 Continue the process with a series of arcs of varying radii until you find an adequate number of points to complete the curve.

of an ellipse. The mathematical equation of an ellipse is

$$\frac{x^2 + y^2}{a^2 + b^2} = 1,$$ where a and b are not 0.

Parabolas

The **parabola** is defined as a plane curve, each point of which is equidistant from a straight line (called a directrix) and a focal point. The parabola is the conic section formed when the cutting plane and an element on the cone's surface make the same angle with the cone's base as shown in **Figure 5.32D.**

Figure 5.37 shows construction of a parabola by using its mathematical definition as is done in analytical geometry. A second method of drawing a parabola, which involves the use of a rectangle or parallelogram, is shown in **Figure 5.38.** The mathematical equation of the parabola is:

$$y = ax^2 + bx + c,$$ where a is not 0

Hyperbolas

The hyperbola is a two-part conic section defined as the path of a point that moves in such a way that the difference of its distances from

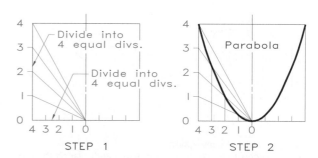

STEP 1 STEP 2

5.38 A parabola by the parallelogram method:
Step 1 Draw a parallelogram or rectangle to contain the parabola; draw its axis parallel to the sides through 0. Divide the sides into equal segments; draw rays from 0.
Step 2 Draw lines parallel to the sides (vertical in this case) to locate points along the rays from 0 and draw a smooth curve through them.

two focal points is a constant **(Figure 5.32C).** **Figure 5.39** shows construction of a hyperbola according to this definition. By selecting a series of radii until enough points have been located, the hyperbolic curve can be accurately drawn.

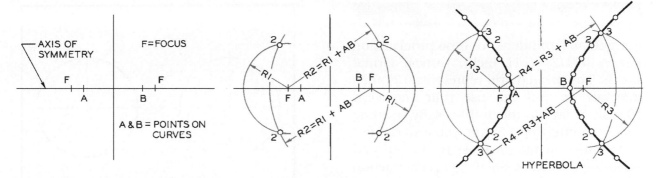

5.39 Hyperbola:

Step 1 Draw a perpendicular through the axis of symmetry. Locate focal points F equidistant from it on both sides. Locate points A and B equidistant from the perpendicular at a distance of your choice but between the focal points.

Step 2 Use radius R1 to draw arcs using focal points F as the centers. Add R1 to AB (the nearest distance between the hyperbolas) to find R2. Draw arcs using radius R2 and the focal points as centers. Rl and R2 locate point 2.

Step 3 Select other radii and add them to AB to locate additional points as shown in step 2. Draw a smooth curve through the points with an irregular curve to draw the two curves.

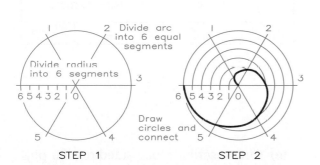

5.40 Spiral:

Step 1 Draw a circle and divide it into equal parts. Divide the radius into the same number of equal parts (six in this case).
Step 2 Begin inside and draw arc 0-1 to intersect radius 0-1. Then swing arc 0-2 to radius 0-2, continue to point 6 on the original circle, and connect the points.

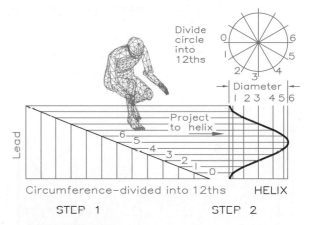

5.41 Cylindrical helix:

Step 1 Divide the top view of the cylinder into equal parts and project them to the front view. Layout the circumference and the lead (pronounced *leed*) of the cylinder. Divide the circumference into the same number of equal parts transferred from the top view.
Step 2 Project the points along the inclined rise to their respective elements on the diameter and connect them with a smooth curve.

5.12 Spirals

The **spiral** is a coil lying in a single plane that begins at a point and becomes larger as it travels around the origin. **Figure 5.40** shows the steps for constructing a spiral. The number of divisions selected in this construction depends on the degree of accuracy desired.

5.13 Helixes

The **helix** is a three-dimensional curve that coils around a cylinder or cone at a constant angle of inclination. Applications of helixes are corkscrews and the threads on a screw. **Figure 5.41** shows the construction of a cylindrical helix.

Problems

Present your solutions to these problems on size A (8-1/2 × 11 inch), vertical format sheets. The printed grid represents 0.20-inch intervals, so you can use your engineers' 10 scale to lay out the problems. By equating each grid interval to 5 mm, you also can use your full-sized metric scale to lay out and solve the problems. Show your construction and mark all points of tangency, as recommended in the chapter.

1. **(Sheet 1):** Basic constructions.
   (a) Draw triangle ABC using the given sides.
   (b–c) Inscribe an angle in the semicircles with the vertexes at point P.
   (d) Inscribe a three-sided regular polygon inside the circle.

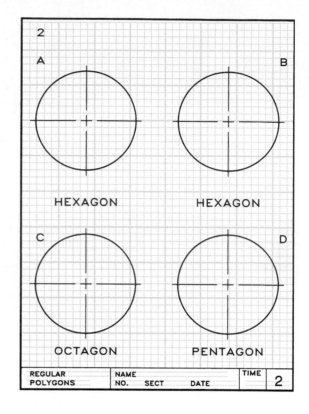

(e) Circumscribe a four-sided regular polygon about the circle.

2. **(Sheet 2):** Construction of regular polygons.
   (a) Circumscribe a hexagon about the circle.
   (b) Inscribe a hexagon in the circle.
   (c) Circumscribe an octagon about the circle.
   (d) Construct a pentagon inside the circle using the compass method.

3. **(Sheet 3):** Basic constructions.
   (a) Bisect the lines.
   (b) Bisect the angles.
   (c) Bisect the sides of the triangle.

4. **(Sheet 4):** Line division and tangencies.
   (a) Divide AB into seven equal parts using a construction line through point A.

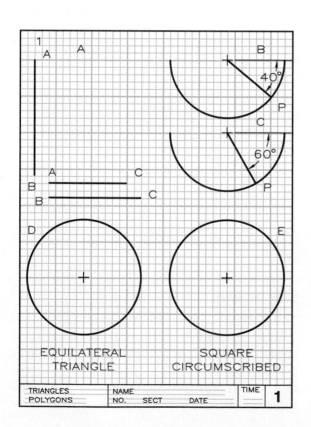

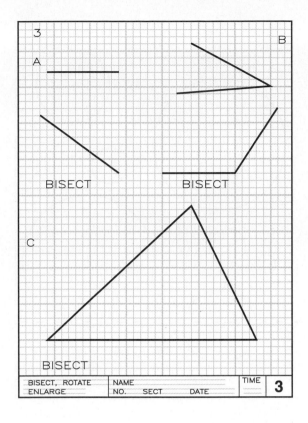

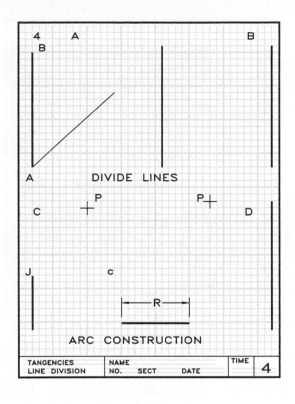

(b) Divide the space between the two vertical lines into four equal segments. Draw three vertical lines at the division points that are equal in length to the given lines.

(c) Construct an arc with radius R that is tangent to the line at J and that passes through point P.

(d) Construct an arc with radius R that is tangent to the line and passes through P.

5. (Sheet 5): Tangency construction.

(a) Using the compass method draw a line from P tangent to the arc. Mark the points of tangency.

(b–d) Construct arcs with the given radii tangent to the lines.

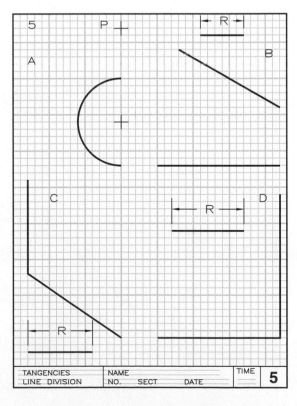

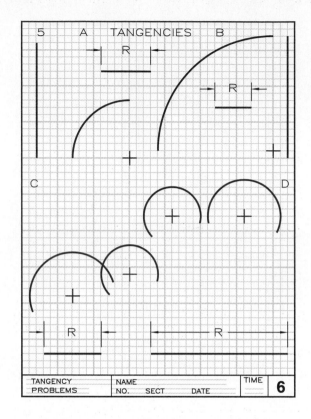

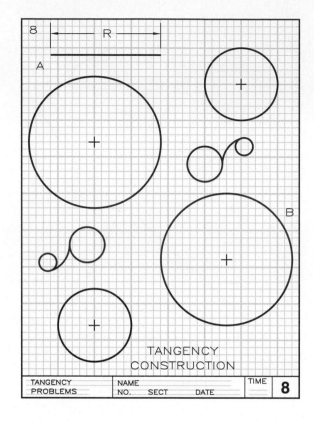

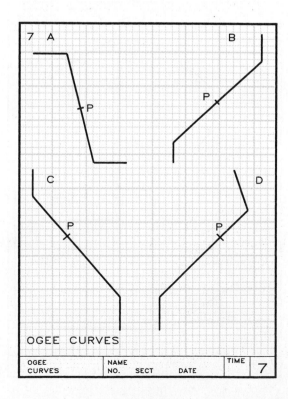

6. **(Sheet 6):** Tangency construction.
   (a–d) Construct arcs that are tangent to the arcs or lines shown. The radii are given for each problem.

7. **(Sheet 7):** Ogee curve construction.
   (a–d) Construct ogee curves that connect the ends of the given lines and pass through points P.

8. **(Sheet 8):** Tangency construction.
   (a–b) Using the given radii, connect the circles with the tangent arcs as indicated in the sketches.

9. **(Sheet 9):** Ellipse construction.
   (a–c) Construct ellipses inside the rectangles given. Use enough divisions to make it possible to draw accurate ellipses with your irregular curve.

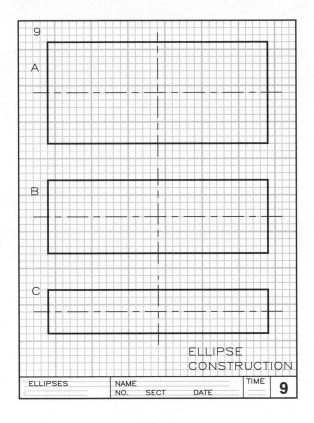

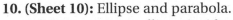

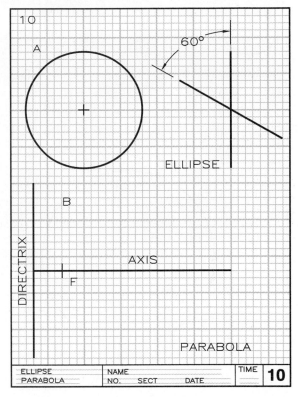

10. **(Sheet 10):** Ellipse and parabola.
    (a) Construct an ellipse inside the circle when the edge view has been rotated as shown.
    (b) Using the focal point F and the directrix, plot and draw the parabola formed by these elements.

11. **(Sheet 11):** Hyperbola and spiral.
    (a) Using the focal point F, points A and B on the curve, and the axis of symmetry, construct the hyperbola.
    (b) Construct a parabola by using the parallelogram method and five divisions along each axis for each half of the parabola.

12. Helix construction **(Sheet 12):**
    (a) Construct ahelix that has a rise equal to the height of the cylinder.
    (b) Construct a spiral by using the six divisions marked along the radius.

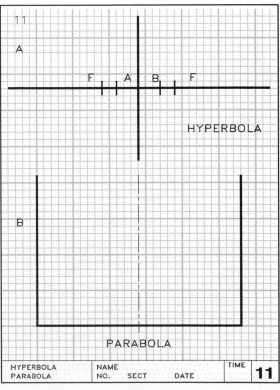

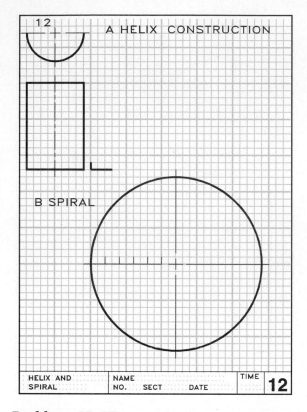

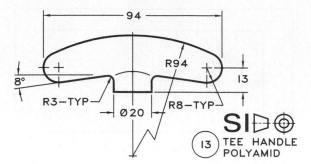

5.43 Problem 14.

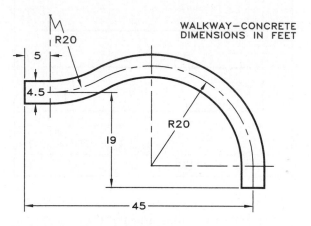

5.44 Problem 15.

Problems 13–22:

Figures 5.42–5.51: Practical applications. Construct the given views on size A sheets, one problem per sheet. Select the scale that will best fit the problem to the sheet. Mark all points of tangency and use good line quality.

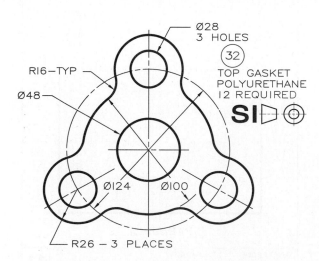

5.42 Problem 13.

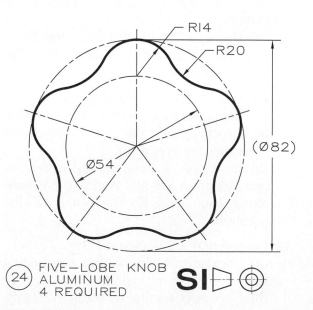

5.45 Problem 16.

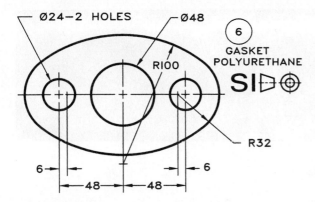

Ø24—2 HOLES | Ø48
R100
R32
6
GASKET
POLYURETHANE
SI⬤
6
48
48
6

5.46 Problem 17.

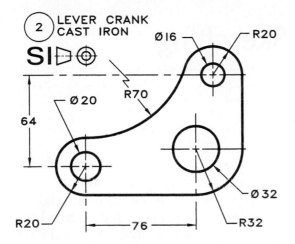

2 LEVER CRANK
CAST IRON
SI⬤
Ø16 | R20
R70
Ø20
64
Ø32
R20 | 76 | R32

5.47 Problem 18.

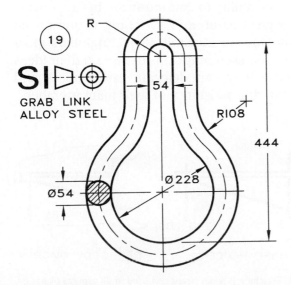

19
SI⬤
GRAB LINK
ALLOY STEEL
R
54
R108
444
Ø54
Ø228

5.48 Problem 19.

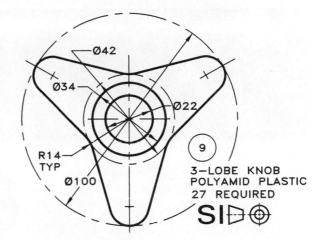

Ø42
Ø34
Ø22
R14
TYP
Ø100
9
3-LOBE KNOB
POLYAMID PLASTIC
27 REQUIRED
SI⬤

5.49 Problem 20.

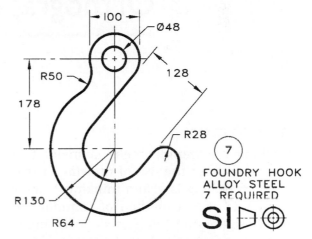

100 | Ø48
R50
128
178
R28
7
FOUNDRY HOOK
ALLOY STEEL
7 REQUIRED
SI⬤
R130
R64

5.50 Problem 21.

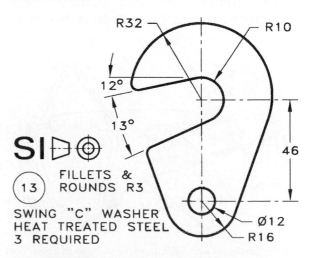

R32 | R10
12°
13°
SI⬤
46
13
FILLETS &
ROUNDS R3
SWING "C" WASHER
HEAT TREATED STEEL
3 REQUIRED
Ø12
R16

5.51 Problem 22.

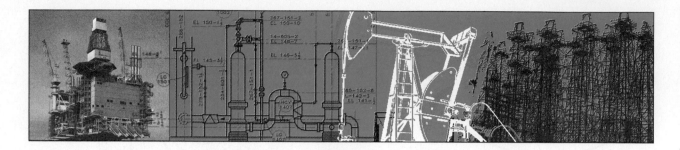

6

Orthographic Sketching

6.1 Introduction

Sketching is a rapid, freehand method of drawing without the use of drawing instruments. Sketching is also a thinking process as much as it is a method of communication. Designers and engineers make many sketches as a method of developing ideas before arriving at the final solution. Many new products and projects have begun as sketches made on the back of an envelope or on a napkin at a restaurant (**Figure 6.1**). Sketching is used by the engineer throughout the engineering process including free-body diagrams during the analysis step of the design process (**Figure 6.2**).

The ability to communicate by any means is a great asset, and sketching is one of the best ways to transmit ideas. Engineers must use their sketching skills to explain their ideas before they can delegate assignments and obtain the assistance of their team members.

6.1 Designs begin with rapid, freehand sketches as a means of developing concepts.

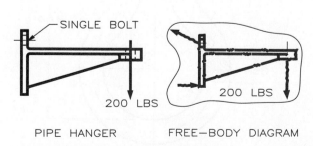

6.2 Sketching is a necessary skill used in all aspects of design, from free-body diagrams through design documentation.

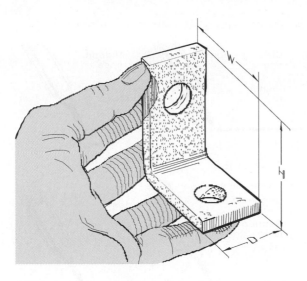

6.3 How can you sketch this angle bracket to convey its shape effectively?

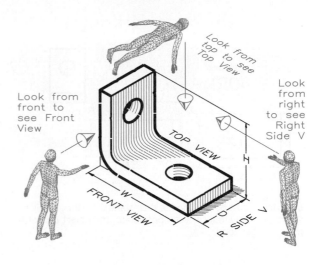

6.4 These positions give the viewpoints for three orthographic views of the angle bracket: top, front, and right side.

6.2 Shape Description

Although the angle bracket in **Figure 6.3** is a simple three-dimensional object, describing it with words is difficult. Most untrained people would think that drawing it as a three-dimensional pictorial would be a challenge. To make drawing such objects easier, engineers devised a standard system called **orthographic projection** for showing objects in multiple views.

In orthographic projection, separate views represent the object at 90° intervals as the viewer moves about it **(Figure 6.4)**. **Figure 6.5** shows two-dimensional views of the bracket from the front, top, and right side. The top view is drawn above the front view and both share the dimension of width. The right-side view is drawn to the right of the front view and both share the dimension of height.

The views of the bracket are drawn with three types of lines: **visible lines, hidden lines,** and **centerlines.** Visible lines are the thickest. Thinner hidden lines (dashed lines) represent features that are invisible or hidden in a view. The thinnest lines are centerlines, which are imaginary lines composed of

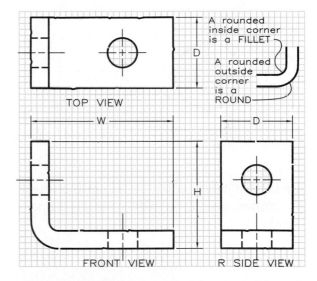

6.5 This sketch shows three orthographic views of the angle bracket.

long and short dashes to show the centers of arcs and the axes of cylinders.

The space between views may vary, but the views must be positioned as shown here. This arrangement is logical, the views are easiest to interpret in this order, and the drawing process is most efficient because the views project from each other. **Figure 6.6** illustrates

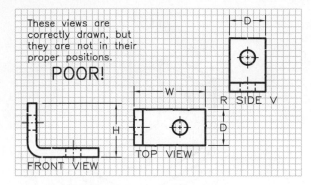

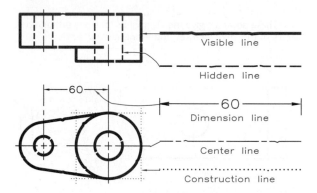

6.6 Views must be sketched in their standard orthographic positions. If they are incorrectly positioned, the object cannot be readily understood.

6.7 The alphabet of lines for sketching are shown here. The lines at the right are full size.

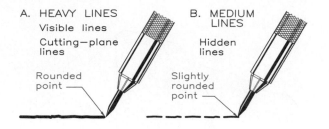

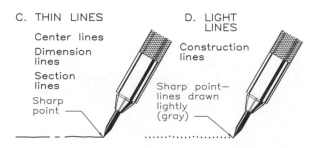

6.8 An F pencil is a good choice for sketching all lines if you sharpen it for varying line widths.

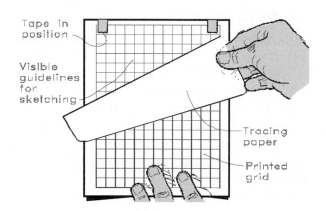

6.9 A grid placed under a sheet of tracing paper will provide guidelines as an aid in freehand sketching.

the lack of clarity when views are incorrectly positioned, even though each view is properly drawn.

6.3 Sketching Techniques

You need to understand the application of line types used in sketching (freehand) orthographic views before continuing with the principles of projection. The "alphabet of lines" for sketching is illustrated in **Figure 6.7.** All lines, except construction lines, should be black and dense. Construction lines, which serve as guides, are drawn lightly so that they need not be erased. The other lines are distinguished by their line widths (line thicknesses), but they are equal in darkness.

Medium-weight pencils, such as H, F, or HB grades, are best for sketching the lines shown in **Figure 6.8.** By sharpening the pencil point to match the desired line width, you may use the same grade of pencil for all these lines. Lines sketched freehand should have a freehand appearance; do not attempt to make them appear mechanical. Using a printed grid or laying translucent paper over a printed grid can aid your sketching technique **(Figure 6.9).**

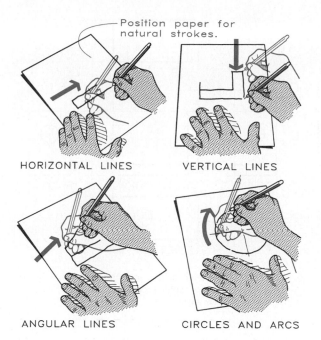

Position paper for natural strokes.

HORIZONTAL LINES

VERTICAL LINES

ANGULAR LINES

CIRCLES AND ARCS

6.10 Sketch lines as shown here for the best results; rotate your drawing sheet for comfortable sketching positions.

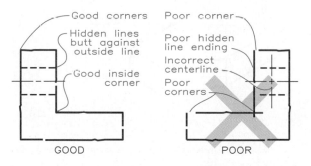

Good corners

Hidden lines butt against outside line

Good inside corner

Poor corner

Poor hidden line ending

Incorrect centerline

Poor corners

GOOD

POOR

6.11 For good sketches, follow the examples of good technique and avoid the common errors of poor technique shown.

When you make a freehand sketch, lines will be vertical, horizontal, angular, and/or circular. By not taping your drawing to the table top, you can easily position the sheet that is most comfortable for each drawing stroke, usually from left to right **(Figure 6.10).** Examples of correctly sketched lines are contrasted with incorrectly sketched ones in **Figure 6.11.** Notice that gaps should not be left at the corners of a view.

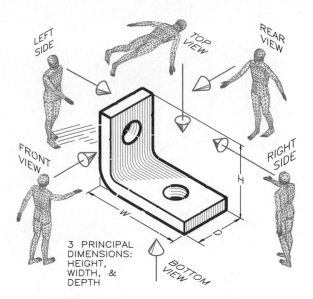

LEFT SIDE

TOP VIEW

REAR VIEW

FRONT VIEW

RIGHT SIDE

BOTTOM VIEW

3 PRINCIPAL DIMENSIONS: HEIGHT, WIDTH, & DEPTH

6.12 Six principal views of the angle bracket can be sketched from the viewpoints shown.

6.4 Six-View Sketching

The maximum number of principal views that can be drawn in orthographic projection is six, as the viewer changes position at 90° intervals **(Figure 6.12).** In each view, two of the three dimensions of **height, width,** and **depth** are seen.

These views must be sketched in their standard positions **(Figure 6.13).** The width dimension is shared by the top, front, and bottom views. The height dimension is shared with the right-side, front, left-side, and rear views. Note the simple and effective dimensioning of each view with two dimensions. Seldom is an object so complex that it requires six orthographic views.

6.5 Three-View Sketching

You can adequately describe most objects with three orthographic views; usually the top, front, and right-side views. **Figure 6.14** shows a typical three-view sketch of a T-block with height, width, and depth dimensions and the front, top, and right-side views labeled.

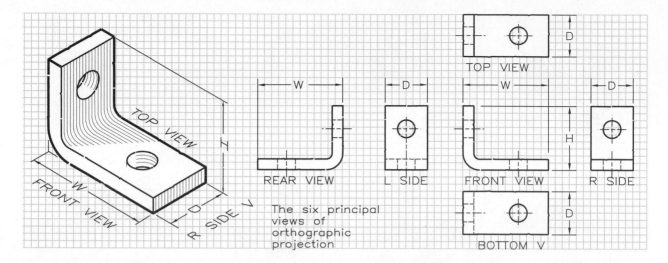

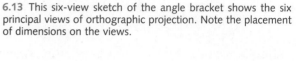

6.13 This six-view sketch of the angle bracket shows the six principal views of orthographic projection. Note the placement of dimensions on the views.

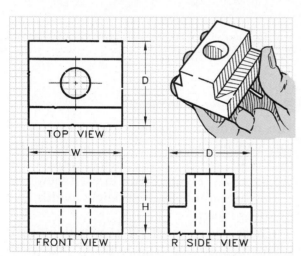

6.14 This sketch shows the standard orthographic arrangement for three views of a T-block, with dimensions and labels.

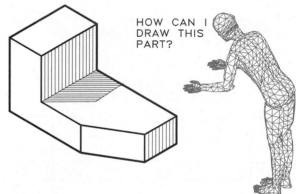

6.15 Sketches of three orthographic views describe this fixture block in Figure 6.16.

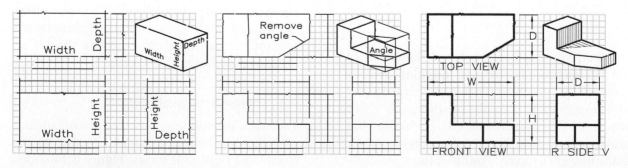

6.16 Three-view sketching:
Step 1 Block in the views with light construction lines. Allow proper spacing for labeling and dimensioning the views.

Step 2 Remove the notches and project the resulting lines of intersection to the adjacent views.

Step 3 Check for correctness, darken the lines, add the dimensions, and letter the labels for each view.

The object shown in **Figure 6.15** is represented by three orthographic views on a grid in **Figure 6.16.** To obtain those views, block in the views of the object with its overall dimensions, then sketch the slanted surface in the top view and project it to the other two views. Finally, darken the lines, label the views, and letter the overall dimensions of height, width, and depth.

Slanted surfaces will appear as **edges** or **foreshortened** (not-true-size) planes in the principal views of orthographic projection (**Figure 6.17**). In **Figure 6.17C,** two intersecting planes of the object slope in two directions; thus both appear foreshortened in the front, top, and right-side views.

A good way to learn orthographic projection is to construct a missing third view (the front view in **Figure 6.18**) when two views are given. In **Figure 6.19,** we construct the missing right-side view from the given top and front views. To obtain the depth dimension for the right-side view, transfer it from the top view with dividers; to obtain the height dimension, project it from the front view.

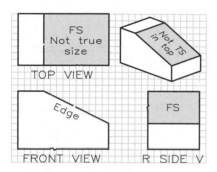

FORESHORTENED IN TOP

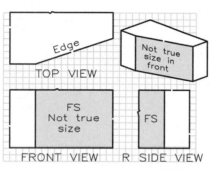

FORESHORTENED IN FRONT

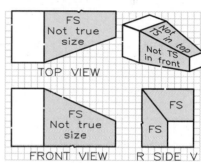

FORESHORTENED IN ALL

6.17 Views of planes:

A The plane appearing as an angular edge in the front view is foreshortened in the top and side views.

B The plane appearing as an angular edge in the top view is foreshortened in the front and side views.

C Two sloping planes appear foreshortened in all views and neither appears as an edge in any of the three views.

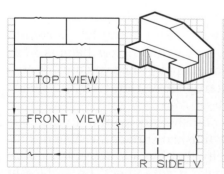

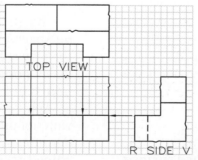

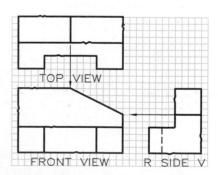

6.18 Sketching a missing view:

Step 1 Begin by blocking in the front view with light construction lines that will not need to be erased.

Step 2 Project the notch from the top view and side view to the front and darken these lines as final lines.

Step 3 Project the ends of the angular notch from the top and right-side views, check the views, and darken the lines.

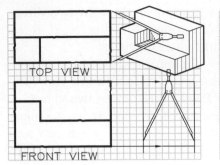

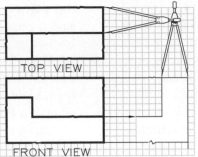

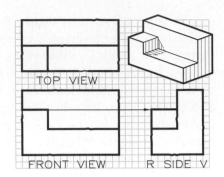

6.19 Sketching a missing side view:
Step 1 Transfer the depth with dividers and project the height from the front. Block in the side view with construction lines.

Step 2 Locate the notch in the side view with your dividers and project its base from the front view. Use light construction lines.

Step 3 Project the top of the notch from the front view, check for correctness, darken the lines, and label the views.

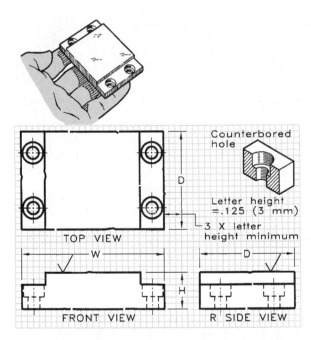

6.20 Three orthographic views adequately describe the rest pad. Space dimension lines at least three letter heights from the views. Finish marks (V marks) indicate that the top surface has been machined to a smooth finish. Counterbored holes allow bolt heads to be recessed.

Figure 6.20 shows a rest pad sketched in three views. The pad has a **finished surface,** indicated by V marks in the two views where the surface appears as an edge, and four **counterbored** holes. Dimension lines for

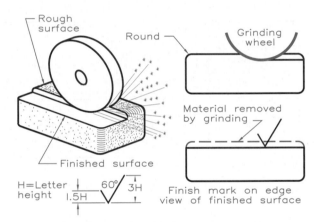

6.21 Place a finish mark on all edge views of a surface (visible or hidden) that is to be smoothed by machining. Grinding is one of the methods used to finish a surface.

the height, width, and depth labels should be spaced at least three letter heights from the views. For example, when you use 1/8-inch letters, position dimension lines at least 3/8-inch from the views.

Apply the finish mark symbol to the edge views of any finished surfaces, visible or hidden, to specify that the surface is to be machined to make it smoother. The surface in **Figure 6.21** is being finished by grinding, which is one of many methods of smoothing a surface.

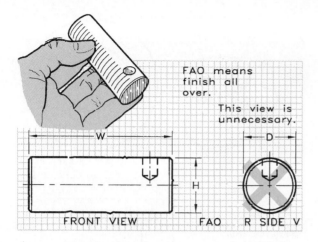

6.22 This pulley shaft is a typical cylindrical part that can be represented adequately by one view.

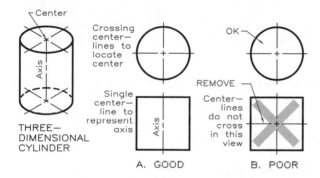

6.23 Centerlines identify the centers of circles and axes of cylinders. Centerlines cross only in the circular view and extend about 1/8 inch beyond the outside lines.

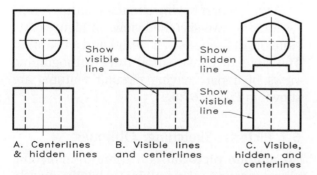

6.24 When visible lines coincide with hidden lines, show the visible lines. When hidden lines coincide with centerlines, show the hidden lines.

6.6 Circular Features

The pulley shaft depicted in **Figure 6.22** in two views is composed of circular features. **Centerlines** are added to better identify these cylindrical features. **Figure 6.23** shows how to apply centerlines to indicate the center of the circular ends of a cylinder and its vertical axis. Perpendicular centerlines cross in circular views to locate the center of the circle and extend beyond the arc by about 1/8 inch. Centerlines consist of alternating long and short dashes, about 1 inch and 1/8 inch in length, respectively.

When centerlines coincide with visible or hidden lines, the centerline should be omitted because object lines are more important and centerlines are imaginary lines. **Figure 6.24** shows the precedence of lines.

The centerlines shown in **Figure 6.25** clarify whether the circles and arcs are concentric (share the same centers). **Figure 6.26** shows the correct manner of applying centerlines to

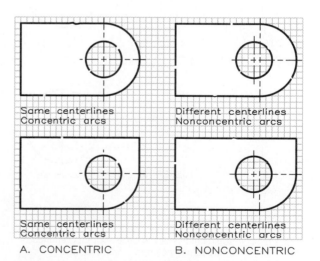

6.25 Centerlines:
A Extend centerlines beyond the last arc that has the same center.
B Sketch separate centerlines when the arcs are not concentric.

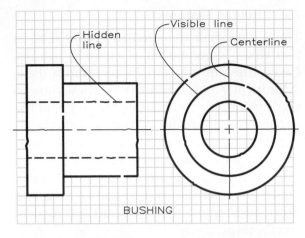

6.26 This orthographic sketch depicts the application of center-lines to concentric cylinders and the relative weights of various lines.

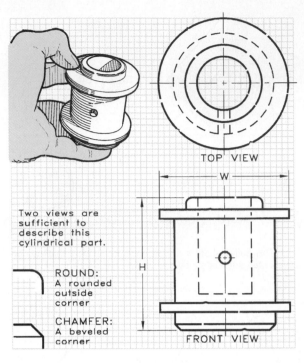

6.28 Two views adequately describe this cylindrical pivot base.

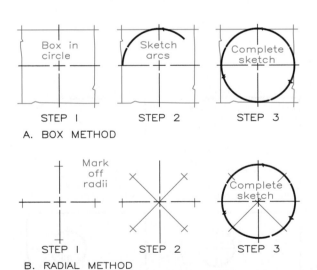

6.27 Sketching circles:
Box Method
Step 1 Block in the diameter of the circle about its centerlines.
Step 2 Sketch an arc tangent to two tangent points.
Step 3 Complete the circle with other arcs.
Radial Method
Step 1 Mark off radii on the centerlines.
Step 2 Mark off radii on two construction lines drawn at 45°.
Step 3 Sketch the circles with arcs passing through the marks.

orthographic views of an object composed of concentric cylinders.

Sketching-Circles

Circles can be sketched by either of the methods shown in **Figure 6.27.** Use light guidelines and dark centerlines to block in the circle. Drawing a freehand circle in one continuous arc is difficult, so draw arcs in segments with the help of the guidelines.

A typical part having circular features is represented by two sketched views in **Figure 6.28.** Note the definitions of a **round** and a **chamfer.** The steps of constructing three orthographic views of a part having circular features are shown in **Figure 6.29.**

6.7 Pictorial Sketching: Obliques

An **oblique pictorial** is a three-dimensional representation of an object's height, width, and depth. It approximates a photograph of an object, making the sketch easier to under-

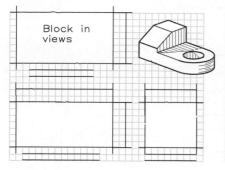
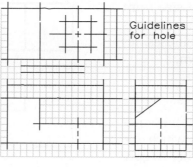
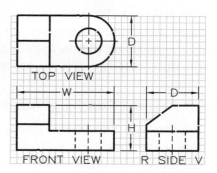

6.29 Sketching circular features:
Step 1 Block in the overall dimensions with construction lines. Leave room for labels and dimensions.

Step 2 Draw the centerlines and the squares that block in the diameter of the circle. Find the slanted surface in the side view.

Step 3 Sketch the arcs, darken the lines, label the views, and show the dimensions of W, D, and H between the views.

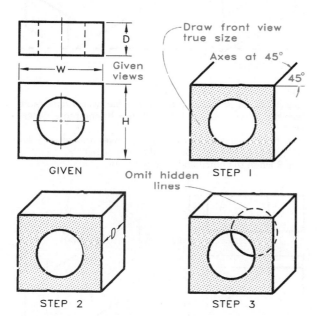

6.30 Sketching oblique pictorials:
Step 1 Sketch the front of the part as an orthographic front view and the receding lines at 45° to show the depth dimension.
Step 2 Measure the depth along the receding axes and sketch the back of the part.
Step 3 Locate the circle on the rear plane, show the visible portion of it, and omit the hidden lines.

stand at a glance than do orthographic views. Sketch the front of the object as a true-shape orthographic view **(Figure 6.30)**. Sketch the receding axes at an angle of between 20° and

60° oblique with the horizontal in the front view. Lay off the depth dimension as its true length along the receding axes. **Figure 6.31** shows an oblique sketch of a saddle. When the depth is laid off true length, the oblique is a **cavalier** oblique.

The major advantage of an oblique pictorial is the ease of sketching circular features

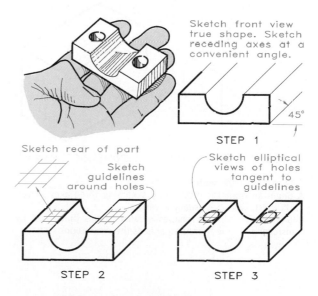

6.31 Sketching arcs in oblique pictorials:
Step 1 Sketch the front view of the saddle as a true front view. Sketch the receding axes from each corner.
Step 2 Sketch the rear of the part by measuring its depth along the receding axes. Sketch guidelines that will enclose the holes.
Step 3 Sketch the circular features as ellipses on the upper planes tangent to the guidelines.

with true circular arcs on the true-size front plane rather than as ellipses. Circular features on the receding planes appear as ellipses, requiring slanted guidelines, as shown.

6.8 Pictorial Sketching: Isometrics

Another type of three-dimensional representation is the **isometric pictorial,** in which the axes make 120° angles with each other (**Figure 6.32**). Specially printed isometric grids with lines intersecting at 60° angles make isometric sketching easier as shown in **Figure 6.33** by transferring dimensions from the square grid in the orthographic views to the isometric grid.

You cannot measure angles in isometric pictorials with a protractor; you must find them by connecting coordinates of the angle laid off along the isometric axes. In **Figure 6.34,** locate the ends of the angular plane by using

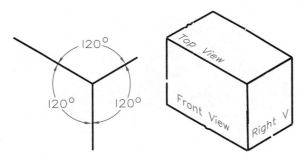

A. THE ISOMETRIC AXES   B. ISOMETRIC DRAWING

6.32 An isometric sketch:
A Begin an isometric pictorial by sketching three axes spaced 120° apart. One axis usually is vertical.
B Sketch the isometric shapes parallel to the three axes by transferring its true measurements as the dimensions.

6.34 Sketching angles in isometric pictorials:
Step 1 Sketch a box using the overall dimensions given in the orthographic views.
Step 2 Angles cannot be measured with a protractor. Find each end of the angle with coordinates measured along the axes.
Step 3 Connect the ends of the angle and darken the lines.

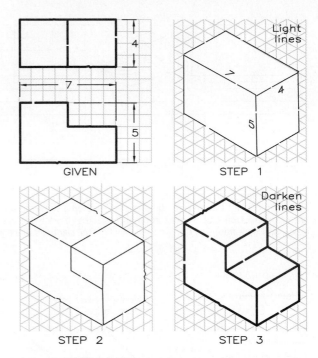

6.33 Sketching isometric pictorials:
Step 1 Use an isometric grid, transfer the overall dimensions from the given views, and sketch a box having those dimensions.
Step 2 Locate the notch by measuring over four squares and down two squares, as shown in the given views.
Step 3 Finish the notch and darken the lines.

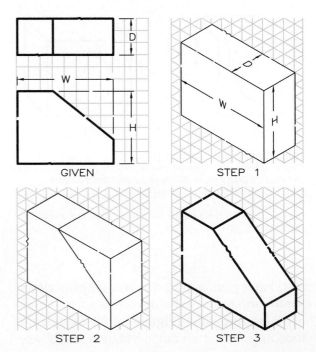

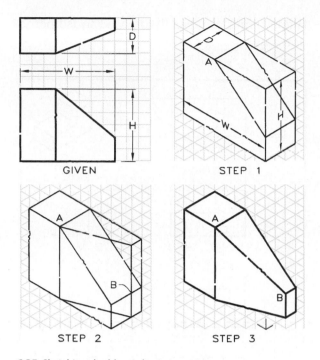

6.35 Sketching double angles in isometric:
Step 1 This object has two sloping angles that intersect. Begin by sketching the overall box and draw one of the angles.
Step 2 Find the second angle, which locates point B, the intersection line between the planes.
Step 3 Connect points A and B and darken the lines. Line AB is the line of intersection between the two sloping planes.

6.36 Sketching circles in isometric pictorials:
Step 1 Sketch a box using the overall dimensions given. Sketch the centerlines and a rhombus blocking in the circular hole.
Step 2 Sketch the isometric arcs tangent to the box. These arcs are elliptical rather than circular.
Step 3 Sketch the hole and darken the lines. Hidden lines usually are omitted in isometric pictorial sketches.

the coordinates for width and height. When a part has two sloping planes that intersect **(Figure 6.35),** you must sketch them one at a time to find point B. Line AB is found as the line of intersection between the planes. A more thorough coverage of isometric drawing is given in Chapter 17.

Circles in Isometric

Circles appear as ellipses in isometric pictorials. When you sketch them, begin with their centerlines and construction lines enclosing their diameters as shown in **Figure 6.36.** The end of the block is semicircular in the front view, so its center must be equidistant from

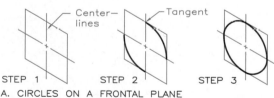

A. CIRCLES ON A FRONTAL PLANE

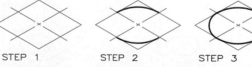

B. CIRCLES ON A HORIZONTAL PLANE

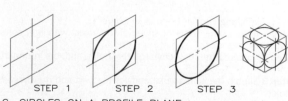

C. CIRCLES ON A PROFILE PLANE

6.37 Sketching circular features in isometrics:
Step 1 Sketch the centerlines and guidelines.
Step 2 Sketch the two opposite arcs of the ellipse.
Step 3 Sketch the two small arcs at the opposite ends of the ellipse.

the top, bottom, and end of the front view. Circles and ellipses are easier to sketch when you use construction lines.

Figure 6.37 shows how to use centerlines and construction lines to draw ellipses in the three isometric planes: frontal, horizontal (top), and profile (side) views. This technique is used to sketch a cylinder in **Figure 6.38** and an object having semicircular ends in **Figure 6.39.**

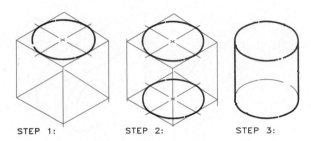

STEP 1:   STEP 2:   STEP 3:

6.38 Sketching a cylinder in isometric:
Step 1 Block in the cylinder with guidelines and sketch the upper ellipse.
Step 2 Sketch the lower ellipse.
Step 3 Connect the ellipses with lines tangent to the elliptical ends and darken the lines.

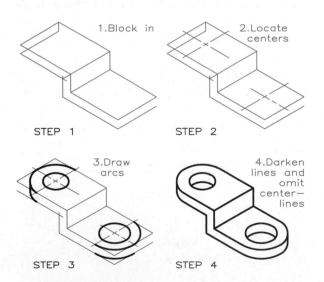

1.Block in

STEP 1

2.Locate centers

STEP 2

3.Draw arcs

STEP 3

4.Darken lines and omit center-lines

STEP 4

6.39 Sketching circular features in isometric:
Step 1 Block in the isometric shape of the object with light construction lines.
Step 2 Locate the centerlines of the holes and the rounded ends.
Step 3 Sketch the semicircular ends of the part and the circular holes.
Step 4 Draw the bottoms of the holes and darken all lines.

Hidden lines are usually omitted in isometric drawings.

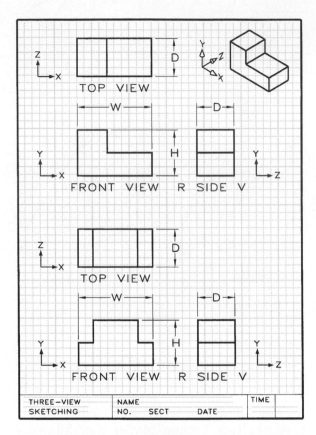

6.40 Use this layout of a size A sheet for sketching problems. You may sketch two problems on each sheet.

Problems

Sketch three views of the problems in **Figures 6.41–6.43** on size A sheets with or without a printed grid as shown in **Figure 6.40.** Each grid is equal to 0.20 inch or 5 mm. Label the views and give the overall dimensions as W, D, and H.

Design Sketching General dimensions are given on the following problems. Determine all missing information as if you were the original designer. Sketch your solutions on size A sheets.

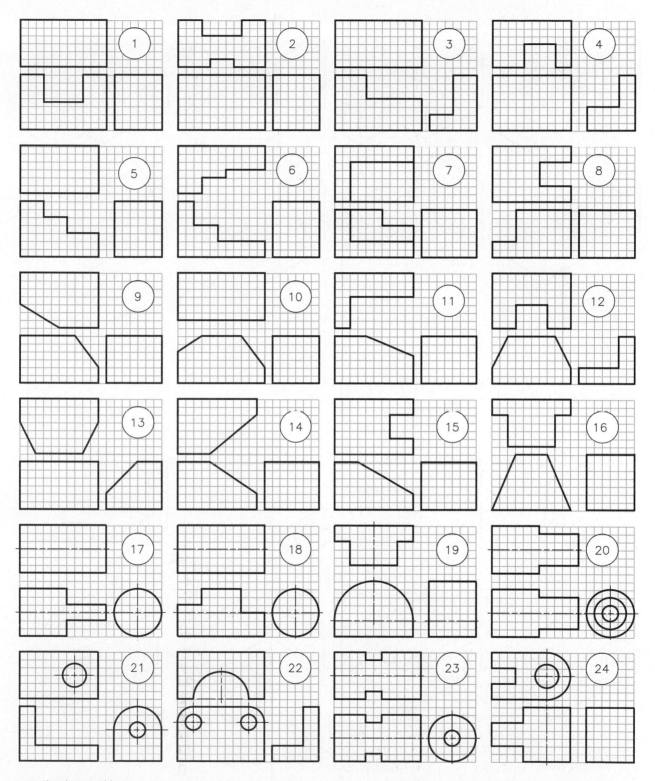

6.41 Sketching problems.

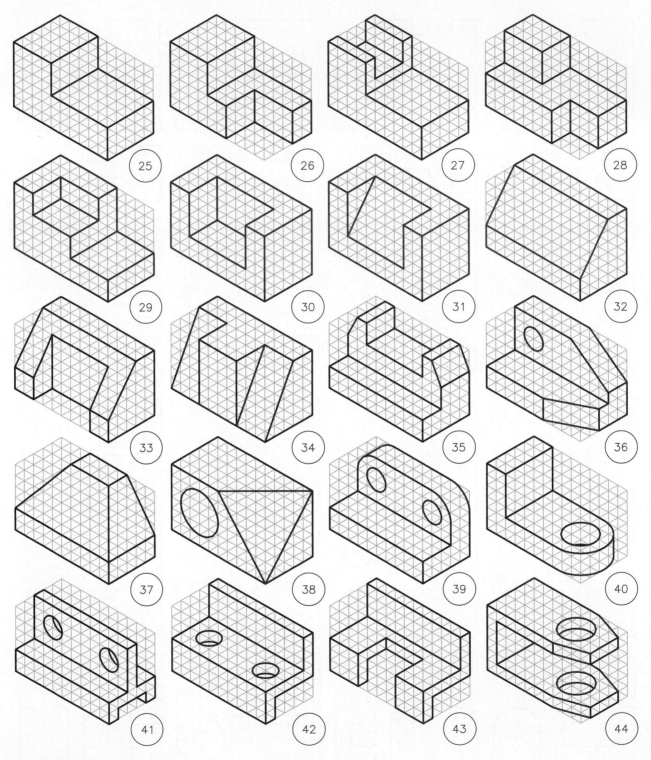

6.42 Sketching problems.

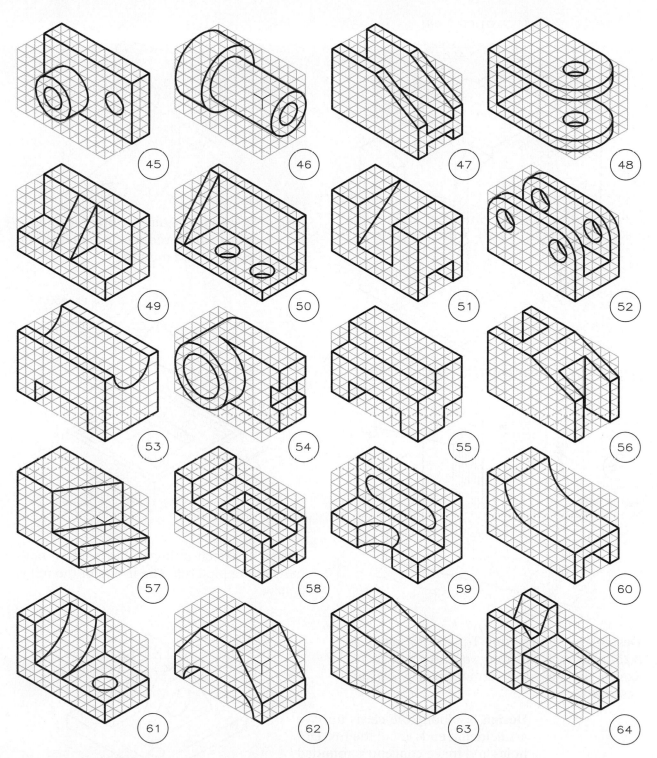

6.43 Sketching problems.

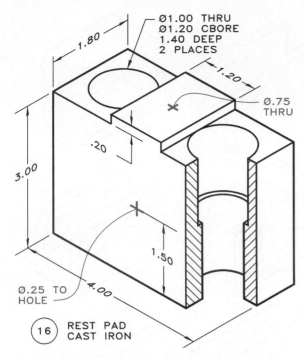

Ø1.00 THRU
Ø1.20 CBORE
1.40 DEEP
2 PLACES

1.80

1.20

Ø.75
THRU

3.00

.20

1.50

Ø.25 TO
HOLE

4.00

16 REST PAD
    CAST IRON

Design 1: Add the two new holes as specified.

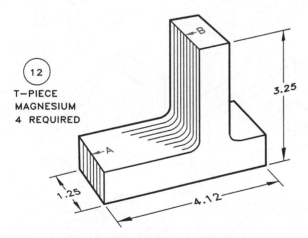

12

T−PIECE
MAGNESIUM
4 REQUIRED

B

3.25

A

1.25

4.12

Design 2: Modify the T-piece to have a 0.25-inch-thick rib from point A to point B. Sketch the necessary orthographic views to describe the part.

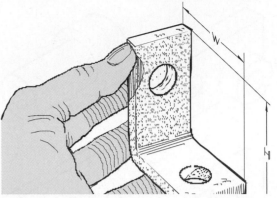

Design 3: Modify the aircraft bracket to have the triangular rib moved to the center of the part. Show fillets and rounds.

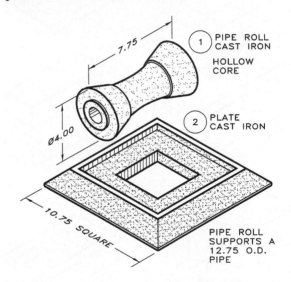

7.75

1 PIPE ROLL
   CAST IRON

   HOLLOW
   CORE

Ø4.00

2 PLATE
   CAST IRON

10.75 SQUARE

PIPE ROLL
SUPPORTS A
12.75 O.D.
PIPE

Design 4: Sketch orthographic views of both parts of the pipe roll. The inside of the roll is hollow.

Design 5: Modify the clevis to have semicircular ends about the 1.60 DIA holes and make concentric rounded corners at each 0.80 DIA hole. Sketch the necessary views.

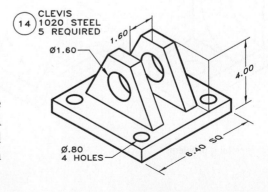

14 CLEVIS
    1020 STEEL
    5 REQUIRED

Ø1.60

1.60

4.00

Ø.80
4 HOLES

6.40 SQ

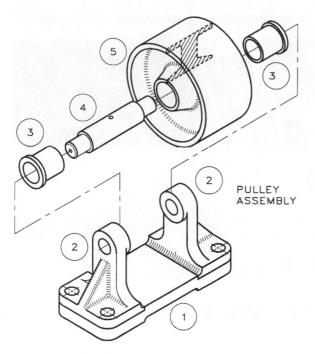

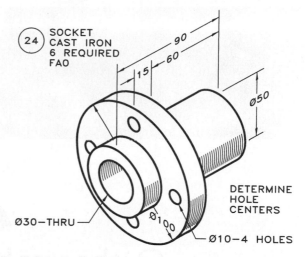

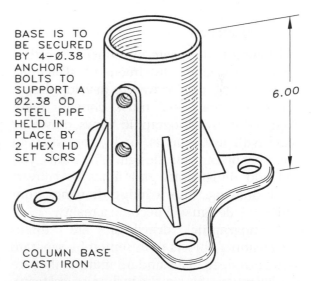

PULLEY
ASSEMBLY

Design 6:
There are five parts in the pulley assembly in Design 6, which has a 1-inch DIA shaft (Part 4). Sketch one part per A size sheet following the instructions below.

Option 1: Make a two-view sketch of the shaft (Part 4). Do you know why there are holes in the shaft?

Option 2: Make a two-view drawing of the bushing (Part 3). What are bushings and why are they used and what type of material is best for a bushing?

Option 3: Make a two-view drawing of the base (Part 1). How are the brackets (Part 2) to be connected to the base?

Option 4: Make a three-view drawing of the shaft bracket (Part 2). How do the brackets support the shaft?

Option 5: Make a two-view drawing of the pulley (Part 5). Why does it have a boss (hub) protruding from it through which the shaft, part 4, passes?

Option 6: Redesign the bracket (Part 2) and show your proposed modification in a three-view sketch.

Design 7: Modify the socket by moving the flange to the center of the part's length. Sketch the necessary views to describe the part on a size A sheet.

Design 8: Column base.
Option 1 Draw the necessary freehand orthographic views to describe the column base.
Option 2 Draw orthographic sketches of your own design of a column base to support a pipe column of this size.

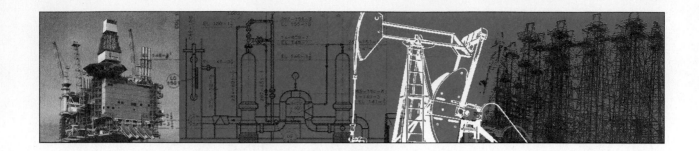

7

Orthographic Projection
with Instruments

7.1 Introduction

In Chapter 6, you were introduced to ortho-graphic projection by freehand sketching, which is an excellent way to develop design concepts. Now, you must convert these sketches into orthographic views drawn to scale with instruments, or by computer, to precisely define your design. Afterwards, you will add dimensions and notes to convert these drawings into working drawings from which the design will become a reality.

Orthographic drawings are three-dimensional objects represented by separate views arranged in a standard manner that are readily understood by the technological team. Because multiview drawings usually are exe-cuted with instruments and drafting aids, they are often called **mechanical drawings.** They are called **working drawings,** or **detail drawings,** when sufficient dimensions, notes, and specifications are added to enable the product to be manufactured or built from the drawings.

7.2 Orthographic Projection

An artist has the option of representing ob-jects impressionistically, but the engineer must represent them precisely. Orthographic pro-jection is used to prepare accurate, scaled, and clearly presented drawings from which the project depicted can be produced.

Orthographic projection is the system of drawing views of an object by projecting them perpendicularly onto projection planes with parallel projectors. **Figure 7.1** illustrates this concept by imagining that the object is inside a glass box and three of its views are projected onto planes of the box.

Figure 7.2 illustrates the principle of or-thographic projection where the front view is projected perpendicularly onto a vertical projection plane, the **frontal plane,** with par-allel projectors. The projected front view is two-dimensional because it has only width and height and lies in a single plane. Simi-larly, the top view is projected onto a **hori-zontal** projection plane, and the side view is

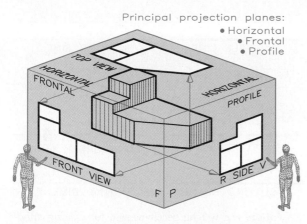

7.1 Orthographic projection is the system of projecting views onto an imaginary glass box with parallel projectors that are perpendicular to the three mutually perpendicular projection planes.

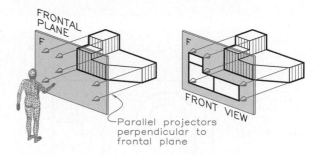

7.2 An orthographic view is found by projecting from the object to a projection plane with parallel projectors that are perpendicular to the projection plane.

projected onto a second vertical projection plane, a profile plane.

Imagine that the box is opened into the plane of the drawing surface as shown in **Figure 7.3A** to yield the three-view arrangement of views shown in **Figure 7.3B.** This layout of views gives the standard positions for three orthographic views that describe the object. These views are the **front, top,** and **right-side** views.

The principal projection planes of orthographic projection are the **horizontal (H), frontal (F),** and **profile (P)** planes. Views projected onto these principal planes are principal views. The dimensions used to give the sizes of principal views are **height (H), width (W),** and **depth (D).**

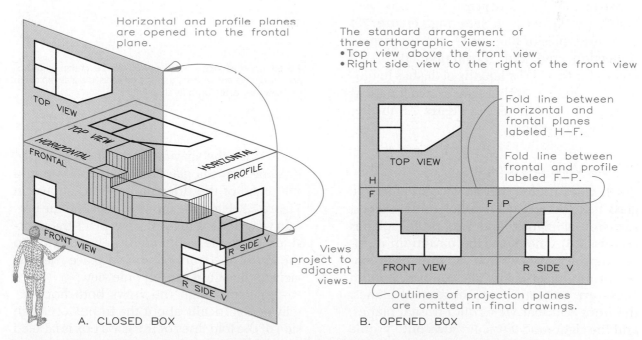

7.3 When the glass box is opened, the orthographic views are drawn and labeled in the format shown at B.

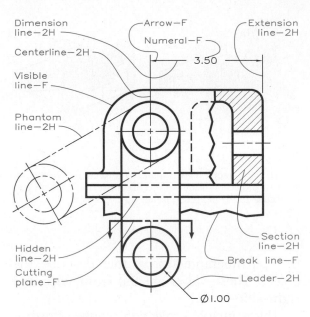

7.4 The alphabet of lines and recommended pencil grades for drawing orthographic views are shown here.

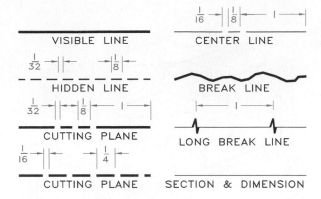

7.5 Full-size line weights recommended for drawing orthographic views are shown above.

7.3 Alphabet of Lines

Draw all orthographic views with dark and dense lines as if drawn with ink. Only the line widths should vary except for guidelines and construction lines, which are drawn very lightly for layout and lettering. **Figure 7.4** gives examples of lines used in orthographic projection and the recommended pencil grades for them. The lengths of dashes in hidden lines and centerlines are drawn longer as a drawing's size increases. **Figure 7.5** further describes these lines.

7.4 Six-View Drawings

When you imagine that an object is inside a glass box you will see two horizontal planes, two frontal planes, and two profile planes **(Figure 7.6).** Therefore, the maximum number of principal views that can be used to represent an object is six. The top and bottom views are projected onto horizontal planes, the front and rear views onto frontal planes, and the right- and left-side views onto profile planes.

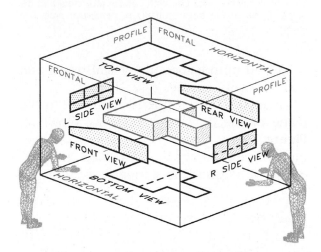

7.6 Six principal views of an object can be drawn in orthographic projection. Imagine that the object is in a glass box with the views projected onto its six planes.

To draw the six views on a sheet of paper, imagine that the glass box is opened up into the plane of the drawing paper as shown in **Figure 7.7.** Place the top view over and the bottom view under the front view; place the right-side view to the right and the left-side view to the left of the front view; and place the rear view to the left of the left-side view.

Projectors align the views both horizontally and vertically about the front view. Each side of the fold lines of the glass box is labeled **H, F,** or **P** (horizontal, frontal, or profile) to

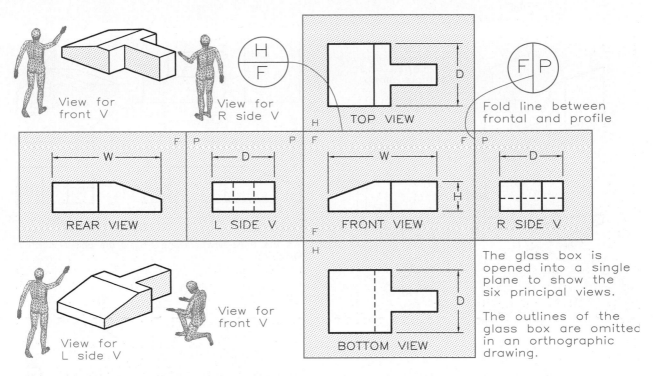

7.7 Opening the box into a single plane positions the six views as shown to describe the object.

identify the projection planes on each side of the imaginary fold lines **(Figure 7.7)**.

Height (H), width (W), and depth (D)—the three dimensions necessary to dimension an object—are shown in their recommended positions in **Figure 7.7.** The standard arrangement of the six views allows the views to share dimensions by projection. For example, the height dimension, which is shown only once between the front and right-side views, applies to the four horizontally aligned views. The width dimension is placed between the top and front views, but applies to the bottom view also.

7.5 Three-View Drawings

The most commonly used orthographic arrangement of views is the three-view drawing, consisting of front, top, and right-side views. Imagine that the views of the object are projected onto the planes of the glass box **(Figure 7.8)** and the three planes are opened into a single plane, the frontal plane. **Figure 7.9** shows the resulting three-view drawing where the views are labeled and dimensioned with H, W, and D.

In addition to using the glass-box approach for visualizing orthographic projection, imagine that the object is revolved until the desired view is obtained. For example, if the front view of the part in **Figure 7.10** is revolved 90°, you will see its side view true shape (TS). Placed in its proper position, the side view aligns with the projectors from the front view at its right.

7.6 Arrangement of Views

Figure 7.11 shows the standard positions for a three-view drawing: The top and side views are projected from and aligned with the front

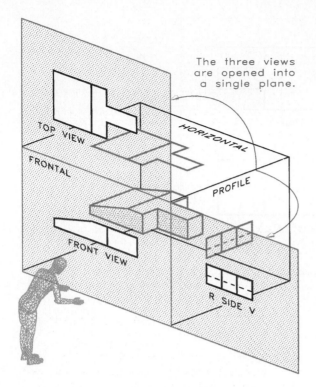

7.8 Three-view drawings are usually adequate to describe most small objects such as machine parts.

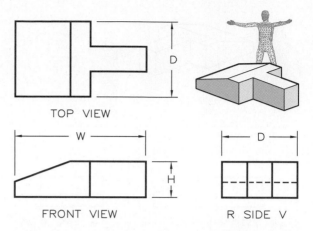

TOP VIEW

FRONT VIEW

R SIDE V

7.9 This three-view drawing depicts the object shown in Figure 7.8.

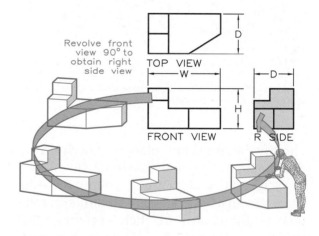

Revolve front view 90° to obtain right side view

TOP VIEW

FRONT VIEW

R SIDE

7.10 By rotating the front view 90° about a vertical axis, the side view is found.

view. Improperly arranged views that do not project from view to view are also shown. **Figure 7.12** illustrates the rules of projection and shows the proper alignment of dimension lines. Orthographic projection shortens layout time, improves readability, and reduces the number of dimensions required because they are placed between and are shared by the views to which they apply.

7.7 Selection of Views

Select the sequence of orthographic views with the fewest hidden lines. **Figure 7.13A** shows that the right-side view is preferable to the left-side view because it has fewer hidden lines. Although the three-view arrangement of top, front, and right-side views is more commonly used, the top, front, and left-side view arrangement is acceptable **(Figure 7.13B)** if

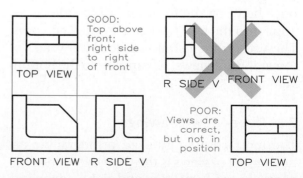

GOOD: Top above front; right side to right of front

TOP VIEW

FRONT VIEW   R SIDE V

R SIDE V   FRONT VIEW

POOR: Views are correct, but not in position

TOP VIEW

7.11 Orthographic views must be arranged in their proper positions in order for them to be interpreted correctly.

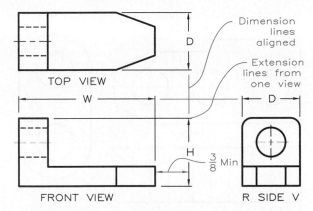

7.12 Dimension and extension lines used in three-view orthographic projection should be aligned. Draw extension lines from only one view when dimensions are placed between views.

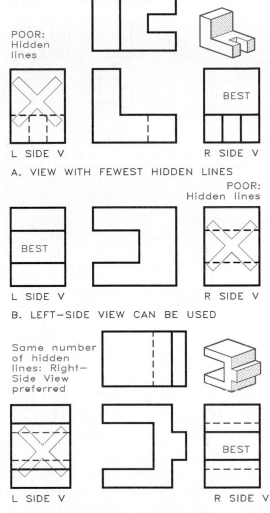

7.13 Selection of views:
A Select the set of views with the fewest hidden lines.
B Select the left-side view because it has fewer hidden lines than the right-side view.
C When both views have an equal number of hidden lines, select the right-side view.

the left-side view has fewer hidden lines than the right-side view.

The most descriptive view usually is selected as the front view. If an object, such as a chair, has predefined views that are generally recognize as the front or top views, you should label the accepted front view as the orthographic front view.

Although the right-side view usually is placed to the right of the front view, the side view can be projected from the top view **(Figure 7.14).** This alternative position saves space when the object has a much larger depth than height.

7.8 Line Techniques

Figure 7.15 illustrates techniques for handling most types of intersecting lines, hidden lines, and arcs in combination. Proper application of these principles improves the readability of orthographic drawings.

Become familiar with the order of importance (precedence) of lines **(Figure 7.16).** The most important line, the visible object line, is shown regardless of any other line lying behind it. Of next importance is the hidden line, which is more important than the centerline.

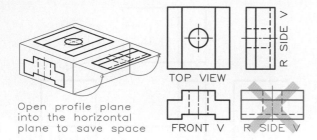

Open profile plane into the horizontal plane to save space

TOP VIEW

FRONT V

R SIDE V

R SIDE V

7.14 The side view can be projected from the top view instead of the front view. This alternate position saves space when the depth of an object is considerably greater than its height.

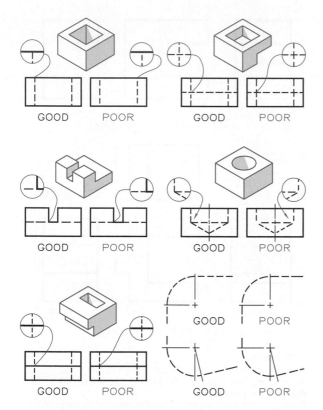

GOOD POOR GOOD POOR

GOOD POOR GOOD POOR

GOOD POOR GOOD POOR

GOOD POOR

GOOD POOR

7.15 These drawings show proper intersections and other line techniques in orthographic views.

7.9 Point Numbering

Some orthographic views are difficult to draw due to their complexity. By numbering the end-points of the lines of the parts in each view as you construct it **(Figure 7.17),** the location of the object's features will be easier. For example, using numbers on the top and side views of this

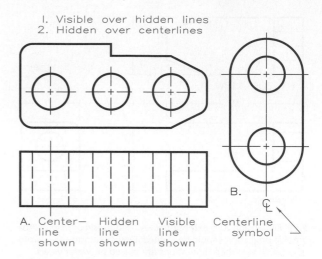

1. Visible over hidden lines
2. Hidden over centerlines

A. Center— Hidden Visible Centerline
   line   line   line   symbol
   shown   shown   shown

B.

7.16 When lines conincide with each other, the more important lines take precedence (cover up) the other lines. The order of importance is: visible lines, hidden lines, and center lines.

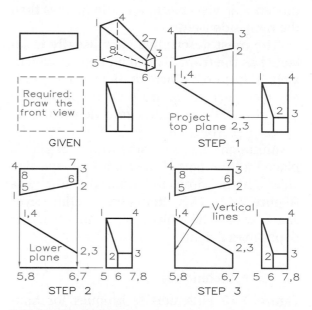

GIVEN STEP 1

Required: Draw the front view

Project top plane 2,3

Lower plane

STEP 2 STEP 3

Vertical lines

7.17 Point numbering:
Required: Find the front view.
Step 1 Number the corners of plane 1-2-3-4 in the top and side views and project them to the front view.
Step 2 Number the corners of plane 5-6-7-8 in the top and side views and project them to the front view.
Step 3 Connect the numbered lines to complete the front view.

object aids in the construction of the missing front view. Projecting points from the top and side views to the intersections of the projectors locates the object's front view.

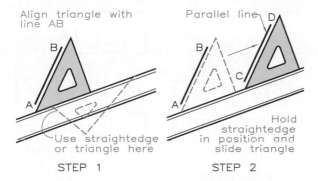

STEP 1   STEP 2

7.18 Drawing parallel lines:
Step 1 Align the upper triangle with AB and in contact with the lower triangle (or straightedge).
Step 2 Hold the lower straightedge in position and slide the upper triangle to where CD is drawn parallel to AB.

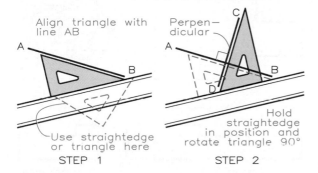

STEP 1   STEP 2

7.19 Drawing perpendiculars:
Step 1 Align your triangle with line AB and in contact with the lower triangle (or straightedge).
Step 2 Hold the lower straightedge in position, rotate your triangle, and draw CD perpendicular to AB.

7.10 Drawing with Triangles

Two triangles can be effectively used to make instrument drawings on 8-1/2 × 11 sheets without taping the sheet to the drawing surface. It is better that it can be moved about to comfortably position the triangles.

Parallel lines The 45° triangle or the 30°–60° triangle can be used to draw parallel lines as shown in **Figure 7.18.** One triangle or a straightedge is held in position while the other triangle is moved to where the parallel line is drawn.

Perpendiculars To draw a line perpendicular to AB in **Figure 7.19,** align the 30°–60° triangle's

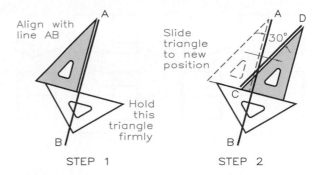

STEP 1   STEP 2

7.20 Drawing a 30° angle:
Step 1 Hold the 30°–60° triangle aligned with AB and in contact with the lower triangle (or straightedge).
Step 2 Hold the lower triangle in position and slide the triangle and draw CD at 30° to AB.

hypotenuse side with AB and against the lower triangle or straightedge. Hold the lower triangle in position and rotate it so that its hypotenuse is perpendicular to AB and line CD can be drawn.

Angles To draw a line making 30° with AB use the steps shown in **Figure 7.20.** Two triangles can be used in other combinations as a means of making instrument drawings in this informal manner, and yet with a sufficient degree of accuracy.

7.11 Views by Subtraction

Figure 7.21 illustrates how three views of a part are drawn by beginning with a blocked in outline having the overall dimensions of height, width, and depth of the finished part and removing volumes from it. This drawing procedure is similar to the steps of making the part in the shop.

7.12 Three-View Drawing Layout

The depth dimension applies to both the top and side views, but these views usually are positioned where depth does not project between them **(Figure 7.23).** The depth dimension must be transferred between the top and side

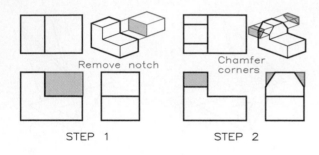

STEP 1   STEP 2

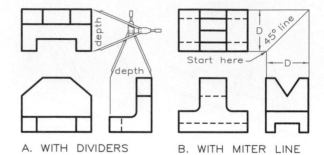

A. WITH DIVIDERS   B. WITH MITER LINE

7.22 Transferring depth:
A Transfer the depth dimension from the top view to the side view with your dividers.
B Use a 45° miter line to transfer the depth dimension between the top and side views by projection.

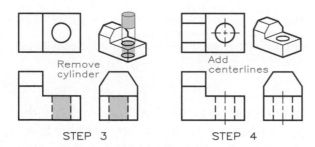

STEP 3   STEP 4

7.21 Views by subtraction:
Step 1 Block in the views of the object using overall dimensions of H, W, and D. Remove the notch.
Step 2 Remove the triangular solids at the corners.
Step 3 Remove the cylindrical volume of the hole.
Step 4 Add centerlines to complete the views.

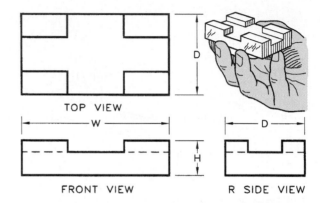

TOP VIEW

FRONT VIEW   R SIDE VIEW

7.23 A three-view drawing depicts an object with only horizontal and vertical planes.

views with dividers or by using a 45° miter line, as illustrated in **Figure 7.22.**

Layout Rules The basic rules of making orthographic drawings are summarized below. Refer to the examples in **Figure 7.23** through **Figure 7.27** and observe how these rules have been used. Notice how the dimensions have been applied and how the views have been labeled.

1. Draw orthographic views in their proper positions.

2. Select the most descriptive view as the front view, if the object does not have a predefined front view.

3. Select the sequence of views with the fewest hidden lines.

4. Label the views; for example, top view, front view, and right-side view.

5. Place dimensions between the views to which they apply.

6. Use the proper alphabet of lines.

7. Leave adequate room between the views for labels and dimensions.

8. Draw the views necessary to describe a part. Sometimes fewer or more views are required.

7.13 Two-View Drawings

Time and effort can be saved by drawing only the views and features that are necessary to describe a part. **Figure 7.28** shows typical objects that require only two views necessary to

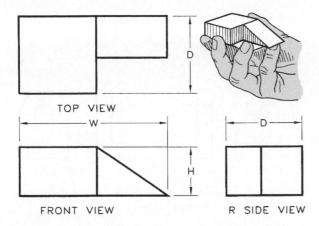

TOP VIEW

W ← → ← D →

FRONT VIEW   R SIDE VIEW

H

7.24 A three-view drawing showing an object with a sloping plane.

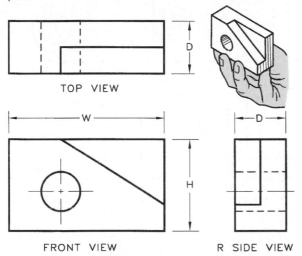

TOP VIEW

W ← → ← D →

FRONT VIEW   R SIDE VIEW

H

7.25 A three-view drawing showing an object with a sloping plane and a cylindrical hole through it.

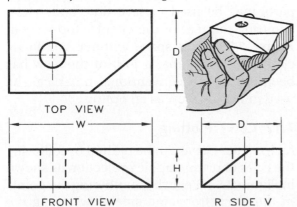

TOP VIEW

W ← → ← D →

FRONT VIEW   R SIDE V

H

7.26 This three-view drawing depicts an object that has a corner plane with a compound slope.

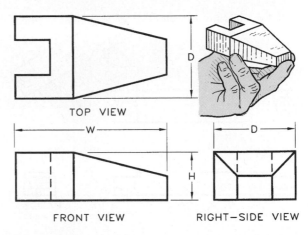

TOP VIEW

W ← → ← D →

FRONT VIEW   RIGHT–SIDE VIEW

H

7.27 This three-view drawing shows an object that has multiple planes with compound slopes.

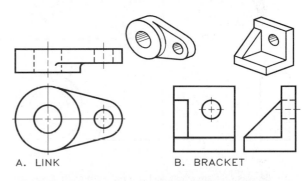

A. LINK   B. BRACKET

7.28 These objects can be adequately described with two orthographic views.

be described. The fixture block in **Figure 7.29** is another example of a part needing only two views to be adequately described. You can see that the top and front views are the best for this part. The front and side views would not be as good.

7.14 One-View Drawings

Simple cylindrical parts and parts of a uniform thickness can be described by only one view as shown in **Figure 7.30.** Supplementary notes clarify features that would have been shown in the omitted views. Diameters are labeled with diameter signs and thicknesses are given in notes.

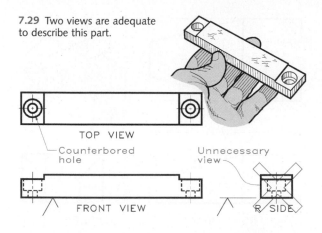

7.29 Two views are adequate to describe this part.

TOP VIEW

Counterbored hole

Unnecessary view

FRONT VIEW

R SIDE

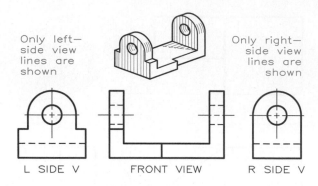

Only left—side view lines are shown

Only right—side view lines are shown

L SIDE V FRONT VIEW R SIDE V

7.31 Use simplified views with unnecessary and confusing hidden lines omitted to improve clarity.

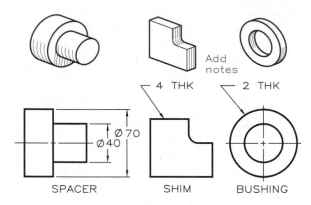

Add notes

4 THK   2 THK

Ø70
Ø40

SPACER   SHIM   BUSHING

7.30 Objects that are cylindrical or of a uniform thickness can be described with only one orthographic view and supplementary notes.

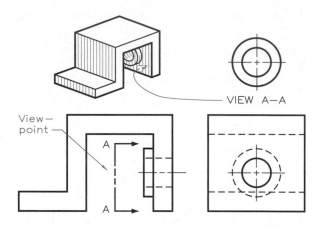

VIEW A–A

View—point

A   A

7.32 Use a removed view, indicated by the directional arrows, to show hard-to-see views in removed locations.

7.15 Simplified and Removed Views

The right- and left-side views of the part in **Figure 7.31** would be harder to interpret if all hidden lines were drawn by rigorously following the rules of orthographic projection. Simplified views in which confusing and unnecessary lines have been omitted are better and more readable.

When it is difficult to show a feature with a standard orthographic view because of its location, a **removed view** can be drawn (**Figure 7.32**). The removed view, indicated by the directional arrows, is clearer and less confusing when moved to an isolated position.

7.16 Partial Views

Partial views of symmetrical or cylindrical parts may be used to save time and space. Omitting the rear of the circular top view in **Figure 7.33** saves space without sacrificing clarity. To clarify that a part of the view has been omitted, a conventional break can be used in the top view as an option.

7.17 Curve Plotting

An irregular curve can be plotted by following the rules of orthographic projection as shown in **Figure 7.34.** Begin by numbering the points in the given front and side views along the curve. Next, project from the points having the same numbers in the front and side views

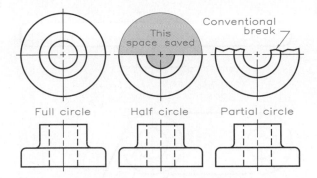

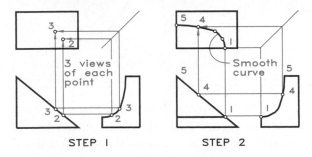

7.33 Save space and time by drawing the circular view of a cylindrical part as a partial view.

7.34 Curve plotting:
Step 1 Locate points 2 and 3 in the front and side views by projection. Project points 2 and 3 to the top view.
Step 2 Locate the remaining points in the three views and connect them in the top view with a smooth curve.

to the top to where the projectors intersect. Continue projecting in this manner and connect the points in the top view with a smooth curve drawn with an irregular curve. **Figure 7.35** shows an ellipse plotted in the top view by projecting points from front and side views. It is best to number the points as they are transferred one at a time to avoid getting lost in your construction.

7.18 Conventional Practices

The readability of an orthographic view may be improved if the rules of projection are violated. Violations of rules customarily made for the sake of clarity are called **conventional practices.**

Symmetrically spaced holes in a circular plate (**Figure 7.36**) are drawn at their true

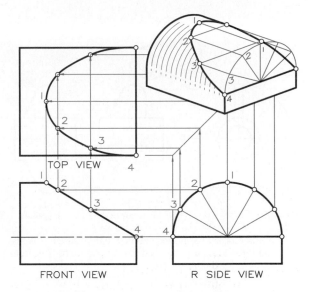

FRONT VIEW   R SIDE VIEW

7.35 The ellipse in the top view was found by numbering points in the front and side views and projecting them to the top view.

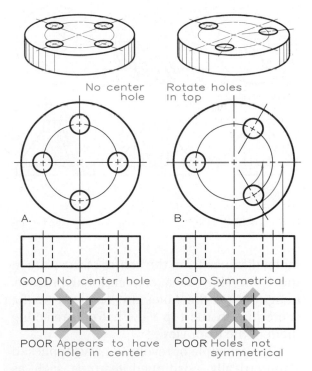

7.36 Placement of holes:
A Omit the center hole found by true projection. It makes the hole appear to pass through the center of the plate.
B Use a conventional view to show the holes located at their true radial distances from the center. Imagine they are rotated to the centerline in the top view.

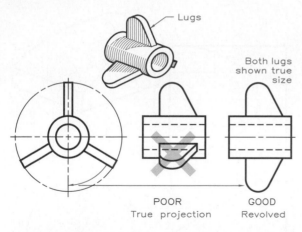

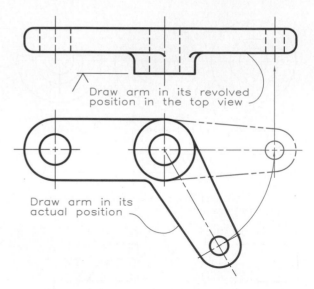

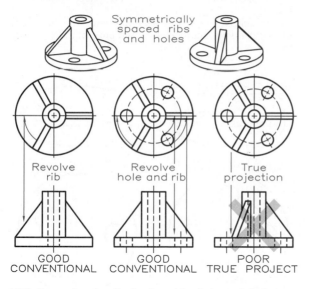

7.37 Symmetrically positioned external features, such as webs, ribs, and these lugs, are imagined to be revolved to their true-size positions for the best views.

7.38 Conventional methods of revolving holes and ribs in combination improve clarity.

7.39 Imagine that the front view of the arm is revolved so its true length can be drawn in the top view as a conventional practice.

radial distance from the center of the plate in the front view as a conventional practice. Imagine that the holes are revolved to the centerline in the top view before projecting them to the front view.

This principle of revolution also applies to symmetrically positioned features such as ribs, webs, and the three lugs on the outside of the part shown in **Figure 7.37. Figure 7.38** shows the applications of conventional practices to holes and ribs in combination.

Another conventional revolution is illustrated in **Figure 7.39** where the front view of an inclined arm is revolved to a horizontal position so that it can be drawn true size in the top view. The arm in its revolved position in the front view is not drawn because the revolution is imaginary.

Figure 7.40 shows how to clarify views of parts by conventional revolution. By revolving the top views of these parts 45°, slots and holes no longer coincide with the centerlines and can be seen more clearly. Draw the front views of the slots and holes true size by imagining that they have been revolved 45°.

Another type of conventional view is the true-size development of a curved sheet-metal part drawn as a flattened-out view (**Figure 7.41**). The top view shows the part's curvature.

7.19 Conventional Intersections

In orthographic projection, lines are drawn to represent the intersections (fold lines) between planes of object. Wherever planes intersect, forming an edge, this line of intersection is projected to its adjacent view. Examples showing where lines are required are given in **Figure 7.42.**

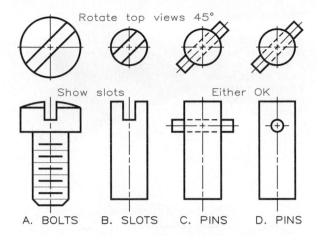

7.40 It is conventional practice to draw the slots and holes at 45° in the top view and true size in the front view.

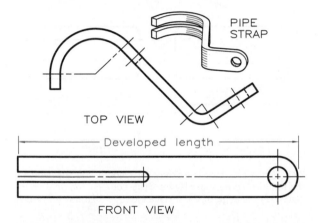

7.41 It is conventional practice to use true-size developed (flattened-out) views of parts made of bent sheet metal.

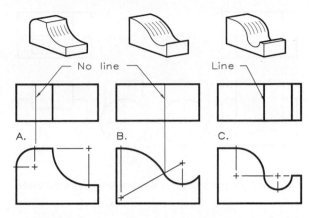

7.42 Object lines are drawn only where there are sharp intersections or where arcs are tangent at their centerlines as at C.

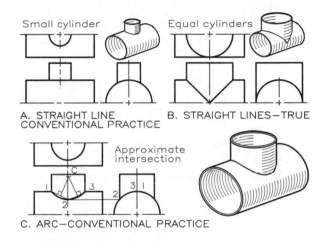

7.43 Intersections between cylinders:
A and **C** Use these methods of representation.
B Equal size cylinders have straight-line intersections.

Figure 7.43 shows how to draw intersections between cylinders rather than plotting more complex, orthographically correct lines of intersection. **Figures 7.43A** and **C** show conventional intersections, which means they are approximations drawn for ease of construction while being sufficiently representative of the object. **Figure 7.43B** shows an easy-to-draw intersection between cylinders of equal diameters, and this is a true intersection as well. **Figures 7.44** and **7.45** show other cylindrical intersections, and **Figure 7.46**

shows conventional practices for depicting intersections formed by holes in cylinders.

7.20 Fillets and Rounds

Fillets and **rounds** are rounded intersections between the planes of a part that are used on castings, such as the body of the pillow block in **Figure 7.47**. A *fillet* is an inside rounding and a *round* is an external rounding on a part. The radii of fillets and rounds usually are small, about 1/4 inch. Fillets give added strength at

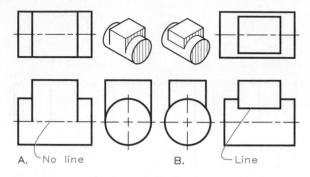

A. No line   B. Line

7.44 These are true intersections between cylinders and prisms.

7.47 The edges of this pillow block are rounded with fillets and rounds. The surface of the casting is rough except where it has been machined. (*Courtesy of Rockwell Automation.*)

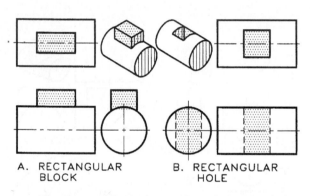

A. RECTANGULAR BLOCK   B. RECTANGULAR HOLE

7.45 These are conventional intersections between cylinders and prisms.

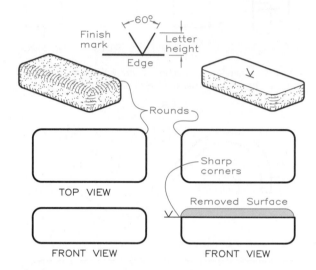

7.48 When a surface is finished (machined), the cut removes the rounded corners and leaves sharp corners. The finish mark placed on the edge of the surface indicates that it is finished.

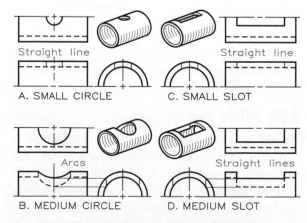

A. SMALL CIRCLE   C. SMALL SLOT

B. MEDIUM CIRCLE   D. MEDIUM SLOT

7.46 Conventional methods of depicting holes in cylinders are easier to draw and to understand than true projections.

inside corners, rounds improve appearance, and both remove sharp edges.

A casting will have square corners after its surface has been finished, which is the process of machining away part of the surface to a smooth finish **(Figure 7.49).** Finished surfaces are indicated by placing a finish mark (V) on all edge views of finished surfaces whether the edges are visible or hidden. **Figure 7.50** shows four types of finish marks. A more detailed surface texture symbol is presented in Chapter 13. **Figure 7.51** illustrates

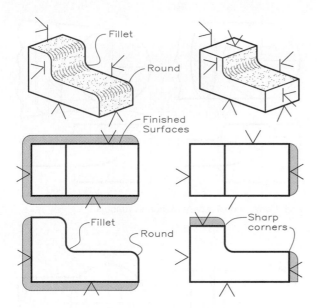

7.49 Fillets and rounds are rounded inside and outside corners, respectively, that are standard features on castings. When surfaces are finished, fillets and rounds are removed as shown here.

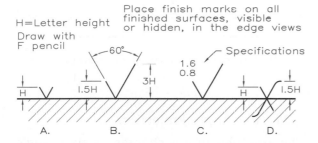

7.50 Any of these finish marks are placed on all edge views of finished surfaces whether visible or hidden.

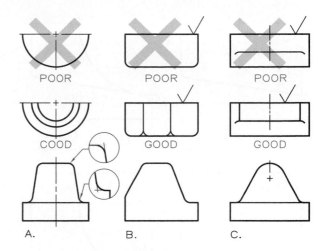

7.51 These examples show both poorly drawn and conventionally drawn fillets and rounds.

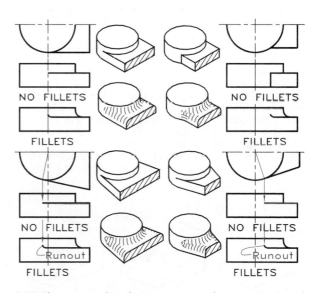

7.52 These examples show conventional intersections and runouts on cylindrical features. Runouts result from fillets and rounds intersecting cylinders.

several techniques for showing fillets and rounds on orthographic views with a circle template.

Figure 7.52 gives a comparison of intersections and **runouts** of parts with and without fillets and rounds. Large runouts are constructed as an eighth of a circle with a compass as shown in **Figure 7.53.** Small runouts are drawn with a circle template. Runouts on

orthographic views reveal much about the details of an object. For example, the runout in the top view of **Figure 7.54A** tells us that the rib has rounded corners, whereas the top view of **Figure 7.54B** tells us the rib is completely round. **Figures 7.55** and **7.56** illustrate other types of filleted and rounded intersections.

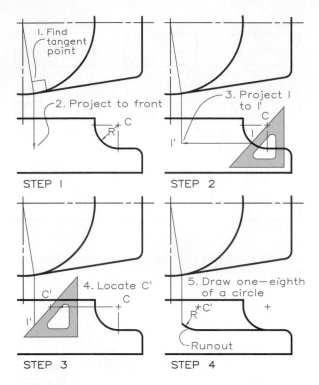

STEP 1

STEP 2

STEP 3

STEP 4

7.53 Plotting runouts:
Step 1 Find the tangency point in the top view and project it to the front view.
Step 2 Find point 1 with a 45° triangle; project it to 1'.
Step 3 Move the 45° triangle to locate point C' on the horizontal projector from center C.
Step 4 Use the radius of the fillet to draw the runout with C' as its center. The runout arc is drawn as one-eighth of a circle.

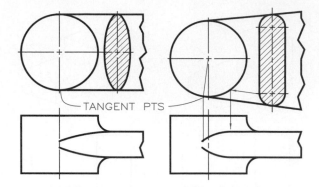

7.55 Conventional representation of runouts.

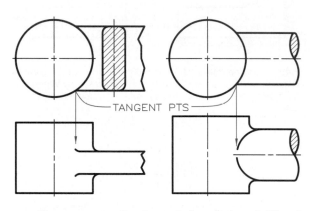

7.56 These are conventional runouts for cylinders of different cross sections.

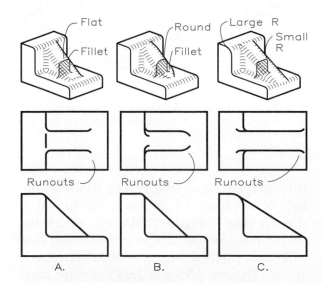

7.54 These are conventional drawings of runouts.

7.21 First-Angle Projection

The examples in this chapter are third-angle projections in which the top view is placed over the front view and the right-side view is placed to the right of the front view as shown in **Figure 7.57.** This method is used in the United States, Great Britain, and Canada. However, most of the world uses first-angle projection.

The first-angle system is illustrated in **Figure 7.58** in which an object is placed above the horizontal plane and in front of the frontal plane. When these projection planes are opened onto the surface of the drawing paper, the front view projects over the top view, and the right-side view to the left of the front view.

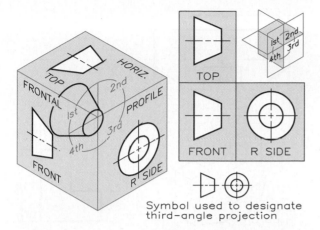

Symbol used to designate
third-angle projection

7.57 Third-angle projection is used for drawing orthographic views in the United States, Great Britain, and Canada. The top view is placed over the front view and the right-side view is placed to the right of the front view. The truncated cone is the symbol used to designate third-angle projection.

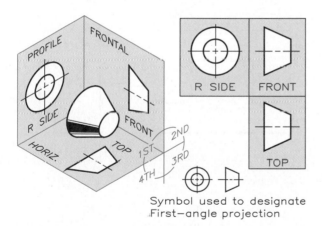

Symbol used to designate
First—angle projection

7.58 First-angle projection is used in most of the world. It shows the right-side view to the left of the front view and the top view under the front view. The truncated cone designates first-angle projection.

The angle of projection used in making a drawing is indicated by placing the truncated cone in or near the title block (**Figure 7.59**). When metric units of measurement are used, the SI symbol is given in combination with the cone on the drawing.

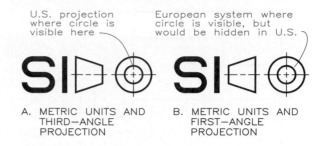

U.S. projection where circle is visible here

European system where circle is visible, but would be hidden in U.S.

A. METRIC UNITS AND THIRD—ANGLE PROJECTION

B. METRIC UNITS AND FIRST—ANGLE PROJECTION

7.59 These symbols are placed on drawings to specify first-angle or third-angle projection and metric units of measurement.

7.22 Summary

The techniques of orthographic projection introduced in this chapter can be used to develop and communicate designs from the simple to the complex. Orthographic projection is an important medium of design and the language of engineering without which the marvels of the twentieth and twenty-first centuries could not have come into being.

With the advent of computer graphics, many designs are depicted as three-dimensional pictorials to show objects in views as if they had been photographed by a camera. In spite of these powerful computer capabilities, a precisely drawn and dimensioned orthograhic view in many cases may give the fabrication details with greater clarity.

You will use orthographic projection throughout your career as a valuable means of developing design concepts by sketching, and finally with instrument-drawn working drawings from which designs will be fabricated.

Problems

1–7. (**Figures 7.60–7.66**) Draw the given views on size A sheets, two per sheet, using the dimensions given and draw the missing top, front, or right-side views. Lines may be missing in the given views.

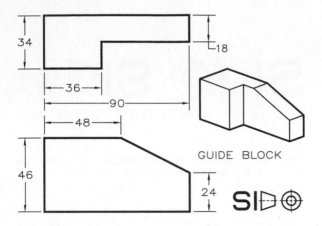

7.60 Problem 1.

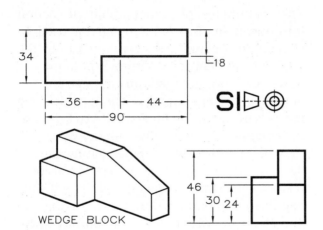

7.61 Problem 2.

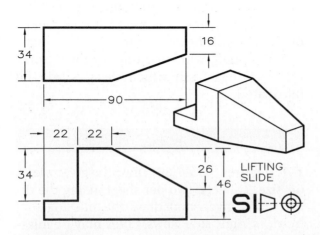

7.62 Problem 3.

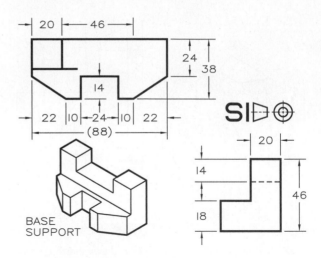

7.63 Problem 4.

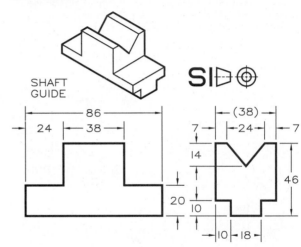

7.64 Problem 5.

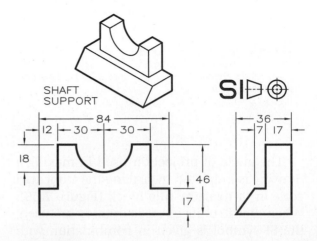

7.65 Problem 6.

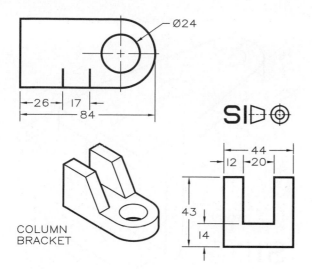

Ø24

├─26─┼─17─┤
├──────84──────┤

SI ▷⊕

COLUMN
BRACKET

├──44──┤
├12┤├20┤

43
14

7.66 Problem 7.

8–28. (Figures 7.67–7.87) Construct the necessary orthographic views to describe the objects on B-size sheets at an appropriate scale. Label the views and show the overall dimensions of W, D, and H.

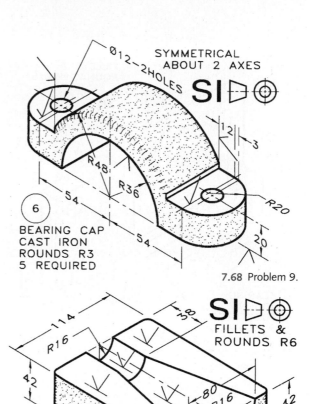

Ø12–2HOLES
SYMMETRICAL
ABOUT 2 AXES

SI ▷⊕

12
3

R48
R36

54

R20

20

54

⑥ BEARING CAP
CAST IRON
ROUNDS R3
5 REQUIRED

7.68 Problem 9.

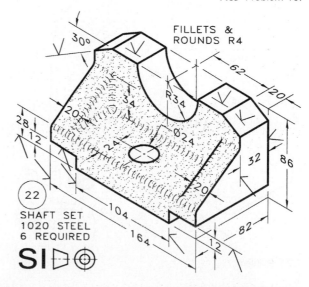

SI ▷⊕
FILLETS &
ROUNDS R6

114
R16
42
80
R16
42

146
R114

㉑ FORMING PLATE
CAST IRON
8 REQUIRED

7.69 Problem 10.

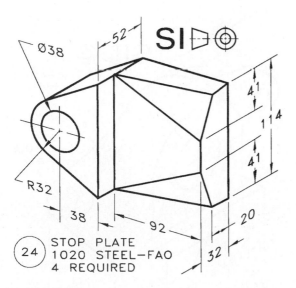

Ø38
52

SI ▷⊕

41
114
41

R32
38
92
20
32

㉔ STOP PLATE
1020 STEEL—FAO
4 REQUIRED

7.67 Problem 8.

7.70 Problem 11.

FILLETS &
ROUNDS R4

30°
62
R34
20
Ø24
28
12
32
86
20
㉒ SHAFT SET
1020 STEEL
6 REQUIRED
104
164
82
12

SI ▷⊕

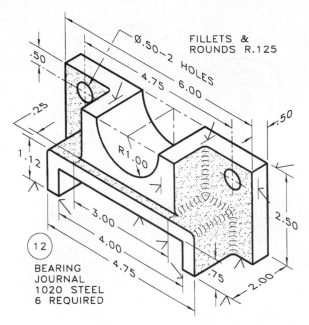

FILLETS & ROUNDS R.125

Ø.50–2 HOLES

.50

.25

4.75

6.00

R1.00

.50

1.12

.50

3.00

4.00

4.75

2.50

.75

2.00

⑫

BEARING JOURNAL
1020 STEEL
6 REQUIRED

7.71 Problem 12.

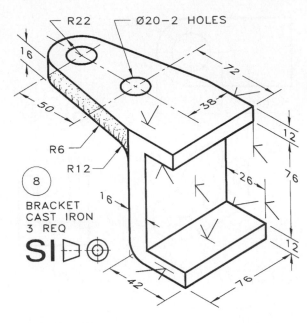

R22

Ø20–2 HOLES

16

50

72

38

R6

R12

12

26

76

16

⑧

BRACKET
CAST IRON
3 REQ

SI ⟊⊙

42

76

12

7.73 Problem 14.

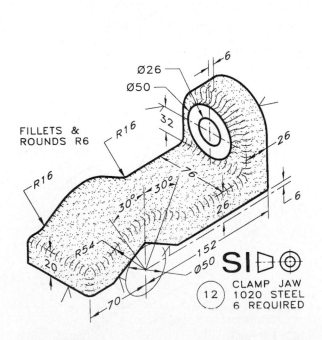

Ø26

Ø50

6

FILLETS & ROUNDS R6

R16

32

26

R16

30° 30°

6

26

6

R54

152

20

Ø50

70

SI ⟊⊙

CLAMP JAW
1020 STEEL
6 REQUIRED

⑫

7.72 Problem 13.

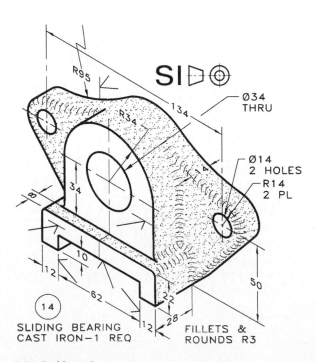

R95

SI ⟊⊙

Ø34
THRU

R34

134

Ø14
2 HOLES
R14
2 PL

8

34

10

12

62

12

28

22

50

⑭

SLIDING BEARING
CAST IRON–1 REQ

FILLETS & ROUNDS R3

7.74 Problem 15.

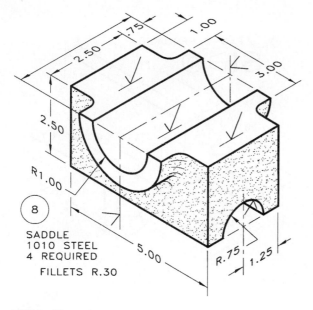

2.50
.75
1.00
3.00
2.50
R1.00
8
SADDLE
1010 STEEL
4 REQUIRED
FILLETS R.30
5.00
R.75
1.25

7.75 Problem 16.

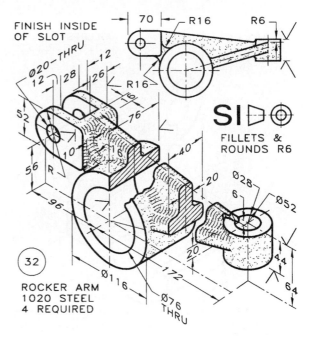

FINISH INSIDE
OF SLOT
70
R16
R6
Ø20–THRU
12
12
28
26
R16
52
76
SI
FILLETS &
ROUNDS R6
56
R
10
16
40
20
Ø28
6
Ø52
96
44
32
Ø116
20
64
ROCKER ARM
1020 STEEL
4 REQUIRED
172
Ø76
THRU

7.77 Problem 18.

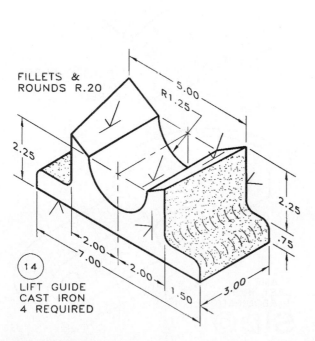

FILLETS &
ROUNDS R.20
5.00
R1.25
2.25
2.25
.75
2.00
7.00
2.00
14
LIFT GUIDE
CAST IRON
4 REQUIRED
1.50
3.00

7.76 Problem 17.

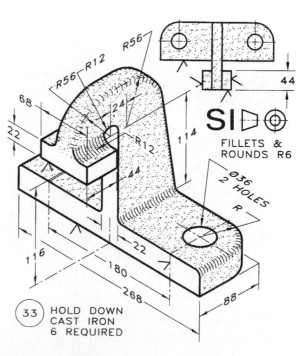

R56
R56
R12
R56
R12
68
24
114
22
R12
44
SI
FILLETS &
ROUNDS R6
Ø36
2 HOLES
R
16
22
180
268
33
HOLD DOWN
CAST IRON
6 REQUIRED
88
44

7.78 Problem 19.

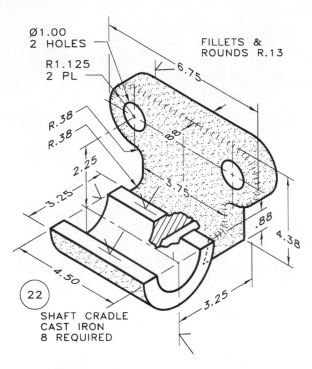

Ø1.00
2 HOLES

R1.125
2 PL

FILLETS &
ROUNDS R.13

6.75

R.38
R.38

2.25

3.25

.88

3.75

.88

4.38

4.50

3.25

22

SHAFT CRADLE
CAST IRON
8 REQUIRED

7.79 Problem 20.

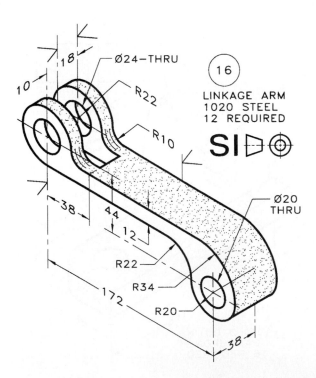

18
10

Ø24–THRU

R22

16

LINKAGE ARM
1020 STEEL
12 REQUIRED

SI ▷⊙

R10

Ø20
THRU

38

44

12

R22

R34

172

R20

38

7.80 Problem 21.

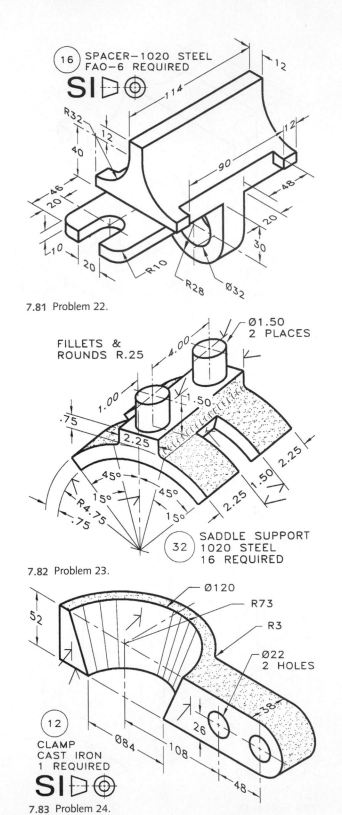

16

SPACER–1020 STEEL
FAO–6 REQUIRED

SI ▷⊙

12

R32

12

114

40

12

46

90

20

48

10

20

20

30

R10

R28

Ø32

7.81 Problem 22.

FILLETS &
ROUNDS R.25

Ø1.50
2 PLACES

4.00

1.00

1.50

.75

2.25

45°

15°

45°

R4.75

15°

.75

2.25

1.50

2.25

SADDLE SUPPORT
1020 STEEL
16 REQUIRED

32

7.82 Problem 23.

Ø120

R73

R3

Ø22
2 HOLES

52

26

38

12

CLAMP
CAST IRON
1 REQUIRED

SI ▷⊙

Ø84

108

48

7.83 Problem 24.

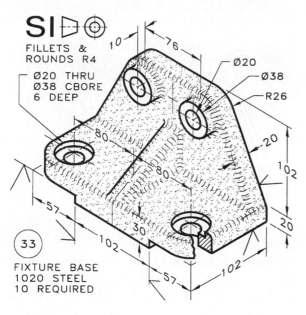

SI ▷⊕

FILLETS & ROUNDS R4

Ø20 THRU
Ø38 CBORE
6 DEEP

10
76
Ø20
Ø38
R26
80
80
20
102
57
102
30
20
102
57

(33)

FIXTURE BASE
1020 STEEL
10 REQUIRED

7.84 Problem 25.

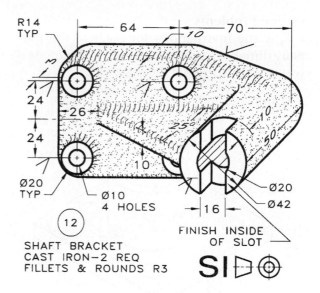

R14
TYP
64
10
70
3
24
26
25°
24
10
10
50
Ø20
TYP
Ø10
4 HOLES
Ø20
Ø42
16

(12)

SHAFT BRACKET
CAST IRON—2 REQ
FILLETS & ROUNDS R3

FINISH INSIDE
OF SLOT

SI ▷⊕

7.86 Problem 27.

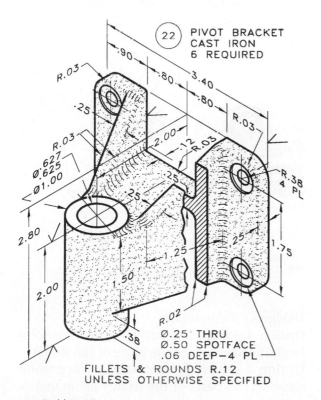

(22) PIVOT BRACKET
CAST IRON
6 REQUIRED

.90
.80
3.40
.80
R.03
R.03
R.03
.25
R.03
.627
.625
Ø.627
Ø.625
Ø1.00
2.00
.12
R.03
.25
R.38
4 PL
.25
.25
2.80
1.25
.25
1.75
1.50
2.00
R.02
.38
Ø.25 THRU
Ø.50 SPOTFACE
.06 DEEP—4 PL

FILLETS & ROUNDS R.12
UNLESS OTHERWISE SPECIFIED

7.85 Problem 26.

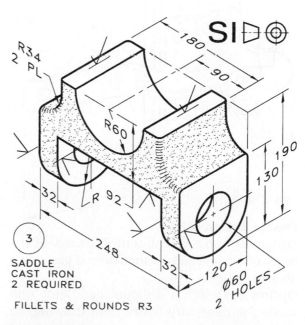

SI ▷⊕

R34
2 PL
180
90
R60
190
R 92
130
32
248
32
120
Ø60
2 HOLES

(3)

SADDLE
CAST IRON
2 REQUIRED

FILLETS & ROUNDS R3

7.87 Problem 28.

Design Problems

Follow the instructions for each of the partially dimensioned problems given below as if you were the original designer.

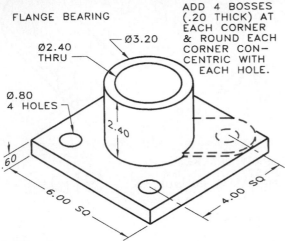

FLANGE BEARING

Ø2.40 THRU

Ø3.20

ADD 4 BOSSES (.20 THICK) AT EACH CORNER & ROUND EACH CORNER CONCENTRIC WITH EACH HOLE.

Ø.80 4 HOLES

2.40

.60

6.00 SQ

4.00 SQ

Design 1

Redesign the flange bearing to have 4 bosses and round each corner as noted. Draw the necessary views to describe the part.

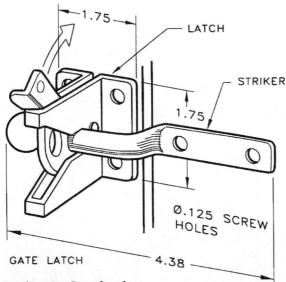

1.75

LATCH

STRIKER

1.75

Ø.125 SCREW HOLES

GATE LATCH   4.38

Design 2: Gate latch

Option 1: Make instrument drawings of the parts of the partially dimensioned gate latch on size A sheets.

Option 2: Design a different gate latch and make the necessary instrument drawings to describe your design.

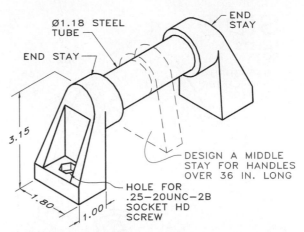

Ø1.18 STEEL TUBE

END STAY

END STAY

3.15

1.80

1.00

DESIGN A MIDDLE STAY FOR HANDLES OVER 36 IN. LONG

HOLE FOR .25-20UNC-2B SOCKET HD SCREW

Design 3: Handle assembly.

Option 1: Make the views of the end stays of the door handle to adequately describe them.

Option 2: Design a middle stay that can be used for handles over 36 inches and draw the necessary orthographic views to describe your design.

Option 3: Design a completely different handle and make the necessary instrument drawings (orthographic views) to describe it.

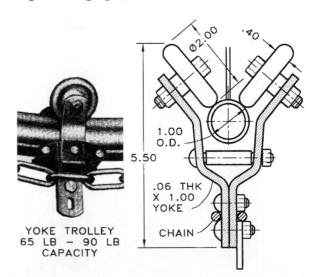

Ø2.00

.40

1.00 O.D.

5.50

.06 THK X 1.00 YOKE

CHAIN

YOKE TROLLEY 65 LB — 90 LB CAPACITY

Design 4: Yoke trolley.

Option 1: Make the necessary orthographic freehand sketches of the parts of the trolley.

Option 2: Make the necessary orthographic views of the parts of the trolley using instruments.

Thought Questions

1. What is the definition of *orthographic projection* and how many principal views are there in this system?

2. How does a surface appear in the front view if it is true size in the top view? How does this surface appear in the side view?

3. What are the differences in line symbols and line weights for the following types of lines: visible, hidden, phantom, extension, dimension, section?

4. What is the general rule of selecting which views are to be the front, top, and side views?

5. Sketch examples of the preferred intersection of crossing hidden lines, visible and hidden lines, and intersecting hidden lines.

6. What determines the space that must be provided between orthographic views?

7. How many views are necessary for representing and describing an object in orthographic projection?

8. What is the minimum number of views required to represent the following: sphere, cylinder, pyramid, bolt, coffee cup? Make sketches of each.

9. What is meant when an orthographic view is referred to as having been drawn using conventional practices? Sketch examples.

10. What are the purposes of finish marks and to which views are they to be applied?

11. What are fillets and rounds and what is their purpose?

12. Define a *simplified orthographic view* and explain its purpose.

13. Sketch the top and front views of a circular plate that has three equally spaced holes located the same radial distance from the plate's center. Repeat for four holes.

14. Sketch the top and front views of the perpendicular intersections between the following sheet-metal shapes: two cylinders of the same diameter, two cylinders of different diameters, a cylinder and square prism with a width smaller than the diameter, and a cylinder and square prism with a width equal to the diameter.

15. Instead of representing an object with orthographic views, write out its description verbally as the sole means of communicating its specifications (ex: coffee cup, drawing triangle, tape dispenser, desk lamp, door knob).

16. What is the difference between ribs and lugs? What is the purpose of ribs?

17. Sketch the front and top views of a slotted bolt head using conventional practice.

18. What is a runout and what is its purpose when drawing views in orthographic projection?

19. What does SI on a drawing stand for and what is its origin?

20. Why is it good practice to sketch the orthographic views of a part before beginning an instrument drawing of it?

21. What is the minimum number of views required to show an object's three-dimensionality?

22. What are the letters placed on each side of the reference line between the top and front views? The front and side views?

23. Why is the arrangement of orthographic views important? Explain.

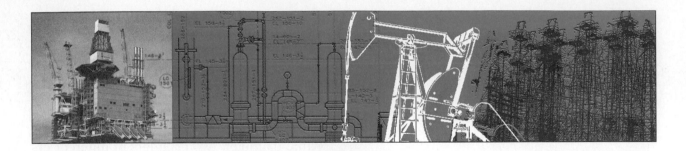

8

Primary Auxiliary Views

8.1 Introduction

Objects often are designed to have sloping or inclined surfaces that do not appear true size in principal orthographic views. A plane of this type is not parallel to a principal projection plane (horizontal, frontal, or profile) and is, therefore, a **nonprincipal plane.** Its true shape must be projected onto a plane that is parallel to it. This view is called an **auxiliary view.**

An auxiliary view projected from a primary view (principal view) is called a **primary auxiliary view.** An auxiliary view projected from a primary auxiliary view is a **secondary auxiliary view.** By the way, get out your dividers; you must use them all the time in drawing auxiliary views.

The inclined surface of the part shown in **Figure 8.1A** does not appear true size in the top view because it is not parallel to the horizontal projection plane. However, the inclined surface will appear true size in an auxiliary view projected perpendicularly from its edge view in the front view **(Figure 8.1B).**

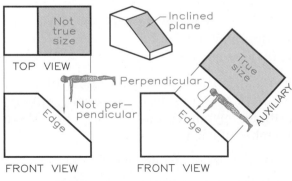

8.1 A surface that appears as an inclined edge in a principal view can be found true size by an auxiliary view. (A) The top view is foreshortened, but the inclined plane is true size in the auxiliary view at (B).

The relationship between an auxiliary view and the view it was projected from is the same as that between any two adjacent orthographic views. **Figure 8.2A** shows an auxiliary view projected perpendicularly from the edge view of the sloping surface. By rotating these

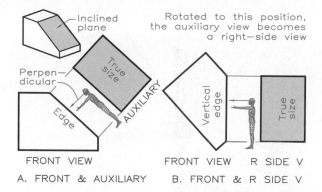

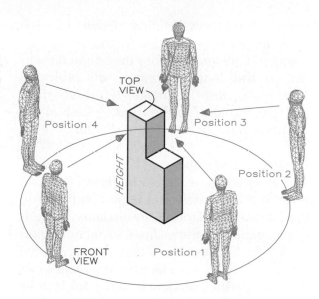

8.2 An auxiliary view has the same relationship with the view it is projected from as that of any two adjacent principal views.

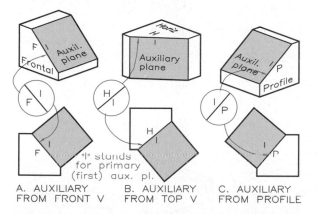

8.3 A primary auxiliary plane can be folded from the frontal, horizontal, or profile planes. The fold lines are labeled F-1, H-1, and P-1, with 1 on the auxiliary plane side and P on the principal-plane side.

8.4 By moving your viewpoint around the top view of an object, you will see a series of auxiliary views in which the height dimension (H) is true length in all of them.

views (the front and auxiliary views) so that the projectors are horizontal, the views have the same relationship as regular front and right-side views **(Figure 8.2B).**

8.2 Folding-Line Principles

The three principal orthographic planes are the **frontal** (F), **horizontal** (H), and **profile** (P) planes. An auxiliary view is projected from a principal orthographic view (a top, front, or side view), and a primary auxiliary plane is perpendicular to one of the principal planes and oblique to the other two.

Think of auxiliary planes as planes that fold into principal planes along a folding line **(Figure 8.3).** The plane in **Figure 8.3A** folds at a 90° angle with the frontal plane and is labeled F-1 where F is an abbreviation for frontal, and 1 represents first, or primary, auxiliary plane. **Figures 8.3B** and **8.3C** illustrate the positions for auxiliary planes that fold from the horizontal and profile planes, labeled H-1 and P-1, respectively.

It is important that reference lines be labeled as shown in **Figure 8.3,** with the numeral 1 placed on the auxiliary side and the letter H, F, or P on the principal-plane side.

8.3 Auxiliaries from the Top View

By moving your position about the top view of a part as shown in **Figure 8.4,** each line of sight is perpendicular to the height dimension. One of the views, the front view, is a principal view while the other positions see nonprincipal views called **auxiliary views.**

Figure 8.5 illustrates how these five views (one of which is a front view) are projected from the top view. The line of sight for each auxiliary view is parallel to the horizontal projection plane; therefore the height dimension is true length in each view projected from a top view. The height (H) dimensions are transferred from the front to each of the auxiliary views by using your dividers.

Folding-Line Method

The inclined plane shown in **Figure 8.6** is an edge in the top view and is perpendicular to the horizontal plane. If an auxiliary plane is drawn parallel to the inclined surface, the view projected onto it will be a true-size view of the inclined surface. **A surface must appear as an edge in a principal view in order for it to be found true size in a primary auxiliary view.** When the auxiliary view is projected from the top view, the height dimensions in the front

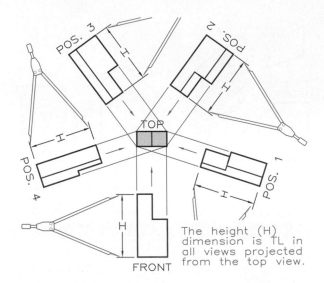

8.5 Auxiliaries from the top view.
The views shown in Figure 8.4 would be drawn as shown here with the same height dimension common to each view.

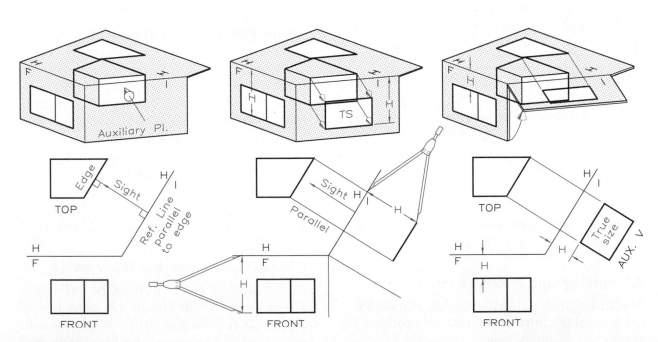

8.6 Auxiliary view from the top view.
Step 1 Draw the line of sight perpendicular to the edge view of the inclined surface. Draw the H-1 line parallel to its edge and draw the H-F reference line between the top and front views.

Step 2 Project from the edge view of the inclined surface parallel to the line of sight. Transfer the H dimensions from the front to locate a line in the auxiliary view.

Step 3 Locate the other corners of the inclined surface by projecting to the auxiliary view and locating the points by transferring the height (H) from the front view.

view must be transferred to the auxiliary view with your dividers.

8.4 Auxiliaries from the Top: Application

Figure 8.7 illustrates how the folding-line method is used to find an auxiliary view of a part that is imagined to be in a glass box. The semicircular end of the part does not appear true size in front or side views, making these views difficult to draw and interpret if they were drawn. However, because the inclined surface appears as an edge in the top view, it can be found true size in a primary auxiliary view projected from the top view. The height dimension (H) in the frontal view will be the same as in the auxiliary plane because both planes are perpendicular to the horizontal projection plane. Height is transferred from the front to the auxiliary view with your dividers. In **Figure 8.8** the auxiliary plane is rotated about the H-1 fold line into the plane of the top view, the horizontal projection plane. This rotation illustrates how the placement of the views is determined when drawing the views.

When drawn on a sheet of paper, the views of this object appear as shown in **Figure 8.9.** The top view is a complete view, but the front view is drawn as a partial view because the omitted portion would have been hard to draw and would not have been true size. The auxiliary view also is drawn as a partial view because the front view shows the omitted features better, which saves drawing time and space on a drawing.

Reference-Plane Method

A second method of locating an auxiliary view uses reference planes instead of the folding-line method. **Figure 8.10A** shows a horizontal reference plane (HRP) drawn through the center of the front view. Because this view is symmetrical, equal height dimensions on both sides of the HRP can be conveniently transferred from the front view to the auxiliary view and laid off on both sides of the HRP.

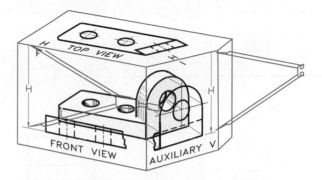

8.7 By imagining that the object is inside a glass box, you can see the relationship of the auxiliary plane on which the true-size view is projected and the horizontal projection plane.

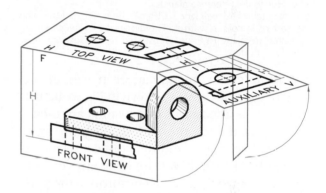

8.8 Fold the auxiliary plane into the horizontal projection plane by revolving it about the H-1 fold line.

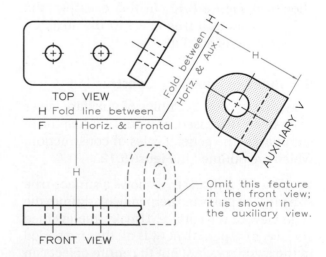

8.9 The front and auxiliary views are shown as partial views. The omitted portions of these views are unnecessary because their details are explained by the adjacent views.

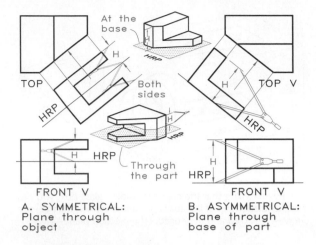

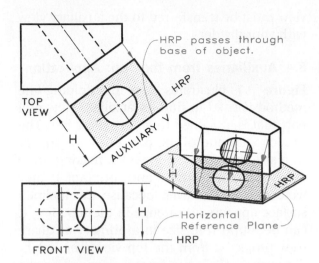

8.10 Horizontal reference plane:
A horizontal reference plane (HRP) can be positioned through the part or in contact with it. The dimension of height (H) is measured from the HRP and transferred to the auxiliary view with your dividers.

The reference plane can be placed at the base of the front view as shown in **Figure 8.10B.** In this case, the height dimensions are measured upward from the HRP in both the front and auxiliary views. You may draw a reference plane (the HRP in this example) in any convenient position in the front view: through the part, above it, or below it.

A similar example of an auxiliary view drawn with a horizontal reference plane is shown in **Figure 8.11.** In this example, the hole appears as a true circle in the auxiliary view instead of an ellipse.

8.5 Rules of Auxiliary Construction

Now that several examples of auxiliary views have been discussed, it would be helpful to summarize the general rules of construction, which are outlined in **Figure 8.12.**

1. An auxiliary view that shows a surface true size must be projected perpendicularly from the edge view of the surface. Usually, the inclined surface, or a partial view, is all that is needed in the auxiliary view, but the entire object can be drawn in the auxiliary view, if desired, as shown here.

8.11 Horizontal reference plane: application.
An auxiliary view projected from the top view is used to draw a true-size view of the inclined surface using a horizontal reference plane. The HRP is drawn through the bottom of the front view.

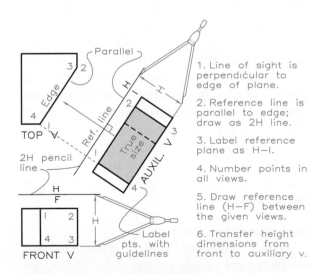

1. Line of sight is perpendicular to edge of plane.

2. Reference line is parallel to edge; draw as 2H line.

3. Label reference plane as H—I.

4. Number points in all views.

5. Draw reference line (H—F) between the given views.

6. Transfer height dimensions from front to auxiliary v.

8.12 Rules of auxiliary view construction:
Step 1 Draw a line of sight perpendicular to the edge of the inclined surface. Draw the H-1 fold line parallel to the edge of the inclined surface and draw an H-F fold line between the given views.
Step 2 Find points 3 and 4 by transferring the height (H) dimensions with your dividers from the front view to the auxiliary view.
Step 3 Find points 1 and 2 in the same manner by transferring the H dimensions.

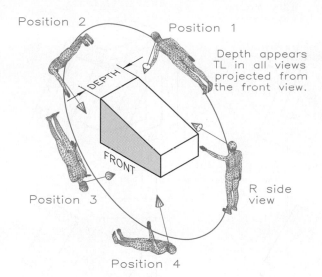

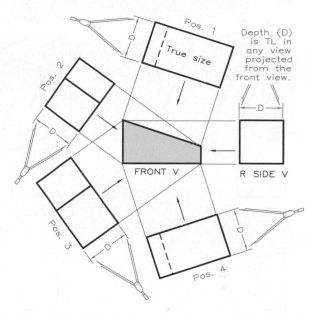

8.13 Viewpoints around the front view.
By moving your viewpoint around the front view of an object, you will see a series of auxiliary views in which the depth dimension (D) is true length in all views.

2. Draw the sight line perpendicular to the inclined edge of the plane you wish to find true size (TS).

3. Draw the reference line, H-1 for example, parallel to the edge view of the inclined plane, which will be perpendicular to the line of sight.

4. Draw and label a reference line between the given views (front and top in this example). Draw reference lines (fold lines) as thin dark lines with a 2H or 3H pencil.

5. If an auxiliary is projected from the front view it will have an F-1 reference line; if it is projected from the horizontal view (top view) it will have an H-1 reference line; and if projected from the side view (profile view) it will have a P-1 reference plane.

6. Transfer measurements from the other given view with your dividers (not the view you are projecting from), height in the front view in this example.

7. It would be helpful to number the points one at a time in the primary views and the auxiliary views as they are plotted.

8.14 Auxiliaries from the front view.
The auxiliary views shown in Figure 8.13 would be seen in this arrangement when viewing the front view.

8. Do your lettering in a professional manner with guidelines.

9. Connect the points with light construction lines and use light gray projectors that do not have to be erased with a pencil in the 2H–4H range.

10. Draw the outlines of the auxiliary view as thick visible lines, the same weight as visible lines in principal views, with an F or HB pencil.

8.6 Auxiliaries from the Front View

By moving about the front view of the part as shown in **Figure 8.13,** you will be looking parallel to the edge view of the frontal plane. Therefore the depth dimension (D) will appear true size in each auxiliary view projected from the front view. One of the positions gives a principal view, the right-side view, and position 1 gives a true-size view of the inclined plane. **Figure 8.14** illustrates the relationship between the auxiliary views projected from the front view.

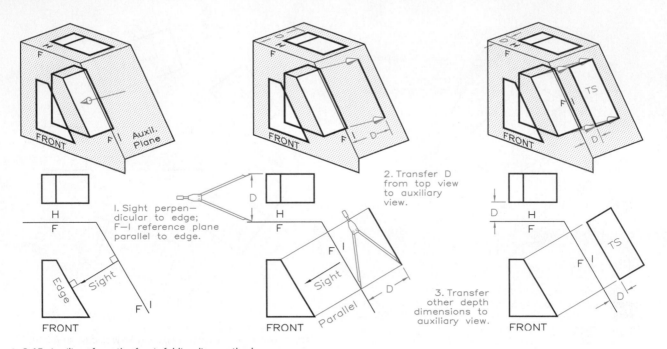

8.15 Auxiliary from the front: folding-line method.

Step 1 Draw the line of sight perpendicular to the edge of the plane and draw the F-1 line parallel to it. Draw the H-F fold line between the top and front views.

Step 2 Project perpendicularly from the edge view of the inclined surface and parallel to the line of sight. Transfer the depth dimensions (D) from the top to the auxiliary view with dividers.

Step 3 Locate the other corners of the inclined surface in the auxiliary view. Locate the points by transferring the depth dimensions (D) from the top view.

Folding-Line Method

A plane of an object that appears as an edge in the front view (**Figure 8.15**) is true size in an auxiliary view projected perpendicularly from it. Draw fold line F-1 parallel to the edge view of the inclined plane in the front view at a convenient location.

Draw the line of sight perpendicular to the edge view of the inclined plane in the front view. When observed from this direction, the frontal plane appears as an edge; therefore measurements perpendicular to the frontal plane depth dimensions (D) will be seen true length. Transfer depth dimensions from the top view to the auxiliary view with your dividers.

The object in **Figure 8.16** is imagined to be enclosed in a glass box and an auxiliary plane is folded from the frontal plane to be parallel to the inclined surface. When drawn on a sheet of paper, the views appear as shown in **Figure 8.17**. The top and side views are drawn as partial views because the auxiliary view eliminates the need for drawing complete views. The auxiliary view, located by transferring the depth dimension measured perpendicularly from the edge view of the frontal plane in the top view and transferred to the auxiliary view, shows the surface's true size.

Reference-Plane Method

The object shown in **Figure 8.18** has an inclined surface that appears as an edge in the front view; therefore this plane can be found true size in a primary auxiliary view. It is helpful to draw a reference plane through the center of the symmetrical top view because

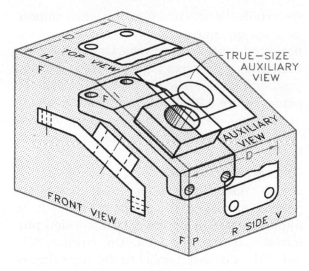

8.16 Glass-box theory.
This part is shown in an imaginary glass box to illustrate the relationship of the auxiliary plane, on which the true-size view of the inclined surface is projected, with the principal planes.

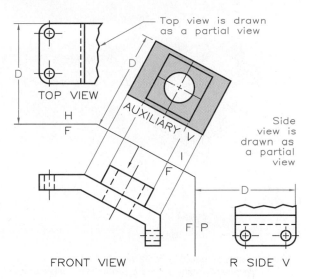

8.17 Auxiliary from the front view: application.
The layout and construction of an auxiliary view of the object shown in Figure 8.16 is shown here.

all depth dimensions can be located on each side of the frontal reference plane. Because the reference plane is a frontal plane, it is labeled FRP in the top and auxiliary views. In the auxiliary view, the FRP is drawn parallel to the edge view of the inclined plane at a convenient distance from it. By transferring depth dimensions from the FRP in the top view to

the FRP in the auxiliary view, the symmetrical view of the part is drawn.

8.7 Auxiliaries from the Profile View

By moving your position about the profile view (side view) of the part as shown in **Figure 8.19,** you will be looking parallel to the edge view of

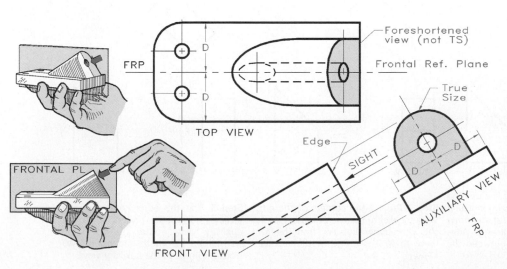

8.18 Because the inclined surface is symmetrical, a frontal reference plane (FRP) that passes through the object is used. Project the auxiliary view perpendicularly from the edge view of the plane. The FRP appears as an edge in the auxiliary view, and depth dimensions (D) are transferred from both sides of it in the top view to find its true-size view of the inclined surface.

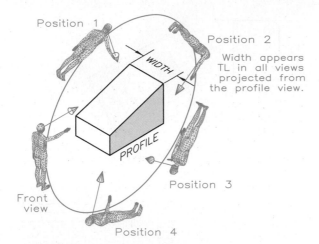

Position 1

Position 2

Width appears
TL in all views
projected from
the profile view.

WIDTH

PROFILE

Front
view

Position 3

Position 4

8.19 Viewpoints around the profile view.
By moving your viewpoint around a profile (side) view of an object, you will obtain a series of auxiliary views in which the width dimension (W) is true length.

the profile plane. Therefore, the width dimension will appear true size in each auxiliary view projected from the side view. One of the positions gives a principal view, the front view, and position 1 gives a true-size view of the inclined plane. **Figure 8.20** illustrates the arrangement of the auxiliary views projected from the side view if they were drawn on a sheet of paper.

Folding-Line Method
Because the inclined surface in **Figure 8.21** appears as an edge in the profile plane, it can be found true size in a primary auxiliary view projected from the side view. The auxiliary fold line, P-1, is drawn parallel to the edge view of the inclined surface. A line of sight perpendicular to the auxiliary plane shows the profile plane

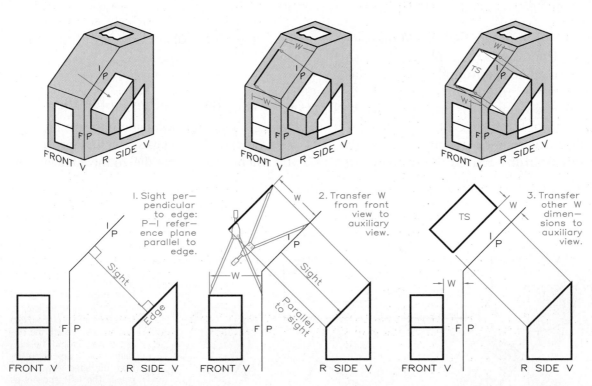

8.21 Profile auxiliary: Folding-line method
Step 1 Draw a line of sight perpendicular to the edge of the inclined surface. Draw the P-1 fold line parallel to the edge view, and draw the F-P fold line between the given views.

Step 2 Project the corners of the edge view parallel to the line of sight. Transfer the width dimensions (W) from the front view to locate a line in the auxiliary view.

Step 3 Find the other corners of the inclined surface by projecting to the auxiliary view. Locate the points by transferring the width dimensions (W) from the front to the auxiliary view.

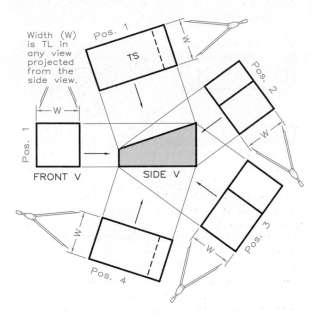

Width (W) is TL in any view projected from the side view.

8.20 Profile auxiliaries.
The auxiliary views shown in Figure 8.19 would be seen in this arrangement when projected from the side view.

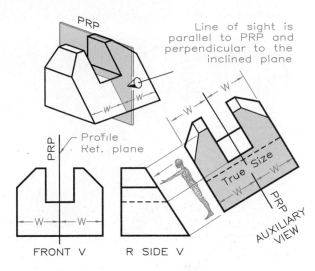

8.22 Profile reference plane: PRP.
An auxiliary view is projected from the right-side view by using a profile reference plane (PRP) to show the true-size view of the inclined surface.

as an edge. Therefore width dimensions (W) transferred from the front view to the auxiliary view appear true length in the auxiliary view.

Reference-Plane Method

The object shown in **Figure 8.22** has an inclined surface that appears as an edge in the right-side view, the profile view. This inclined surface may be drawn true size in an auxiliary view by using a profile reference plane (PRP) that is a vertical edge in the front view. Draw the PRP through the center of the front view because this view is symmetrical. Then find the true-size view of the inclined plane by transferring equal width dimensions (W) with your dividers from the edge view of the PRP in the front view to both sides of the PRP in the auxiliary view.

8.8 Curved Shapes

The cylinder shown in **Figure 8.23** has an inclined surface that appears as an edge in the front view. The true-size view of this plane can

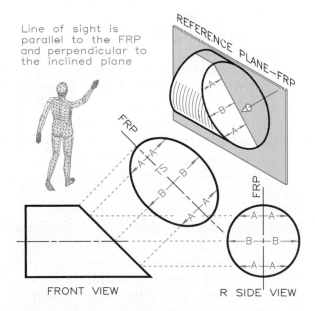

8.23 Curved surface: frontal reference plane.
The auxiliary view of this elliptical surface was found by projecting a series of points on its perimeter. The frontal reference plane (FRP) is drawn through its center in the side view since the object is symmetrical.

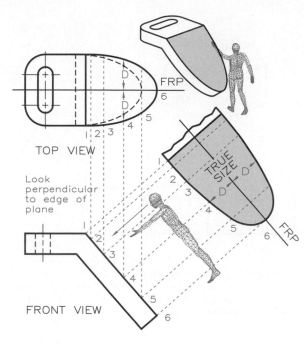

8.24 Curved surfaces:
The auxiliary view of this curved surface required that a series of points located in the top view be projected to the front view, and then projected to the auxiliary view. The FRP was passed through the center of the top view.

be seen in an auxiliary view projected from the front view.

Because the cylinder is symmetrical, a frontal reference plane (FRP) is drawn through the center of the side view so that equal dimensions can be laid off on both sides of it. Points located about the circular side view are projected to its edge view in the front view.

In the auxiliary view, the FRP is drawn parallel to the edge view of the plane in the front view, and the points are projected perpendicularly from the edge view of the plane. Dimensions A and B are shown as examples of depth dimensions used for locating points in the auxiliary view. To construct a smooth elliptical curve, more points than shown in this example are needed.

A true-size auxiliary view of a surface bounded by an irregular curve is shown in **Figure 8.24.** Project points from the curve in

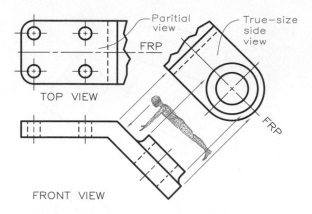

8.25 Partial views.
Partial views with foreshortened portions omitted can be used to represent objects. The FRP reference line is drawn through the center of the object in the top view because the object is symmetrical to make point location easier.

the top view to the front view. Locate these points in the auxiliary view by transferring depth dimensions (D) from the FRP in the top view to the auxiliary view.

8.9 Partial Views

Auxiliary views are used as supplementary views to clarify features that are difficult to depict with principal views alone. Consequently, portions of principal views and auxiliary views may be omitted, provided that the partial views adequately describe the part. The object shown in **Figure 8.25** is composed of a complete front view, a partial auxiliary view, and a partial top view. These partial views are easier to draw and are more descriptive without sacrificing clarity.

8.10 Auxiliary Sections

In **Figure 8.26,** a cutting plane labeled A-A is passed through the part to obtain the auxiliary section labeled Section A-A. The auxiliary section provides a good and efficient way to describe features of the part that could not be as easily described by additional principal views.

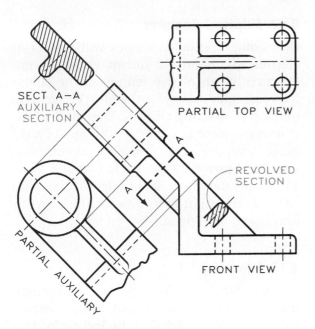

SECT A-A
AUXILIARY SECTION

PARTIAL TOP VIEW

A

REVOLVED SECTION

PARTIAL AUXILIARY

FRONT VIEW

8.26 Auxiliary section.
A cutting plane labeled A-A is passed through the object and the auxiliary section, Section A-A, is drawn as a supplementary view to describe the part. The top and front views are drawn as partial views.

8.11 Secondary Auxiliary Views

Figure 8.27 shows how to project an auxiliary, called a **secondary auxiliary view,** from a primary auxiliary view. An edge view of the oblique plane is found in the primary auxiliary view by finding the point view of a true-length line (2-3) that lies on the oblique surface. A line of sight perpendicular to the edge view of the plane gives a secondary auxiliary view that shows the oblique plane as true size.

Note that the reference line between the primary auxiliary view and the secondary auxiliary view is labeled 1-2 to represent the fold line between the primary plane (1) and the secondary plane (2). The 1 label is placed on the primary side and the 2 label is placed on the secondary side.

Figure 8.28 illustrates the construction of a secondary auxiliary view that gives the true-size view of a surface on a part using

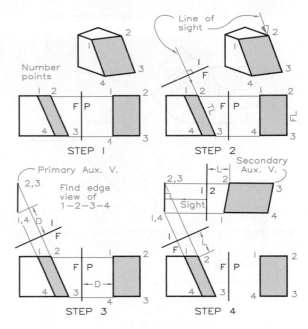

8.27 Secondary auxiliary views.
Step 1 Draw a fold line F-P between the front and side views. Label the corner points in both views.
Step 2 Line 2-3 is a true-length frontal line in the front view. Draw reference line F-1 perpendicular to line 2-3 with a line of sight parallel to line 2-3.
Step 3 Find the edge view of plane 1-2-3-4 by transferring depth dimensions (D) from the side view.
Step 4 Draw a line of sight perpendicular to the edge view of 1-2-3-4 and draw the 1-2 fold line parallel to the edge view. Find the true-size auxiliary view by transferring the dimensions (L) from the front view to the auxiliary view.

these same principles and a combination of partial views. A secondary auxiliary view must be used in this case because the oblique plane does not appear as an edge in a principal view.

Find the point view of a line (AB) on the oblique plane to find the edge view of the plane in the primary auxiliary view. The secondary auxiliary view is projected perpendicularly from the edge view of the plane found in the primary auxiliary view to find a true-size view of the plane. In this example all of the views are drawn as partial views, which are adequate to describe the part.

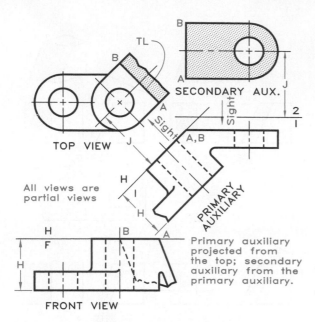

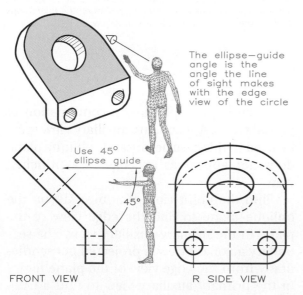

8.12 Elliptical Features

Occasionally, circular shapes will project as ellipses, which must be drawn with an irregular curve or an ellipse template. The ellipse template (guide) is by far the most convenient method of drawing ellipses. The angle of the ellipse template is the angle the line of sight makes with the edge view of the circular feature. In **Figure 8.29** the angle is found to be 45° where the curve is an edge in the front view, so the right-side view of the curve is drawn as a 45° ellipse.

8.13 Summary

Auxiliary views can be projected from principal orthographic views and from auxiliary views in order to achieve the viewpoint that you desire. Auxiliary views are orthographic views but they are not *principal* orthographic views.

The mastery of primary auxiliary views is the first step toward the development of your ability to solve problems in three-dimensional space by descriptive geometry. The engineer and technologist must be skilled with spatial relationships in order to represent them two-dimensionally on working drawings from which designs become realities. Essentially all designs have spatial problems that must be solved.

Problems

1–13. (Figure 8.30) Using the example layout, change the top and front views by substituting the top views given at the right in place of the one given in the example. The angle of inclination in the front view is 45° for all problems, and the height is 38 mm (1.5 inches) in the front view. Construct auxiliary views that show the inclined surface true size. Draw two problems per size A sheet.

8.28 Secondary auxiliaries: partial views.
A secondary auxiliary view projected from a primary auxiliary view that was projected from the top view is shown here. All views are drawn as partial views. Find the edge view of the inclined plane in the primary auxiliary view by finding the point view of line AB which is true length in the top view.

8.29 Ellipse guide angles.
The ellipse guide angle is the angle that the line of sight makes with the edge view of the circular feature. The ellipse angle for the right-side view is 45°.

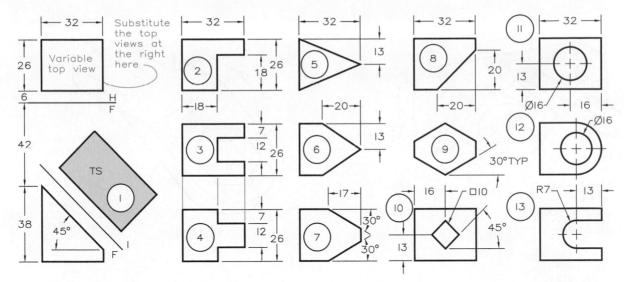

8.30 Problems 1–13. Primary auxiliary views.

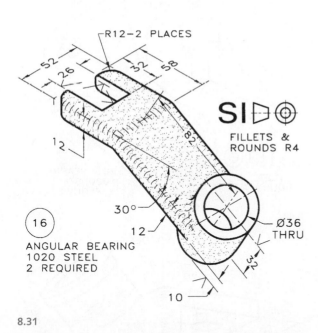

R12–2 PLACES

52
26
32
58
12
82
30°
12
Ø36 THRU
32
10

⑯

ANGULAR BEARING
1020 STEEL
2 REQUIRED

SI⊅⊕
FILLETS &
ROUNDS R4

8.31

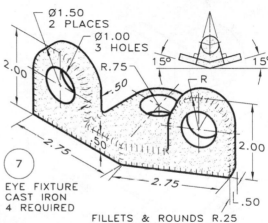

Ø1.50
2 PLACES
Ø1.00
3 HOLES
R.75
2.00
.50
15°   15°
R
2.75
.50
2.00
2.75
.50

⑦

EYE FIXTURE
CAST IRON
4 REQUIRED

FILLETS & ROUNDS R.25

8.32

Figures 8.31–8.49 Draw the necessary primary and auxiliary views to describe the objects assigned. Draw one per size A or size B sheet. Adjust the scale of each to utilize the space on the sheet.

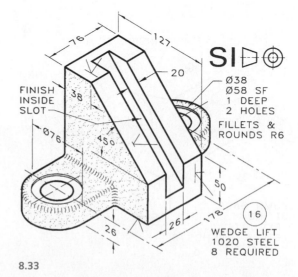

76
127
20
38
45°
Ø76
26
26
178
50

FINISH
INSIDE
SLOT

SI⊅⊕
Ø38
Ø58 SF
1 DEEP
2 HOLES
FILLETS &
ROUNDS R6

⑯

WEDGE LIFT
1020 STEEL
8 REQUIRED

8.33

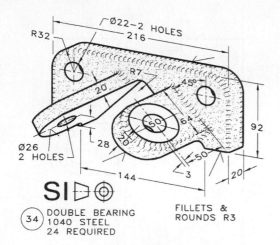

Ø22–2 HOLES
216
R32
R7
45°
20
50
64
20
3
50
92
20
144
28
Ø26
2 HOLES

SI ⊅⊕

(34) DOUBLE BEARING
1040 STEEL
24 REQUIRED

FILLETS &
ROUNDS R3

8.34

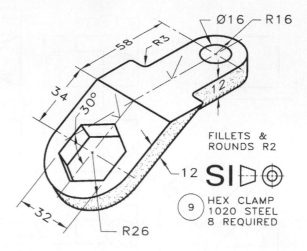

58 R3
34
30°
Ø16 R16
12
FILLETS &
ROUNDS R2
12 SI ⊅⊕
32
R26

(9) HEX CLAMP
1020 STEEL
8 REQUIRED

8.37

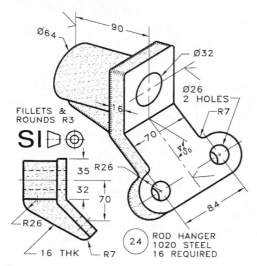

Ø64
90
Ø32
Ø26
2 HOLES
16
R7
70
50°

FILLETS &
ROUNDS R3

SI ⊅⊕

35 R26
32
70
R26
16 THK R7

(24) ROD HANGER
1020 STEEL
16 REQUIRED

8.35

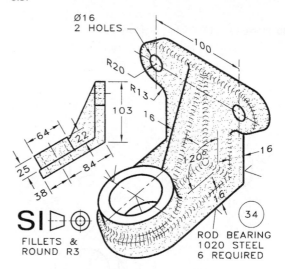

Ø16
2 HOLES
100
R20
R13
103
16
64
22
25
84
38
120°
16
16

SI ⊅⊕
FILLETS &
ROUND R3

(34)
ROD BEARING
1020 STEEL
6 REQUIRED

8.38

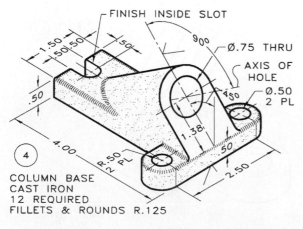

FINISH INSIDE SLOT
1.50
50 50
50
900
Ø.75 THRU
AXIS OF
HOLE
Ø.50
2 PL
50
1.38
50
4.00
R.50 2 PL
R.2
2.50

(4)
COLUMN BASE
CAST IRON
12 REQUIRED
FILLETS & ROUNDS R.125

8.36

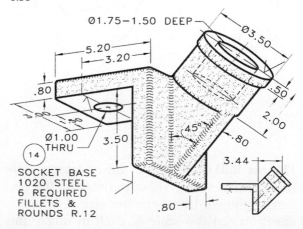

Ø1.75–1.50 DEEP
Ø3.50
5.20
3.20
.80
.50
2.00
45°
Ø1.00
THRU
.80
3.50
3.44
.80

(14)
SOCKET BASE
1020 STEEL
6 REQUIRED
FILLETS &
ROUNDS R.12

8.39

126 • **CHAPTER 8   PRIMARY AUXILIARY VIEWS**

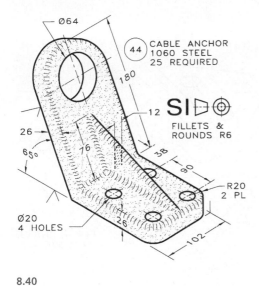

Ø64

44 CABLE ANCHOR
1060 STEEL
25 REQUIRED

180

12 SI ◁ ⊕

FILLETS &
ROUNDS R6

26

65°

76

38

90

R20
2 PL

Ø20
4 HOLES

26

102

8.40

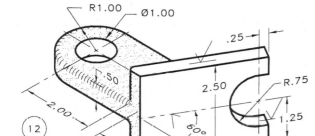

R1.00   Ø1.00

.25

1.50

2.50

R.75

2.00

60°

1.25

.50

12 SOCKET SLIDE
CAST IRON
4 REQUIRED
FILLETS &
ROUNDS R.125

3.50

8.41

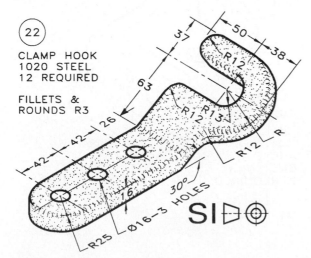

22 CLAMP HOOK
1020 STEEL
12 REQUIRED

FILLETS &
ROUNDS R3

37

50

38

R12

63

R12

R13

26

R12

42

R

42

16

30°

R25

Ø16—3 HOLES

SI ◁ ⊕

8.42

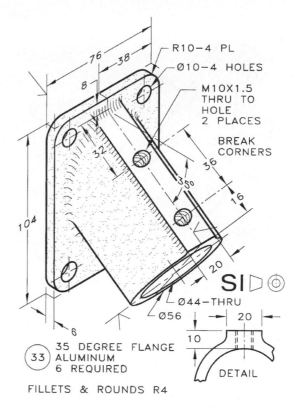

76   38

R10—4 PL

Ø10—4 HOLES

8

M10X1.5
THRU TO
HOLE
2 PLACES

BREAK
CORNERS

32

36

35°

16

104

20

SI ◁ ⊕

Ø44—THRU

Ø56

6

20

10

DETAIL

33 35 DEGREE FLANGE
ALUMINUM
6 REQUIRED

FILLETS & ROUNDS R4

8.43

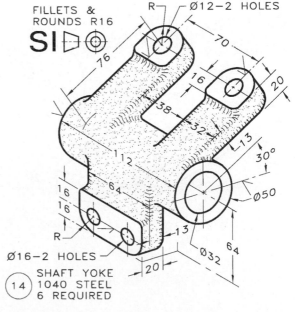

FILLETS &
ROUNDS R16

SI ◁ ⊕

R   Ø12—2 HOLES

76

70

16

20

38

32

L13

112

30°

Ø50

64

16

16

R

13

Ø32

64

Ø16—2 HOLES

20

14 SHAFT YOKE
1040 STEEL
6 REQUIRED

8.44

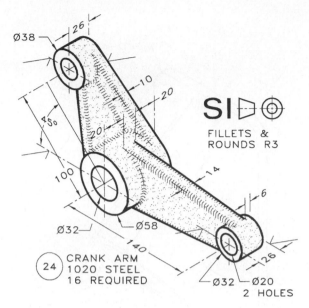

Ø38
26
10
20
50°
20
100
14
6
Ø32
Ø58
Ø32
Ø20
2 HOLES
140
26

SI ▷⊕
FILLETS &
ROUNDS R3

(24) CRANK ARM
1020 STEEL
16 REQUIRED

8.45

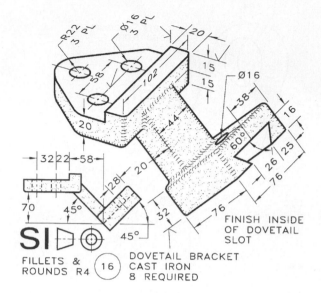

R22
3 PL
Ø16
3 PL
20
15
15
Ø16
38
16
58
102
60°
26 25
44
20
32 22 58
128
76
70
45°
20
32
26
76
76

SI ▷⊕
FILLETS &
ROUNDS R4
45°
45°

FINISH INSIDE
OF DOVETAIL
SLOT

(16) DOVETAIL BRACKET
CAST IRON
8 REQUIRED

8.47

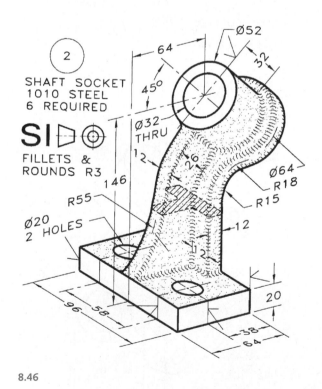

(2)
SHAFT SOCKET
1010 STEEL
6 REQUIRED

64
Ø52
32
45°
Ø32
THRU
12
26
Ø64
R18
R15
146
R55
Ø20
2 HOLES
12
12
20
96
58
38
64

SI ▷⊕
FILLETS &
ROUNDS R3

8.46

SI ▷⊕
FILLETS &
ROUNDS R10

245
26
25°
Ø36
2 HOLES
60
26
100
26
48
60 80
180
Ø32−2 HOLES
2 HOLES

(22)
ANGLE BRACKET
CAST IRON
2 REQUIRED

8.48

128 • **CHAPTER 8 PRIMARY AUXILIARY VIEWS**

8.49 Lay out the necessary orthographic views of the oblique bracket **(Figure 8.49)** on a B size sheet. Construct the true-size auxiliary view that shows the inclined surface true size. Select the appropriate scale.

Design 1: Design a fixture to hold the oblique bracket **(Figure 8.49)** in position on a drill press in order to drill the 1.25 DIA hole in the inclined surface. Show the fixture details in orthographic views.

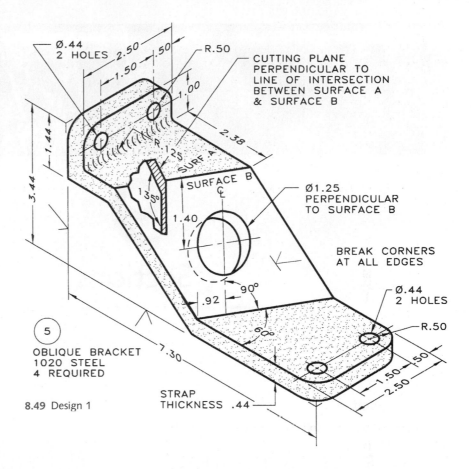

Ø.44
2 HOLES

CUTTING PLANE PERPENDICULAR TO LINE OF INTERSECTION BETWEEN SURFACE A & SURFACE B

Ø1.25 PERPENDICULAR TO SURFACE B

BREAK CORNERS AT ALL EDGES

Ø.44
2 HOLES

R.50

5

OBLIQUE BRACKET
1020 STEEL
4 REQUIRED

STRAP THICKNESS .44

8.49 Design 1

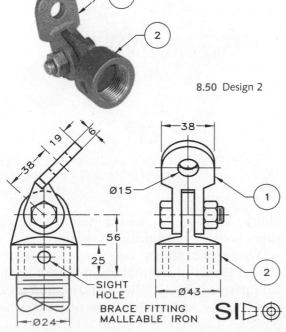

8.50 Design 2

BRACE FITTING
MALLEABLE IRON

SIGHT HOLE

Design 2: The two-piece brace fitting **(Figure 8.50)** is used to brace piping against sway and seismic movement. One end is attached to the building structure and the other end connects to the pipe attachment. Assume the role of the designer and provide the dimensions and details needed to fully describe the parts. *Question:* What is the purpose of the sight hole in part 2?

A. Make freehand, orthographic views of each part on size A sheets as a preliminary step toward making an instrument drawing.

B. Convert your freehand sketches into instrument drawn orthographic views on size A sheets. Part 1 will require a primary auxiliary view to properly depict it.
(Courtesy of Anvil International, Inc.)

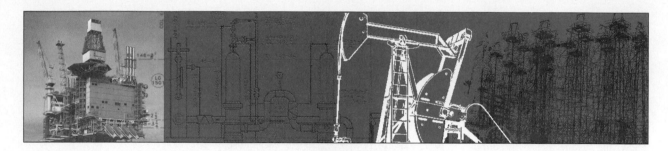

9

Sections

9.1 Introduction

Correctly drawn orthographic views that show all hidden lines may not clearly describe an object's internal details. The gear housing shown in **Figure 9.1** is such an example and is better understood when a section has been cut from it. The technique of constructing imaginary cross-sectional cuts through a drawing of a part results in an orthographic view called a section.

9.2 Basics of Sectioning

In **Figure 9.2A** standard views of a cylinder are shown as top and front views where its interior features are drawn as hidden lines. If you imagined a knife edge cutting through the top view, the front view would become a **section.** This section is a **full section** since the cutting plane passes fully through the part **(Figure 9.2B).** The portion of the part that was cut by the imaginary plane is cross-hatched, and

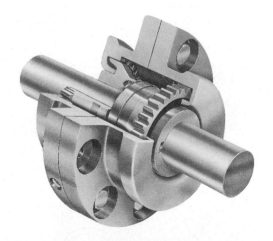

9.1 This gear housing has many internal features that cannot be described clearly in a standard orthographic view. Sections are used to clarify interior parts.

hidden lines usually are omitted in sectional views because they are not needed.

Figure 9.3 shows two types of cutting planes. Either is acceptable although the one with pairs of short dashes is most often used. The spacing and proportions of the dashes

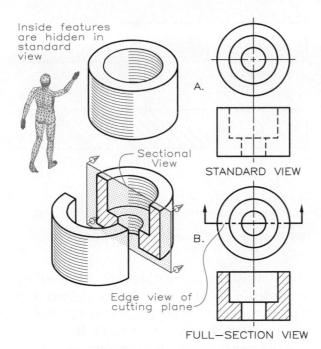

Inside features
are hidden in
standard
view

Sectional
View

STANDARD VIEW

A.

B.

Edge view of
cutting plane

FULL—SECTION VIEW

9.2 A part with internal features can be better shown with sectional views than with standard views with hidden lines.

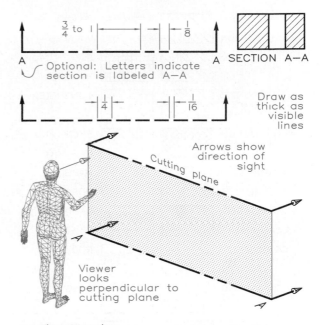

$\frac{3}{4}$ to 1 $\frac{1}{8}$

Optional: Letters indicate
section is labeled A—A

SECTION A—A

$\frac{1}{4}$ $\frac{1}{16}$

Draw as
thick as
visible
lines

Arrows show
direction of
sight

Cutting plane

Viewer
looks
perpendicular to
cutting plane

9.3 The cutting plane.
Cutting planes can be thought of as knife edges that pass through views to reveal interior features in sections. The cutting plane marked A-A results in a section labeled SECTION A-A.

depend on the size of the drawing. The line thickness of the cutting plane is the same as the visible object line. Letters placed at each end of the cutting plane are used to label the sectional view, such as **SECTION A-A.**

The sight arrows at the ends of the cutting plane are always perpendicular to the cutting plane. In the sectional view, the observer is looking in the direction of the sight arrows, perpendicular to the surface of the cutting plane.

Figure 9.4 shows the three basic positions of sections and their respective cutting planes. In each case perpendicular arrows point in the direction of the line of sight. For example, the cutting plane in **Figure 9.4A** passes through and removes the front of the top view, and the line of sight is perpendicular to the remainder of the top view.

The top view appears as a section when the cutting plane passes through the front view and the line of sight is downward **(Figure**

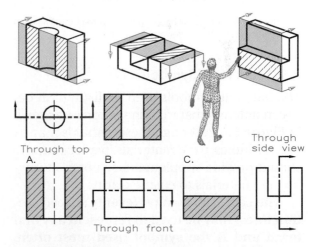

Through top
A.

B.

C.

Through front

Through
side view

9.4 Principal sections.
The three examples of cutting planes that pass through the principal views, (A) top, (B) front, and (C) side views, are shown here. The arrows point in the direction of your line of sight for each section.

9.4B). When the cutting plane passes vertically through the side view **(Figure 9.4C),** the front view becomes a section.

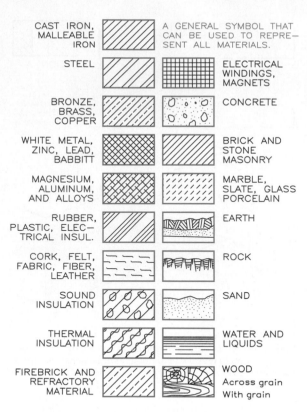

9.5 Use these symbols for hatching parts in sections. The cast-iron symbol may be used for any material.

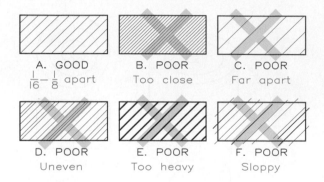

9.6 Hatching techniques.
A Section lines are thin, dense lines drawn from 1/16 to 1/8 inch apart.
B–F Avoid these common errors of section lining.

9.3 Sectioning Symbols

The hatching symbols used to distinguish between different materials in sections are shown in **Figure 9.5.** Although these symbols may be used to indicate the materials in a section, you should provide supplementary notes specifying the materials to ensure clarity.

The cast-iron symbol (evenly spaced section lines) may be used to represent any material and is the symbol used most often. Draw cast-iron symbols with a 2H pencil, slant the lines upward and to the right at 30°, 45°, or 60° angles, and space the lines about 1/16 inch apart (close together in small areas and farther apart in larger areas).

Properly drawn section lines—thin and evenly spaced—are shown in **Figure 9.6.** Common errors of section lining are shown in **Figures 9.6B–F.** Thin parts such as sheet

metal, washers, and gaskets are sectioned by completely blacking in their areas (**Figure 9.7**) because space does not permit the drawing of section lines. On the other hand, large parts are sectioned with an **outline section** to save time and effort.

Sectioned areas should be hatched with symbols that are neither parallel nor perpendicular to the outlines of the parts lest they be confused with serrations or other machining treatments of the surface (**Figure 9.8**).

9.4 Sectioning Assemblies of Parts

When sectioning an assembly of several parts, draw section lines at varying angles to distinguish the parts from each other (**Figure 9.9A**). Using different material symbols in an assembly also helps distinguish between the parts and their materials. Crosshatch the same part at the same angle and with the same symbol even though portions of the part may be separated (**Figure 9.9B**).

9.5 Full Sections

A cutting plane passed fully through an object and then removing half of it forms a **full section** view. **Figure 9.10** shows two orthographic views of an object with all its hidden lines. We can describe the part better by passing a

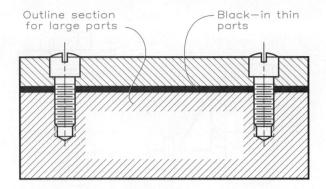

9.7 Large and small sectional areas.
Large sectional areas are hatched with outline sectioning (around their edges) and thin parts are blacked in solid.

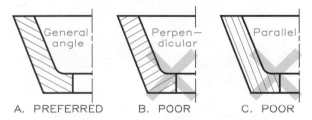

A. PREFERRED   B. POOR   C. POOR

9.8 Directions of section lines.
Draw section lines that are neither parallel nor perpendicular to the outlines of the part, so that they are not misunderstood as features of the part.

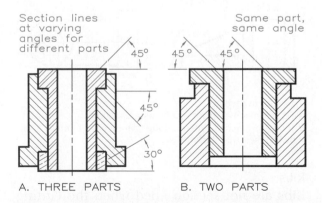

A. THREE PARTS   B. TWO PARTS

9.9 Hatching parts in assembly.
A Draw section lines of different parts in an assembly at varying angles to distinguish between them.
B Draw section lines on separated portions of the same part (both sides of a hole here) in the same direction.

cutting plane through the top view to remove half of it. The arrows on the cutting plane indicate the direction of sight. The front view

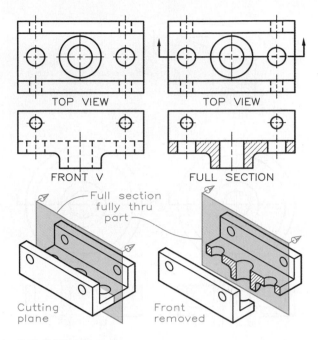

9.10 Full section.
A full section is found by passing a cutting plane fully through this part, removing half of it. The arrows on the cutting plane gives your direction of sight. The sectional view shows the internal features clearly.

becomes a full section, showing the surfaces cut by the cutting plane. **Figure 9.11** shows a full section through a cylindrical part, with half the object removed. **Figure 9.11A** shows the correctly drawn sectional view. A common mistake in constructing sections is the omission of the visible lines behind the cutting plane (**Figure 9.11B).**

Omit hidden lines in sectional views unless you consider them necessary for a clear understanding of the view. Also, cutting planes may be omitted if you consider them unnecessary. **Figure 9.12** shows a full section of a part from which the cutting plane was omitted because its path is obvious.

Parts Not Requiring Section Lining
Many standard parts, such as nuts and bolts, rivets, shafts, and set screws, do not require section lining even though the cutting plane passes through them (**Figure 9.13).** These

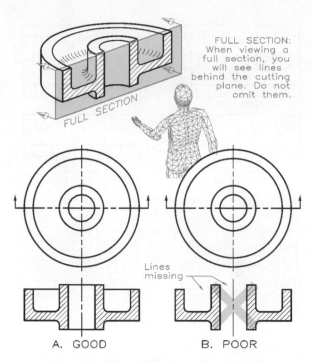

FULL SECTION: When viewing a full section, you will see lines behind the cutting plane. Do not omit them.

A. GOOD   B. POOR

Lines missing

9.11 Full section: Cylindrical part.
A When a front view of a cylinder is shown as a full section as shown here, visible lines will be seen behind the cut surface.
B If only the lines of the cut surface (the section) are shown, the view will be incompete.

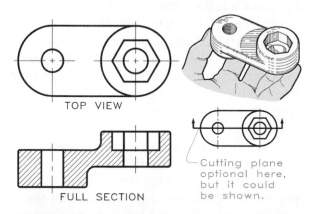

TOP VIEW

FULL SECTION

Cutting plane optional here, but it could be shown.

9.12 Cutting plane omission.
The cutting plane of a sectiona can be omitted if its location is obvious.

parts have no internal features, therefore sections through them would be unnecessary. Other parts not requiring section lining are roller bearings, ball bearings, gear teeth, dowels, pins, and washers.

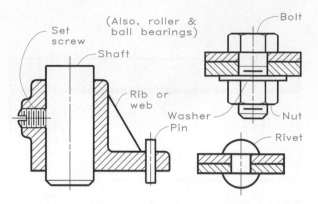

Set screw
Shaft
(Also, roller & ball bearings)
Bolt
Rib or web
Washer
Pin
Nut
Rivet

9.13 Parts not sectioned.
By conventional practice these parts are not section lined even though cutting planes pass through them.

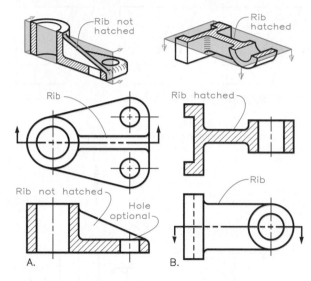

Rib not hatched
Rib hatched
Rib
Rib hatched
Rib not hatched
Hole optional
Rib
A.   B.

9.14 Ribs in section.
A Do not hatch a rib cut in a flatwise direction.
B Ribs are hatched when cutting planes pass through them showing their true thicknesses.

Ribs

Ribs are not section lined when the cutting plane passes flatwise through them (**Figure 9.14A**), because to do so would give a misleading impression of the rib. But ribs do require section lining when the cutting plane passes perpendicularly through them and shows their true thicknesses (**Figure 9.14B**).

By not section lining the ribs in **Figure 9.15A** we provided a descriptive section view

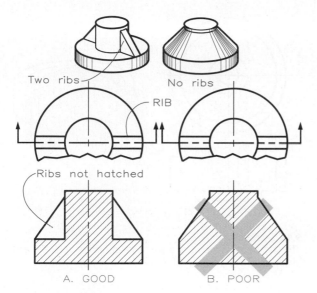

9.15 Ribs in section.
A Ribs are not hatched in section to better describe the part.
B If ribs were hatched in section, a misleading impression of the part would be given.

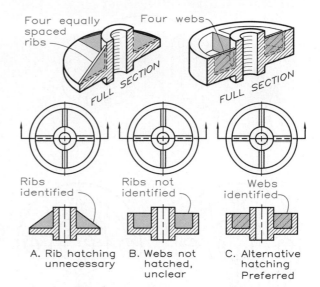

9.16 Ribs and webs in section.
A These ribs are well defined in this section and are not hatched.
B These webs are poorly defined when not hatched in the sectional view.
C Alternate hatch lines are drawn to call attention to poorly-defined webs.

of the part. Had we hatched the ribs, the section would give the impression that the part was solid and conical **(Figure 9.15B).**

Figure 9.16 shows an alternative method of section lining webs and ribs. The outside ribs in **Figure 9.16A** do not require section lining because the cutting plane passes flatwise through them and they are well identified. As a rule, webs do not require cross-hatching, but the webs shown in **Figure 9.16B** are not well identified in the front section and could go unnoticed. Therefore using alternate section lines as shown in **Figure 9.16C** is better. Here, extending every-other section line through the webs ensures that they can be identified easily.

9.6 Partial Views

A conventional method of representing symmetrical views is the half view that requires less space and less time to draw than full views **(Figure 9.17).** A half top view is sufficient when drawn adjacent to the section view or

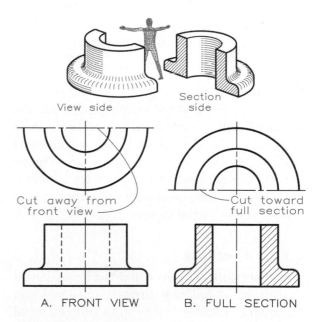

9.17 Half views and sections.
Half views of symmetrical parts are used to conserve space and drawing time as an approved conventional practice. (A) The external portion of the half view is toward the front view. (B) The internal portion of the half view is toward the front view when it is a section. In half sections, the omitted half view can be either toward or away from the section.

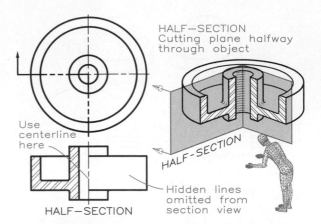

Use
centerline
here

HALF—SECTION

Hidden lines
omitted from
section view

HALF—SECTION

9.19 Half section.
This half section describes a part shown orthographically and pictorially. The parting line for cylinders is a centerline.

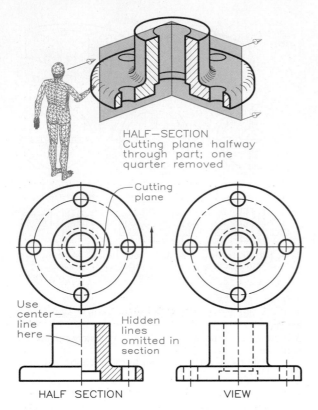

HALF—SECTION
Cutting plane halfway
through part; one
quarter removed

Cutting
plane

Use
center—
line
here

Hidden
lines
omitted in
section

HALF SECTION

VIEW

9.18 Half section.
In a half section the cutting plane passes halfway through the object removing a quarter of it to show half its outside and half its inside. Omit hidden lines in sectional views unless they are needed for clarity.

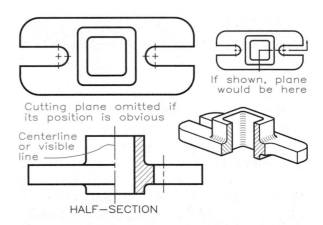

Cutting plane omitted if
its position is obvious

If shown, plane
would be here

Centerline
or visible
line

HALF—SECTION

9.20 Half section options.
Cutting planes can be omitted when their location is obvious. The parting line between the section and view may be a visible line or a centerline if the part is not cylindrical.

front view. For half views (not sections), the removed half is away from the adjacent view **(Figure 9.17A).** For full sections, the removed half is the half nearest the section **(Figure 9.17B).** When drawing partial views with half sections, you may omit either the front or rear halves of the partial views.

9.7 Half Sections

A **half section** is a view obtained by passing a cutting plane halfway through an object and removing a quarter of it to show both external and internal features. Half sections are used with symmetrical parts and with cylinders, in particular, as shown in **Figure 9.18.** By comparing the half section with the standard front view, you can see that both internal and exter-

nal features show more clearly in a half section than in a view. Hidden lines are unnecessary, and we've omitted them to simplify the section. **Figure 9.19** shows a half section of a pulley.

Note omission of the cutting plane from the half section shown in **Figure 9.20** because the cutting plane's location is obvious. Because the parting line of the half section is not at a centerline, you may use a solid line or a centerline to separate the sectional half from the half that appears as an external view.

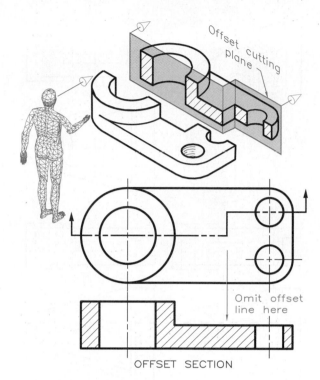

9.21 Offset section.
An offset section is formed by a cutting plane that must be off-set to pass through nonaligned features.

9.8 Offset Sections

An **offset section** is a full section in which the cutting plane is offset to pass through important features that do not lie in a single plane. **Figure 9.21** shows an offset section in which the plane is offset to pass through the large hole and one of the small holes. The cut formed by the offset is not shown in the section because it is imaginary.

9.9 Broken-Out Sections

A **broken-out section** shows a partial view of a part's interior features. The broken-out section of the part shown in **Figure 9.22** reveals details of the wall thickness to describe the part better. The irregular lines representing the break are conventional breaks (discussed later in this chapter).

The broken-out section of the pulley in **Figure 9.23** clearly depicts the keyway and

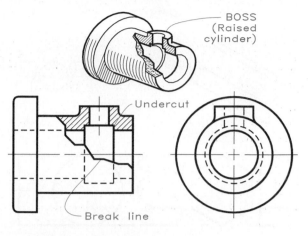

9.22 Broken-out section.
A broken-out section is one in which part of the object has been broken away to show internal features.

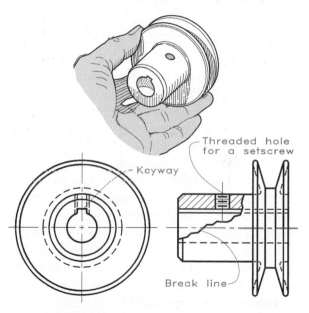

9.23 Broken-out section.
This broken-out section shows the keyway and the threaded hole for the setscrew in the pulley.

threaded hole for a setscrew. This method shows the part efficiently, with the minimum of views.

9.10 Revolved Sections

A **revolved section** describes a part when you revolve its cross section about an axis of revolution and place it on the view where the

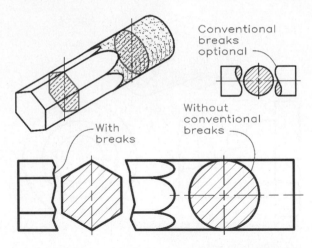

9.24 Revolved sections.
Revolved sections show cross-sectional features of a part to eliminate the need for additional orthographic views. Revolved sections may be superimposed on the given views or conventional breaks can be used to separate them from the given view.

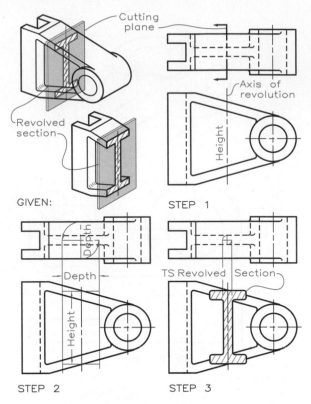

9.25 Revolved section construction.
Given: The part is shown pictorially with a cutting plane showing the cross section that will be revolved.
Step 1 An imaginary cutting plane is located in the top view and the axis of rotation is drawn in the front view.
Step 2 The depth in the top view is rotated and is projected to the front view. The height and depth in the front view give the overall dimensions of the revolved view.
Step 3 The revolved section is drawn, fillets and rounds are added, and the section lines applied to finish the view.

revolution occurred. Note the use of revolved sections to explain two cross sections of the shaft shown in **Figure 9.24** (with and without conventional breaks). Conventional breaks are optional; however, you may draw a revolved section on the view without them.

A revolved section helps to describe the part shown in **Figure 9.25.** Imagine passing a cutting plane through the top view of the part (step 1). Then imagine revolving the cutting plane in the top view and projecting it to the front (step 2). The true-size revolved section is competed in step 3. Conventional breaks could be used on each side of the revolved section.

Figure 9.26 demonstrates how to use typical revolved sections to show cross sections through parts, eliminating the need to draw additional orthographic views.

9.11 Removed Sections

A **removed section** is a revolved section that is shown outside the view in which it was revolved **(Figure 9.27).** Centerlines are used as axes of rotation to show the locations from which the sections are taken. Where space does not permit revolution on the given view **(Figure 9.28A),** removed sections must be used instead of revolved sections **(Figure 9.28B).**

Removed sections do not have to be positioned directly along an axis of revolution adjacent to the view from which they were revolved. Instead, removed sections can be located elsewhere on a drawing if they are properly labeled **(Figure 9.29).** For example, the plane labeled with an A at each end identifies the location of SECTION A-A, and the same applies to SECTION B-B.

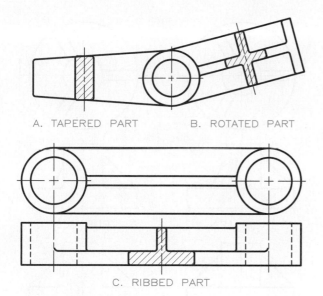

A. TAPERED PART   B. ROTATED PART

C. RIBBED PART

9.26 Revolved sections.
These revolved sections describe the cross sections of the two parts that would be difficult to depict in supplementary orthographic views, such as a side view.

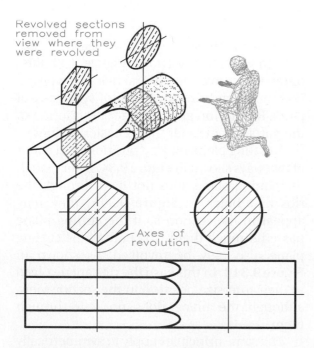

Revolved sections removed from view where they were revolved

Axes of revolution

9.27 Removed sections.
Removed sections are revolved sections that are drawn outside the object and along their axes of revolution.

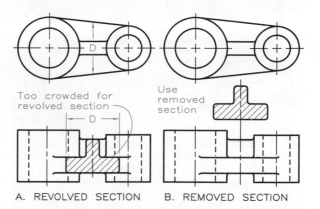

Too crowded for revolved section

Use removed section

A. REVOLVED SECTION   B. REMOVED SECTION

9.28 Removed sections.
Removed sections are necessary when space does not permit a revolved section to be superimposed on the part.

When a set of drawings consists of multiple sheets, removed sections and the views from which they are taken may appear on different sheets. When this method of layout is necessary, label the cutting plane in the view from which the section was taken and the sheet on which the section appears **(Figure 9.30).**

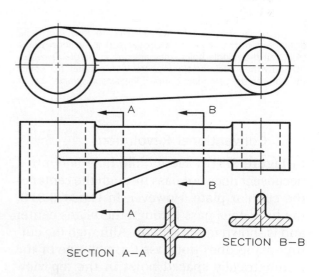

SECTION A–A   SECTION B–B

9.29 Removed sections.
Letters at the ends of a cutting plane (such as A-A) identify the removed section labeled SECTION A-A drawn elsewhere on the drawing.

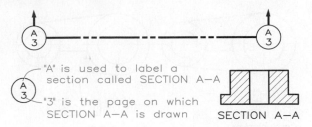

"A" is used to label a section called SECTION A—A

"3" is the page on which SECTION A—A is drawn

SECTION A—A

9.30 When a removed section is placed on another page of a set of drawings, label each end of the cutting plane with a letter and a number. The letters identify the section and the numerals indicate the page on which the removed section is drawn.

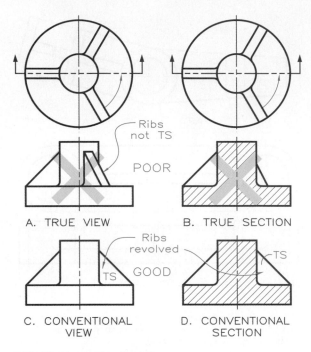

A. TRUE VIEW   Ribs not TS   POOR   B. TRUE SECTION

C. CONVENTIONAL VIEW   Ribs revolved   GOOD   TS   D. CONVENTIONAL SECTION

9.32 Ribs: conventional practice.
Symmetrically spaced ribs are revolved and drawn true size in their orthographic and sectional views as a conventional practice.

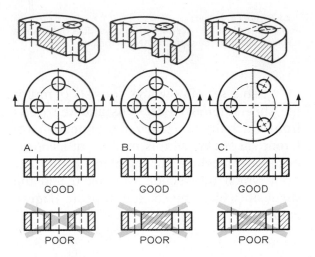

A.   GOOD   POOR

B.   GOOD   POOR

C.   GOOD   POOR

9.31 Symmetrical holes.
Symmetrically spaced holes are revolved to their true radial distance from the center of a circular part in sectional views. (A) Omit the middle holes; they are not at the plate's center. (B) Show the middle hole; it is at the plate's center. (C) Rotate the holes to the centerline to make the section symmetrical.

9.12 Conventional Revolutions

In **Figure 9.31A,** the middle hole is omitted because it does not pass through the center of the circular plate. However, in **Figure 9.31B,** the hole does pass through the plate's center and is shown in the section. Although the cutting plane does not pass through one of the symmetrically spaced holes in the top view **(Figure 9.31C),** the hole is revolved to the cutting plane to show the full section.

When ribs are symmetrically spaced about a hub **(Figure 9.32),** it is conventional practice

to revolve them so that they appear true size in both views and sections. **Figure 9.33** illustrates the conventional practice of revolving both holes and ribs (or webs) of symmetrical parts. Revolution gives a better description of the parts in a manner that is easier to draw.

A cutting plane may be positioned in either of two ways shown in **Figure 9.34.** Even though the cutting plane does not pass through the ribs and holes in **Figure 9.34A,** they may appear in the section as if the cutting plane passed through them. The path of the cutting plane also may be revolved, as shown in **Figure 9.34B.** In this case the ribs are revolved to their true-size position in the section view, although the plane does not cut through them.

The same principles apply to symmetrically spaced spokes **(Figure 9.35).** Draw only the revolved, true-size spokes and do not section line them. If the spokes shown in **Figure 9.36A**

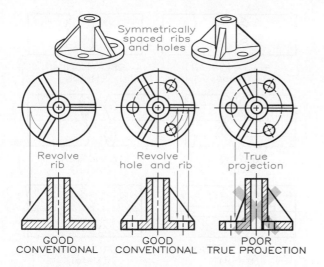

9.33 Ribs and holes.
Symmetrically spaced ribs and holes should be drawn in sections to show the ribs true size and the holes rotated to show them at their true radial distance from the center.

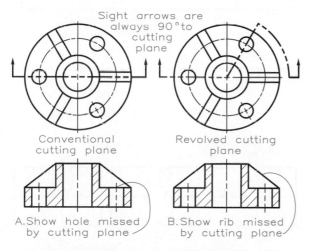

9.34 Cutting plane options.
Rotate symmetrically spaced ribs to show them true size whether or not the cutting plane passes through them. As an alternative, the cutting plane can be drawn to pass through the ribs for clarity.

were hatched, they could be misunderstood as a solid web, as shown in **Figure 9.36B.**

Revolving the symmetrically positioned lugs shown in **Figure 9.37** gives their true size in both the front view and sectional view. The same principles of rotation apply to the part

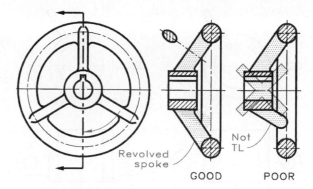

9.35 Spokes in section.
Revolve spokes to show them true size in section. Do not section line (hatch) spokes.

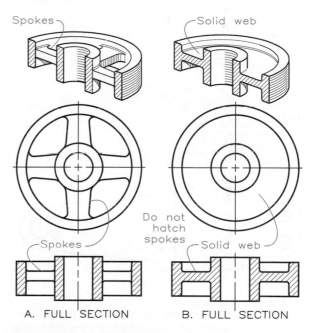

9.36 Spokes and webs in section.
A. Spokes are not hatched even though the cutting plane of the section passes through them.
B. Webs are hatched when cut by the cutting plane.

shown in **Figure 9.38,** where the inclined arm appears in the section as if it had been revolved to the centerline in the top view and then projected to the sectional view. These conventional practices save time and space on a drawing and also make the views more understandable by the reader.

9.12 CONVENTIONAL REVOLUTIONS • 141

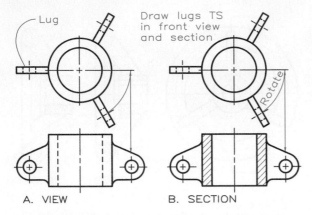

9.37 Revolved lugs.
Lugs are revolved to show their true size in (A) the front view, and also in (B) the sectional view.

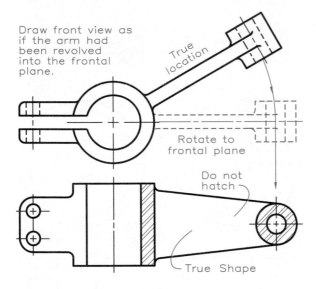

9.38 Revolution of features.
It is conventional practice to revolve parts with inclined features in order to show them true size in both sections and regular orthographic views as well.

9.13 Conventional Breaks

Figure 9.39 shows types of conventional breaks to use when you remove portions of an object. You may draw the "figure-eight" breaks used for cylindrical and tubular parts freehand **(Figure 9.40)** or with a compass when they are larger as shown in **Figure 9.41.**

Conventional breaks can be used to shorten a long piece by removing the portion between the breaks so that it may be drawn at

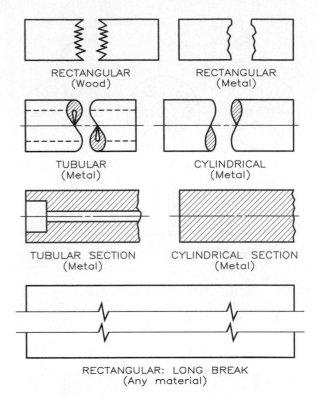

9.39 Conventional breaks.
These conventional breaks indicate that a portion of an object has been omitted.

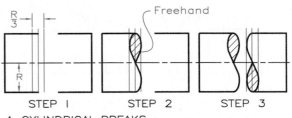

A. CYLINDRICAL BREAKS

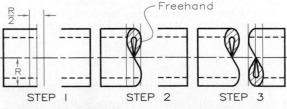

B. TUBULAR BREAKS

9.40 Cylindrical breaks.
It is essential that guidelines are used in drawing conventional breaks freehand for both (A) solid cylinders and (B) tubular cylinders. The radius R is used to determine the widths.

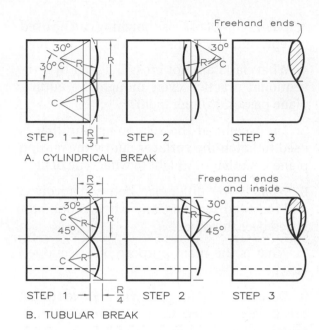

A. CYLINDRICAL BREAK

STEP 1 — $\frac{R}{3}$ — STEP 2 STEP 3

B. TUBULAR BREAK

STEP 1 — $\frac{R}{4}$ — STEP 2 STEP 3

9.41 Cylindrical breaks: Instruments.
These steps can be followed to draw conventional breaks with a compass.

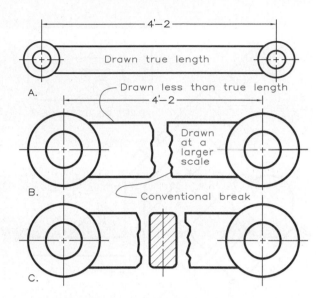

9.42 Conventional breaks: Long parts.
By using conventional breaks, part of the object can be removed so it can be drawn at a larger scale and space can be saved. A revolved section can also be inserted between the breaks to further describe the part.

a larger scale **(Figure 9.42).** The dimension specifies the true length of the part, and the breaks indicate that a portion of the length has been removed.

9.14 Phantom (Ghost) Sections

A **phantom** or **ghost section** depicts parts as if they were being x-rayed. In **Figure 9.43,** the cutting plane is used in the normal manner, but the section lines are drawn as dashed lines. If the object were shown as a regular full section, the circular hole through the front surface could not be shown in the same view. A phantom section lets you show features on both sides of the cutting plane.

9.15 Auxiliary Sections

You may use **auxiliary sections** to supplement the principal views of orthographic projections **(Figure 9.44).** Pass auxiliary cutting plane A-A through the front view and project the auxiliary view from the cutting plane as indicated by the sight arrows. SECTION A-A gives a cross-sectional description of the part

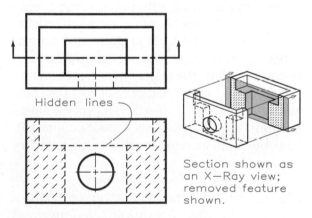

9.43 Phantom sections.
Phantom sections give an "x-ray" view of a part to show features on both sides of the cutting plane. Section lines are drawn as dashed lines.

that would be difficult to depict by other principal orthographic views.

9.16 Summary

Conventional practices are permissable violations of the established principles of orthographic projection that are used to represent

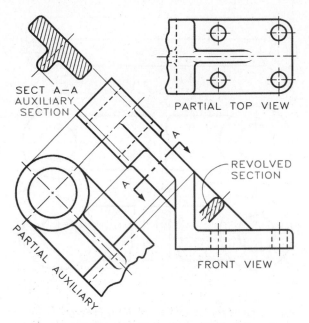

SECT A—A
AUXILIARY
SECTION

PARTIAL TOP VIEW

REVOLVED
SECTION

PARTIAL AUXILIARY

FRONT VIEW

9.44 Auxiliary sections.
Auxiliary sections are helpful in clarifying the details of the inclined features of a part.

objects in a drawing with greater clarity. Sections are examples of conventional practices.

Sections can be used to show internal features, or internal features and external features in combination. The major types of sections are full sections, half sections, offset sections, broken-out sections, revolved sections, removed sections, and phantom sections. Sections can also be used in combination with auxiliary views.

Also, other conventional practices besides sections include the revolution of features such as webs, ribs, spokes, and lugs. Approved types of conventional breaks enable you to omit portions of a drawing to save time and space while improving clarity.

Thought Questions

1. What type of section is formed by passing a cutting plane through a view, removing half of it?

2. In a half section of a symmetrical object, how much is imagined to be removed?

3. What symbol of cross hatching can be used to represent any material in a sectioned view?

4. What parts are not cross hatched by conventional practice even though the cutting plane passes through them?

5. The weight of the sectioning line that is used to hatch the surfaces cut by the cutting plane is similar in weight to what type of line?

6. What is the difference between revolved and removed sections, and when is one preferred over the other?

7. What is the main purpose of a sectioned view?

8. Describe when and why conventional breaks are needed in a section of tubular parts?

9. What are the rules for drawing crosshatch lines of adjacent parts in an assembly view that appears as a section?

10. How are thin parts crosshatched in a sectioned view? Or, overly large areas?

11. How are webs and ribs shown in a section? Explain your answer.

12. What type of section is used to show a series of holes that do not lie in a single cutting plane?

13. Name and explain the types of sections that are available for representing parts.

Problems

Problems 1–24. (Figure 9.45) Solve the problems on size A sheets by drawing two solutions per sheet. Each grid space equals 0.20 in., or 5 mm.

Problems 25–36. (Figures 9.46–9.57) Complete these drawings as full sections. Draw one problem per size A sheet (horizontal format). Each grid space equals 0.20 in., or 5 mm.

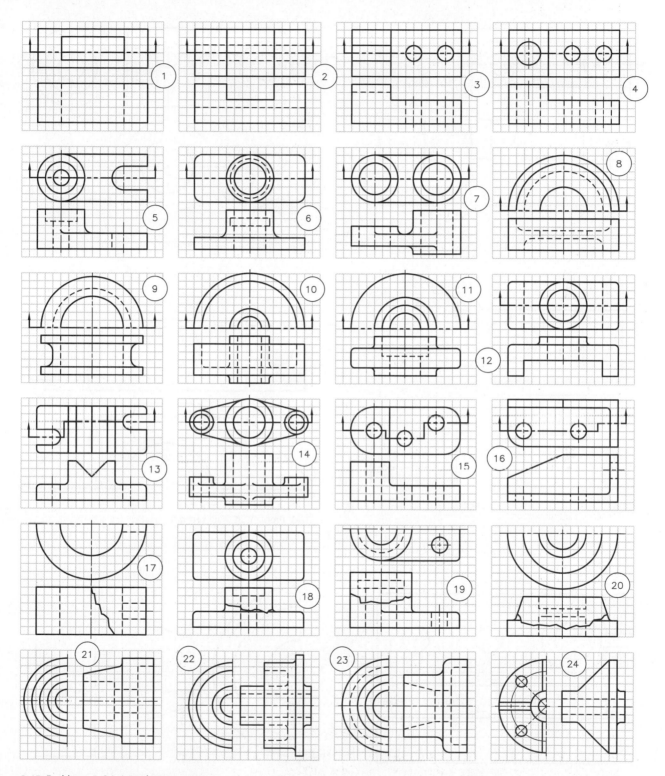

9.45 Problems 1–24. Introductory sections.

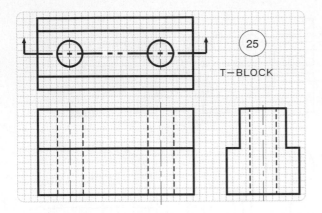

9.46 Problem 25. Full section.

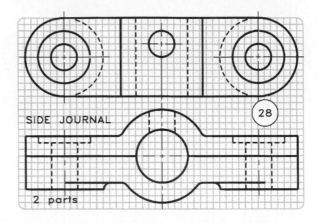

9.49 Problem 28. Offset section.

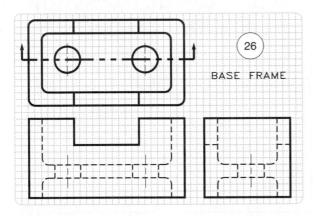

9.47 Problem 26. Full section.

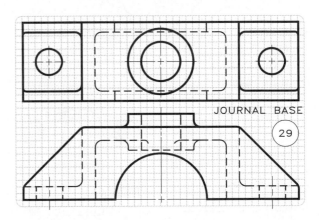

9.50 Problem 29. Full section.

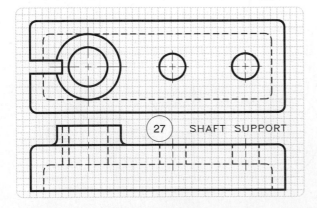

9.48 Problem 27. Full section.

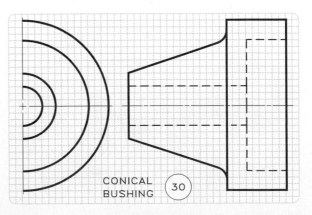

9.51 Problem 30. Half section.

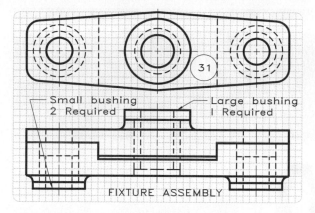

Small bushing
2 Required

Large bushing
1 Required

FIXTURE ASSEMBLY

31

9.52 Problem 31. Half section.

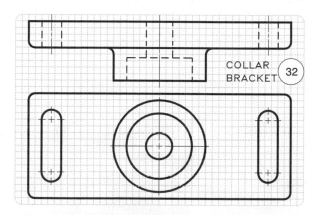

COLLAR
BRACKET 32

9.53 Problem 32. Half section.

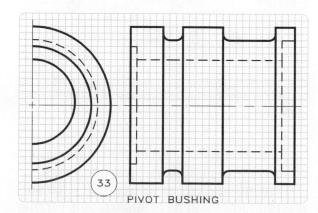

33

PIVOT BUSHING

9.54 Problem 33. Half section.

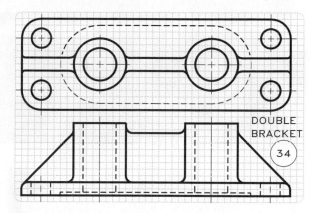

DOUBLE
BRACKET

34

9.55 Problem 34. Half section.

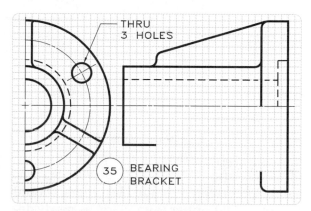

THRU
3 HOLES

35 BEARING
BRACKET

9.56 Problem 35. Offset section.

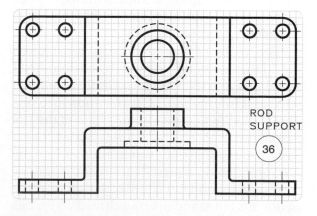

ROD
SUPPORT

36

9.57 Problem 36. Offset section.

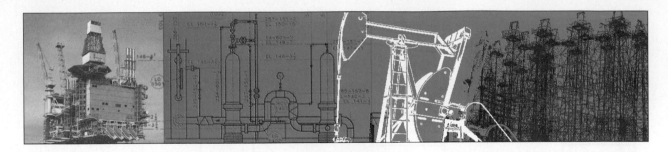

10

Screws, Fasteners, and Springs

10.1 Introduction

Screws provide a fast and easy method of fastening parts together, adjusting the positions of parts, and transmitting power. **Screws, sometimes called threaded fasteners, should be purchased rather than made as newly designed parts for each product.** Screws are available through commercial catalogs in countless forms and shapes for various specialized and general applications. Such screws are cheap, interchangeable, and easy to replace.

The types of threaded parts most often used in industry are covered by current ANSI Standards and include both Unified National (UN) and International Organization for Standardization (ISO) threads. Adoption of the UN thread in 1948 by the United States, Great Britain, and Canada (sometimes called the ABC Standards), a modification of the American Standard and the Whitworth thread, was a major step in standardizing threads. The ISO developed metric thread standards for even

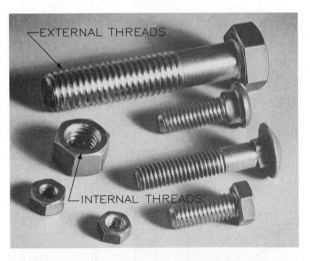

10.1 This photograph shows examples of internal and external threads and three head types. *(Courtesy of RB&W Manufacturing LLC.)*

broader worldwide applications. Other types of fasteners include **keys, pins,** and **rivets.**

Also introduced are **springs** that resist and react to forces and have applications varying from pogo sticks to automobiles. Springs also

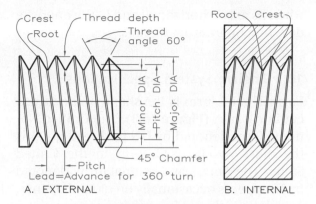

10.2 Most of the definitions of thread terminology are labeled for (A) external and (B) internal threads.

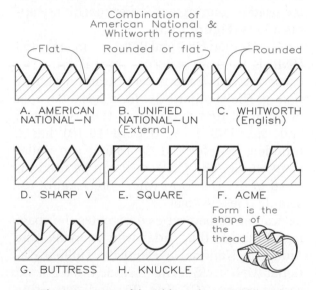

10.3 The various types of thread forms for external threads are shown here.

are available in many forms and styles from specialty manufacturers who supply most of them to industry.

10.2 Thread Terminology

Understanding threaded parts begins with learning their definitions and terminology, which are used throughout this chapter.

External thread: a thread on the outside of a cylinder, such as a bolt **(Figure 10.1).**

Internal thread: a thread cut on the inside of a part, such as a nut **(Figure 10.1).**

Major diameter: the largest diameter on an internal or external thread **(Figure 10.2).**

Minor diameter: the smallest diameter on an internal or external thread **(Figure 10.2).**

Crest: the peak edge of a screw thread **(Figure 10.2).**

Root: the bottom of the thread cut into a cylinder to form the minor diameter **(Figure 10.2).**

Depth: the depth of the thread from the major diameter to the minor diameter; as shown in **Figure 10.2.**

Thread angle: the angle between threads made by the cutting tool, usually 60° **(Figure 10.2).**

Pitch (thread width): the distance between crests of threads, found by dividing 1 inch by the number of threads per inch of a particular thread **(Figure 10.2).**

Pitch diameter: the diameter of an imaginary cylinder passing through the threads at the points where the thread width is equal to the space between the threads **(Figure 10.2** and **Figure 10.4).**

Lead (pronounced leed): the distance a screw will advance when turned 360°.

Form: the shape of the thread cut into a threaded part **(Figure 10.3).**

Series: the number of threads per inch for a particular diameter, grouped into coarse, fine, extra fine, and eight constant-pitch thread series.

Class: the closeness of fit between two mating parts. Class 1 represents a loose fit and Class 3 a tight fit.

Right-hand thread: one that will assemble when turned clockwise. A right-hand external thread slopes downward to the right when its axis is horizontal and in the opposite direction on internal threads.

Left-hand thread: one that will assemble when turned counterclockwise. A left-hand external thread slopes downward to the left

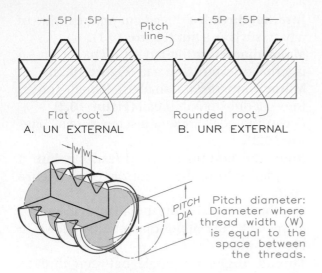

A. UN EXTERNAL   B. UNR EXTERNAL

Pitch diameter: Diameter where thread width (W) is equal to the space between the threads.

10.4 UN and UNR thread forms.
A The UN external thread has a flat root (a round root is optional) and a flat crest.
B The UNR thread has a rounded root formed by rolling rather than by cutting. The UNR form does not apply to internal threads; internal threads cannot be rolled.

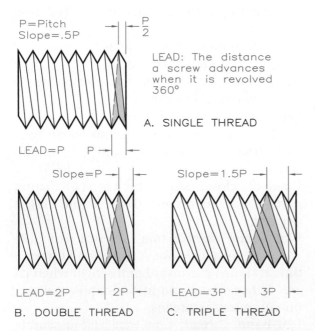

P=Pitch
Slope=.5P

LEAD: The distance a screw advances when it is revolved 360°

A. SINGLE THREAD

LEAD=P

B. DOUBLE THREAD   C. TRIPLE THREAD

10.5 Threads can be (A) single, (B) double, or (C) triple, which represents the number of pitches (P) that each advances when turned 360°.

when its axis is horizontal and in the opposite direction on internal threads.

10.3 English System Specifications

Form Thread form is the shape of the thread cut into a part **(Figure 10.3).** The Unified National form denoted by **UN** in thread notes is the most widely used form in the United States. The American National form, denoted by **N,** appears occasionally on older drawings. The sharp V thread is used for set screws and in applications where friction in assembly is desired. **Acme, square,** and **buttress** threads are used in gearing and other machinery applications **(Figure 10.3).**

The Unified National Rolled form, denoted UNR, is used for external threads only, because internal threads cannot be formed by rolling. The standard UN form has a flat root (a rounded root is optional) **(Figure 10.4A),** and the UNR form **(Figure 10.4B)** has a rounded root formed by rolling a cylinder across a die. The UNR form can be used instead of the UN form where precision of assembly is less critical.

Series The thread series designates the spacing of threads that varies with diameter. The American National (N) and the Unified National (UN/UNR) forms include three graded series: **coarse (C), fine (F),** and **extra fine (EF).** Eight **constant-pitch** series (4, 6, 8, 12, 16, 20, 28, and 32 threads per inch) are also available.

Coarse Unified National threads are denoted **UNC** or **UNRC,** which is a combination of form and series designation. The coarse thread (UNC/UNRC or NC) has the largest pitch of any series and is suitable for bolts, screws, nuts, and general use with cast iron, soft metals, and plastics when rapid assembly is desired. An American National (N) form for a coarse thread is written **NC.**

Fine threads (NF or UNF/UNRF) are used for bolts, nuts, and screws when a high degree of tightening is required. Fine threads are closer together than coarse threads, and their pitch is graduated to be smaller on smaller diameters.

Extra-fine threads (UNEF/UNREF or NEF) are suitable for sheet metal screws and bolts, thin nuts, ferrules, and couplings when the length of engagement is limited and high stresses must be withstood.

Constant-pitch threads (4 UN, 6 UN, 8 UN, 12 UN, 16 UN, 20 UN, 28 UN, and 32 UN) are used on larger diameter threads (beginning near the 1/2-inch size) and have the same pitch size regardless of the diameter size. The most commonly used constant-pitch threads are 8 UN, 12 UN, and 16 UN members of the series, which are used on threads of about 1 inch in diameter and larger. Constant-pitch threads may be specified as UNR or N thread forms. The ANSI table in Appendix 3 shows constant-pitch threads for larger thread diameters instead of graded pitches of coarse, fine, and extra fine.

Class of Fit The class of fit is the tightness between two mating threads as between a nut and bolt, and is indicated in the thread note by the numbers 1, 2, or 3 followed by the letters **A** or **B.** For UN forms, the letter A represents an external thread and the letter B represents an internal thread. The letters A and B do not appear in notes for threads of the American National form (N).

Class 1A and **1B** threads are used on parts that assemble with a minimum of binding and precision. **Class 2A** and **2B** threads are general-purpose threads for bolts, nuts, and screws used in general and mass-production applications. **Class 3A** and **3B** threads are used in precision assemblies where a close fit is required to withstand stresses and vibration.

Single and Multiple Threads A single thread **(Figure 10.5A)** is a thread that advances the distance of its pitch in a revolution of 360°; that is, its pitch is equal to its lead. The crest

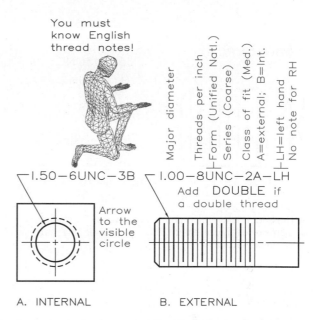

10.6 Thread notes of the English form that are applied to (A) internal and (B) external threads are defined here.

lines have a slope of 1/2 P since only 180° of the revolution is visible in the view.

Multiple threads are used where quick assembly is required. A double thread is composed of two threads that advance a distance of 2P when turned 360° **(Figure 10.5B);** that is, its lead is equal to 2P. The crest lines have a slope of P because only 180° of the revolution is visible in the view. A triple thread advances 3P in 360° with a crest line slope of 1-1/2 P in the view where 180° of the revolution is visible as shown in **Figure 10.5C.**

10.4 English Thread Notes

Drawings of threads are only symbolic depictions and are inadequate representations of threads unless accompanied by notes **(Figure 10.6).** In a thread note, the major diameter is given first, then the number of threads per inch, the form, series, the class of fit, and the letter A or B to denote external or internal threads, respectively. For a double or triple thread, include the word *double* or *triple* in the note, and for left-hand threads add the letters LH.

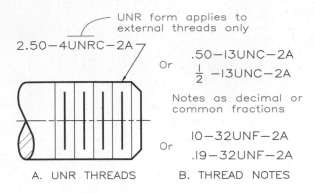

A. UNR THREADS   B. THREAD NOTES

2.50—4UNRC—2A ⌐ UNR form applies to external threads only

Or .50—13UNC—2A

$\frac{1}{2}$ —13UNC—2A

Notes as decimal or common fractions

Or 10—32UNF—2A

.19—32UNF—2A

10.7 Thread notes.
A The UNR thread note is applied to only external threads.
B Diameters in thread notes can be given as decimal fractions or common fractions.

AMERICAN NATIONAL STANDARDS INSTITUTE
UNIFIED INCH SCREW THREADS (UN AND UNR)

Nominal Diameter	Basic Diameter	Coarse NC & UNC		Fine NF & UNF		Extra Fine NEF/UNEF	
		Thds per In.	Tap Drill DIA	Thds per In.	Tap Drill DIA	Thds per In.	Tap Drill DIA
1	1.000	8	.875	12	.922	20	.953
1–1/16	1.063	...	...	...	...	18	1.000
1–1/8	1.125	7	.904	12	1.046	18	1.070
1–3/16	1.188	...	...	...	...	18	1.141
1–1/4	1.250	7	1.109	12	1.172	18	1.188
1–5/16	1.313	...	...	...	...	18	1.266
1–3/8	1.375	6	1.219	12	1.297	18	1.313
1–7/16	1.438	...	...	...	...	18	1.375
1–1/2	1.500	6	1.344	12	1.422	18	1.438

10.8 This is a portion of the ANSI tables for UN and UNR threads in Appendix 2.

Figure 10.7 shows a UNR thread note for the external thread. (UNR does not apply to internal threads.) When inches are the unit of measurement, fractions can be written as decimal or as common fractions. The information for UN thread notes comes from ANSI tables in Appendixes 2 and 3.

Using Thread Tables

A portion of Appendix 2 is shown in **Figure 10.8,** which gives the UN/UNR thread table from which specifications for standardized interchangeable threads can be selected. Note that a 1.50-inch-diameter bolt with fine thread

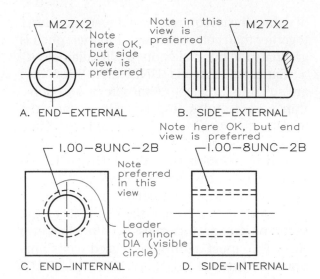

A. END—EXTERNAL   B. SIDE—EXTERNAL

M27X2 — Note here OK, but side view is preferred

Note in this view is preferred — M27X2

Note here OK, but end view is preferred

1.00—8UNC—2B

Note preferred in this view

Leader to minor DIA (visible circle)

1.00—8UNC—2B

C. END—INTERNAL   D. SIDE—INTERNAL

10.9 Notes for external threads are best if they are placed on the rectangular view of the threads. Notes for internal threads are best applied in the circular view if space permits. Metric notes are given at A and B, and English notes are given at C and D.

(UNF) has 12 threads per inch, and its thread note is written as

1.500–12 UNF–2A or 1-1/2–12 UNF–2A.

If the thread were internal (nut), the thread note would be the same but the letter B would be used instead of the letter A. For constant-pitch thread series, selected for larger diameters, write the thread note as

1.750–12 UN–2A or 1-3/4–12 UN–2A.

For the UNR thread form (for external threads only) substitute UNR for UN in the last three columns; for example, UNREF for UNEF (extra fine). **Figure 10.9** shows the preferred placement of thread notes (with leaders) for external and internal threads.

10.5 Metric Thread Notes

Metric thread notes are simpler than English notes since only the major diameter and the pitch are given as shown in A and B of **Figure 10.9.** The letter M (for metric) precedes diameter and pitch: M27 X 2 for example, where 27

COARSE		FINE	
MAJ. DIA & THD PITCH	TAP DRILL	MAJ. DIA & THD PITCH	TAP DRILL
M20 X 2.5	17.5	M20 X 1.5	18.5
M22 X 2.5	19.5	M22 X 1.5	20.5
M24 X 3	21.0	M24 X 2	22.0
M27 X 3	24.0	M27 X 2	25.0
M30 X 3.5	26.5	M30 X 2	28.0
M33 X 3.5	29.5	M33 X 2	31.0
M36 X 4	32.0	M36 X 2	33.0
M39 X 4	35.0	M39 X 2	36.0
M42 X 4.5	37.5	M42 X 2	39.0

10.10 This portion of the metric (ISO) thread tables in Appendix 4 shows specifications metric thread notes.

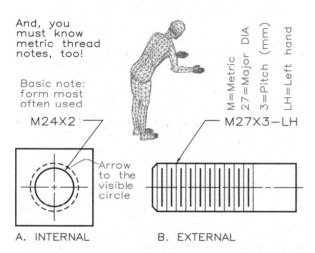

10.11 This is the basic thread note that will be used for (A) internal and (B) external threads.

is the major diameter and 2 is the pitch in millimeters. A portion of the ISO thread table in Appendix 4 is shown in **Figure 10.10** from which thread notes can be selected. Notice that metric threads come in two series, **coarse** and **fine.** The tap drill, the approximate diameter of the minor diameter, is also given.

Basic Designation   Examples of metric thread notes that were taken directly from the tables are shown in **Figure 10.11.** These are basic designation notes that are sufficient for general applications. Metric notes do not distinguish

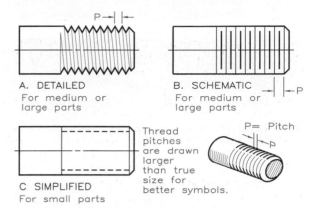

10.12 The three types of thread symbols used for drawing threads are (A) detailed, (B) schematic, and (C) simplified.

between internal and external threads as in the English system where the letters A and B are given. But like the English system, left-hand threads are labeled LH at the end of the note.

10.6 Drawing Threads

Threads may be represented by **detailed, schematic,** and **simplified** symbols (**Figure 10.12**). Detailed symbols represent a thread most realistically, simplified symbols represent a thread least realistically, and schematic symbols are a compromise between the two. Detailed and schematic symbols can be used for drawing larger threads on a drawing (1/2″ dia and larger) and simplified symbols are best for smaller threads.

10.7 Detailed Symbols

UN/UNR-Threads   Detailed thread symbols for external threads in view and in section are shown in **Figure 10.13**. Instead of drawing helical curves, straight lines are used to depict crest and root lines. Variations of detailed thread symbols for internal threads drawn in views and sections are shown in **Figure 10.14.**

The steps of drawing a detailed thread representation, whether English or metric threads, are shown in **Figure 10.15**. When using the

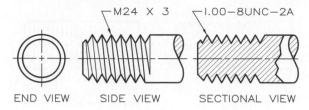

10.13 These detailed symbols represent external threads in view and section.

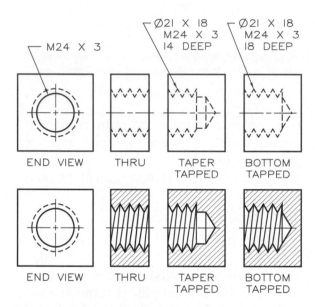

10.14 These detailed symbols represent internal threads. Approximate the minor diameter as 75% of the major diameter. Tap drill diameters are found in Appendix 4.

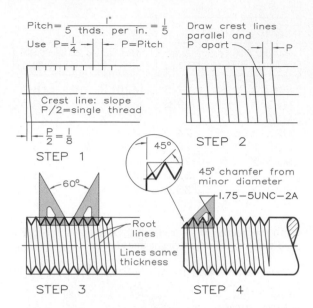

10.15 Detailed representation of threads.
Step 1 For a 1.75–5 UNC–2A thread, find the pitch by dividing 1 inch by the number of threads per inch, 5 in this case. Use a pitch of 1/4 instead of 1/5 to space the threads farther apart. Lay off the pitch along the thread's length and draw a crest line at a slope of P/2=1/8.
Step 2 Draw the other crest lines as dark, visible lines parallel to the first crest line.
Step 3 Draw 60° vees between the crest lines and draw root lines from the bottom of the vees. Root lines are parallel to each other but not to crest lines.
Step 4 Draw a 45° chamfer from the minor diameter at the thread's end. Darken lines and add a thread note.

English tables, calculate the pitch by dividing 1 inch by the number of threads per inch. In the metric system, pitch is given in the tables. **Draw the spacing between crest lines larger than the actual pitch size to avoid "clogged-up" lines.**

Square Threads How to draw and note a detailed drawing of a square thread is shown in **Figure 10.16.** Follow the same basic steps to draw views and sections of square internal threads **(Figure 10.17).** In a section, draw both the internal crest and root lines, but in a view draw only the outline of the threads. Place thread notes for internal threads in the

circular view, whenever possible, with the leader pointing toward the center and stopping at the visible circle.

When a square thread is long, it can be represented by using phantom lines, without drawing all the threads **(Figure 10.18).** This conventional practice saves time and effort without reducing the drawing's effectiveness.

Acme Threads A modified version of the square thread is the Acme thread, which has tapered (15°) sides for easier engagement than square threads. The steps involved in drawing detailed Acme threads are shown in **Figure 10.19.** Acme threads are heavy threads that are

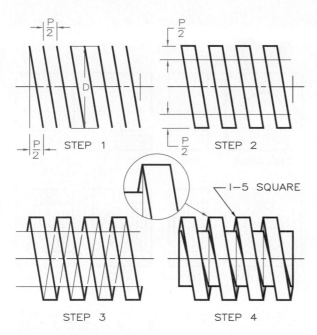

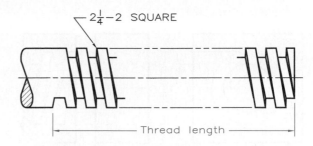

10.18 The conventional method of showing square threads is to draw sample threads at each end and to connect them with phantom lines.

10.16 Drawing the square thread.
Step 1 Lay out the major diameter. Space the crest lines 1/2P or larger apart and slope them downward to the right for right-hand threads.
Step 2 Connect every other pair of crest lines. Find the minor diameter by measuring 1/2P inward from the major diameter.
Step 3 Connect the opposite crest lines with light construction lines to establish the thread's profile.
Step 4 Connect the inside crest lines with light construction lines to locate the points on the minor diameter where the thread wraps around the minor diameter. Darken the final lines.

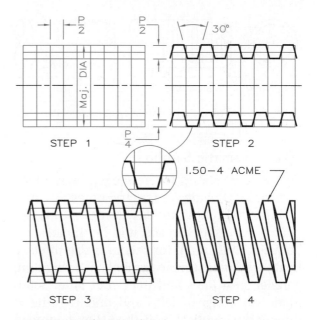

10.19 Drawing the Acme thread.
Step 1 Lay out the major diameter and thread length and divide the shaft into equal divisions 1/2P apart. Locate the minor and pitch diameters by using distances 1/2P and 1/4P.
Step 2 Draw lines at 15° angles with the vertical along the pitch diameter to make a total angle of 30°.
Step 3 Draw the crest lines across the screw.
Step 4 Darken the lines, draw the root lines, and add the thread note to complete the drawing.

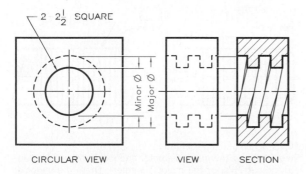

10.17 This drawing shows internal square threads in view and section.

used to transmit force and power in mechanisms such as screw jacks, leveling devices, and lathes. Appendix 5 contains the table for Acme thread specifications and dimensions.

Internal Acme threads are shown in view and section in **Figure 10.20.** Left-hand internal threads in section appear the same as right-hand external threads.

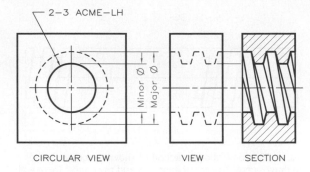

CIRCULAR VIEW   VIEW   SECTION

10.20 This drawing shows internal Acme threads in view and section.

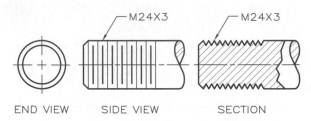

END VIEW   SIDE VIEW   SECTION

10.21 These schematic symbols represent external threads in view and section.

10.8 Schematic Symbols

Figure 10.21 shows schematic representations of external threads with metric notes. Because schematic symbols are easy to draw and adequately represent threads, it is the thread symbol used most often for medium-size threads. Draw schematic thread symbols by using thin, parallel crest lines and thick root lines. Schematic drawings of left-hand and right-hand threads are identical, only the LH in the thread note indicates that a thread is left-handed. Right-hand threads are not marked RH, but are understood to be right-hand threads.

Figure 10.22 shows threaded holes in view and in section drawn with schematic symbols. The size of the tap drill diameter is approximately equal to the major diameter minus the pitch. However, the minor diameter usually is drawn a bit smaller to provide better separation between the lines representing the major and minor diameters.

Figure 10.23 shows how to draw schematic threads using English specifications. Draw

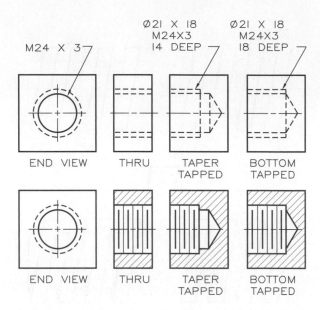

END VIEW   THRU   TAPER TAPPED   BOTTOM TAPPED

END VIEW   THRU   TAPER TAPPED   BOTTOM TAPPED

10.22 These schematic symbols represent internal threads in view and section. Tap drill diameters for their minor diameters are given in Appendix 2.

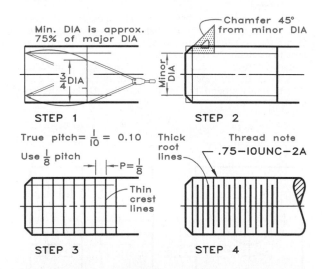

10.23 Schematic representation.
Step 1 Lay out the major diameter and locate the minor diameter (about 75% of the major diameter). Draw the minor diameter with light construction lines.
Step 2 Chamfer the end of the threads with a 45° angle from the minor diameter.
Step 3 Find the pitch of a .75–10UNC–2A thread (0.1) by dividing 1 inch by the number of threads per inch (10). Use a larger pitch, 1/8 inch in this case, for spacing the thin crest lines.
Step 4 Draw root lines as thick as the visible lines between the crest lines to the construction lines representing the minor diameter. Add a thread note.

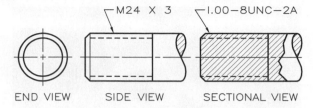

10.24 These simplified thread symbols represent external threads in view and section.

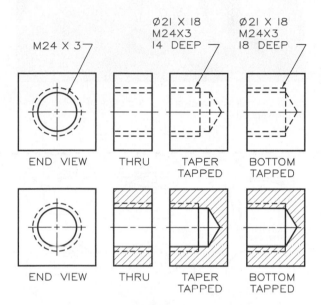

10.25 These simplified thread symbols represent internal threads in view and section. Draw minor diameters at about 75% of the major diameter.

the minor diameter at approximately three-quarters of the major diameter and the chamfer (bevel) 45° from the minor diameter. Draw crest lines as thin lines and root lines as thick visible lines.

10.9 Simplified Symbols

Examples of external threads drawn with simplified symbols and noted with both metric and English formats are shown in **Figure 10.24.** Simplified symbols are the easiest to draw and are the best suited for drawing small threads where drawing using schematic and

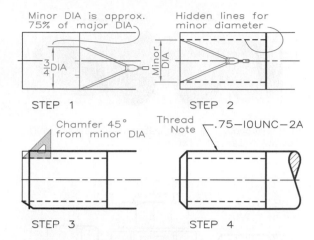

10.26 Drawing simplified threads.
Step 1 Lay out the major diameter. Locate the minor diameter (about 75% of the major diameter).
Step 2 Draw hidden lines to represent the minor diameter.
Step 3 Draw a 45° chamfer from the minor diameter to the major diameter.
Step 4 Darken the lines and add a thread note.

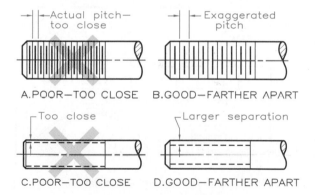

10.27 Most threads must be drawn using exaggerated dimensions instead of actual measurements to prevent the drawing from having lines drawn too closely together.

detailed symbols would be too crowded. Various techniques of applying simplified symbols to threads are shown in **Figure 10.25.** The minor diameter is drawn as hidden lines spaced at about three-quarters the major diameter. **Figure 10.26** shows the steps of drawing simplified threads. With experience, you can approximate the location of the minor diameter of simplified threads by eye.

10.28 This photo shows a nut, bolt, and washers in combination. *(Courtesy of Lamson & Sessions.)*

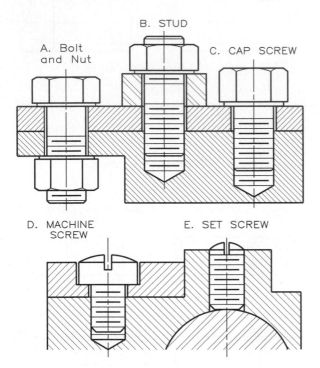

10.29 Types of threaded bolts are illustrated here.

10.30 Examples of types of nuts. *(Courtesy of RB&W Manufacturing LLC.)*

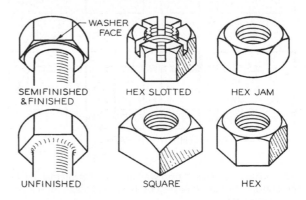

10.31 Finished and semifinished bolts and nuts have raised washer faces. Several types of nuts are shown here also.

Drawing Small Threads

Remember, a drawing of a thread is a pictorial symbol, therefore do not try to draw the true spacing of the threads. True spacing will be too close and hard to read. Instead, select a wider spacing between root and crest lines for a drawing that is easier to read **(Figure 10.27).** This conventional practice of enlarging the thread's pitch to separate thread symbols is applied in drawing all three types of symbols (simplified, schematic, and detailed). Add a thread note to the drawing to give the necessary detailed specifications.

10.10 Nuts and Bolts

Nuts and bolts (Figure 10.28) come in a variety of forms and sizes for many different applications. Some common types of threaded fasteners are shown in **Figure 10.29.** A **bolt** is a threaded cylinder with a head that is used with a **nut** to hold parts together. A **stud** is a headless bolt, threaded at both ends, that is screwed into one part with a nut attached to the other end.

A **cap screw** usually does not have a nut, but passes through a hole in one part and screws into a threaded hole in another part. A

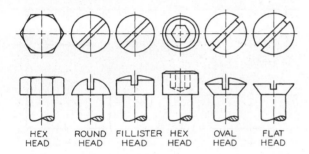

HEX HEAD ROUND HEAD FILLISTER HEAD HEX HEAD OVAL HEAD FLAT HEAD

10.32 These are standard types of bolt and screw heads. See Appendixes 7–16.

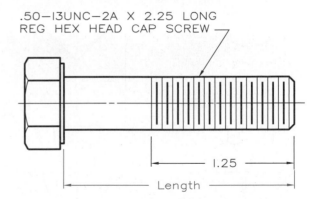

.50—13UNC—2A X 2.25 LONG REG HEX HEAD CAP SCREW

1.25

Length

10.33 A properly dimensioned and noted hexagon-head bolt.

hexagon-head **machine screw** is similar to but smaller than a cap screw. Machine screws also come with other types of heads. A **set screw** is used to hold one part fixed in place with another, usually to prevent rotation, as with a pulley on a shaft.

Types of heads used on **regular** and **heavy** bolts and nuts are shown in **Figure 10.30** and **Figure 10.31.** Heavy bolts have thicker heads than regular bolts for heavier usage. A **finished head** (or nut) has a 1/64-inch-thick washer face (a circular boss) to provide a bearing surface for smooth contact. **Semifinished bolt heads** and **nuts** are the same as finished bolt heads and nuts. Unfinished bolt heads and nuts have no bosses and no machined surfaces.

A **hexagon jam nut** does not have a washer face, but it is chamfered (beveled at its corners) on both sides. **Figure 10.32** shows other standard bolt and screw heads for cap screws and machine screws.

Dimensions

Figure 10.33 shows a properly dimensioned bolt. The ANSI tables in Appendixes 7–10 give nut and bolt dimensions, but you may use the following guides for hexagon-head and square-head bolts.

Overall-Lengths Hexagon-head bolts are available in 1/4-inch increments up to 8 inches

long, in 1/2-inch increments from 8 to 20 inches long, and in 1-inch increments from 20 to 30 inches long. Square-head bolts are available in 1/8-inch increments from 1/2 to 3/4 inch long, in 1/4-inch increments from 3/4 inch to 5 inches long, in 1/2-inch increments from 5 to 12 inches long, and in 1-inch increments from 12 to 30 inches long.

Thread-Lengths For both hexagon-head and square-head bolts up to 6 inches long,

$$\text{Thread length} = 2D + 1/4 \text{ inch}$$

where D is the diameter of the bolt. For bolts more than 6 inches long

$$\text{Thread length} = 2D + 1/2 \text{ inch.}$$

Threads for bolts can be coarse, fine, or 8-pitch threads. The class of fit for bolts and nuts is understood to be 2A and 2B if no class is specified in the note.

Dimension Notes

Designate standard square-head and hexagon-head bolts by notes in one of three forms:

3/8–16 X 1-1/2 SQUARE BOLT–STEEL;

1/2–13 X 3 HEX CAP SCREW
SAE GRADE 8–STEEL;

.75 X 3 5.00 UNC–2A HEX HD
LAG SCREW.

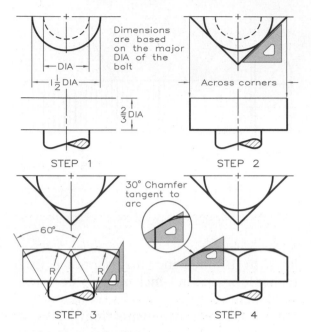

10.34 Drawing a square head.
Step 1 Draw the major diameter, DIA, of the bolt. Use 1.5 DIA to draw the hexagon-head's diameter and 2/3 DIA to establish its thickness.
Step 2 Draw the top view of the square head at a 45° angle to give an across-corners view.
Step 3 Show the chamfer in the front view by using a 30°–60° triangle to find the centers for the radii.
Step 4 Show a 30° chamfer tangent to the arcs in the front view. Darken the lines.

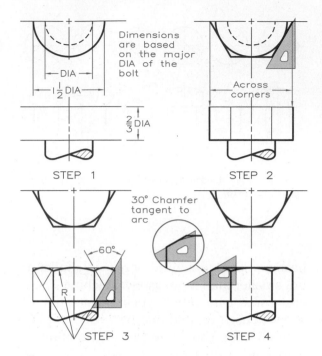

10.35 Drawing a hexagon head.
Step 1 Draw the major DIA of the bolt, find the diameter of the head as 1.5 DIA, and thickness as 2/3 DIA.
Step 2 Construct a hexagon head with a 30°–60° triangle to give an across-corners view.
Step 3 Find arcs in the front view to draw the chamfer of the head.
Step 4 Draw a 30° chamfer tangent to the arcs in the front view. Darken the lines.

The numbers (left to right) represent bolt diameter, threads per inch (omit for lag screws), bolt length, screw name, and material (material designation is optional). When not specified in a note, each bolt is assumed to have a class 2 fit. Three types of notes for designating nuts are

1/2–13 SQUARE NUT–STEEL;

3/4–16 HEAVY HEX NUT;

1.00–8UNC–2B HEX HD THICK
SLOTTED NUT
CORROSION-RESISTANT STEEL.

When nuts are not specified as heavy, they are assumed to be regular. When the class of fit is not specified in a note, it is assumed to be 2B for nuts.

10.11 Drawing Square Heads

Appendixes 7 and 8 give dimensions for square bolt heads and nuts. However, it is conventional practice to draw nuts and bolts by using the general proportions shown in **Figure 10.34.** Your first step in drawing a bolt head or nut is determining whether the view is to be across corners or across flats; that is, whether the lines at either side of the view represent the square's corners or flats. Nuts and bolts drawn across corners is best, but occasionally they are drawn across flats when the head or nut is truly in this orientation.

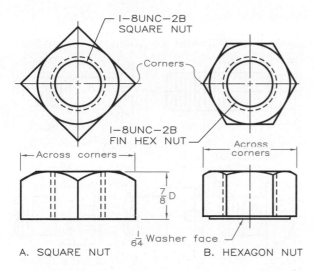

10.36 Drawing square and hexagon nuts across corners involves the same steps used for drawing bolt heads. Add notes to give nut specifications.

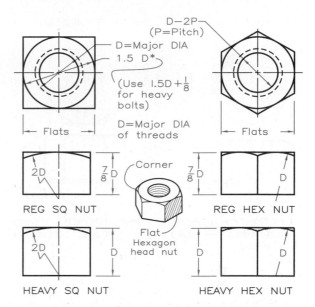

10.37 These square and hexagon nuts are drawn across flats with notes added to give their specifications. Square nuts are unfinished.

10.12 Drawing Hexagon Heads

Figure 10.35 shows the steps of drawing the head of a hexagon bolt across corners by using the bolt's major diameter, D, as the basis for all other proportions. Begin by drawing the top view of the head as a circle of a diameter of 1-1/2 D. For a regular head, the thickness is 2/3 D and for a heavy head it is 7/8 D. Circumscribe a hexagon about the circle. Then draw outside arcs in the rectangular view and tangent chamfers (bevels) to complete the drawing.

Drawing Nuts

Use the same steps to draw a square and a hexagon nut (shown across corners in **Figure 10.36**) that you used to draw bolt heads. The difference is that nuts are thicker than bolt heads: The thickness of a regular nut is 7/8 D, and the thickness of a heavy nut is 1 D, where D is the bolt diameter. Hidden lines may be inserted in the front view to indicate threads, or omitted. Exaggerate the thickness of the 1/64-inch washer face on the finished and semifinished hexagon nuts to about 1/32 inch to make it more noticeable. Place thread notes on circular views with leaders when space permits. Square nuts that are not labeled heavy are assumed to be regular nuts.

Figure 10.37 shows how to construct square and hexagon nuts across flats. For regular nuts, the distance across flats is 1-1/2 X D (D = major diameter of the thread), and 1-5/8 D for heavy nuts. Draw the top views in the same way you did across-corner top views, but rotate them to give across-flat front views. **Figure 10.38** depicts dimensioned and noted hexagon regular and heavy nuts drawn across corners.

Drawing Nut and Bolt Combinations

Nuts and bolts in assembly are drawn in the same manner as they are drawn individually **(Figure 10.39).** Use the major diameter, D, of the bolt as the basis for other dimensions. Here, the views of the bolt heads are across corners, and the views of the nuts are across flats, although both views could have been drawn across corners. The half end views are used to find the front views by projection. Add a note to give the specifications of the nut and bolt.

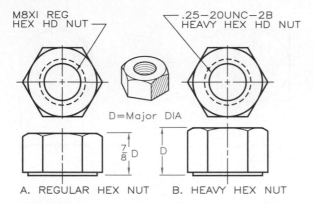

M8XI REG HEX HD NUT

.25−20UNC−2B HEAVY HEX HD NUT

D=Major DIA

$\frac{7}{8}$D

D

A. REGULAR HEX NUT   B. HEAVY HEX NUT

10.38 These regular and heavy hexagon nuts are drawn across corners, with notes added to give their specifications. See Appendix 10.

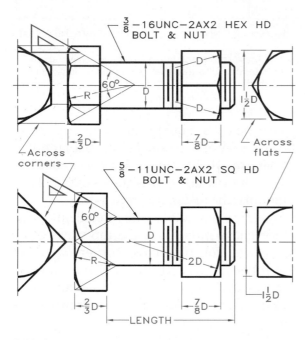

$\frac{3}{8}$−16UNC−2AX2 HEX HD BOLT & NUT

D
D
60°
R
D
$1\frac{1}{2}$D
Across flats
$\frac{2}{3}$D
$\frac{7}{8}$D
Across corners

$\frac{5}{8}$−11UNC−2AX2 SQ HD BOLT & NUT

60°
R
D
2D
$1\frac{1}{2}$D
$\frac{2}{3}$D
$\frac{7}{8}$D
LENGTH

10.39 The proportions and the geometry for drawing square nuts and bolts and hexagon nuts and bolts are shown here.

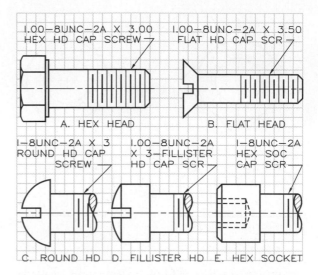

1.00−8UNC−2A X 3.00 HEX HD CAP SCREW

1.00−8UNC−2A X 3.50 FLAT HD CAP SCR

A. HEX HEAD   B. FLAT HEAD

I−8UNC−2A X 3 ROUND HD CAP SCREW

1.00−8UNC−2A X 3−FILLISTER HD CAP SCR

I−8UNC−2A HEX SOC CAP SCR

C. ROUND HD   D. FILLISTER HD   E. HEX SOCKET

10.40 These cap screws are drawn on a grid to give the proportions for drawing them at different sizes. Notes give thread specifications, length, head type, and bolt name (cap screw). See Appendixes 11–15.

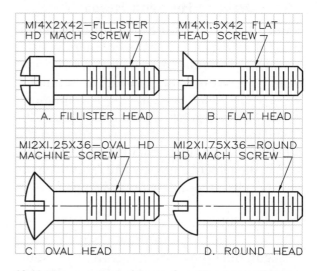

MI4X2X42−FILLISTER HD MACH SCREW

MI4XI.5X42 FLAT HEAD SCREW

A. FILLISTER HEAD   B. FLAT HEAD

MI2XI.25X36−OVAL HD MACHINE SCREW

MI2XI.75X36−ROUND HD MACH SCREW

C. OVAL HEAD   D. ROUND HEAD

10.41 These are standard types of machine screws. The same proportions may be used to draw machine screws of all sizes. See Appendix 16.

10.13 Types of Screws

Cap Screws

The cap screw passes through a hole in one part and screws into a threaded hole in the other part so the two parts can be held together without a nut. Cap screws are usually larger than machine screws and they may also be used with nuts. **Figure 10.40** shows the standard types of cap screw heads drawn on a grid that can be used as a guide for drawing cap screws of other sizes. Appendixes 11–15 give cap screw dimensions, which will aid in drawing them.

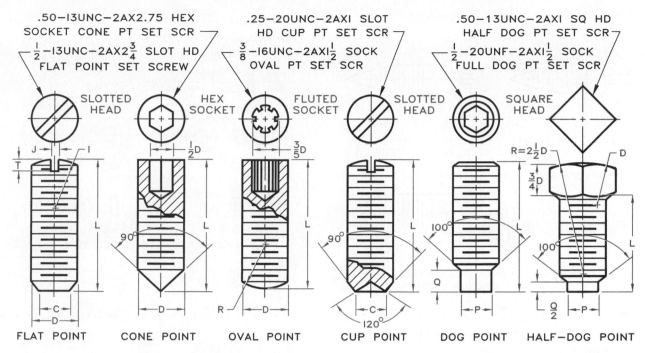

.50–13UNC–2AX2.75 HEX
SOCKET CONE PT SET SCR

½–13UNC–2AX2¾ SLOT HD
FLAT POINT SET SCREW

.25–20UNC–2AXI SLOT
HD CUP PT SET SCR

⅜–16UNC–2AXI½ SOCK
OVAL PT SET SCR

.50–13UNC–2AXI SQ HD
HALF DOG PT SET SCR

½–20UNF–2AXI½ SOCK
FULL DOG PT SET SCR

SLOTTED HEAD   HEX SOCKET   FLUTED SOCKET   SLOTTED HEAD   SQUARE HEAD

FLAT POINT   CONE POINT   OVAL POINT   CUP POINT   DOG POINT   HALF–DOG POINT

10.42 Set screws are available with various combinations of heads and points. Notes give their measurements. (See Appendix 17.)

Machine Screws

Smaller than most cap screws, **machine screws** usually are less than 1 inch in diameter. They screw into a threaded hole in a part or into a nut. Machine screws are fully threaded when their length is 2 inches or less. Longer screws have thread lengths of 2D + 1/4 inch (D = major diameter of the thread). **Figure 10.41** shows four types of machine screws, along with notes, drawn on a grid that may be used as an aid in drawing them without dimensions from a table. Machine screws range in diameter from No. 0 (0.060 inch) to 3/4 inch, as shown in Appendix 16, which gives the dimensions of round-head machine screws.

Set Screws

Set screws are used to hold parts together, such as pulleys and handles on a shaft, and prevent rotation. **Figure 10.42** shows various types of set screws with dimensions denoted

by letters that correspond to the tables of dimensions in Appendix 17.

Set screws are available in combinations of points and heads. The shaft against which the set screw is tightened may have a machined flat surface to provide a good bearing surface for a **dog** or **flat-point** set screw end to press against. The **cup point** gives good gripping when pressed against round shafts. The **cone point** works best when inserted into holes drilled in the part being held. The **headless set screw** has no head to protrude above a rotating part. An **exterior square head** is good for applications in which greater force must be applied with a wrench to hold larger set screws in position.

Wood Screws

A wood screw is a pointed screw having sharp coarse threads that will screw into wood making its own internal threads in the process. **Figure 10.43** shows the three most common

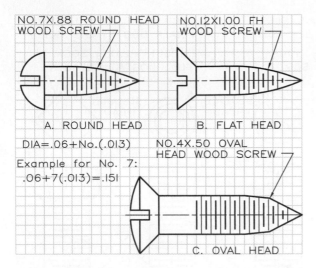

10.43 Standard types of wood screws are drawn on a grid to give their proportions for drawing them at other sizes.

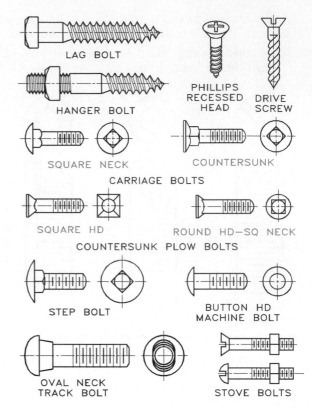

10.44 A few of the many different types of bolts and screws are illustrated here.

types of wood screws drawn on a grid to show their relative proportions.

Sizes of wood screws are specified by single numbers, such as 0, 6, or 16. From 0 to 10, each digit represents a different size. Beginning at 10, only even-numbered sizes are standard, that is, 10, 12, 14, 16, 18, 20, 22, and 24. Use the following formula to translate these numbers into the actual diameter sizes:

Actual DIA = 0.06 + (screw number × 0.013).

For example, the diameter for the No. 7 wood screw shown in **Figure 10.43** is calculated as follows:

$$DIA = 0.06 + 7(0.013) = 0.151$$

10.14 Other Threaded Fasteners

Only the more standard types of nuts and bolts are covered in this chapter. **Figure 10.44** illustrates a few of the many other types of threaded fasteners that have their own special applications. Three types of wing screws that are turned by hand are available in incremental lengths of 1/8 inch (**Figure 10.45**). **Figure 10.46** shows two types of thumb screws, which serve the same purpose as wing screws,

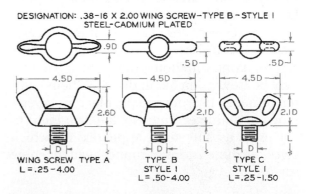

10.45 These wing screw proportions are for screw diameters of about 5/16 inch. The same proportions may be used to draw wing screws of any diameter. Type A screws are available in diameters of 4, 6, 8, 10, 12, 0.25″, 0.313″, 0.375″, 0.438″, 0.50″, and 0.625″. Type B screws are available in diameters of no. 10 to 0.625″. Type C screws are available in diameters of no. 6 to 0.375″.

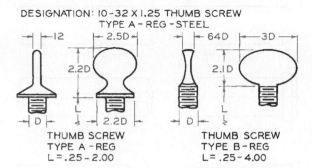

DESIGNATION: 10-32 X 1.25 THUMB SCREW
TYPE A - REG - STEEL

THUMB SCREW
TYPE A - REG
L = .25 - 2.00

THUMB SCREW
TYPE B - REG
L = .25 - 4.00

10.46 These thumb screw proportions are for screw diameters of about 1/4 inch. The same proportions may be used to draw thumb screws of any diameter. Type A screws are available in diameters of 6, 8, 10, 12, 0.25", 0.313", and 0.375". Type B thumb screws are available in diameters of no. 6 to 0.50".

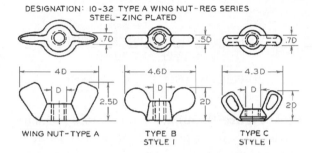

DESIGNATION: 10-32 TYPE A WING NUT - REG SERIES
STEEL - ZINC PLATED

WING NUT - TYPE A

TYPE B
STYLE I

TYPE C
STYLE I

10.47 These wing nut proportions are for screw diameters of 3/8 inch. The same proportions may be used to draw thumb screws of any size. Type A wing nuts are available in screw diameters of 3, 4, 5, 6, 8, 10, 12, 0.25", 0.313", 0.375", 0.438", 0.50", 0.583", 0.625", and 0.75". Type B nuts are available in sizes from no. 5 to 0.75". Type C nuts are available in sizes from no. 4 to 0.50".

and **Figure 10.47** shows wing nuts that can be screwed together by fingertip without wrenches or screwdrivers.

10.15 Tapping a Hole

An internal thread is made by drilling a hole with a tap drill with a 120° point (**Figure 10.48**). The depth of the drilled hole is measured to the shoulder of the conical point, not to the point. The diameter of the drilled hole is approximately equal to the root diameter, calculated as the major diameter of the screw

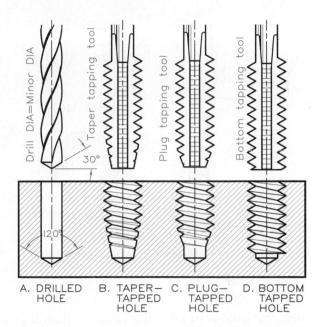

A. DRILLED HOLE B. TAPER—TAPPED HOLE C. PLUG—TAPPED HOLE D. BOTTOM TAPPED HOLE

10.48 Three types of tapping tools are used to thread internal drilled holes: taper tap, plug tap, and bottom tap.

thread minus its pitch (Appendixes 2–4). The hole is **tapped,** or threaded, with a tool called a **tap** of one of the types shown.

The **taper, plug,** and **bottoming** hand taps have identical measurements except for the chamfered portion of their ends. The taper tap has a long chamfer (8 to 10 threads), the plug tap has a shorter chamfer (3 to 5 threads), and the bottoming tap has the shortest chamfer (1 to 1-1/2 threads).

When tapping is to be done by hand in open or "through" holes, the taper tap should be used for coarse threads and in harder metals because it ensures straighter alignment and starting. The plug tap may be used in soft metals and for fine-pitch threads. When a hole is tapped to its bottom, all three taps—taper, plug, and bottoming—are used in that sequence on the same internal threads.

Notes are added to specify the depth of a drilled hole and the depth of the threads within it. For example, a note reading 7/8 DIA–3 DEEP X 1/8 UNC–2A X 2 DEEP means

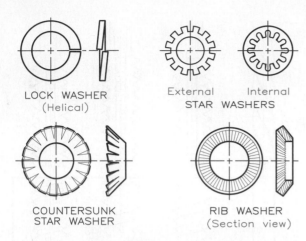

LOCK WASHER
(Helical)

External   Internal
STAR WASHERS

COUNTERSUNK
STAR WASHER

RIB WASHER
(Section view)

10.49 Lock washers are used to keep threaded parts from vibrating apart. See Appendix 24.

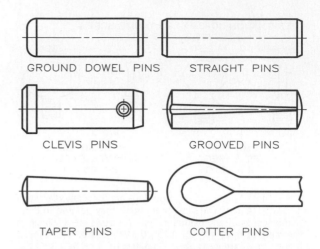

GROUND DOWEL PINS   STRAIGHT PINS

CLEVIS PINS   GROOVED PINS

TAPER PINS   COTTER PINS

10.50 Pins are used to hold parts together in assembly. See Appendixes 18 and 19.

that the hole is to be drilled deeper than it is threaded and that the last usable thread will be 2 inches deep in the hole.

10.16 Washers and Pins

Various types of washers are used with nuts and bolts to improve their assembly and increase their fastening strength.

Plain washers are noted on a drawing as

.938 X 1.750 X 0.134 TYPE A PLAIN WASHER

where the numbers (left to right) represent the washer's inside diameter, outside diameter, and thickness (Appendixes 22 and 23).

Lock washers reduce the likelihood that threaded parts will loosen because of vibration and movement. **Figure 10.49** shows several common types of lock washers. Appendix 24 contains a table of dimensions for regular and extra-heavy-duty helical-spring lock washers. Designate them with a note in the form:

HELICAL-SPRING LOCK WASHER
1/4 REGULAR–PHOSPHOR BRONZE,

where the 1/4 is the washer's inside diameter. You can designate tooth lock washers with a note in one of two forms:

INTERNAL-TOOTH LOCK WASHER
1/4–TYPE A–STEEL;

EXTERNAL-TOOTH LOCK WASHER
.562–TYPE B–STEEL.

Pins (Figure 10.50) are used to hold parts together in a fixed position. Appendix 19 gives dimensions for straight pins. The **cotter pin** is another locking device that you will remember held the wheels on your toy wagon. Appendix 18 contains a table of dimensions for cotter pins.

10.17 Pipe Threads and Fittings

Pipe threads are used for connecting pipes, tubing, and various fittings including lubrication fittings. The most commonly used pipe thread is tapered at a ratio of 1:16 on its diameter, but straight pipe threads also are available **(Figure 10.51)**. Tapered pipe threads will engage only for an effective length of

$$L = (0.80 \ D + 6.8) \ P$$

where D is the outside diameter of the threaded pipe and P is the pitch of the thread.

The pipe threads shown in **Figure 10.51** have a taper exaggerated to 1:16 on radius (instead of on diameter) to emphasize it.

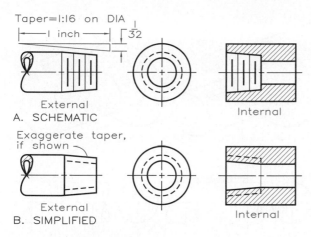

A. SCHEMATIC

B. SIMPLIFIED

Exaggerate taper, if shown

External   Internal

10.51 Pipe threads are shown with schematic and simplified symbols here. See Appendix 6.

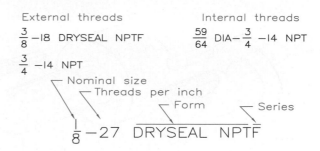

10.52 These are typical pipe thread notes.

Thread size	$\frac{1}{8}$ 3mm		$\frac{1}{4}$ 6mm		$\frac{3}{8}$ 10mm	
Overall length	L=in.	mm	L=in.	mm	L=in.	mm
Straight	.625	16	1.000	25	1.200	30
90° Elbow	.800	20	1.250	32	1.400	36
45° Angle	1.000	25	1.500	38	1.600	41

GREASE FITTINGS
Threads may be NPT or UN form

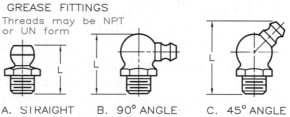

A. STRAIGHT   B. 90° ANGLE   C. 45° ANGLE

10.53 Three standard types of grease fittings used to lublicate moving parts with a grease gun are shown here.

Drawing them with no taper obviously is easier. You may use either schematic or simplified symbols to show the threaded features.

Use the following ANSI abbreviations in pipe thread notes. All begin with NP (for National Pipe thread).

NPT: national pipe taper
NPTF: national pipe thread (dryseal, for pressure-tight joints)
NPS: straight pipe thread
NPSC: straight pipe thread in couplings
NPSI: national pipe straight internal thread
NPSF: straight pipe thread (dryseal)
NPSM: straight pipe thread for mechanical joints
NPSL: straight pipe thread for lock nuts and lock nut pipe threads
NPSH: straight pipe thread for hose couplings and nipples
NPTR: taper pipe thread for railing fittings

To specify a pipe thread in note form, give the nominal pipe diameter (the common-fraction size of its internal diameter), the number of threads per inch, and the thread-type symbol:

1-1/4–11–1/2 NPT or 3–8 NPTR

Appendix 6 gives a table of dimensions for pipe threads. **Figure 10.52** shows how to present specifications for external and internal threads in note form. Dryseal threads, either straight or tapered, provide a pressure-tight joint without the use of a lubricant or sealer.

Grease Fittings
Grease fittings (Figure 10.53) allow the application of a lubricate to moving parts. Threads of grease fittings are available as tapered and straight pipe threads. The ends where grease is inserted with a grease gun are available straight or at 90° and 45° angles. A one-way valve, formed by a ball and spring, permits grease to enter the fitting (forced through by a grease gun) but prevents it from escaping.

10.17 PIPE THREADS AND FITTINGS • 167

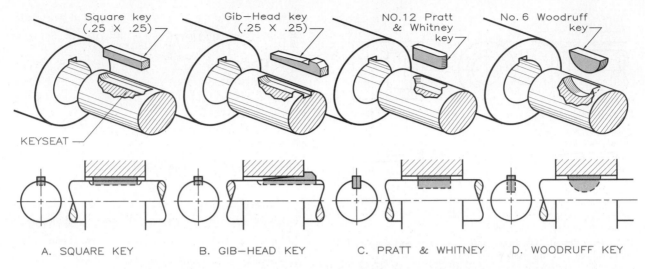

10.54 Standard types of keys used to hold parts on a shaft. See Appendixes 20–21.

10.18 Keys

Keys are used to attach pulleys, gears, or crank handles to shafts to hold them securely in place while moving and transmitting power. The four types of keys shown in **Figure 10.54** are the most commonly used ones. Appendixes 20 and 21 contain tables of dimensions for keyways, keys, and keyseats.

10.19 Rivets

Rivets are fasteners that permanently join thin, overlapping materials **(Figure 10.55).** The rivet is inserted in a hole slightly larger than the diameter of the rivet, and the application of pressure to the projecting end forms the headless end into shape. Forming may be done with either hot or cold rivets depending on the application.

    Figure 10.56 shows typical shapes and proportions of small rivets that vary in diameter from 1/16 to 1-3/4 inches. Rivets are used extensively in pressure-vessel fabrication, heavy construction (such as bridges and buildings), and sheet-metal construction.

    Figure 10.57 shows some of the standard ANSI symbols for representing rivets. Rivets that are driven in the shop are called **shop**

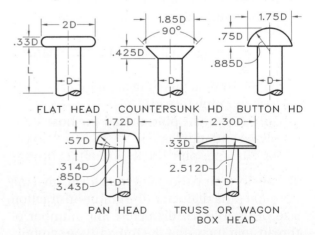

10.55 Rivets are used to permanently fasten structural elements together. *(Courtesy of RB&W Manufacturing Company.)*

10.56 The proportions of small rivets with shanks up to 1/2 inch are shown here.

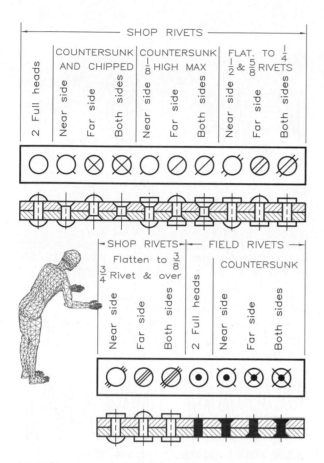

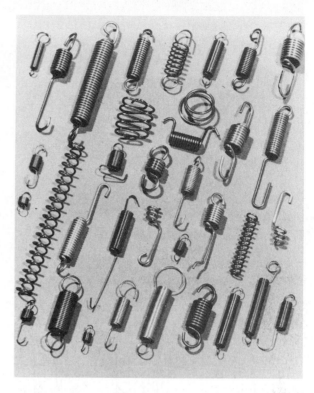

10.58 Springs are available for numerous special applications.

10.57 Rivets are represented by these symbols in a drawing.

rivets, and those assembled at the job site are called **field rivets.**

10.20 Springs

Springs are devices that absorb energy and react with an equal force (**Figure 10.58**). Most springs are **helical,** as are bed springs, but they can also be **flat** (leaf), as in an automobile chassis. Some of the more common types of springs are **compression, torsion, extension, flat,** and **constant force** springs. **Figures 10.59A–C** shows single-line conventional representations of the first three types. **Figures 10.59D–F** represent the types of ends used on compression springs.

Plain ends of springs simply end with no special modification of the coil. **Ground plain**

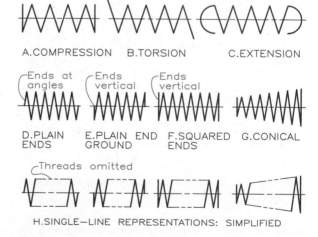

A. COMPRESSION   B. TORSION   C. EXTENSION

D. PLAIN ENDS   E. PLAIN END GROUND   F. SQUARED ENDS   G. CONICAL

H. SINGLE–LINE REPRESENTATIONS: SIMPLIFIED

10.59 Single-line spring drawings.
A–C These are single-line representations of various types of springs.
D–G These single-line representations of springs show various types of ends.
H These are simplified single-line representations of the springs depicted in D–G.

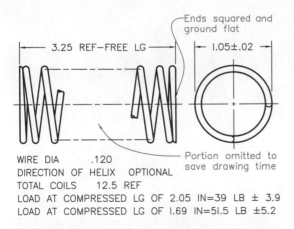

WIRE DIA   .120
DIRECTION OF HELIX   OPTIONAL
TOTAL COILS   12.5 REF
LOAD AT COMPRESSED LG OF 2.05 IN=39 LB ± 3.9
LOAD AT COMPRESSED LG OF 1.69 IN=51.5 LB ±5.2

10.60 This conventional double-line drawing is of a compression spring and includes its specifications.

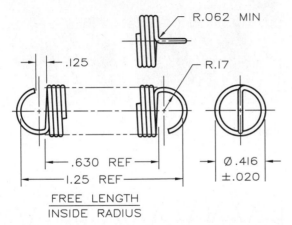

FREE LENGTH
INSIDE RADIUS

WIRE DIA	0.42
DIRECTION OF HELIX OPTIONAL	
TOTAL COILS	14 REF
RELATIVE POSITION OF ENDS	180° ±20°
EXTENDED LENGTH INSIDE ENDS WITHOUT PERMANENT SET	2.45 IN (MAX)
INITIAL TENSION	1.00 LB ±.10 LB
LOAD	4.0 LB ±.4 LB AT 1.56 IN
EXTENDED LG INSIDE ENDS	
LOAD	6.30 LB ±.63 LB AT 1.95

10.61 This conventional double-line drawing shows an extension spring and its specifications.

ends are coils that have been machined by grinding to flatten the ends perpendicular to their axes. **Squared ends** are inactive coils that have been closed to form a circular flat coil at the spring's end, which may also be ground.

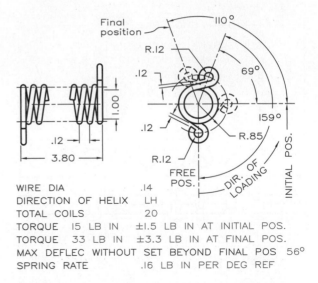

WIRE DIA   .14
DIRECTION OF HELIX   LH
TOTAL COILS   20
TORQUE   15 LB IN   ±1.5 LB IN AT INITIAL POS.
TORQUE   33 LB IN   ±3.3 LB IN AT FINAL POS.
MAX DEFLEC WITHOUT SET BEYOND FINAL POS 56°
SPRING RATE   .16 LB IN PER DEG REF

10.62 This conventional double-line drawing is of a helical torsion spring and includes its specifications.

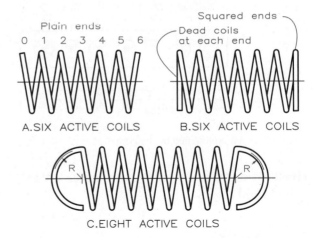

A.SIX ACTIVE COILS   B.SIX ACTIVE COILS

C.EIGHT ACTIVE COILS

10.63 Double-line drawings of springs.
A This double-line drawing shows a spring with six active coils.
B This double-line drawing shows a spring with six coils and "dead" coil (inactive coil) at each end.
C This double-line drawing shows an extension spring with eight active coils.

In **Figure 10.59G** a conical helical spring is shown in its simplified form. **Figure 10.59H** shows schematic single-line representations of the same types of springs depicted in **Figures 10.59D–G** with phantom outlines instead of all the coils. This is conventional method of drawing springs that saves time and effort.

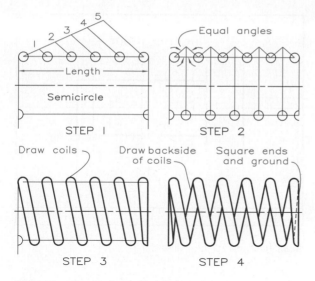

10.64 Drawing a spring in detail.
Step 1 Lay out the diameter and length of the spring and locate the five coils by the diagonal-line technique.
Step 2 Locate the coils on the lower side along the bisectors of the spaces between the coils on the upper side.
Step 3 Connect the coils on each side. This is a right-hand coil; a left-hand spring would slope in the opposite direction.
Step 4 Construct the back side of the spring and the end coils to complete the drawing. The spring has a square end that is to be ground.

Working drawing specifications of a compression spring drawn as a double-line representation are shown in **Figure 10.60.** Two coils are drawn at each end of the spring and phantom lines are drawn between them in order to save drawing time. A dimension is given with the diameter and free length of the spring on the drawing. The remaining specifications are given in a table placed near the drawing.

A working drawing of an extension spring **(Figure 10.61)** is similar to that of a compression spring. An extension spring is designed to resist stretching, whereas a **compression spring** is designed to resist squeezing. In a drawing of a helical **torsion spring,** which resists and reacts to a **twist**ing motion **(Figure 10.62),** angular dimensions specify the initial and final positions of the spring as torsion is applied. Again, dimension the drawing and add specifications to describe their details.

10.21 Drawing Springs

Springs may be represented with single-line drawings **(Figure 10.59)** or as more realistic double-line drawings **(Figure 10.63).** Draw each type shown by first laying out the diameters of the coils and lengths of the springs and then dividing the lengths into the number of active coils **(Figure 10.63A).** In **Figure 10.63B,** both end coils are "dead" (inactive) coils, and only six coils are active. **Figure 10.63C** depicts an extension spring with eight active coils.

The steps of drawing a double-line detailed representation of a compression spring are shown in **Figure 10.64.** Springs can be drawn right-hand or left-hand, but like threads, most are drawn as right-hand coils. The ends of the spring in this case are to be squared by grinding the ends to make them flat and perpendicular to the axis of the spring.

Problems

Solve and draw these problems on size A sheets. Each grid space equals 0.20 inch, or 5 mm.

1. (Figure 10.65) Draw detailed representations of Acme threads with 2 inch major diameters. Show both external and internal threads as views and sections. Give a thread note by referring to Appendix 5.

2. Repeat Problem 1 using detailed representations of square threads. Appendix 5.

3. Repeat Problem 1 using detailed representations of UN threads with a class 2 fit.

4. (Figure 10.66) Draw detailed representations of the internally threaded holes in a section that complies with the given notes.

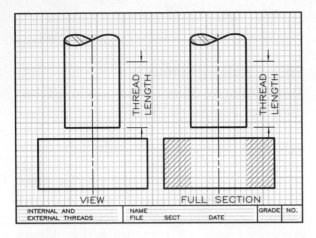

10.65 Problems 1–3.

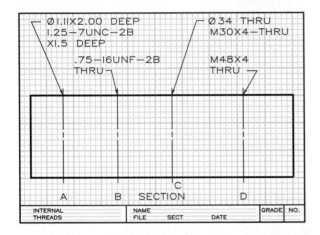

10.66 Problems 4–6.

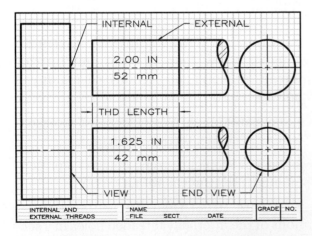

10.67 Problems 7–9.

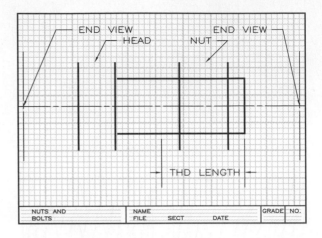

10.68 Problems 10–12. Full size.

5. Repeat Problem 4 with schematic symbols.

6. Repeat Problem 4 with simplified symbols.

7. (**Figure 10.67**) Complete the partial views using detailed thread symbols, draw external, internal, and end views of the full-size threaded parts. Note the UNC threads with a class 2 fit.

8. Repeat Problem 7 with schematic symbols.

9. Repeat Problem 7 with simplified symbols.

10. (**Figure 10.68**) Complete the drawing of the finished hexagon-head bolt and a heavy hexagon nut. Draw the bolt head and nut across corners using detailed thread symbols. Provide metric or English thread notes.

11. Repeat Problem 10 but draw the nut and bolt with unfinished square heads. Use schematic thread symbols.

12. Repeat Problem 10 but draw the bolt with a regular finished hexagon head across flats, using simplified thread symbols. Draw the nut across flats and provide thread notes for both.

13. (**Figure 10.69**) Draw the screws with detailed symbols in accordance with the notes.

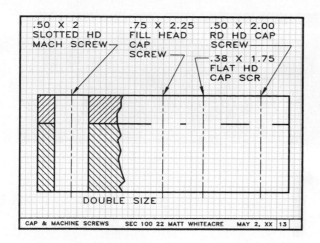

10.69 Problems 13–15.

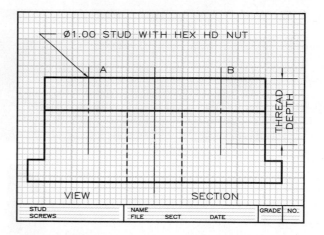

10.70 Problem 16.

Complete the broken-out section, and show all crosshatching. Provide thread notes.

14. Repeat Problem 13 with schematic symbols.

15. Repeat Problem 13 with simplified symbols.

16. (**Figure 10.70**) On axes A and B, construct hexagon-head cap screws (across flats) with UNC threads and a class 2 fit. The cap screws should not reach the bottoms of the threaded holes. Convert the view to a half section.

17. (**Figure 10.71**) On axes A and B, draw studs with hexagon-head nuts (across flats) that hold the two parts together. Note the studs to have a fine series and a class 2 fit. They should not reach the bottom of the part. Show the view as a half section.

18. (**Figure 10.72**) Draw a 2.00-in. (50-mm) diameter hexagon-head bolt (head across flats) using schematic symbols. Draw a plain washer and regular nut (across corners) at the right end. Size the recess in the part at the left to hold the bolt head to prevent turning. Note the bolt to have UNC thread and a class 2 fit.

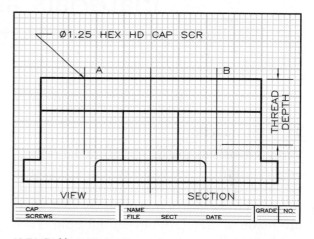

10.71 Problem 17.

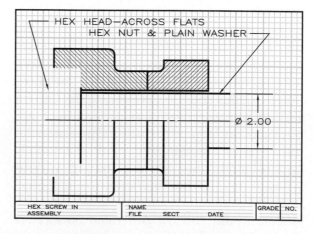

10.72 Problem 18.

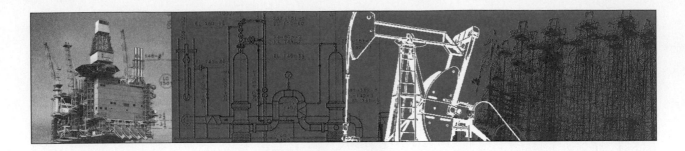

11

Materials and Processes

11.1 Introduction

Various materials and manufacturing processes are commonly used to make parts similar to those discussed in this textbook. A large proportion of parts designed by engineers are made of metal, but other materials such as plastics, fibers, and ceramics are available to the designer in increasingly useful applications.

Metallurgy, the study of metals, is a field that is constantly changing as new processes and alloys are developed **(Figure 11.1).** These developments affect the designer's specification of metals and their proper application for various purposes and applications from appliances to large structures. Three associations have standardized and continually update guidelines for designating various types of metals: the American Iron and Steel Institute (AISI), the Society of Automotive Engineers (SAE), and the American Society for Testing Materials (ASTM).

11.1 These examples of forged products that offer great strength, lightness of weight, and attractiveness are typical of the parts required by industry. *(Courtesy of the Waltec Forgings Inc.)*

11.2 Commonly Used Metals

Iron*

Metals that contain iron, even in small quantities, are called **ferrous** metals. Three common

DESIGNATION OF GRAY IRON (450 LBS/CF)

ATSM Grade (1000 psi)	SAE Grade	Typical Uses
ASTM 25 CI	G 2500 CI	Small engine blocks, pump bodies, clutch plates, transmission cases
ASTM 30 CI	G 3000 CI	Auto engine blocks, heavy castings, flywheels
ASTM 35 CI	G 3500 CI	Diesel engine blocks, tractor transmission cases, heavy and high—strength parts
ASTM 40 CI	G 4000 CI	Diesel cylinders, pistons, camshafts

11.2 These are the numbering designations of gray iron and their typical uses.

types of iron are **gray iron, white iron,** and **ductile iron.**

Gray iron contains flakes of graphite that result in low strength and low ductility, which makes it easy to machine. Gray iron resists vibration better than other types of iron. **Figure 11.2** shows designations of and typical applications for gray iron.

White iron contains carbide particles that are extremely hard and brittle, enabling it to withstand wear and abrasion. Because the composition of white iron differs from one supplier to another, there are no designated grades of white iron. It is used for parts on grinding and crushing machines, digging teeth on earthmovers and mining equipment, and wear plates on reciprocating machinery used in textile mills.

Ductile iron (also called nodular or spheroidized iron) contains tiny spheres of graphite, making it stronger and tougher than most types of gray iron and more expensive to produce. Three sets of numbers **(Figure 11.3)** describe the most important features of ductile iron. **Figure 11.4** shows the designations of and typical applications for the commonly used alloys of ductile iron.

*This section on iron was developed by Dr. Tom Pollock, a metallurgist at Texas A&M University

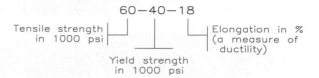

11.3 Ductile iron is specified with numerical notes in this format.

DESIGNATIONS OF DUCTILE IRON (490 LBS/CF)

Grade	Typical Uses
60—40—18 CI	Valves, steam fittings, chemical plant equipment, pump bodies
65—45—12 CI	Machine components that are shock loaded, disc brake calipers
80—55—6 CI	Auto crankshafts, gears, rollers
100—70—3 CI	High—strength gears and machine parts
120—90—2 CI	Very high—strength gears, rollers, and slides

11.4 These are the numbering designations of ductile iron and their typical uses.

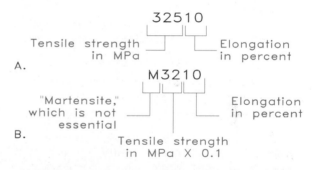

11.5 These notes illustrate the numbering designations for malleable iron.

Malleable iron is made from white iron by a heat-treatment process that converts carbides into carbon nodules (similar to ductile iron). The numbering system for designating grades of malleable iron is shown in **Figure 11.5.** Some of the commonly used grades of malleable iron and their typical applications are shown in **Figure 11.6.**

Cast iron is iron that is melted and poured into a mold to form it by casting, a commonly used process for producing machine parts. Although cheaper and easier to machine than steel, iron does not have steel's ability to withstand shock and force.

DESIGNATIONS OF MALLEABLE IRON (490 LBS/CF)

ASTM Grade	Typical Uses
35018 CI	Marine and railroad valves and fittings, "black—iron" pipe fittings (similar to 60—40—18 ductile CI)
45006 CI	Machine parts (similar to 80—55—6 ductile CI)
M3210 CI	Low—stress components, brackets
M4504 CI	Crankshafts, hubs
M7002 CI	High—strength parts, connecting rods, universal joints
M8501 CI	Wear—resistant gears and sliding parts

11.6 These are the numbering designations of malleable iron and their typical uses.

DESIGNATIONS OF STEEL (490 LBS/CF)

Type of steel	Number	Applications
Carbon steels		
Plain carbon	10XX	Tubing, wire, nails
Resulphurized	11XX	Nuts, bolts, screws
Manganese steel	13XX	Gears, shafts
Nickel steel	23XX	Keys, levers, bolts
	25XX	Carburized parts
	31XX	Axles, gears, pins
	32XX	Forgings
	33XX	Axles, gears
Molybdenum	40XX	Gears, springs
Chromium—moly.	41XX	Shafts, tubing
Nickel—chromium	43XX	Gears, pinions
Nickel—moly.	46XX	Cams, shafts
	48XX	Roller bearings, pins
Chromium steel	51XX	Springs, gears
	52XX	Ball bearings
Chrom. vanadium	61XX	Springs, forgings
Silicon manganese	92XX	Leaf springs

11.7 These are the numbering designations of steel and their applications.

Steel

Steel is an alloy of iron and carbon, which often contains other constituents such as manganese, chromium, or nickel. Carbon (usually between 0.20% and 1.50%) is the ingredient having the greatest effect on the grade of steel. The three major types of steel

ALUMINUM DESIGNATIONS (169 LBS/CF)

Composition	Alloy Number	Application
Aluminum (99% pure)	1XXX	Tubing, tank cars
Aluminum alloys		
Copper	2XXX	Aircraft parts, screws, rivets
Manganese	3XXX	Tanks, siding, gutters
Silicon	4XXX	Forging, wire
Magnesium	5XXX	Tubes, welded vessels
Magnesium and silicon	6XXX	Auto body, pipes
Zinc	7XXX	Aircraft structures
Other elements	8XXX	

11.8 These are the numbering designations of aluminum and aluminum alloys and their applications.

are **plain carbon steels, free-cutting carbon steels,** and **alloy steels. Figure 11.7** gives the types of steels and their SAE designations by four-digit numbers. The first digit indicates the type of steel: 1 is carbon steel, 2 is nickel steel, and so on. The second digit gives content (as a percentage) of the material represented by the first digit. The last two or three digits give the percentage of carbon in the alloy: 100 equals 1%, and 50 equals 0.50%.

Steel weighs about 490 pounds per cubic foot. Some frequently used SAE steels are 1010, 1015, 1020, 1030, 1040, 1070, 1080, 1111, 1118, 1145, 1320, 2330, 2345, 2515, 3130, 3135, 3240, 3310, 4023, 4042, 4063, 4140, and 4320.

Copper

One of the first metals discovered, copper is easily formed and bent without breaking. Because it is highly resistant to corrosion and is highly conductive, it is used for pipes, tubing, and electrical wiring. It is an excellent roofing and screening material because it withstands the weather well. Copper weighs about 555 pounds per cubic foot.

ALUMINUM CASTINGS AND INGOT DESIGNATIONS

Composition	Alloy Number
Aluminum (99% pure)	1XX.X
Aluminum alloys	
Copper	2XX.X
Silicon with copper and/or magnesium	3XX.X
Silicon	4XX.X
Magnesium	5XX.X
Magnesium and silicon	6XX.X
Zinc	7XX.X
Tin	8XX.X
Other elements	9XX.X

11.9 These are the numbering designations of cast aluminum, ingots, and aluminum alloys.

Copper has several alloys, including brasses, tin bronzes, nickel silvers, and copper nickels. Brass (about 530 pounds per cubic foot) is an alloy of copper and zinc, and bronze (about 548 pounds per cubic foot) is an alloy of copper and tin. Copper and copper alloys are easily finished by buffing or plating; joined by soldering, brazing, or welding; and machined.

Wrought copper has properties that permit it to be formed by hammering. A few of the numbered designations of wrought copper are C11000, C11100, C11300, C11400, C11500, C11600, C10200, C12000, and C12200.

Aluminum

Aluminum is a corrosion-resistant, lightweight metal (approximately 169 pounds per cubic foot) that has numerous applications. Most materials called aluminum actually are aluminum alloys, which are stronger than pure aluminum.

The types of wrought aluminum alloys are designated by four digits (**Figure 11.8**). The first digit (2 through 8) indicates the alloying element that is combined with aluminum. The second digit indicates modifications of the original alloy or impurity limits. The last two digits identify other alloying materials or indicate the aluminum's purity.

Figure 11.9 shows a four-digit numbering system used to designate types of cast aluminum and alloys. The first digit indicates the alloy group, and the next two digits identify the aluminum alloy or aluminum purity. The number to the right of the decimal point represents the aluminum form: XX.0 indicates castings, XX.1 indicates ingots with a specified chemical composition, and XX.2 indicates ingots with a specified chemical composition other than the XX.1 ingot. **Ingots** are blocks of cast metal to be remelted, and **billets** are castings of aluminum to be formed by forging.

Magnesium

Magnesium is a light metal (109 pounds per cubic foot) available in an inexhaustible supply because it is extracted from seawater and natural brines. Magnesium is an excellent material for aircraft parts, clutch housings, crankcases for air-cooled engines, and applications where lightness is desirable.

Magnesium is used for die and sand castings, extruded tubing, sheet metal, and forging. Magnesium and its alloys may be joined by bolting, riveting, or welding. Some numbered designations of magnesium alloys are M10100, M11630, M11810, M11910, M11912, M12390, M13320, M16410, and M16620.

11.3 Properties of Metals

All materials have properties that designers must utilize to the best advantage. The following terms describe these properties.

Ductility: a softness in some materials, such as copper and aluminum, which permits them to be formed by stretching (drawing) or hammering without breaking.

Brittleness: a characteristic that will not allow metals such as cast irons and hardened steels to stretch without breaking.

Malleability: the ability of a metal to be rolled or hammered without breaking.

Hardness: the ability of a metal to resist being dented when it receives a blow.

Toughness: the property of being resistant to cracking and breaking while remaining malleable.

Elasticity: the ability of a metal to return to its original shape after being bent or stretched.

Modifying Properties by Heat Treatment

The properties of metals can be changed by various types of heat treating. Although heat affects all metals, steels are affected to a greater extent than others.

Hardening: heating steel to a prescribed temperature and quenching it in oil or water.

Quenching: rapidly cooling heated metal by immersing it in liquids, gases, or solids (such as sand, limestone, or asbestos).

Tempering: reheating previously hardened steel and then cooling it, usually by air, to increase its toughness.

Annealing: heating and cooling metals to soften them, release their internal stresses, and make them easier to machine.

Normalizing: heating metals and letting them cool in air to relieve their internal stresses.

Case hardening: hardening a thin outside layer of a metal by placing the metal in contact with carbon or nitrogen compounds that it absorbs as it is heated; afterward, the metal is quenched.

Flame hardening: hardening by heating a metal to within a prescribed temperature range with a flame and then quenching the metal.

11.4 Forming Metal Shapes

Casting

One of the two major methods of forming shapes is casting, which involves preparing a mold in the shape of the part desired, pouring molten metal into it, and cooling the metal to form the part. The types of casting,

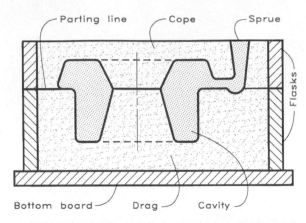

11.10 A two-section sand mold is used for casting a metal part.

11.11 This pattern is held in the bottom half (the drag) of a sand mold to form a mold for a casting.

which differ in the way the molds are made, are **sand casting, permanent-mold casting, die casting,** and **investment casting.**

Sand-Casting   In the first step of sand casting, a wood or metal form or pattern is made in the shape of the part to be cast. The pattern is placed in a metal box called a flask and molding sand is packed around the pattern. When the pattern is withdrawn from the sand, it leaves a void forming the mold. Molten metal is poured into the mold through sprues or gates. After cooling, the casting is removed and cleaned (**Figure 11.10**).

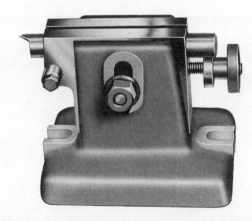

11.12 The tailstock casting for a lathe has raised bosses and contact surfaces that were finished to improve the effectiveness of nuts and bolts. Fillets and rounds were added to the inside and outside corners. *(L. W. Chuck Company.)*

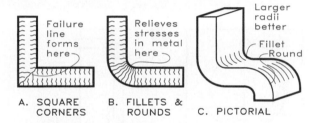

11.13 Fillets and Rounds:
A Square corners cause a failure line to form, causing a weakness at this point.
B Fillets and rounds make the corners of a casting stronger and more attractive.
C The larger the radii of fillets and rounds, the stronger the casting will be.

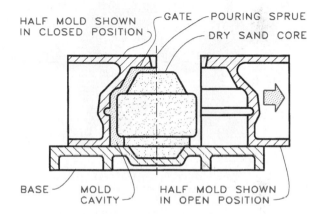

11.14 Permanent molds are made of metal for repetitive usage. Here, a sand core made from another mold is placed in the permanent mold to create a void within the casting.

Cores formed from sand may be placed in a mold to create holes or hollows within a casting. After the casting has been formed, the cores are broken apart and removed, leaving behind the desired void within the casting.

Because the patterns are placed in and removed from the sand before the metal is poured, the sides of the patterns must be tapered, called **draft,** for ease of withdrawal from the sand. The angle of draft depends on the depth of the pattern in the sand and varies from 2° to 8° in most applications. **Figure 11.11** shows a pattern held in the sand by a lower flask. Patterns are made oversize to compensate for shrinkage that occurs when the casting cools.

Because sand castings have rough surfaces, features that come into contact with other parts must be machined by drilling, grinding, finishing, or shaping. The tailstock base of a lathe shown in **Figure 11.12** illustrates raised bosses that have been finished. The casting must be made larger than finished size where metal is to be removed by machining.

Fillets and **rounds** are used at the inside and outside corners of castings to increase their strength by relieving the stresses in the cast metal **(Figure 11.13).** Fillets and rounds also are used because forming square corners by the sand-casting process is difficult and because rounded edges make the finished product more attractive **(Figure 11.12).**

Permanent-Mold Casting Permanent molds are made for the mass production of parts. They are generally made of cast iron and coated to prevent fusing with the molten metal poured into them **(Figure 11.14).**

Die Casting Die castings are used for the mass production of parts made of aluminum, magnesium, zinc alloys, copper, and other materials. Die castings are made by forcing

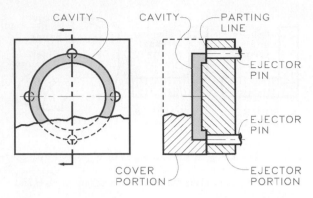

11.15 This die is used for casting a simple part. The metal is forced into the die to form the casting.

11.16 An investment casting (lost-wax process) is used to produce complex metal objects and art pieces.

molten metal into dies (or molds) under pressure. They are inexpensive, meet close tolerances, and have good surface qualities. The same general principles of sand castings—using fillets and rounds, allowing for shrinkage, and specifying draft angles—apply to die castings (**Figure 11.15**).

Investment Casting   Investment casting is used to produce complicated parts or artistic sculptures that would be difficult to form by other methods (**Figure 11.16**). A new pattern must be used for each investment casting, so a mold or die is made for casting a wax master pattern. The wax pattern, identical to the casting, is placed inside a container and plaster or sand is poured (invested) around it. Once the investment has cured, the wax pattern is melted, leaving a hollow cavity to serve as the mold for

11.17 This aircraft landing-gear component was formed by forging. *(Courtesy of Cameron Division, Cooper Cameron Corporation.)*

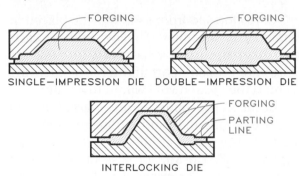

11.18 This drawing shows three types of forging dies.

the molten metal. After the casting has set, the plaster or sand is broken away from it.

Forgings

The second major method of forming shapes is forging, which is the process of shaping or forming heated metal by hammering or forcing it into a die. Drop forges and press forges are used to hammer metal billets into forging dies. Forgings have the high strength and resistance to loads and impacts required for applications such as aircraft landing gears (**Figure 11.17**).

Figure 11.18 shows three types of dies. A single-impression die gives an impression on one side of the parting line between the mating dies; a double-impression die gives an

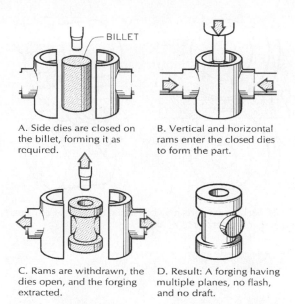

A. Side dies are closed on the billet, forming it as required.

B. Vertical and horizontal rams enter the closed dies to form the part.

C. Rams are withdrawn, the dies open, and the forging extracted.

D. Result: A forging having multiple planes, no flash, and no draft.

11.19 These are the steps involved in forging a part with external dies and an internal ram.

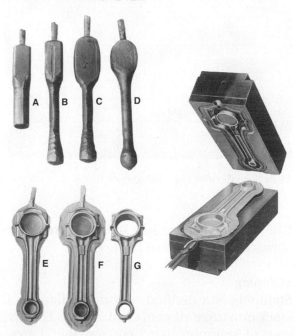

11.20 Steps A through G are required to forge a billet into a finished connecting rod. *(Courtesy of Forging Industry Association.)*

impression on both sides of the parting line; and the interlocking dies give an impression that may cross the parting line on either side. **Figure 11.19** shows how an object is forged

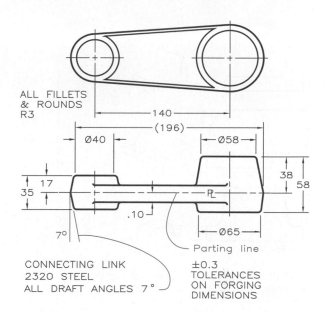

ALL FILLETS & ROUNDS R3

CONNECTING LINK
2320 STEEL
ALL DRAFT ANGLES 7°

±0.3
TOLERANCES ON FORGING DIMENSIONS

Parting line

11.21 This working drawing for a forging shows draft angles and the parting line (PL) where the dies come together.

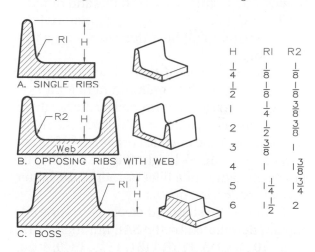

A. SINGLE RIBS

B. OPPOSING RIBS WITH WEB

C. BOSS

H	RI	R2
$\frac{1}{4}$	$\frac{1}{8}$	$\frac{1}{8}$
$\frac{1}{2}$	$\frac{1}{8}$	$\frac{1}{8}$
1	$\frac{1}{4}$	$\frac{3}{8}$
2	$\frac{1}{2}$	$\frac{3}{8}$
3	$\frac{3}{8}$	1
4	1	$1\frac{3}{8}$
5	$1\frac{1}{4}$	$1\frac{3}{4}$
6	$1\frac{1}{2}$	2

11.22 These guidelines are for determining the minimum radii for fillets (inside corners) on forged parts.

with horizontal dies and a vertical ram to hollow the object.

Figure 11.20 illustrates the sequence of forging a part from a billet by hammering it into different dies. It is then machined to its proper size within specified tolerances.

Figure 11.21 shows a working drawing for making a forged part. When preparing forging drawings, you must consider (1) draft

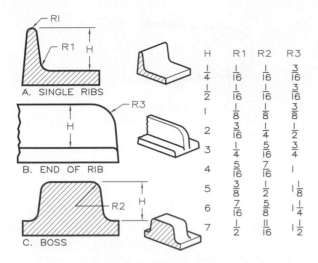

11.23 These guidelines are for determining the minimum radii of rounds (outside corners) on forged parts.

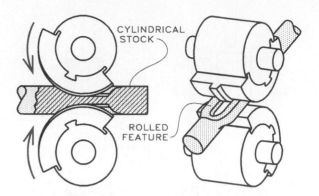

11.24 Features on parts may be formed by rolling. Here a part is being rolled parallel to its axes.

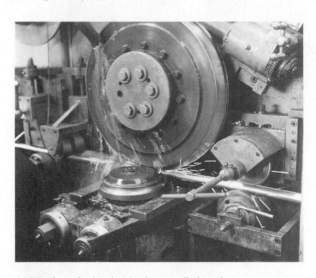

11.25 This cylindrical rod is being rolled to shape.

angles and parting lines, (2) fillets and rounds, (3) forging tolerances, (4) extra material for machining, and (5) heat treatment of the finished forging.

Draft, the angle of taper, is crucial to the forging process. The minimum radii for inside corners (fillets) are determined by the height of the feature **(Figure 11.22).** Similarly, the minimum radii for the outside corners (rounds) are related to a feature's height **(Figure 11.23).** The larger the radius of a fillet or round, the better it is for the forging process.

Some of the standard steels used for forging are designated by the SAE numbers 1015, 1020, 1025, 1045, 1137, 1151, 1335, 1340, 4620, 5120, and 5140. Iron, copper, and aluminum also can be forged.

Rolling Rolling is a type of forging in which the stock is rolled between two or more rollers to shape it. Rolling can be done at right angles or parallel to the axis of the part **(Figure 11.24).** If a high degree of shaping is required, the stock usually is heated before rolling. If the forming requires only a slight change in shape, rolling can be done without heating the metal, which is called **cold rolling** (CR);

CRS means cold-rolled steel. **Figure 11.25** shows a cylindrical rod being rolled.

Stamping

Stamping is a method of forming flat metal stock into three-dimensional shapes. The first step of stamping is to cut out the shapes, called blanks, which are formed by bending and pressing them against forms. **Figure 11.26** shows three types of box-shaped parts formed by stamping, and **Figure 11.27** shows a design for a flange to be formed by stamping. Holes in stampings are made by punching, extruding, or piercing **(Figure 11.28).**

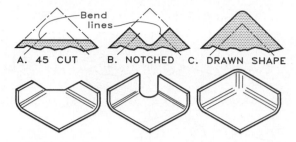

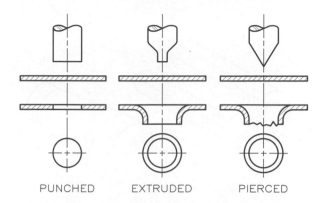

11.26 Box-shaped parts formed by stamping:
A A corner cut of 45° permits flanges to be folded with no further trimming.
B Notching has the same effect as the 45° cut and is often more attractive.
C A continuous corner flange requires that the blank be developed so that it can be drawn into shape.

11.28 These three methods are used to form holes in sheet metal are punching, extrusion, and piercing.

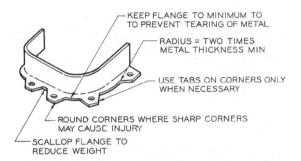

11.27 This drawing shows a sheet metal flange design with notes that explain design details.

11.5 Machining Operations

After metal parts have been formed, machining operations must be performed to complete them. The machines used most often are: **lathe, drill press, milling machine, shaper,** and **planer.** Some of these machines require manual operation; others are computer programmed to run at high speeds automatically, and require minimal or no operator attention.

Lathe

The **lathe** shapes cylindrical parts while rotating the work piece between its centers **(Figure 11.29).** The fundamental operations performed on the lathe are **turning, facing,**

11.29 This typical metal lathe holds and rotates the work piece between its centers for machining. (*Courtesy of the Clausing Industrial Inc.*)

drilling, boring, reaming, threading, and **undercutting (Figure 11.30).**

Turning forms a cylinder with a tool that advances against, and moves parallel to the cylinder being turned between the centers of the lathe **(Figure 11.31). Facing** forms flat surfaces perpendicular to the axis of rotation of the part being rotated.

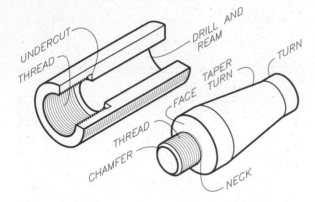

11.30 The basic operations that are performed on a lathe are shown here.

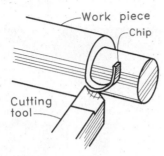

11.31 The most basic operation performed on the lathe is turning, whereby a continuous chip is removed by a cutting tool as the part rotates.

Drilling is performed by mounting a drill in the tail stock of the lathe and rotating the work while the bit is advanced into the part (**Figures 11.32A–C**). **Boring** makes large holes by enlarging smaller drilled holes with a tool mounted on a boring bar (**Figure 11.33A**). **Undercutting** is a groove cut inside a cylindrical hole with a tool mounted on a boring bar. The groove is cut as the tool advances from the center of the axis of revolution into the part (**Figure 11.33B**). **Reaming** removes only thousandths of an inch of material inside cylindrical and conical holes to enlarge them to their required tolerances (**Figure 11.34**).

Threading of internal holes can be done on the lathe as shown in **Figure 11.35**. The die used for cutting internal holes is called a **tap.** A threading die can be used to cut external threads on a shaft held in the chuck of a lathe as illustrated in **Figure 11.36**. Other external cuts made on a lathe are shown in **Figure 11.37**.

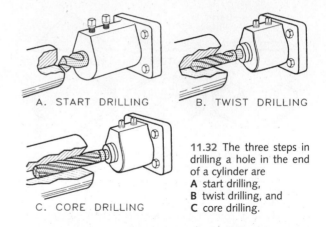

11.32 The three steps in drilling a hole in the end of a cylinder are
A start drilling,
B twist drilling, and
C core drilling.

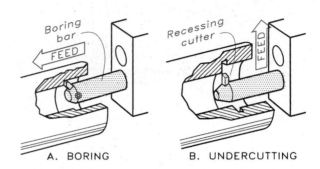

11.33 (A) This hole is being bored with a cutting tool attached to a boring bar of a lathe. (B) An undercut is being cut by the tool and boring bar.

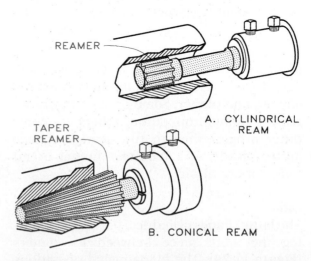

11.34 Fluted reamers can be used to finish inside (A) cylindrical and (B) conical holes within a few thousandths of an inch.

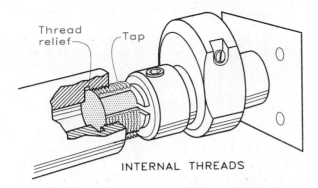

11.35 Internal threads can be cut on a lathe with a die called a **tap.** A recess, called a **thread relief,** was formed at the end of the threaded hole.

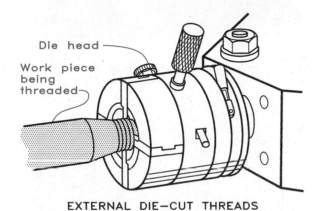

11.36 This work piece (a shaft) is being threaded by a die head.

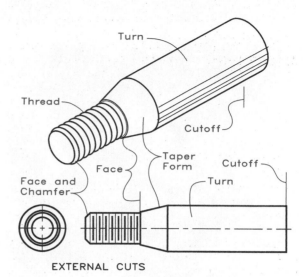

11.37 The various external cuts made on the lathe are shown here.

11.38 A turret lathe performs a sequence of operations by revolving the turret on which are mounted various tools. *(Courtesy of Clausing Industrial, Inc.)*

The **turret lathe** is a programmable lathe that can perform sequential operations, such as drilling a series of holes, boring them, and then reaming them. The turret is a multisided tool holder that sequentially rotates each tool into position for its particular operation **(Figure 11.38).**

Drill Press

The **drill press** is used to drill small- and medium-sized holes **(Figure 11.39).** The stock being drilled is held securely by fixtures or clamps. The drill press can be used for **counterdrilling, countersinking, counterboring, spotfacing,** and **threading (Figure 11.40).**

Multiple-tool drilling systems can be programmed to perform a series of drilling operations for mass production applications **(Figure 11.41).**

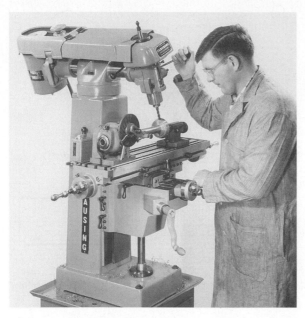

11.39 This small drill press is used to make holes in parts. *(Courtesy of Clausing Industrial Inc.)*

11.41 This heavy duty radial drill is of the type used for large capacity, high productivity output. *(Courtesy of Clausing Industrial, Inc.)*

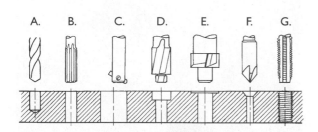

11.40 The basic operations performed on the drill press are (A) drilling, (B) reaming, (C) boring, (D) counterboring, (E) spotfacing, (F) countersinking, and (G) tapping (threading).

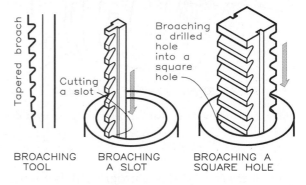

11.42 A broaching tool can be used to cut slots and holes with square corners on the interior and exterior of parts. Other shapes may be broached also.

Measuring Cylinders The diameters, not the radii, of cylindrical features of parts made on a drill press or a lathe are measured to determine their sizes. Internal and external micrometer calipers are used for this purpose to measure to within one ten-thousandth of an inch.

Broaching Machine
Cylindrical holes can be converted into square, rectangular, or hexagonal holes with a **broach** mounted on a special machine **(Figure 11.42).** A broach has a series of teeth graduated in size along its axis, beginning with teeth that are nearly the size of the hole to be broached and tapering to the final size of the hole. The broach is forced through the hole by pushing or pulling in a single pass, with each tooth cutting more from the hole as it passes through. Broaches can be used to cut external grooves, such as keyways or slots, in a part.

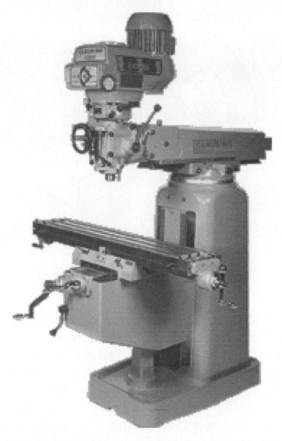

11.43 The milling machine operates by mounting the work on a bed that moves beneath revolving cutters. *(Courtesy of Clausing Industrial, Inc.)*

Milling Machine

The **milling machine** uses a variety of cutting tools, rotated about a shaft **(Figure 11.43),** to form different grooved slots, threads, and gear teeth. The milling machine can cut irregular grooves in cams and finish surfaces on a part within a high degree of tolerance. The cutters revolve about a stationary axis while the work pieces on the work table are passed beneath them.

Shaper

The shaper is a machine that holds a work piece stationary while the cutter passes back and forth across it to shape the surface or to cut a groove one stroke at a time **(Figure 11.44).** With each stroke of the cutting

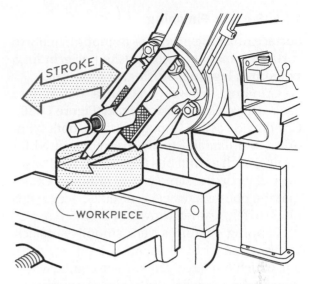

11.44 The shaper moves back and forth across the part, removing metal as it advances, to shape surfaces, cut slots, and perform other operations.

tool, the material is shifted slightly to align the part for the next overlapping stroke.

Planer (Mill)

Unlike the shaper, which holds the work piece stationary, the **planer** passes the piece under the cutters to machine large flat surfaces However, the planer has been replaced with the more efficient planer-type milling machine. Planer-type milling machines are especially efficient for surfacing large areas as shown in **Figure 11.45.**

11.45 This bed-type milling machine (Model 30 KF) is capable of milling surfaces of parts weighing up to 10,000 pounds *(Courtesy of Zayer, S. A.)*

11.6 Surface Finishing

Surface finishing produces a smooth, uniform surface. It may be accomplished by **grinding, polishing, lapping, buffing,** and **honing.**

Grinding involves holding a flat surface against a rotating abrasive wheel (**Figure 11.46**). Grinding is used to smooth surfaces, both cylindrical and flat, and to sharpen edges used for cutting, such as drill bits (**Figure 11.47**). Polishing is done in the same way as grinding, except that the polishing wheel is flexible because it is made of felt, leather, canvas, or fabric.

Lapping produces very smooth surfaces. The surface to be finished is held against a lap, which is a large, flat surface coated with a fine abrasive powder that finishes a surface as the lap rotates. Lapping is done only after the surface has been previously finished by a less accurate technique, such as grinding or polishing. Cylindrical parts can be lapped by using a lathe with the lap.

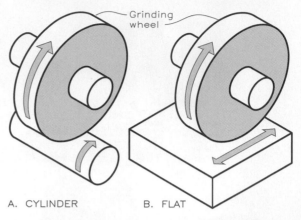

A. CYLINDER   B. FLAT

11.47 Grinding may be used to finish (A) cylindrical and (B) flat surfaces.

11.46 The operator is grinding the upper surface of this part to a smooth finish with a grinding wheel. *(Courtesy of the Clausing Industrial Inc.)*

Buffing removes scratches from a surface with a belt or rotating buffer wheel made of wool, cotton, felt, or other fabric. To enhance the buffing, an abrasive mixture is applied to the buffed surface during the process.

Honing finishes the outside or inside of holes within a high degree of tolerance. The honing tool is rotated as it is passed through the holes to produce the types of finishes found in gun barrels, engine cylinders, and other products requiring a high degree of smoothness.

11.7 Plastics and Other Materials

Plastics (polymers) are widely used in numerous applications ranging from clothing, containers, and electronics to automobile bodies and components. Plastics are easily formed into irregular shapes, have a high resistance to weather and chemicals, and are available in limitless colors. The three basic types of plastics are **thermoplastics, thermosetting plastics,** and **elastomers.**

Thermoplastics may be softened by heating and formed to the desired shape. If a polymer returns to its original hardness and

	MACHINABILITY	FORMABILITY	CASTABILITY	WELDABILITY	CORROSION RES.	ABRASION RES.	LB/CU FT	YIELD: 1000 PSI	Typical Applications
THERMOPLASTICS									
ACRYLIC	G	G	E	A	E	F	74	9	Aircraft windows, TV parts, Lenses, skylights
ABS	G	G	G	A	E	G	66	66	Luggage, boat hulls, tool handles, pipe fittings
POLYMIDES (NYLON)	E	G	G	—	G	E	73	15	Helmets, gears, drawer slides, hinges, bearings
POLYETHYLENE	G	F	G	A	F	F	58	2	Chemical tubing, containers, ice trays, bottles
POLYPROPYLENE	G	G	G	A	E	G	56	5.3	Card files, cosmetic cases, auto pedals, luggage
POLYSTYRENE	G	E	G	A	P	G	67	7	Jugs, containers, furniture, lighted signs
POLYVINYL CHLORIDE	E	E	G	A	G	G	78	4.8	Rigid pipe and tubing, house siding, packaging
THERMOSETS									
EPOXY	F	G	G	—	E	G	69	17	Circuit boards, boat bodies, coatings for tanks
SILICONE	F	G	G	—	G	G	109	28	Flexible hoses, heart valves, gaskets
ELASTOMERS									
POLYURETHANE	G	G	G	A	G	E	74	6	Rigid: Solid tires, bumpers; Flexible: Foam, sponges
SBR RUBBER	—	—	E	—	F	E	39	3	Belts, handles, hoses, cable coverings
GLASSES									
GLASS	F	G	—	—	F	F	160	10+	Bottles, windows, tumblers, containers
FIBERGLASS	G	—	E	A	G	G	109	20+	Boats, shower stalls, auto bodies, chairs, signs

E=Excellent
G=Good
F=Fair
P=Poor
A=Adhesives

11.48 These are the characteristics of and typical applications for commonly used plastics and other materials.

strength after being heated, it is classified as a **thermoplastic.** In contrast, thermosetting plastics cannot be changed in shape by reheating after they have permanently set. **Elastomers** are rubberlike polymers that are soft, expandable, and elastic, which permits them to be deformed greatly and then returned to their original size.

Figure 11.48 shows commonly used plastics and other materials, including **glass** and **fiberglass.** The weights and yields of the materials are given, along with examples of their applications.

The motorized golf cart shown in **Figure 11.49** is made of plastic. It has fewer parts and weighs less than carts made of metal. It has rounded corners and fewer joints which make it easy to fabricate and clean.

11.49 The use of DuraShield®, a thermoplastic elastomer, in constructing the E-Z-GO® golf car results in a body that withstands extremes of heat, cold, sunlight, and impact. *(Courtesy of E-Z-GO – A Textron Company.)*

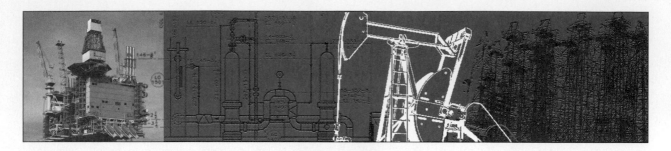

12

Dimensioning

12.1 Introduction

Working drawings have dimensions and notes that convey sizes, specifications, and other information necessary to build a project. Drawings with their details, dimensions, and specifications serve as construction documents that will become legal contracts.

The techniques of dimensioning presented here are based primarily on the standards of the American National Standards Institute (ANSI), especially Y14.5M, *Dimensioning and Tolerancing for Engineering Drawings*. Standards of companies such as the General Motors Corporation also are referenced.

12.2 Terminology

The strap shown in **Figure 12.1** is described in **Figure 12.2** with orthographic views to which dimensions were added. This example introduces the basic terminology of dimensioning.

Dimension lines: thin lines (2H–4H pencil) with arrows at each end and numbers placed near their midpoints to specify size.

Extension lines: thin lines (2H–4H pencil) extending from the part and between which dimension lines are placed.

Centerlines: thin lines (2H–4H pencil) used to locate the centers of cylindrical parts such as holes.

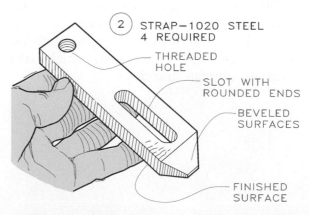

12.1 This tapered strap is a part of a clamping device that is dimensioned in Figure 12.2.

Leaders: thin lines (2H–4H pencil) drawn from a note to the feature to which it applies.

Arrowheads: drawn at the ends of dimension lines and leaders and the same length as the height of the letters or numerals, usually 1/8 inch, as shown in **Figure 12.3.**

Dimension numbers: placed near the middle of the dimension line and usually 1/8 inch high, with no units of measurement shown (", in., or mm).

12.3 Units of Measurement

The two commonly used units of measurement are the **decimal inch** in the English (imperial) system, and the **millimeter** in the metric system (SI) **(Figure 12.4).** Giving fractional inches as decimals rather than common fractions makes arithmetic easier.

Figure 12.5 demonstrates the proper and improper dimensioning techniques with millimeters, decimal inches, and fractional inches. In general, round off dimensions in millimeters to whole numbers without fractions. However, when you must show a metric dimension of less than a millimeter, use a zero before the decimal point. Do not use a zero before the decimal point when using inch units.

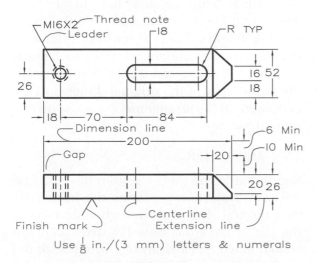

12.2 This dimensioned drawing of the tapered strap shown in Figure 12.1 introduces the terminology of dimensioning.

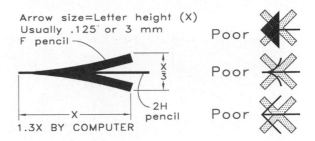

12.3 Draw arrowheads as long as the height of the letters used on the drawing and one-third as wide as they are long. When drawn by computer they can be drawn about 30% longer.

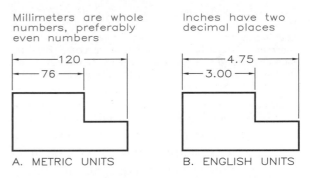

12.4 For the metric system, round millimeters to the nearest whole number. For the English system, show inches with two decimal places even for whole numbers such as 3.00.

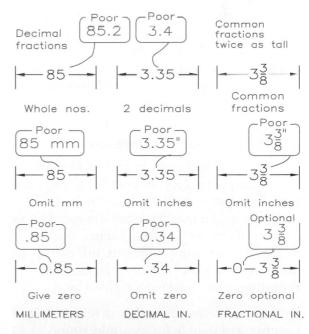

12.5 Basic principles of specifying measurements in SI and English units on a drawing.

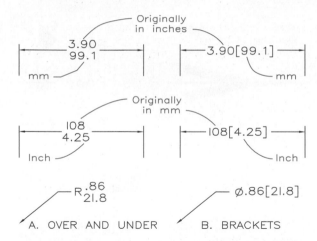

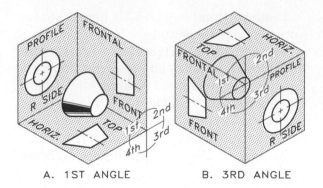

12.6 In dual dimensioning, place size equivalents in millimeters under or to the right of the inches (in brackets). Place the equivalent measurement in inches under or to the right of millimeters (in brackets). Show millimeters converted from inches as decimal fractions.

12.7 Projection systems.
A The SI system uses the first angle of orthographic projection in which the top view is placed under the front view.
B The American system uses the third angle of projection, which places the top view over the front view.

Show decimal inch dimensions with two-place decimal fractions, even if the last numbers are zeros. Omit units of measurement from the dimensions because they are understood to be in millimeters or inches. For example, use 112 (not 112 mm) and 67 (not 67″ or 5′-7″).

Architects use combinations of feet and inches in dimensioning and show foot marks, but usually omit inch marks, for example 7′-2. Engineers use feet and decimal fractions of feet to dimension large-scale projects such as road designs, for example 252.7′.

12.4 English/Metric Conversions

To convert dimensions in inches to millimeters, multiply by 25.4. To convert dimensions in millimeters to inches, divide by 25.4.

When millimeter fractions are required as a result of conversion from inches, one-place fractions usually are sufficient, but two-place fractions are used in some cases. Rules of rounding of dimensions are given here:

• Retain the last digit if it is followed by a number less than 5; for example, round 34.43 to 34.4.

• Increase the last digit retained by 1 if it is followed by a number greater than 5; for example, round 34.46 to 34.5.
• Leave the last digit unchanged if it is an even numbers and is followed by the digit 5; for example, round 34.45 to 34.4.
• Increase the last digit retained by 1 if it is an odd number and is followed by the digit 5; for example, round 34.75 to 34.8.

12.5 Dual Dimensioning

On some drawings you may have to give both metric and English units, called **dual dimensioning (Figure 12.6).** Place the millimeter equivalent either under or over the inch units, or place the converted dimension in brackets to the right of the original dimension. Be consistent in the arrangement you use on any set of drawings.

12.6 Metric Units

Recall that in the metric system (SI), the first angle of projection positions the front view over the top view and the right-side view to the left of the front view **(Figure 12.7).** You should label metric drawings with one of the symbols shown in **Figure 12.8** to designate the angle of projection. Display either the letters

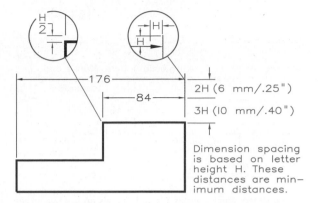

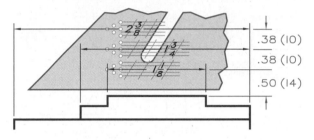

12.10 Place dimensions on a view as shown here, where all dimensioning geometry is based on the letter height (H) used.

12.11 Draw guidelines for common fractions in dimensions by aligning the center holes in the Braddock-Rowe triangle with the dimension line.

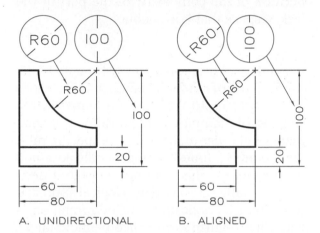

U.S. projection, where circle is visible here

First angle, where circle is visible, but would be hidden in U.S. system

A. METRIC UNITS AND THIRD—ANGLE PROJECTION

B. METRIC UNITS AND FIRST—ANGLE PROJECTION

12.8 The SI symbol.
A The SI symbol indicates that the millimeter is the unit of measurement and this truncated cone specifies that third-angle projection was used to position the orthographic views.
B Again, the SI symbol denotes use of the millimeter, but this truncated cone designates that first-angle projection was used.

A. UNIDIRECTIONAL

B. ALIGNED

12.9 Unidirectional and aligned dimensions.
A Dimensions are unidirectional when they are horizontal in both vertical and horizontal dimension lines.
B Dimensions are aligned when they are lettered parallel to angular and vertical dimension lines to read from the right side of the drawing (not from the left).

SI or the word METRIC prominently in or near the title block to indicate that the measurements are metric.

12.7 Numerals and Symbols

Vertical Dimensions

Numerals on vertical dimension lines may be **aligned** or **unidirectional.** When the unidirectional method is used, all dimensions are given in the standard horizontal position **(Figure 12.9A).** When the aligned method is used, numerals are given parallel with vertical

and angular dimension lines and read from the right-hand side of the drawing, never from the left-hand side **(Figure 12.9B).** Aligned dimensions are used almost entirely in architectural drawings where dimensions composed of feet, inches, and fractions are too long to fit unidirectionally (for example, 22'-10 1/2).

Placement

Dimensions should be placed on the most descriptive views of the part being dimensioned. The first row of dimensions should be at least three times the letter height (3H) from the object **(Figure 12.10).** Successive rows of dimensions should be spaced equally at least two times the letter height apart (0.25 inch or 6 mm, when 1/8-inch letters are used). Use the Braddock-Rowe lettering guide triangle to space the dimension lines **(Figure 12.11).**

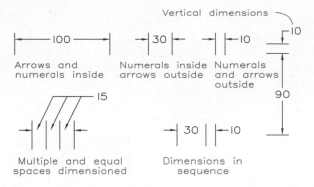

12.12 When space permits place numerals and arrows inside extension lines. For smaller spaces use other placements, as shown.

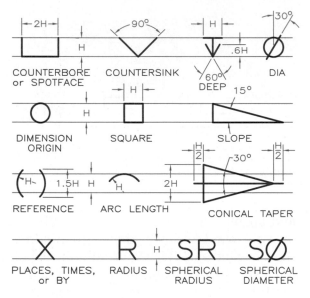

12.13 These symbols can be used instead of words to dimension parts. Their proportions are based on the letter height, H, which usually is 1/8 inch.

Figure 12.12 illustrates how to place dimensions in limited spaces. Regardless of space limitations, do not make numerals smaller than they appear elsewhere on the drawing.

Dimensioning Symbols
Figure 12.13 shows standard dimensioning symbols and their sizes based on the letter height, usually 1/8 inch. By using these symbols, lengthy notes can be replaced and drawing time saved.

12.8 Dimensioning Rules

There are many rules of dimensioning that you should become familiar with in order to place dimensions and notes on drawings in the conventionally approved manner. Each geometrical shape has its own set of rules: prisms, angular surfaces, cylindrical features, pyramids, cones, spheres, and arcs.

Dimensioning rules should be considered more as guidelines rather than as rigid rules. Quite often, rules of dimensioning must be violated or applied in a different manner because of the complexity of the part or the lack of space that is available.

Prisms
Beginning with **Figures 12.14** through **12.26,** the fundamental rules of dimensioning prisms are illustrated. All rules are presented in the simplest of examples in the most fundamental manner in order to focus on the specific rules one point at a time. These generally accepted dimensioning rules were not arrived at arbitrarily, but they are based on a logical approach to aid in their application and interpretation as you will recognize as you become more familiar with dimensioning.

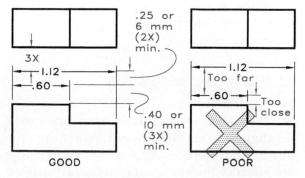

12.14 Place the first row of dimensions at least three times the letter height from the object. Successive rows should be at least two times the letter height apart.

RULE 2: Place dimensions between the views.

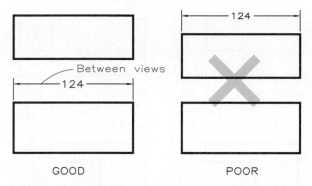

12.15 Place dimensions between the views sharing these dimensions.

RULE 3: Dimension the most descriptive views.

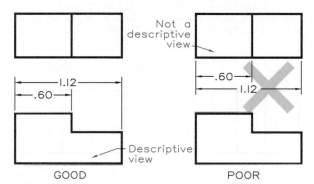

12.16 Place dimensions on the most descriptive views of an object.

RULE 4: Dimension from visible lines, not hidden lines.

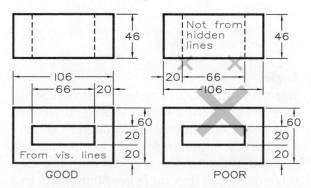

12.17 Dimension visible features, not hidden features.

RULE 5: Give an overall dimension and omit one of the chain dimensions.

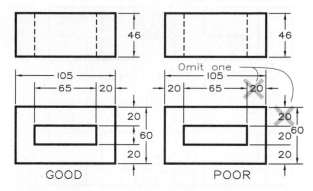

12.18 Leave the last dimension blank in a chain of dimensions and give an overall dimension.

RULE 5 (DEVIATION): When one chain dimension is not omitted, mark one dimension as a reference dimension.

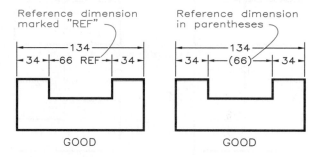

12.19 If you give all dimensions in a chain, mark the reference dimension (the one that would be omitted) with REF or place it in parentheses. Giving a reference dimension is a way of eliminating mathematical calculations in the shop.

RULE 6: Organize and align dimensions for ease of reading.

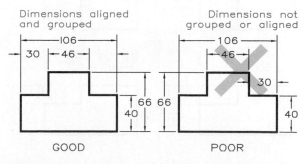

12.20 Place dimensions in well-organized lines for uncluttered drawings.

RULE 7: Do not repeat dimensions.

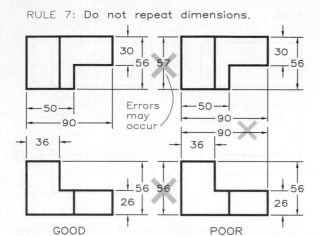

12.21 Do not duplicate dimensions on a drawing to avoid errors or confusion.

RULE 8: Dimension lines should not cross other lines.

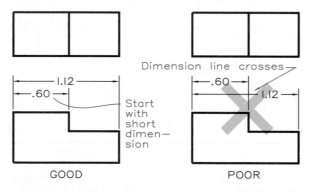

12.22 Dimension lines should not cross any other lines unless absolutely necessary.

RULE 9: Extension lines may cross other lines if they must.

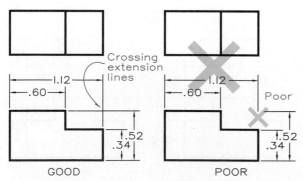

12.23 Extension lines may cross other extension lines or object lines if necessary.

RULE 9 (Cont.): Extension lines may cross other lines.

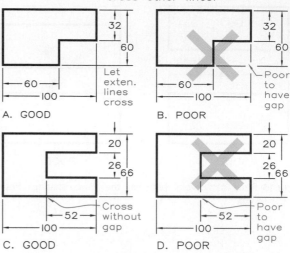

12.24 Leave a small gap from the edges of an object to extension lines that extend from them. Do not leave gaps where extension lines cross object lines or other extension lines.

RULE 10: Do not place dimensions within the views unless necessary.

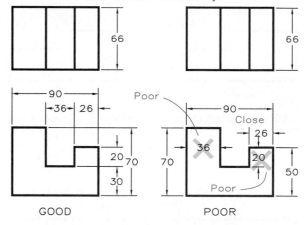

12.25 Whenever possible place dimensions outside objects rather than inside their outlines.

Angles

You may dimension angles either by coordinates locating the ends of sloping surfaces or by angular measurements in degrees (**Figure 12.27).** Fractional angles can be specified in decimal units or in degrees, minutes, and seconds. Recall that there are 60 minutes in a degree and 60 seconds in a minute. It is

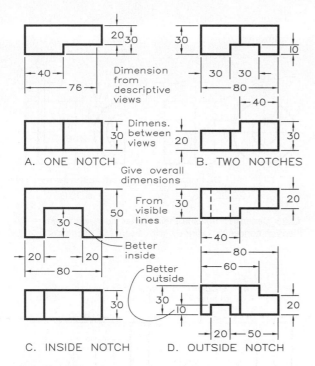

A. ONE NOTCH   B. TWO NOTCHES

Dimension from descriptive views

Dimens. between views

Give overall dimensions

From visible lines

Better inside

Better outside

C. INSIDE NOTCH   D. OUTSIDE NOTCH

12.26 Dimensioning prisms.
A–B Dimension prisms from descriptive views and between views.
C You may dimension a notch inside the object if doing so improves clarity.
D Dimension visible lines, not hidden lines.

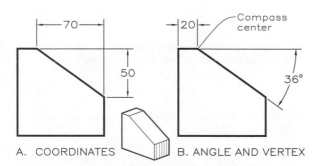

A. COORDINATES   B. ANGLE AND VERTEX

12.27 Dimensioning angles.
A Dimension angular planes by using coordinates.
B Measure angles by locating the vertex and measuring the angle in degrees. When accuracy is essential, specify angles in degrees, minutes, and seconds.

seldom that you will you need to measure angles to the nearest second. **Figures 12.28** and **12.29** illustrate basic rules for dimensioning angles.

RULE 11: Place angular dimensions outside the angle.

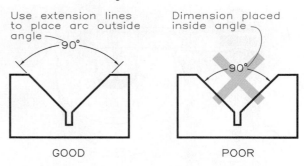

Use extension lines to place arc outside angle

Dimension placed inside angle

GOOD   POOR

12.28 Place angular dimensions outside the object by using extension lines.

RULE 12: Dimension rounded corners to the theoretical intersection.

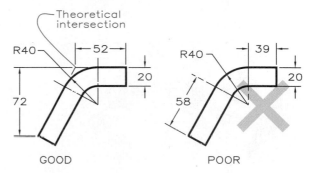

Theoretical intersection

GOOD   POOR

12.29 Dimension a bent surface rounded corner by locating its theoretical point of intersection with extension lines.

Cylindrical Parts and Holes

The diameters of cylinders are measured with a micrometer **(Figures 12.30** and **12.31).** Therefore, dimension cylinders with diameters in their rectangular views (the views in which diameters are measured) as shown in **Figures 12.32** and **12.33.** Place diameter symbols in front of the diametral dimensions to indicate that the dimension is a diameter. You will recall that the diametric symbol is a circle with a slash through it. In the English system the abbreviation DIA placed after the diametral dimension is still sometimes used, but the metric diameter symbol is preferred.

Space is almost always a problem in dimensioning, and all means of conserving space must be used. Stagger dimensions for

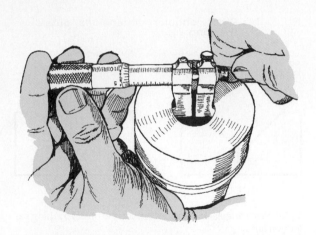

12.30 This internal micrometer caliper measures internal cylindrical diameters (radii cannot be measured).

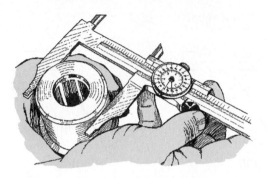

12.31 This external micrometer caliper measures the diameter of a cylinder.

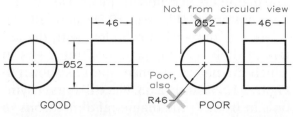

12.32 Dimension the diameter (not the radius) of a cylinder and in the rectangular view.

concentric cylinders to avoid crowding as shown in **Figure 12.34.** Dimension cylindrical holes in their circular view with leaders **(Figure 12.35).** The circular view is the view that would be used when the hole is located and drilled. Draw leaders specifying hole sizes

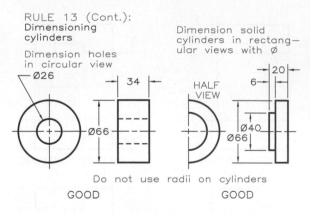

12.33 Dimension holes in their circular views with leaders. Dimension concentric cylinders with a series of diameters.

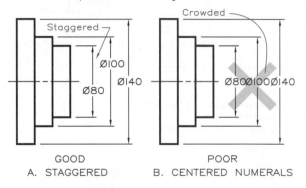

12.34 Numerals on concentric cylinders are easier to read if they are staggered within their dimension lines.

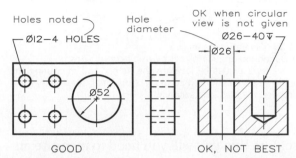

12.35 Dimension holes in their circular view with leaders whenever possible, but dimension them in their rectangular views if necessary.

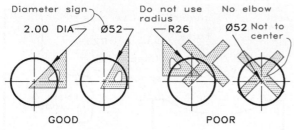

RULE 16: Leaders have horizontal elbows and point toward the hole centers.

GOOD   POOR

12.36 Draw leaders pointing toward the centers of holes.

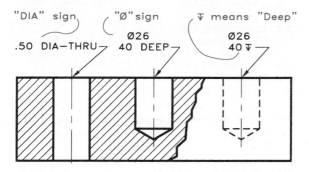

12.37 These are examples of holes noted in their rectangular views.

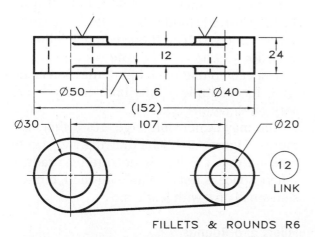

FILLETS & ROUNDS R6

12.38 This drawing illustrates the application of dimensions to cylindrical features. (F&R R6 means that fillets and rounds have a 6 mm radius.)

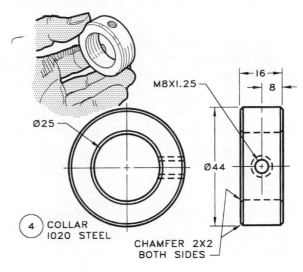

12.39 This cylindrical part, a collar, is shown drawn and dimensioned below.

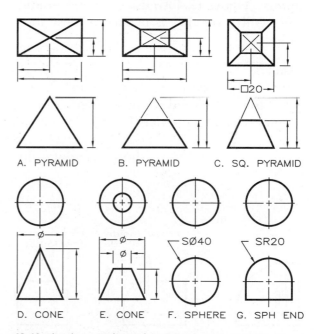

A. PYRAMID   B. PYRAMID   C. SQ. PYRAMID

D. CONE   E. CONE   F. SPHERE   G. SPH END

12.40 This drawing shows the proper way to dimension pyramids, cones, and spheres.

as shown in **Figure 12.36.** When you must place diameter notes for holes in the rectangular view instead of the circular view, draw them as shown in **Figure 12.37.** Examples of dimensioned parts with cylindrical features are shown in **Figures 12.38** and **12.39.**

Pyramids, Cones, and Spheres
Figures 12.40A–C show three methods of dimensioning pyramids. **Figures 12.40D** and **E** show two acceptable methods of dimensioning cones. Spheres are dimensioned by giving its diameter as shown in **Figure 12.40F.**

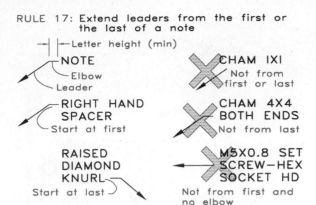

RULE 17: Extend leaders from the first or the last of a note

|—Letter height (min)

NOTE
Elbow
Leader

RIGHT HAND SPACER
Start at first

RAISED DIAMOND KNURL
Start at last

GOOD

CHAM 1X1
Not from first or last

CHAM 4X4 BOTH ENDS
Not from last

M5X0.8 SET SCREW—HEX SOCKET HD
Not from first and no elbow

POOR

12.41 Extend leaders from the first or the last word of a note with a horizontal elbow.

If the spherical shape is less than a hemisphere **(Figure 12.40G)** use a spherical radius (SR). Only one view is needed to describe a sphere.

Leaders
Leaders are used to apply notes and dimensions to features and are most often drawn at standard angles of triangles **(Figure 12.36).** Leaders should begin at either the first or last word of a note with a short horizontal line

RULE 18: Dimension arcs (less than 180°) with radii.

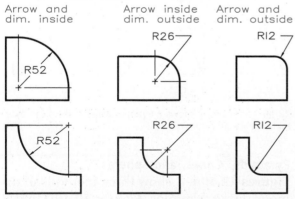

Arrow and dim. inside

Arrow inside dim. outside

Arrow and dim. outside

R52

R26

R12

R52

R26

R12

12.42 When space permits, place dimensions and arrows between the center and the arc. When the number will not fit, place it outside and the arrow inside. If the arrow will not fit inside, place both the dimension and arrow outside the arc with a leader.

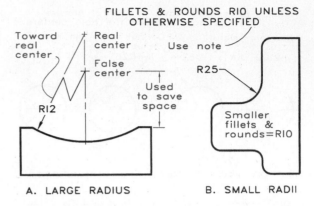

FILLETS & ROUNDS R10 UNLESS OTHERWISE SPECIFIED

Toward real center

Real center

Use note

False center

Used to save space

R12

R25

Smaller fillets & rounds=R10

A. LARGE RADIUS

B. SMALL RADII

12.43 Dimensioning radii.
A Show a long radius with a false radius (a line with a zigzag) to indicate that it is not true length. Show its false center on the centerline of the true center.
B Specify fillets and rounds with a note to reduce repetitive dimensions of small arcs.

(elbow) from the note and extend to the feature being described as shown in **Figure 12.41.**

Arcs and Radii
Full circles are dimensioned with diameters but arcs are dimensioned with radii **(Figure 12.42).** Current standards specify that radii be dimensioned with an R preceding the dimension (for example, R10). The previous standard specified that the R follow the dimension (10R, for example). Thus both methods are seen on drawings.

You may dimension large arcs with a false radius **(Figure 12.43)** by drawing a zigzag to indicate that the line is not the true radius. Where space is not available for radii, dimension small arcs with leaders.

Fillets and Rounds
When all fillets and rounds are equal in size, you may place a note on the drawing stating that condition or use separate notes **(Figure 12.44).** If most, but not all, of the fillets and rounds have equal radii, the note may read ALL FILLETS AND ROUNDS R6 UNLESS OTHERWISE SPECIFIED (or abbreviated as F&R R6), with the fillets and rounds of different radii dimensioned separately.

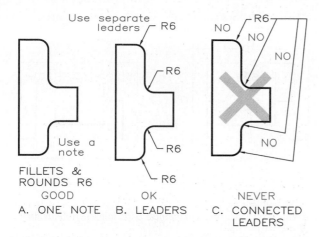

Use separate leaders — R6

Use a note

FILLETS & ROUNDS R6

GOOD
A. ONE NOTE

OK
B. LEADERS

NEVER
C. CONNECTED LEADERS

12.44 Indicate fillets and rounds by (A) notes, or (B) separate leaders and dimensions. Never use confusing leaders (C).

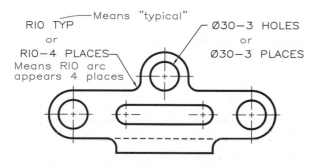

Means "typical"

R10 TYP
or
R10—4 PLACES
Means R10 arc appears 4 places

Ø30—3 HOLES
or
Ø30—3 PLACES

12.45 Use notes to indicate that identical features and dimensions are repeated to simplify dimensioning.

You may note repetitive features as shown in **Figure 12.45** by using the notes TYPICAL, or TYP, which means that the dimensioned feature is typical of those not dimensioned. You may use the note PLACES, or PL, to specify the number of places that identical features appear although only one is dimensioned.

Figure 12.46 shows how the views of a pulley are dimensioned. This example demonstrates proper application of many of the rules discussed in this section.

12.9 Curved and Symmetrical Parts

Curved Parts

An irregular shape comprised of tangent arcs of varying sizes can be dimensioned by using a series of radii as shown in **Figure 12.47.**

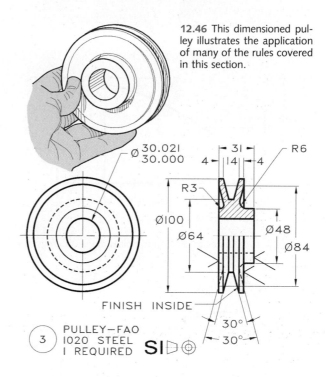

12.46 This dimensioned pulley illustrates the application of many of the rules covered in this section.

FINISH INSIDE

PULLEY—FAO
1020 STEEL
1 REQUIRED

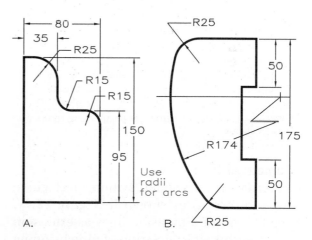

12.47 Use radii to dimension parts composed of arcs of partial circles.

Irregular curves are dimensioned by using coordinates to locate a series of points along the curve **(Figure 12.48).** You must use your judgement in determining how many located points are necessary to define the curve. Placing extension lines at an angle provides additional space for showing dimensions.

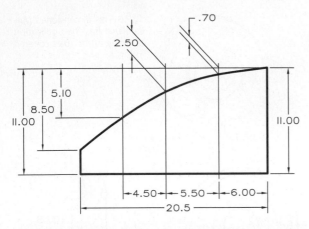

12.48 Use coordinates to dimension points along an irregular curve on a part.

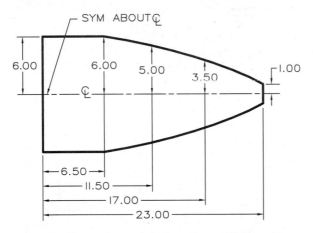

12.49 Use coordinates to dimension points along curves of a symmetrical part.

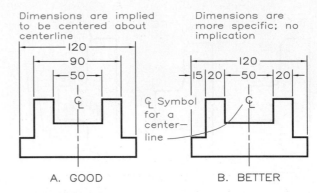

A. GOOD   B. BETTER

12.50 Symmetrical parts.
A You may dimension symmetrical parts implicitly about their centerlines.
B The better way to dimension symmetrical parts is explicitly about their centerlines.

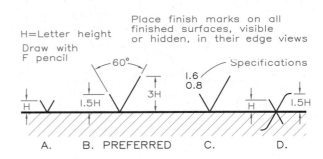

A.   B. PREFERRED   C.   D.

12.51 Finish marks indicate that a surface is to be machined to a smooth surface.
A The traditional V can be used for general applications.
B The unequal finish mark is the best for general applications.
C Where surface texture must be specified, this finish mark is used with texture values.
D The f-mark is the oldest and least-used symbol.

Symmetrical Parts

Dimension an irregular symmetrical curve with coordinates as shown in **Figure 12.49.** Note the use of dimension lines as extension lines, a permissible violation of dimensioning rules in this case.

Dimension symmetrical objects by using coordinates to imply that the dimensions are symmetrical about the centerline (abbreviated CL) as shown in **Figure 12.50A. Figure 12.50B** shows a better method of dimensioning this type of object, where symmetry is dimensioned with no interpretation required on the part of the reader.

12.10 Finished Surfaces

Parts formed in molds, called **castings,** have rough exterior surfaces. If these parts are to assemble with and move against other parts, they will not function well unless their contact surfaces are machined to a smooth finish by grinding, shaping, lapping, or similar processes.

To indicate that a surface is to be finished, finish marks are drawn on edge views of surfaces to be finished **(Figure 12.51).** Finish marks should be shown in every view where

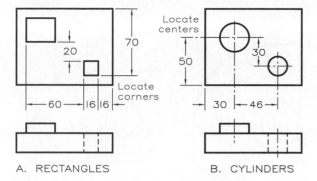

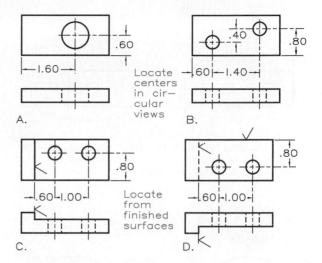

12.52 Location dimensions give the positions of geometric features with respect to other geometric features, but not their sizes.

12.54 Location rules in summary:
A Locate cylindrical holes in their circular views from two surfaces of the object.
B Locate multiple holes from center to center.
C Locate holes from finished surfaces, even if the finished surfaces are hidden as in **D**.

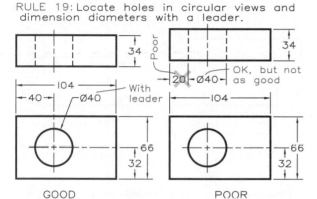

12.53 Locate cylindrical holes in their circular views by coordinates to their centers.

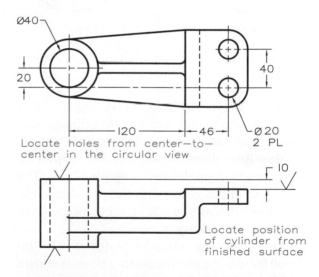

12.55 This example of a dimensioned shaft arm shows the application of location dimensions, with feature sizes omitted.

finished surfaces appear as edges even if they are hidden lines.

The preferred finish mark for the general cases is the uneven mark shown in **Figure 12.51B.** When an object is finished on all surfaces, the note FINISHED ALL OVER (abbreviated as FAO) can be placed on the drawing.

12.11 Location Dimensions

Location dimensions give the positions, not the sizes, of geometric shapes **(Figure 12.52).** Locate rectangular shapes by using coordinates of their corners and cylindrical shapes by using coordinates of their centerlines. In

each case, dimension the view that shows both measurements. Always extend coordinates from any finished surface (even if the finished surface is a hidden line) because smooth machined surfaces allow the most accurate measurements. Locate and dimension single holes as shown in **Figure 12.53** and multiple

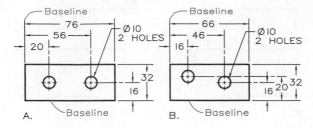

A.   B.

12.56 Measuring holes from two datum planes is a more accurate way to locate them since the accumulation of errors that occur in chain dimensioning is reduced.

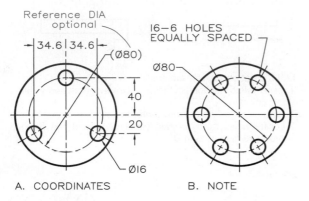

A. COORDINATES   B. NOTE

12.57 Locate holes on a bolt circle by using (A) coordinates or (B) notes.

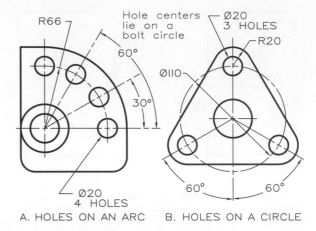

A. HOLES ON AN ARC   B. HOLES ON A CIRCLE

12.58 Locate centers of holes by using (A) a combination of radii and degrees or (B) a circle of centers.

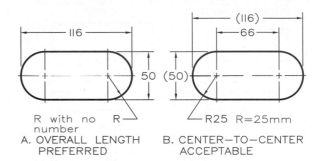

A. OVERALL LENGTH PREFERRED   B. CENTER—TO—CENTER ACCEPTABLE

12.59 Rounded ends.
A Dimension objects having rounded ends from end-to-end and give the diameter.
B A less desirable choice is to dimension the rounded ends from center-to-center and give the radius.

holes as shown in **Figure 12.54. Figure 12.55** shows the application of location dimensions to a typical part with size dimensions omitted.

Baseline dimensions extend from two baselines in a single view **(Figure 12.56).** The use of baselines eliminates the possible accumulation of errors in size that can occur from chain dimensioning.

Holes through circular plates can be located by using coordinates or a note as shown in **Figure 12.57.** Dimension the diameter of the imaginary circle passing through the centers of the holes in the circular view as a reference dimension and locate the holes by using coordinates **(Figure 12.57A)** or a note **(Figure 12.57B).** This imaginary circle is called the **bolt circle** or **circle of centers.**

You may also locate holes with radial dimensions and their angular positions in degrees **(Figure 12.58).** Holes may be located

on their bolt circle even if the shape of the object is not circular.

Objects with Rounded Ends

Dimension objects with rounded ends from one rounded end to the other **(Figure 12.59A)** and show their radii as R without dimensions to specify that the ends are arcs. Obviously, the end radius is half the height of the part. If you dimension the object from center to center **(Figure 12.59B)** you must give the radius size. You may specify the overall width as a reference dimension (116) to eliminate the need for calculations.

Dimension parts with partially rounded ends as shown in **Figure 12.60A.** Dimension

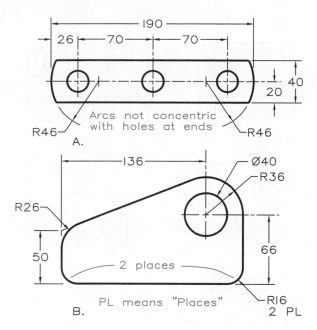

A.

Arcs not concentric
with holes at ends

R46

R46

B.

R26

Ø40

R36

50

2 places

PL means "Places"

66

R16
2 PL

12.60 These examples show dimensioned parts having (A) rounded ends that are not concentric with the holes and (B) rounds and cylinders.

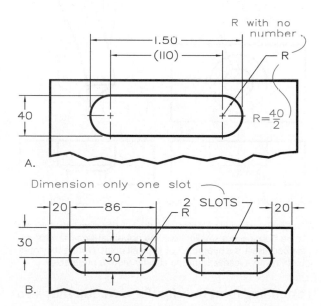

R with no number

R

R=$\frac{40}{2}$

A.

Dimension only one slot

2 SLOTS
R

B.

12.61 These drawings illustrate methods of dimensioning parts that have (A) one slot and (B) more than one slot.

objects with rounded ends that are smaller than a semicircle with a radius and locate the arcs' centers (**Figure 12.60B**).

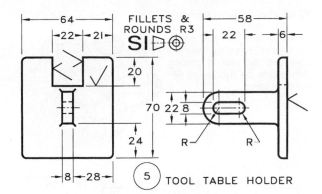

FILLETS & ROUNDS R3

SI ▷ ⊕

TOOL TABLE HOLDER

12.62 This dimensioned part has both slots and arcs.

Dimension a single slot with its overall width and height (**Figure 12.61A**). When there are two or more slots, dimension one slot and use a note to indicate that there are other identical slots (**Figure 12.61B**).

The tool holder table shown in **Figure 12.62** illustrates dimensioning of arcs and slots. To prevent dimension lines from crossing, several dimensions are placed on a less descriptive view. Notice that the diameters of the semicircular features are given; therefore, notes of R are given to indicate radii but the radius sizes are unnecessary since that can be easily calculated as half of the diameters.

12.12 Outline Dimensioning

Now that you are familiar with most of the rules of dimensioning, you can better understand outline dimensioning, which is a way of applying dimensions to a part's outline (silhouette). By taking this approach, you have little choice but to place dimensions in the most descriptive views of the part. For example, imagine that the T-block shown in **Figure 12.63** has no lines inside its outlines. It is dimensioned beginning with its location dimensions. When the inside lines are considered, additional dimensions are seldom needed.

Figure 12.64 shows an example of outline dimensioning. Note the extension of all dimensions from the outlines of well-defined features.

12.63 Outline dimensioning is the placement of dimensions on views as if they had no internal lines. Applying this concept will aid you in placing dimensions on their most descriptive views.

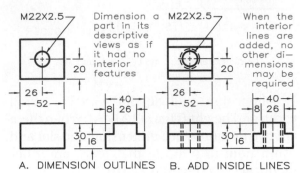

Dimension a part in its descriptive views as if it had no interior features

When the interior lines are added, no other dimensions may be required

A. DIMENSION OUTLINES   B. ADD INSIDE LINES

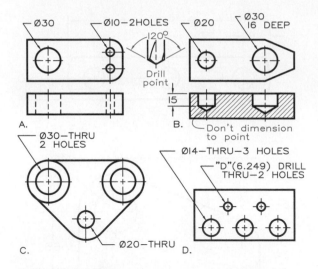

A.

B. Don't dimension to point

C.

D.

12.65 Cylindrical holes:
A–B Dimension cylindrical holes by either of these methods.
C–D When only one view is used, notes must be used to specify THRU holes or specify their depths.

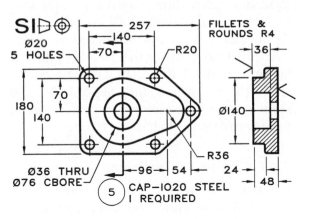

12.64 This drawing illustrates use of the outline method to dimension the cap in its most descriptive views.

12.13 Machined Holes

Machined holes are formed by machine operations, such as drilling, boring, or reaming **(Figure 12.65)**. Occasionally a machining operation is specified in the note, such as ∅32 DRILL, but it is preferred to omit the specific maching operation. Give the diameter of the hole with its diameter symbol (for example, ∅32) in front of its dimension with a leader

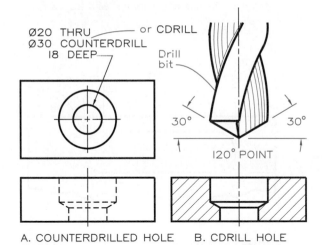

A. COUNTERDRILLED HOLE   B. CDRILL HOLE

12.66 Counterdrilling notes give the specifications for drilling a larger hole inside a smaller hole. Do not dimension the 120° angle because it is a byproduct of the drill point. Noting the counterdrill with a leader from the circular view is preferable.

extending from the circular view. You may also note hole diameters with DIA after their size (for example, 2.00 DIA).

Drilling is the basic method of making holes. Dimension the size of a drilled hole with a leader pointing to the circumference and extending from its circular view. The hole's

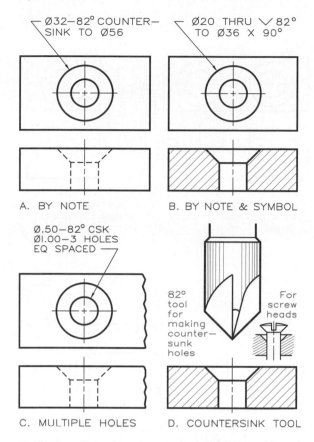

A. BY NOTE

⌀32–82° COUNTER-
SINK TO ⌀56

B. BY NOTE & SYMBOL

⌀20 THRU ⌵ 82°
TO ⌀36 X 90°

⌀.50–82° CSK
⌀1.00–3 HOLES
EQ SPACED

82°
tool
for
making
counter-
sunk
holes

For
screw
heads

C. MULTIPLE HOLES

D. COUNTERSINK TOOL

12.67 These illustrations show methods of noting and specifying countersunk holes for receiving screw heads.

depth can be given in the note from the circular view or specified in the rectangular view with a conventional dimension **(Figure 12.65B).** Dimension the depth of a drilled hole to the usable part of the hole, not to its conical point.

Counterdrilling involves drilling a large hole inside a smaller hole to enlarge it **(Figure 12.66).** The drill point leaves a 120° conical shoulder as a by-product of counterdrilling.

Countersinking is the process of forming conical holes for receiving screw heads **(Figure 12.67).** Give the diameter of a countersunk hole (the maximum diameter on the surface) and the angle of the countersink in a note. Countersunk holes also are used as guides in shafts, spindles, and other cylindrical parts held between the centers of a lathe.

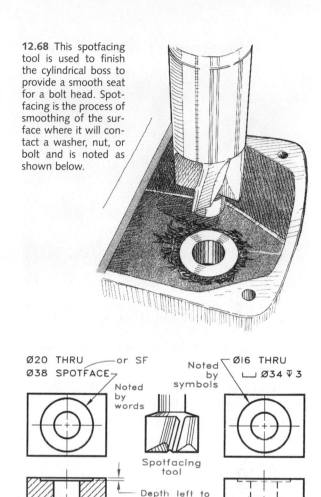

12.68 This spotfacing tool is used to finish the cylindrical boss to provide a smooth seat for a bolt head. Spotfacing is the process of smoothing of the surface where it will contact a washer, nut, or bolt and is noted as shown below.

⌀20 THRU
⌀38 SPOTFACE

or SF

Noted
by
words

Noted
by
symbols

⌀16 THRU
⌴ ⌀34 ⍌ 3

Spotfacing
tool

Depth left to
shop; purpose
is to smooth
the surface

A. SECTION

B. VIEW

Spotfacing is the process of finishing the surface around holes to provide bearing surfaces for washers or bolt heads **(Figure 12.68A).** Figure 12.68B shows the method of spotfacing a boss (a raised cylindrical element).

Boring is used to make large holes and it is usually done on a lathe with a bore or a boring bar **(Figure 12.69).**

Counterboring is the process of enlarging the diameter of a drilled hole **(Figure 12.70)** to give a flat bottom to the hole without the taper as in counterdrilled holes.

12.69 This photo shows the use of a lathe to bore a large hole with a boring bar. *(Courtesy of Clausing Industrial, Inc.)*

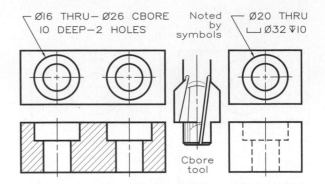

12.70 Counterbored holes are similar to counterdrilled holes but have flat bottoms instead of tapered sides. Dimension them as shown.

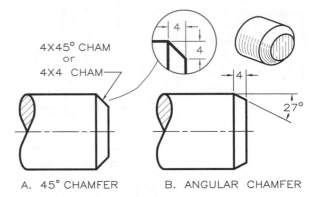

12.71 Chamfers.
A Dimension 45° chamfers by either of the notes shown.
B Dimension chamfers of all other angles as shown.

Reaming is the operation of finishing or slightly enlarging drilled or bored holes within their prescribed tolerances. A ream is similar to a drill bit.

12.14 Chamfers

Chamfers are beveled edges cut on cylindrical parts, such as shafts and threaded fasteners, to eliminate sharp edges and to make them easier to assemble. When the chamfer angle is 45°, use a note in either of the forms shown in **Figure 12.71A.** Dimension chamfers of other angles as shown in **Figure 12.71B.** When inside openings of holes are chamfered, they are dimensioned as shown in **Figure 12.72.**

12.15 Keyseats

A keyseat is a slot cut into a shaft for aligning and holding a pulley or a collar on a shaft. **Figure 12.73** shows how to dimension keyways and keyseats with dimensions taken from the tables in the Appendix. The double dimensions on the diameter are tolerances (discussed in Chapter 13).

12.16 Knurling

Knurling is the operation of cutting diamond-shaped or parallel patterns on cylindrical surfaces for gripping, decoration, or press fits between mating parts that are permanently assembled. Draw and dimension diamond knurls and straight knurls as shown in **Figure 12.74,** with notes specifying type, pitch, and diameter.

The abbreviation DP means diametral pitch, or the ratio of the number of grooves on the circumference (N) to the diameter (D) expressed as DP = N/D. The preferred diametral pitches for knurling are 64 DP, 96 DP, 128 DP, and 160 DP.

For diameters of 1 inch, knurling of 64 DP, 96 DP, 128 DP, and 160 DP will have 64, 96,

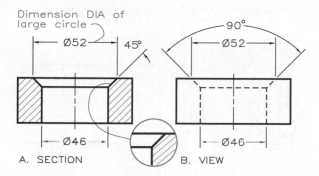

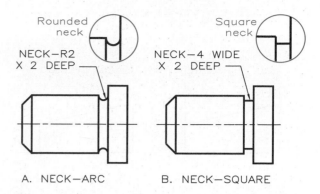

12.72 Dimension chamfers on the insides of cylinders using either of the methods shown here.

12.75 Necks are recessed grooves cut into cylinders with rounded or square bottoms, usually at the intersections of concentric cylinders. Dimension necks as shown.

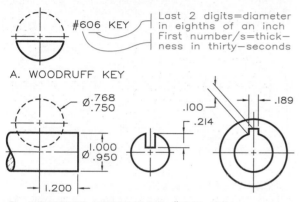

12.73 These drawings show methods of dimensioning (A) Woodruff keys and (B) keyways used to hold a part on a shaft. Appendix 20 gives their tables of sizes.

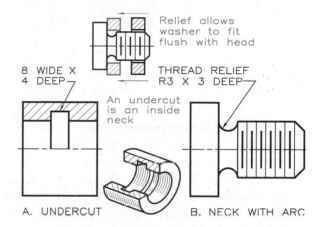

12.76 Undercuts and necks.
A An undercut is a groove cut inside a cylinder.
B A thread relief is a groove cut at the end of a thread to improve the screw's assembly. Dimension both types of necks as shown.

128, and 160 teeth, respectively, on the circumference. The note P0.8 means that the knurling grooves are 0.8 mm apart. Make knurling calculations in inches and then convert them to millimeters for metric equivalents. Specify knurls for press fits with the diameter size before knurling and with the minimum diameter size after knurling.

12.17 Necks and Undercuts

A **neck** is a groove cut around the circumference of a cylindrical part. If cut where cylinders of different diameters join **(Figure 12.75)**, a

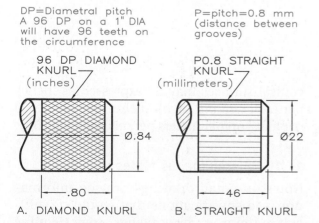

12.74 This diamond knurl (A) has a diametral pitch, DP, of 96 and the straight knurl. (B) has a linear pitch, P, of 0.8 mm. Pitch is the distance between the grooves on the circumference.

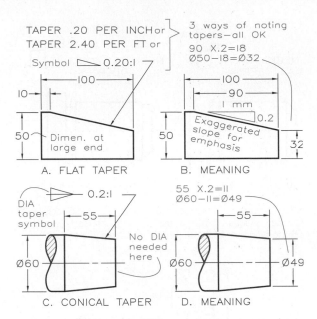

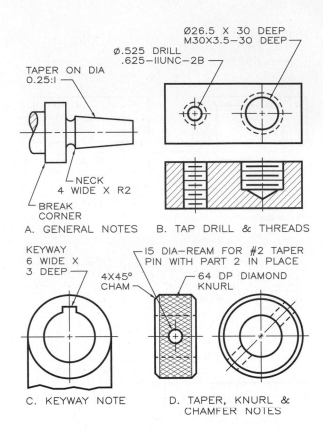

12.77 Tapers may be specified for either flat or conical surfaces: dimensioning and interpretation for (A and B) a flat taper and (C and D) for a conical taper.

12.78 Miscellaneous dimensioning.
A Notes indicate a neck, a taper, and a break corner, which is a slight round to remove corner sharpness.
B Tap drill sizes are sometimes specified in addition to the thread specifications, but selection of the tap drill size usually is left to the shop.
C The keyway is dimensioned by a note.
D The notes for this collar call for chamfering, knurling, and drilling for a #2 taper pin.

neck ensures that the assembled parts fit flush at the shoulder of the larger cylinder and allows trash, which would cause binding, to drop out of the way into the neck.

An **undercut** is a recessed neck inside a cylindrical hole (**Figure 12.76A**). A thread relief is a neck that has been cut at the end of a thread to ensure that the head of the threaded part will fit flush against the part it screws into (**Figure 12.76B**).

12.18 Tapers

Tapers for both flat planes and conical surfaces may be specified with either notes or symbols. Flat taper is the ratio of the difference in the heights at each end of a surface to its length (**Figures 12.77A** and **B**). Tapers on flat surfaces may be expressed as inches per inch (.20 per inch), inches per foot (2.40 per foot), or millimeters per millimeter (0.20:1).

Conical taper is the ratio of the difference in the diameters at each end of a cone to its length (**Figures 12.77C** and **D**). Tapers on

conical surfaces may be expressed as inches per inch (.25 per inch), inches per foot (3.00 per foot), or millimeters per millimeter (0.25:1).

12.19 Miscellaneous Notes

Notes on detail drawings provide information and specifications that would be difficult to represent by drawings alone (**Figures 12.78–12.80**). Place notes horizontally on the sheet whenever possible because they

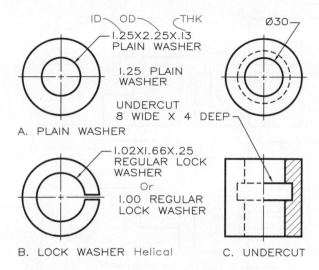

A. PLAIN WASHER

ID — OD — THK
1.25X2.25X.13
PLAIN WASHER

1.25 PLAIN
WASHER

UNDERCUT
8 WIDE X 4 DEEP

B. LOCK WASHER Helical

1.02X1.66X.25
REGULAR LOCK
WASHER
Or
1.00 REGULAR
LOCK WASHER

Ø30

C. UNDERCUT

12.79 Washers and undercuts.
A and **B** Dimension washers and lock washers with specifications taken from Appendixes 22–24. **C** Dimension an undercut with a note.

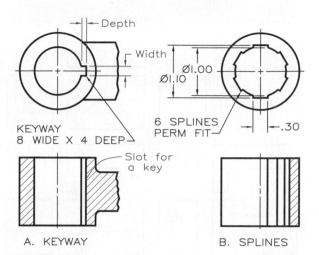

Depth

Width

KEYWAY
8 WIDE X 4 DEEP

Ø1.00
Ø1.10

6 SPLINES
PERM FIT

.30

Slot for
a key

A. KEYWAY

B. SPLINES

12.80 These drawings illustrate how to dimension (A) keyways and (B) splines.

are easier to letter and read in that position. Several notes in sequence on the same line should be separated with short dashes between them (for example, 15 DIA-30 DIA SPOTFACE). Use standard abbreviations in notes to save space and time.

Problems

1–36. (Figures 12.81 and **12.82)** Solve these problems on size A paper, one per sheet when drawn full size. Use size B sheets if you draw them double size. The problems are drawn on a 0.20-in. (5 mm) grid. Vary the spacing between views to provide adequate room for the dimensions. Sketching the views and dimensions to determine the required spacing before laying out the solutions with instruments would be helpful. Supply lines that may be missing in all views.

Supplementary problems:
Problems at the ends of Chapters 6 and 7 can be used as dimensioning exercises.

Thought Questions

1. What is the standard measurement unit when dimensioning in the metric system?

2. Which are the best views on which to place dimensions?

3. What is the size of gap between an extension line and the view being dimensioned?

4. Which is the best view for locating a drilled hole in a view?

5. How far should the first row of dimensions be placed from the view?

6. When dimensioned with inches, how many decimal places are used in standard practice?

7. What is it called when a large hole is drilled inside a smaller one for part of its depth?

8. What is the term for a bevel at the end of a cylindrical shaft?

9. What is the machining operation that smooths the surface at the top of a drilled hole?

10. Finished marks should be placed on what views of finished surfaces?

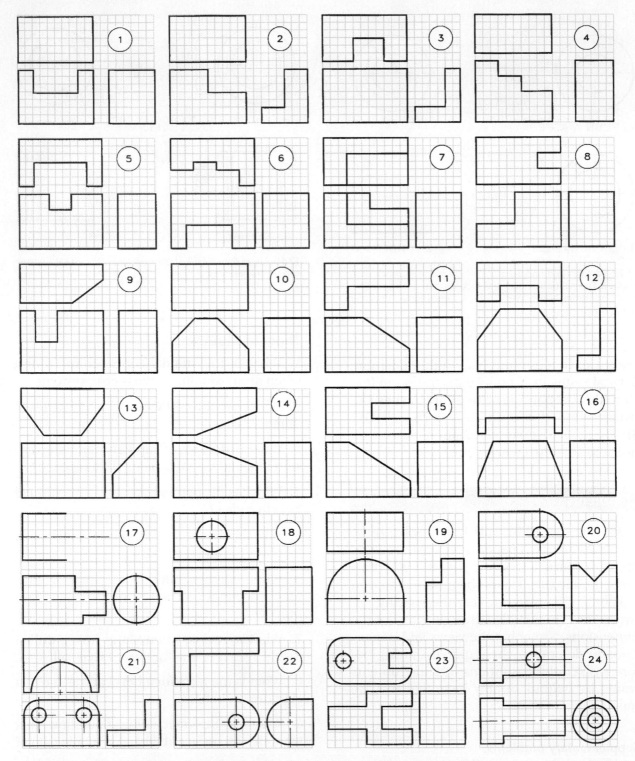

12.81 Problems 1–24.

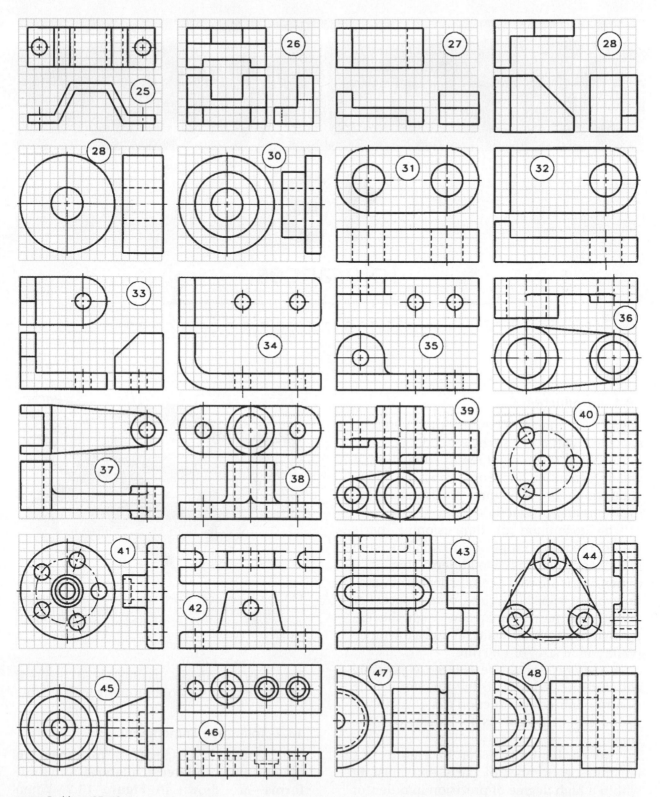

12.82 Problems 25–48.

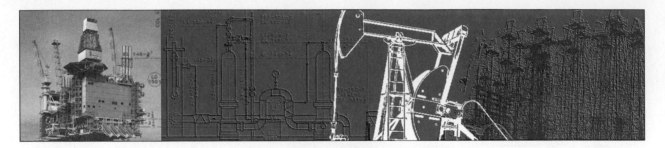

13

Tolerances

13.1 Introduction

Today's technology requires that parts be specified with increasingly exact dimensions. Many parts, made by different companies at widely separated locations, must be interchangeable, which requires precise size specifications and production.

The technique of dimensioning parts within a required range of variation to ensure interchangeability is called **tolerancing.** Each dimension is allowed a certain degree of variation within a specified zone, or tolerance. For example, a part's dimension might be expressed as 20 ± 0.50, which allows a tolerance (variation in size) of 1.00 mm.

A tolerance should be as **large as possible** without interfering with the function of the part to minimize production costs. Manufacturing costs increase as tolerances become smaller.

The cutting-tool holder in **Figure 13.1** illustrates a number of parts that must fit within a high degree of precision in order for

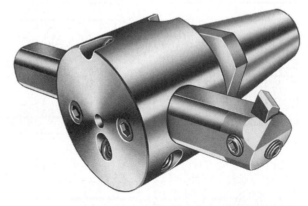

13.1 This cutting-tool holder would not function precisely while being used to make cuts on a lathe unless tolerances were used. All of its parts have been finished within close tolerances. *(Courtesy of Rockwell Automation.)*

it to work properly. It must hold the tool exactly in order to work properly with a lathe.

13.2 Tolerance Dimensions

Three methods of specifying tolerances on dimensions—**unilateral, bilateral,** and **limit forms**—are shown in **Figure 13.2.** When

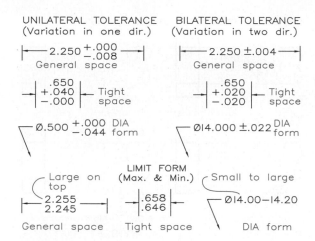

UNILATERAL TOLERANCE
(Variation in one dir.)

BILATERAL TOLERANCE
(Variation in two dir.)

2.250 +.000 / −.008   General space

2.250 ±.004   General space

.650 +.040 / −.000   Tight space

.650 +.020 / −.020   Tight space

Ø.500 +.000 / −.044 DIA form

Ø14.000 ±.022 DIA form

Large on top

LIMIT FORM
(Max. & Min.)

Small to large

2.255 / 2.245   General space

.658 / .646   Tight space

Ø14.00−14.20   DIA form

13.2 These examples show properly applied tolerances in unilateral, bilateral, and limit forms for general and tight spaces.

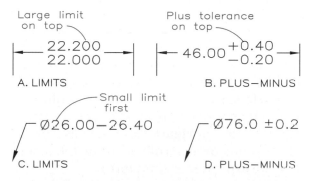

Large limit on top

Plus tolerance on top

22.200 / 22.000

A. LIMITS

46.00 +0.40 / −0.20

B. PLUS−MINUS

Small limit first

Ø26.00−26.40

C. LIMITS

Ø76.0 ±0.2

D. PLUS−MINUS

13.3 Place upper limits either above or to the right of lower limits. In plus-or-minus tolerancing, place the plus limits above the minus limits.

plus-or-minus tolerancing is used, it is applied to a theoretical dimension called the **basic dimension.** When dimensions can vary in only one direction from the basic dimension (either larger or smaller) tolerancing is **unilateral.** Tolerancing that permits variation in both directions from the basic dimension (larger and smaller) is **bilateral.**

Tolerances may be given in limit form, with dimensions representing the largest and smallest sizes for a feature. When tolerances are shown in limit form, the basic dimension will be unknown.

The customary methods of applying tolerance values on dimension lines are shown in

H = 1/8

Same no. of decimal places

H/2 = 1/16

H Max

2.0000 +.0040 / −.0020

A. PLUS−MINUS TOLERANCES

H/2 = 1/16

2.0400 / 1.9980

B. LIMIT−FORM TOLERANCES

13.4 The spacing and proportions of numerals used to specify tolerances on dimensions are shown here.

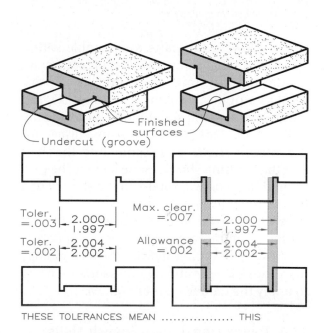

Undercut (groove)

Finished surfaces

THESE TOLERANCES MEAN .... THIS

Toler. =.003   2.000 / 1.997

Toler. =.002   2.004 / 2.002

Max. clear. =.007   2.000 / 1.997

Allowance =.002   2.004 / 2.002

13.5 These mating parts have tolerances (variations in size) of 0.003″ and 0.002″, respectively. The allowance (tightest fit) between the assembled parts is 0.002″.

Figure 13.3. The spacing and proportions of the tolerance dimensions are shown in **Figure 13.4.**

13.3 Mating Parts

Mating parts must be toleranced to fit within a prescribed degree of accuracy **(Figure 13.5).** The upper part is dimensioned with limits indicating its maximum and minimum sizes. The slot in the lower part is toleranced to be slightly

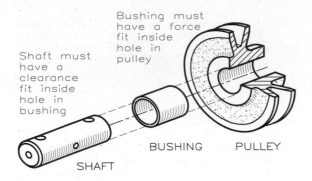

13.6 Mating parts.
These parts must be assembled with cylindrical fits that give a clearance and an interference fit.

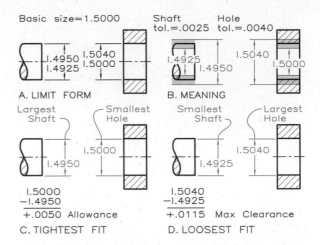

13.7 Cylindrical fits: English units.
The allowance (tightest fit) between these assembled parts is +0.005″. The maximum clearance is +0.0115″.

larger, allowing the parts to assemble with a clearance fit that allows freedom of movement.

Mating parts may be cylindrical forms, such as a pulley, bushing, and shaft **(Figure 13.6).** The bushing should force fit inside the pulley to provide a good bearing surface for the rotating shaft. At the same time, the shaft and the bushing should mate so that the pulley and bushing will rotate on the shaft with a free running fit.

ANSI tables (see Appendixes 25–29) prescribe cylindrical-fit tolerances for different applications. Familiarity with the terminology of cylindrical tolerancing is essential in order to apply the data from these tables.

13.4 Tolerancing Terms: English Units

The following terminology and definitions of tolerancing are illustrated in **Figure 13.7,** which shows two mating cylindrical parts.

Tolerance: The difference between the limits of size prescribed for a single feature, or 0.0025 inch for the shaft and 0.0040 for the hole in **Figure 13.7A.**

Limits of tolerance: The maximum and minimum sizes of a feature, or 1.4925 and 1.4950 for the shaft and 1.5000 and 1.5040 for the hole in **Figure 13.7B.**

Allowance: The tightest fit between two mating parts, or +0.0050 in **Figure 13.7C.** Allowance is negative for an interference fit.

Nominal size: A general size of a shaft or hole, usually expressed with common fractions, 1-1/2 inch or 1.50 inch as in **Figure 13.7.**

Basic size: The size to which a plus-or-minus tolerance is applied to obtain the limits of size, 1.5000 in **Figure 13.7.** The basic diameter cannot be determined from tolerances that are expressed in limit form.

Actual size: The measured size of the finished part.

Fit: The degree of tightness or looseness between two assembled parts, which can be one of the following: **clearance, interference, transition,** and **line.**

Clearance fit gives a clearance between two assembled mating parts. The shaft and the hole in **Figures 13.7C** and **D** have a minimum clearance of 0.0050″ and a maximum clearance of 0.0115″.

Interference fit results in a binding fit that requires the parts to be forced together much as if they are welded **(Figure 13.8A).**

Transition fit may range from an interference to a clearance between the assembled parts. The shaft may be either smaller or larger than the hole and still be within the prescribed tolerances as in **Figure 13.8B.**

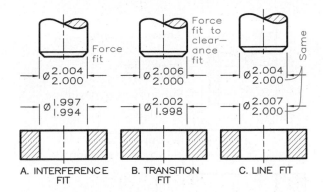

A. INTERFERENCE FIT   B. TRANSITION FIT   C. LINE FIT

13.8 Types of cylindrical fits.
Three types of fits between mating parts are shown here in addition to the clearance fit shown in the Figure 13.7.

Line fit results in surface contact or clearance when the limits are reached (**Figure 13.8C).**

Selective assembly: A method of selecting and assembling parts by hand by trial and error, which allows parts to be made with larger tolerances at less cost as a compromise between a high manufacturing accuracy and ease of assembly.

Single-limits: Dimensions designated by either minimum (MIN) or maximum (MAX) as shown in **Figure 13.9.** Depths of holes, lengths, threads, corner radii, and chamfers are sometimes dimensioned in this manner.

13.5 Basic Hole System: Inches

The **basic hole system** uses the smallest hole size as the **basic diameter** for calculating tolerances and allowances. The basic hole system is best when drills, reamers, and machine tools are used to give precise hole sizes.

The smallest hole size is the basic diameter because a hole can be enlarged by machining but not reduced in size. If the smallest diameter of the hole is 1.500″, subtract the allowance, 0.0050″, from it to find the diameter of the largest shaft, 1.4950″. To find the smallest limit for the shaft diameter, subtract the tolerance from 1.4950 inches.

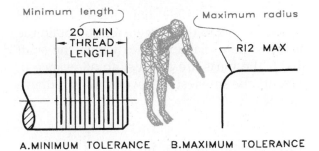

A. MINIMUM TOLERANCE   B. MAXIMUM TOLERANCE

13.9 Single tolerances in maximum (MAX) or minimum (MIN) form can be given in applications of this type.

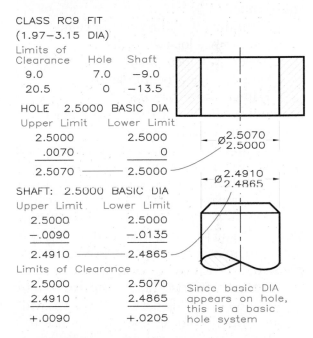

13.10 Calculations: Basic hole system.
This example shows how to calculate basic hole limits and allowances for an RC9 fit between a shaft and hole with a basic diameter of 2.5000″. Values are taken from Appendix 25.

13.6 Basic Shaft System: Inches

The **basic shaft system** uses the largest diameter as the **basic diameter** to which the tolerances are applied. This system is applicable when shafts are available in uniform standard sizes.

The largest shaft size is used as the basic diameter because shafts can be machined to a smaller size but not enlarged. For example, if the largest permissible shaft size is 1.500″, add the allowance to this dimension to obtain the smallest hole diameter into which the shaft fits. If the parts are to have an allowance of 0.0040″, the smallest hole would have a diameter of 1.5040″.

13.7 Cylindrical Fits: Inches

The *ANSI B4.1* standard gives a series of fits between cylindrical features in inches for the basic hole system. The types of fit covered in this standard are:

RC: running or sliding clearance fits
LC: clearance locational fits
LT: transition locational fits
LN: interference locational fits
FN: force and shrink fits

Appendixes 25–29 list these five types of fit, each of which has several classes.

Running or **sliding clearance fits** (RC) provide a similar running performance with suitable lubrication allowance. The clearance for the first two classes (RC 1 and RC 2), which are used chiefly as slide fits, increases more slowly with diameter size than other classes to maintain an accurate location even at the expense of free relative motion.

Locational fits (LC, LT, LN) determine only the location of mating parts; they may provide nonmoving rigid locations (interference fits) or permit some freedom of location (clearance fits). The three locational fits are: **clearance fits (LC), transition fits (LT),** and **interference fits (LN).**

Force fits (FN) are interference fits characterized by a constant bore pressure throughout the range of sizes. There are five types of force fits: FN1 through FN5 varying from light drive to heavier drives, respectively.

The method of applying tolerance values from the tables in Appendix 25 for an RC9 (basic hole system) fit is shown in **Figure 13.10.** The basic diameter of 2.5000″ falls between 1.97″ and 3.15″ in the *Size* column of the table.

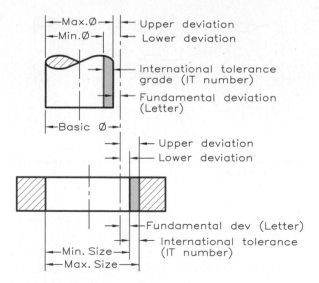

13.11 Cylindrical fits: Metric units.
The terminology and definitions of the metric system of cylindrical fits are given here.

Limits are in thousandths, which requires that the decimal point be moved three places to the left. For example, +7 is +0.0070″.

Add the limits to the basic diameter when a **plus sign** preceeds the values and **subtract** when a **minus sign** is given. Add the limits (+0.0070″ and 0.0000″) to the basic diameter to find the upper and lower limits of the hole (2.5070″ and 2.5000″). Subtract the limits (−.0090″ and −.0135″) from the basic diameter to find the limits of the shaft (2.4910″ and 2.4865″).

The tightest fit (the allowance) between the assembled parts (+0.0090″) is the difference between the largest shaft and the smallest hole. The loosest fit, or maximum clearance, (+0.0205″) is the difference between the smallest shaft and the largest hole. These values appear in the *Limits* column of Appendix 25.

This method of extracting tolerancing dimensions is applied to other types of fits by using their respective tables: force fit, interference fit, transition fit, and locational fit. Subtract negative limits from the basic diameter and add positive limits to it. A **minus sign** preceding limits of clearance in

PREFERRED BASIC SIZES (Millimeters)					
First Choice	Second Choice	First Choice	Second Choice	First Choice	Second Choice
I		10		100	
	1.1		11		110
1.2	1.4	12	14	120	140
1.6	1.8	16	18	160	180
2	2.2	20	22	200	220
2.5	2.8	25	28	250	280
3	3.5	30	35	300	350
4	4.5	40	45	400	450
5	5.5	50	55	500	550
6	7	60	70	600	700
8	9	80	90	800	900
				1000	

13.12 Basic sizes for metric fits selected first from the first-choice column are preferred over those in the second-choice column.

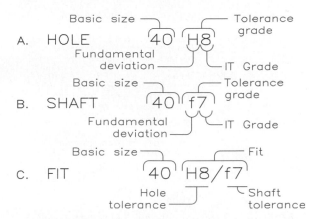

13.13 Tolerance notes.
These tolerance symbols and their definitions apply to holes and shafts.

the tables indicates an **interference fit** between the assembled parts, and a **positive sign** preceding the **limits** of clearance indicates a **clearance fit.**

13.8 Tolerancing: Metric Units

The system recommended by the *International Standards Organization (ISO)* in *ANSI B4.2* for metric measurements are fits that usually apply to cylinders—holes and shafts—but these tables can be used to specify fits between parallel contact surfaces such as a key in a slot.

Basic size: The size, usually a diameter from which limits or deviations are calculated (**Figure 13.11**). Select basic sizes from the *First Choice* column in the table in **Figure 13.12.**

Deviation: The difference between the hole or shaft size and the basic size.

Upper deviation: The difference between the maximum permissible size of a part and its basic size (**Figure 13.11**).

Lower deviation: The difference between the minimum permissible size of a part and its basic size (**Figure 13.11**).

Fundamental deviation: The deviation closest to the basic size (**Figure 13.11**). In the note 40 H8 in **Figure 13.13,** H represents the fundamental deviation for a hole, and in

the note 40 f7, the f represents the fundamental deviation for a shaft.

Tolerance: The difference between the maximum and minimum allowable sizes of a part.

International tolerance (IT) grade: A series of tolerances that vary with basic size to provide a uniform level of accuracy within a given grade (**Figure 13.11**). In the note 40 H8 in **Figure 13.13,** the 8 represents the IT grade. There are 18 IT grades: IT01, IT0, IT1, . . . , IT16.

Tolerance zone: A combination of the fundamental deviation and the tolerance grade. The H8 portion of the 40 H8 note in **Figure 13.13** is the tolerance zone.

Hole basis: A system of fits based on the minimum hole size as the basic diameter. The fundamental deviation letter for a hole-basis system is the **uppercase** H. Appendixes 31 and 32 give hole-basis data for tolerances.

Shaft basis: A system of fits based on the maximum shaft size as the basic diameter. The fundamental deviation letter for a shaft-basis system is the **lowercase** h. Appendixes 33 and 34 give shaft-basis data for tolerances.

Clearance fit: A fit resulting in a clearance between two assembled parts under all tolerance conditions.

Interference fit: A force fit between two parts, requiring that they be driven together.

Hole Basis	Shaft Basis	Description
H11/c11	C11/h11	Loose Running Fit for wide commerical tolerances on external members
H9/d9	D9/h9	Free Running Fit for large temperature variations, high running speeds, or high journal pressures
H8/f7	F8/h7	Close Running Fit for accurate location and moderate speeds and journal pressures
H7/g6	G7/h6	Sliding Fit for accurate fit and location and free moving and turning, not free running
H7/h6	H7/h6	Locational Clearance for snug fits for parts that can be freely assembled
H7/k6	K7/h6	Locational Transition Fit for accurate locations
H7/n6	N7/h6	Locational Transition Fit for more accurate locations and greater interference
H7/p6	P7/h6	Locational Interference Fit for rigidity and alignment without special bore pressures
H7/s6	S7/h6	Medium Drive Fit for shrink fits on light sections; tightest fit usable for cast iron
H7/u6	U7/h6	Force Fit for parts that can be highly stressed and for shrink fits.

(Clearance Fits: H11/c11, H9/d9, H8/f7, H7/g6. Transition Fits: H7/h6, H7/k6, H7/n6. Interference Fits: H7/p6, H7/s6, H7/u6.)

13.14 This list gives the preferred hole-basic and shaft-basis fits for the metric system.

Transition fit: A fit that can result in either a clearance or an interference between assembled parts.

Tolerance symbols: Notes giving the specifications of tolerances and fits (**Figure 13.13**). The basic size is a number, followed by the fundamental deviation letter and the IT number, that gives the tolerance zone.

Uppercase letters: (H) indicate the fundamental deviations for **holes; lowercase letters** (h) indicate fundamental deviations for **shafts.**

Preferred Sizes and Fits

The table in **Figure 13.12** shows the preferred basic sizes for computing tolerances. Under the *First Choice* heading, each number increases by about 25% from the preceding

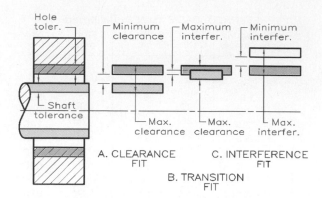

13.15 Types of fits: (A) clearance fit, where there is space between the parts, (B) transition fit, where there can be either interference or clearance, and (C) interference fit, where the parts must be forced together.

value. Each number in the *Second Choice* column increases by about 12%. To minimize cost, select basic diameters from the first column because they correspond to standard stock sizes for round, square, and hexagonal metal products.

Figure 13.14 shows preferred clearance, transition, and interference fits for the hole-basis and shaft-basis systems. Appendixes 30–36 contain the complete tables. **Figure 13.15** illustrates the differences between clearance, interference, and transition fits.

Preferred Fits: Hole-Basis System The preferred fits for the hole-basis system in which the smallest hole is the basic diameter are shown in **Figure 13.15.** Variations in fit between parts range from a clearance fit of H11/c11 to an interference fit of H7/u6 (see **Figure 13.14**).

Preferred Fits: Shaft-Basis System The preferred fits of the shaft-basis system in which the largest shaft is the basic diameter are shown in **Figure 13.14.** Variations in fit range from a clearance fit of C11/h11 to an interference fit of U7/h6 (see **Figure 13.14**).

Standard Cylindrical Fits

The following examples demonstrate how to calculate and apply tolerances to cylindrical

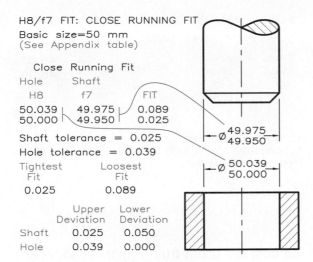

H8/f7 FIT: CLOSE RUNNING FIT
Basic size=50 mm
(See Appendix table)

Close Running Fit

Hole	Shaft	
H8	f7	FIT
50.039	49.975	0.089
50.000	49.950	0.025

Shaft tolerance = 0.025
Hole tolerance = 0.039

Tightest Fit	Loosest Fit
0.025	0.089

	Upper Deviation	Lower Deviation
Shaft	0.025	0.050
Hole	0.039	0.000

Ø 49.975 / 49.950

Ø 50.039 / 50.000

13.16 Example 1—Close running fit, hole basis. This drawing shows how to calculate and apply metric limits and fits to a shaft and hole (Appendix 31).

LOCATIONAL TRANSITION FIT—H7/k6
Basic DIA=60 mm

Ø 60k6   Ø 60H7

A. NOTE FORM

Ø 60.021 / 60.002   Ø 60.030 / 60.000

B. LIMIT FORM

13.17 Example 2—Location transition fit, hole basis. These methods are used to note metric tolerances to a hole and a shaft with a transition fit (Appendix 32).

parts. You must use **Figure 13.11, Figure 13.14,** and data from Appendixes 30–36.

Example 1 (Figure 13.16) *Required:* Use the hole-basis system, a close running fit, and a basic diameter of 49 mm.

Solution: Use a preferred basic diameter of 50 mm (**Figure 13.11**) and fit of H8/f 7 (**Figure 13.14**).

Hole: Find the upper and lower limits of the hole in Appendix 31 under H8 and across from 50 mm; 50.000 mm and 50.039 mm.

Shaft: From Appendix 31 find the upper and lower limits of the shaft under f 7 and across from 50 mm; 49.950 mm and 49.975 mm.

Symbols: **Figure 13.16** shows how to apply toleranced dimensions to the hole and shaft.

Example 2 (Figure 13.17) *Required:* Use the hole-basis system, a location transition fit of medium accuracy, and a basic diameter of 57 mm.

Solution: Use a preferred basic diameter of 60 mm (**Figure 13.11**) and a fit of H7/k6 (**Figure 13.14**).

Hole: Find the upper and lower limits of the hole in Appendix 32 under H7 and across from 60 mm; 60.000 mm and 60.030 mm.

Shaft: From Appendix 32 find the upper and lower limits of the shaft under k6 and across from 60 mm; 60.002 and 60.021 mm.

Symbols: **Figure 13.17** shows two methods of applying the tolerance symbols to a drawing: note and limit form.

Example 3 (Figure 13.18) *Required:* Use the hole-basis system, a medium drive fit, and a basic diameter of 96 mm.

Solution: Use a preferred basic diameter of 100 mm (**Figure 13.11**) and a fit of H7/s6 (**Figure 13.14**).

Hole: Find the upper and lower limits of the hole in Appendix 32 under H7 and across

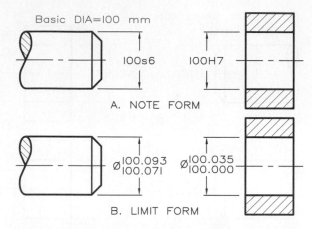

Basic DIA=100 mm

100s6   100H7

A. NOTE FORM

⌀100.093   ⌀100.035
100.071   100.000

B. LIMIT FORM

13.18 Example 3—Interference fit, hole basis. Either of these formats can be used to apply metric tolerances to a hole and shaft that have an interference fit (Appendix 32).

CALCULATION OF NONSTANDARD LIMITS

FIT: H8/f7   ⌀45 BASIC DIA

From Appendix			
Hole	Shaft	Hole Limits	45.039
H8	f7		45.000
0.039	−0.025	Shaft Limits	44.975
0.000	−0.050		44.950

13.19 Nonstandard limits.
This calculation is for an H8/f7 nonstandard diameter of 45 mm (Appendixes 35 and 36).

from 100 mm to obtain limits of 100.035 and 100.000 mm.

Shaft: Find the upper and lower limits of the shaft under s6 and across from 100 mm in Appendix 32. These limits are 100.093 and 100.071 mm. Appendix 32 gives the tightest fit as an interference of −0.093 mm, and the loosest fit as an interference of −0.036 mm. Minus signs in front of these numbers in the fit column indicate interference fits.

Symbols: **Figure 13.18** shows two methods of applying toleranced dimensions to the hole and shaft: note and limit form.

Nonstandard Fits: Nonpreferred Sizes
Limits of tolerances for nonstandard sizes of the fits in **Figure 13.14** that do not appear in Appendixes 31–34 can be calculated as shown in **Figure 13.19.** Limits of tolerances for nonstandard hole sizes are in Appendix 35, and limits of tolerances for nonstandard shaft sizes are in Appendix 36.

Figure 13.19 shows the hole and shaft limits for an H8/f7 fit and a 45-mm DIA. The tolerance limits of 0.000 and 0.039 mm for an

H8 hole are taken from Appendix 35, across from the size range of 40–50 mm. The f7 tolerance limits of −0.025 and −0.050 mm for the shaft are from Appendix 36. Calculate the hole limits by adding the positive tolerances to the 45 mm basic diameter and the shaft limits by subtracting the negative tolerances from the 45 mm basic diameter.

13.9 Chain versus Datum Dimensions

When parts are dimensioned to locate surfaces or geometric features by a chain of dimensions laid end-to-end **(Figure 13.20A),** variations may accumulate in excess of the specified tolerance. For example, the tolerance between surfaces A and B is 0.02, between A and C it is 0.04, and between A and D it is 0.06.

Tolerance accumulation can be eliminated by measuring from a single plane called a **datum plane** or **baseline.** A datum plane is usually on the object, but it can also be on the machine used to make the part. Because each plane in **Figure 13.20B** is located with respect to a datum plane, the tolerances between the intermediate planes do not exceed the maximum tolerance of 0.02. Always base the application of tolerances on the function of a part in relationship to its mating parts.

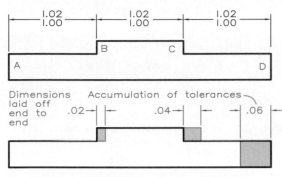

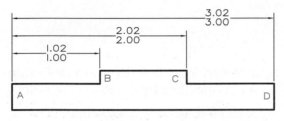

A. CHAIN DIMENSIONS

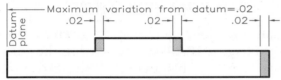

B. DATUM PLANE (BASELINE) DIMENSIONS

13.20 Chain vs. datum dimensioning.
A Dimensions given end to end in a chain fashion may result in an accumulation of tolerances of up to 0.06″ at D instead of the specified 0.02″.
B When dimensioned from a single datum, the variations of B, C, and D cannot deviate more than the specified 0.02″ from the datum.

13.10 Tolerance Notes

You should tolerance all dimensions on a drawing either by using the rules previously discussed or by placing a note in or near the title block. For example, the note

TOLERANCE ±1/64

might be given on a drawing for less critical dimensions. Some industries give dimensions in inches with two, three, and four decimal place fractions. A note for dimensions with

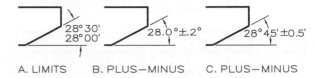

A. LIMITS   B. PLUS—MINUS   C. PLUS—MINUS

13.21 Angular tolerances.
Tolerances on angles can be specified by one of these methods.

two and three decimal places might be given on the drawing as

TOLERANCES XX.XX ±0.10
XX.XXX ±0.005.

Tolerances of four places would be given directly on the dimension lines. The most common method of noting tolerances is to give as large a tolerance as feasible in a note, such as

TOLERANCES ±0.05

and to give tolerances on the dimension lines for dimensions requiring closer tolerances. Give angular tolerances in a general note in or near the title block, such as

ANGULAR TOLERANCES ±0.50° or 30′

Use one of the formats shown in **Figure 13.21** to give specific angular tolerances directly on angular dimensions.

13.11 General Tolerances: Metric

All dimensions on a drawing must be specified within certain tolerance ranges when they are not shown on dimension lines. Tolerances not shown on dimension lines should be specified by a general tolerance note on the drawing.

Linear-Dimensions: Tolerance linear dimensions by indicating plus and minus (±) one half of an international tolerance (IT) grade as given in Appendix 30. You may select the IT grade from the chart in **Figure 13.22,** where IT grades for mass-produced items range from

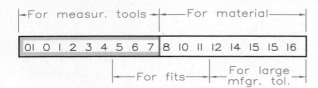

13.22 International tolerance (IT) grades and their applications are shown here. See Appendix 30 to obtain IT tolerance grade numerical values.

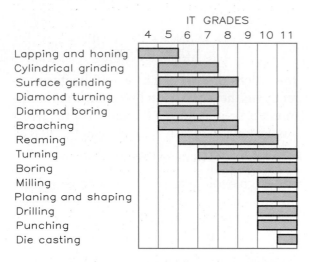

13.23 International tolerance (IT) values may be selected from this table, which is based on the general capabilities of various maching processes.

GENERAL TOLERANCES:
LINEAR DIMENSIONS (mm)

Basic Dimensions	Fine Series	Medium Series	Coarse Series
0.5 to 3	± 0.05	± 0.1	— —
Over 3 to 6	± 0.05	± 0.1	± 0.2
Over 6 to 30	± 0.1	± 0.2	± 0.5
Over 30 to 120	± 0.15	± 0.3	± 0.8
Over 120 to 315	± 0.2	± 0.5	± 1.2
Over 315 to 1000	± 0.3	± 0.8	± 2
Over 1000 to 2000	± 0.5	± 1.2	± 3

13.24 Select general tolerance values from this table for fine, medium, and coarse series. Tolerances vary with the sizes of dimensions.

IT12 through IT16. IT grades can be selected from **Figure 13.23** for a particular machining process.

General tolerances using IT grades may be expressed in a note as follows:

> UNLESS OTHERWISE SPECIFIED
> ALL UNTOLERANCED
> DIMENSIONS ARE IT14

This note means that a tolerance of ±0.700 mm is allowed for a dimension between 315 and 400 mm. The value of the tolerance 1.400 mm is extracted from Appendix 30.

Figure 13.24 shows recommended tolerances for fine, medium, and coarse series for ranges of size. A medium tolerance, for example, can be specified by the following note:

> GENERAL TOLERANCES SPECIFIED
> IN ANSI B4.3-MEDIUM SERIES APPLY

Equivalent tolerances may be given in table form **(Figure 13.25)** on the drawing, the grade (medium, in this example) selected from **Figure 13.24.** General tolerances may be given in a table for dimensions expressed with one or no decimal places **(Figure 13.26).** General tolerances may also be notated in the following form:

> UNLESS OTHERWISE SPECIFIED
> ALL UNTOLERANCED DIMENSIONS
> ARE ±0.8 mm

Use this method only when the dimensions on a drawing are similar in size.

Angular Tolerances: Express angular tolerances as (1) an angle in decimal degrees or in degrees and minutes, (2) a taper expressed in percentage (mm per 100 mm), or (3) milliradians. (To find milliradian, multiply the degrees of an angle by 17.45.) **Figure 13.27** shows the suggested tolerances for decimal degrees and taper, based on the length of the shorter leg of the angle. General angular tolerances may be notated on the drawing as follows:

GENERAL TOLERANCES (mm) UNLESS OTHERWISE SPECIFIED, THE FOLLOWING TOLERANCES ARE APPLICABLE							
LINEAR	Over to	0.5 6	6 30	30 120	120 315	315 1000	1000 2000
TOL.	±	0.1	0.2	0.3	0.5	0.8	1.2

13.25 This table for a medium series of values was extracted from Figure 13.24 for insertion on a working drawing to provide the tolerances for a medium series of sizes.

Medium series for numbers with one decimal place | Coarse series for numbers with no decimal places

GENERAL TOLERANCES (mm) UNLESS OTHERWISE SPECIFIED, THE FOLLOWING TOLERANCES ARE APPLICABLE					
LINEAR		OVER TO	— 120	120 315	315 1000	1000 —
TOL.	ONE DECIMAL ±	0.3	0.5	0.8	1.2
	NO DECIMALS ±	0.8	1.2	2	3

13.26 Placed on a drawing, this table of tolerances would indicate the tolerances for dimensions having one or no decimal places, such as 24.0 and 24, denoting medium and coarse series.

Length of shorter leg (mm)	Up to 10	Over 10 to 50	Over 50 to 120	Over 120 to 400
Degrees	± 1°	± 0° 30'	± 0°20'	± 0°10'
mm per 100	± 1.8	± 0.9	± 0.6	± 0.3

13.27 General tolerances for angular and taper dimensions may be taken from this table of values.

UNLESS OTHERWISE SPECIFIED
THE GENERAL TOLERANCES
IN ANSI B4.3 APPLY

A second method involves showing a portion of the table from **Figure 13.27** as a table of tolerances on the drawing (**Figure 13.28**). A

ANGULAR TOLERANCES				
LENGTH OF SHORTER LEG (mm)	UP TO 10	OVER 10 TO 50	OVER 50 TO 120	OVER 120 TO 400
TOLERANCE	±1°	±0°30'	±0°20'	±0°10'

Values in degrees and minutes taken from previous table

13.28 This table, extracted from Figure 13.27, is placed on the drawing to indicate the general tolerance for angles in degrees and minutes.

third method is a note with a single tolerance such as:

UNLESS OTHERWISE SPECIFIED
ANGULAR TOLERANCES ARE ±0°30'

13.12 Geometric Tolerances

Geometric tolerancing is a system that specifies tolerances that control **form, profile, orientation, location,** and **runout** on a dimensioned part as covered by the *ANSI Y14.5M Standards* and the *Military Standards* (Mil-Std) of the U.S. Department of Defense. Before discussing those types of tolerancing, however, we need to introduce you to symbols, size limits, rules, three-datum-plane concepts, and applications.

Symbols

The most commonly used symbols for representing geometric characteristics of dimensioned drawings are shown in **Figure 13.29.** The proportions of feature control symbols in relation to their feature control frames, based on the letter height, are shown in **Figure 13.30.** On most drawings, a 1/8-inch or 3-mm letter height is recommended. Examples of feature control frames and their proportions are shown in **Figure 13.31.**

Size Limits

Three conditions of size are used when geometric tolerances are applied: **maximum material condition, least material condition,** and **regardless of feature size.**

GEOMETRIC SYMBOLS

	Tolerance	Characteristic	Symbol
INDIVIDUAL FEATURES	Form	Straightness	—
		Flatness	▱
		Circularity	○
		Cylindricity	⌭
BOTH	Profile	Profile: Line	⌒
		Profile: Surface	⌓
RELATED FEATURES	Orientation	Angularity	∠
		Perpendicularity	⊥
		Parallelism	//
	Location	Position	⊕
		Concentricity	◎
		Symmetry	≡
	Runout	Runout: Circular	↗
		Runout: Total	↗↗

13.29 Geometric symbols.
These symbols specify the geometric characteristics of a part's features.

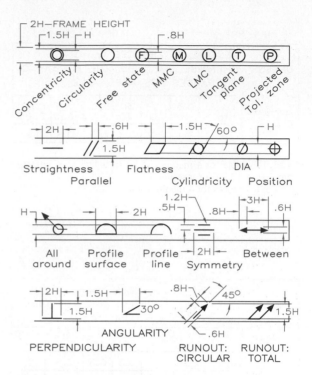

13.30 Feature control symbols.
The proportions of these feature control symbols are based on the letterheight, usually 1/8 inch high.

Maximum material condition (MMC) indicates that a feature contains the maximum amount of material. For example, the shaft shown in **Figure 13.32** is at MMC when it has the largest permitted diameter of 24.6 mm. The hole is at MMC when it has the most material, or the smallest diameter of 25.0 mm.

Least material condition (LMC) indicates that a feature contains the least amount of material. The shaft in **Figure 13.32** is at LMC when it has the smallest diameter of 24.0 mm. The hole is at LMC when it has the least material, or the largest diameter of 25.6 mm.

Regardless of feature size (RFS) indicates that tolerances apply to a geometric feature regardless of its size ranging from MMC to LMC.

13.13 Rules for Tolerancing

Two general rules of tolerancing geometric features should be followed.

Rule 1 Individual Feature of Size: When only a tolerance of size is specified on a feature, the limits of size control the variation in its geometric form. The forms of the shaft and hole shown in **Figure 13.33** are permitted to vary within the tolerance ranges of the dimensions.

Rule 2 All Applicable Geometric Tolerances: Where no modifying symbol is specified, RFS (regardless of features size) applies with respect to the individual tolerance, datum reference, or both. Where required on a drawing, the modifiers MMC or LMC must be specified.

Alternate Practice: For a tolerance of position, RFS may be specified on the drawing

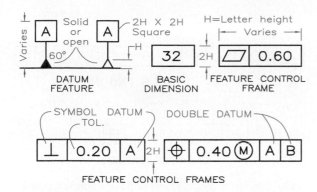

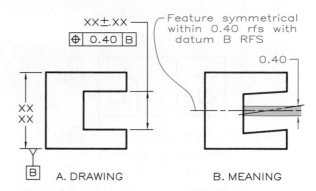

13.31 These are examples of geometric tolerancing frames and feature control symbols used to indicate geometric tolerances. H is the letter height.

13.34 Tolerance of position, when dimensioned in this manner, indicate that a tolerance of 0.40 applies regardless of feature size (RFS).

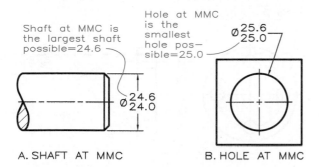

13.32 A shaft is at maximum material condition (MMC) when it is at the largest size permitted by its tolerance. A hole is at MMC when it is at its smallest size.

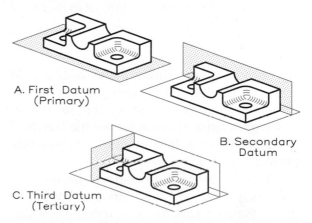

13.35 When an object is referenced to a primary datum plane, it contacts the datum at at least three points. The vertical surface contacts the secondary datum plane at at least two points. The third surface contacts the third datum at at least one point. Datum planes are listed in order of priority in the feature control frame.

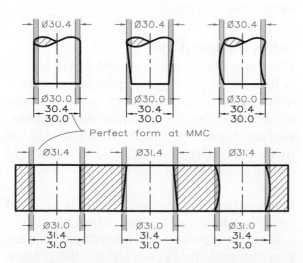

13.33 When only a tolerance of size is specified on a feature, the limits prescribe the form of the features, as shown for these shafts and holes having identical limits.

with respect to the individual tolerance, datum, reference, or both.

The specification of symmetry of the part in **Figure 13.34** is based on a tolerance at RFS from the datum.

Three-Datum-Plane Concept

A datum plane is used as the origin of a part's features that have been toleranced. Datum planes usually relate to manufacturing equipment, such as machine tables or locating pins.

Three mutually perpendicular datum planes are required to dimension a part accurately. For example, the part shown in **Figure 13.35**

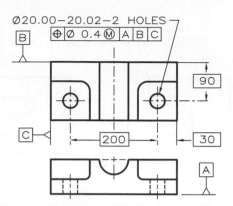

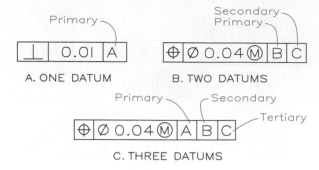

13.37 Use feature control frames to indicate from one to three datum planes in order of priority.

13.36 The three planes of the reference system are noted where they appear as edges. The primary datum plane (A) is given first in the feature control frame; the secondary plane (B) second; and the tertiary plane (C) third. Single numbers in frames are basic dimensions.

sits on the primary datum plane, with at least three points of its base in contact with the datum. The part is related to the secondary plane by at least two contact points. The third (tertiary) datum is in contact with at least one point on the object.

The priority of datum planes is presented in sequence in feature control frames. For example, in **Figure 13.36,** the primary datum is surface A, the secondary datum is surface B, and the tertiary datum is surface C. **Figure 13.37** lists the order of priority of datum planes A–C sequentially in the feature control frames.

13.14 Cylindrical Datum Features

A part with a cylindrical datum feature that is the axis of a cylinder is illustrated in **Figure 13.38.** Datum K is the primary datum. Datum M is associated with two theoretical planes—the second and third in a three-plane relationship. The two theoretical planes are represented in the circular view by perpendicular centerlines that intersect at the point view of the datum axis. All dimensions originate from the datum axis perpendicular to datum K; the other two intersecting datum planes are used for measurements in the *x* and *y* directions.

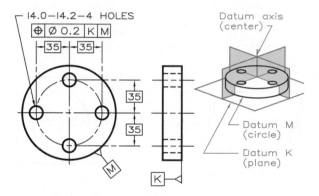

13.38 These true-position holes are located with respect to primary datum K and secondary datum M. Because datum M is a circle, the holes are located about two intersecting datum planes at the crossing centerlines in the circular view, satisfying the three-plane concept.

Datum Features at RFS

When size dimensions are applied to a feature at RFS, the processing equipment that comes into contact with surfaces of the part establishes the datum. Variable machine elements, such as chucks or center devices, are adjusted to fit the external or internal features and establish datums.

Primary-Diameter Datums: For an external cylinder (shaft) at RFS, the datum axis is the axis of the smallest circumscribed cylinder that contacts the cylindrical feature (**Figure 13.39A**). That is, the largest diameter of the part making contact with the smallest cylinder

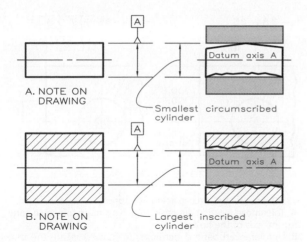

13.39 The datum axis of a shaft is the smallest circumscribed cylinder in contact with the shaft. The datum axis of a hole is the centerline of the largest inscribed cylinder in contact with the hole.

of the machine element holding the part is the datum axis.

For an internal cylinder (hole) at RFS, the datum axis is the axis of the largest inscribed cylinder making contact with the hole. That is, the smallest diameter of the hole making contact with the largest cylinder of the machine element inserted in the hole is the datum axis **(Figure 13.39B).**

Primary External Parallel Datums: The datum for external features at RFS is the center plane between two parallel planes—at minimum separation—that contact the planes of the object **(Figure 13.40A).** These are planes of a viselike device at minimum separation that holds the part.

Primary Internal Parallel Datums: The datum for internal features is the center plane between two parallel planes—at their maximum separation—that contact the inside planes of the object **(Figure 13.40B).**

Secondary-Datums: The secondary datum (axis or center plane) for both external and internal diameters (or distances between parallel planes) has the additional requirement

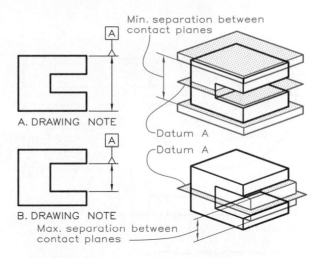

13.40 The datum plane for external parallel surfaces is the center plane between two contact parallel planes at their minimum separation. The datum plane for internal parallel surfaces is the center plane between two contact parallel surfaces at their maximum separation.

that the cylinder in contact with the parallel elements of the hole be perpendicular to the primary datum **(Figure 13.41).** Datum axis B is the axis of cylinder B.

Tertiary-Datums: The third datum (axis or center plane) for both external and internal features has the further requirement that either the cylinder or parallel planes be oriented angularly to the secondary datum. Datum C in **Figure 13.41** is the tertiary datum plane.

13.15 Location Tolerancing

Tolerances of location specify **position, concentricity,** and **symmetry.**

Position: Location dimensions that are toleranced result in a square (or rectangular) tolerance zone for locating the center of a hole **(Figure 13.42A).** In contrast, untoleranced location dimensions, called **basic dimensions,** locates the **true position** of a hole's center about which a circular tolerance zone is specified **(Figure 13.42B).**

In both methods the size of the hole's diameter is toleranced by identical notes. In the true-position method, a feature control frame

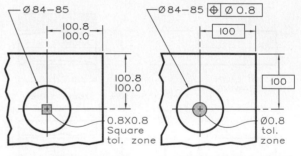

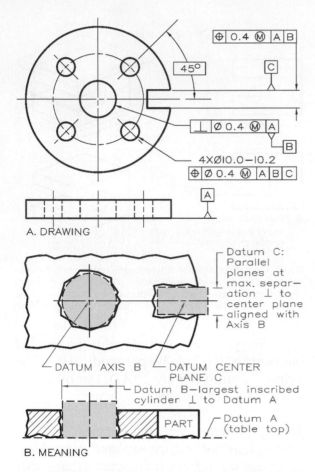

A. DRAWING

Datum C:
Parallel
planes at
max. separ-
ation ⊥ to
center plane
aligned with
Axis B

DATUM AXIS B — DATUM CENTER PLANE C

Datum B—largest inscribed cylinder ⊥ to Datum A

PART

Datum A (table top)

B. MEANING

13.41 The features of this part have been dimensioned with respect to primary, secondary, and tertiary datum planes.

A. SQUARE TOL. ZONE   B. CIRCULAR TOL. ZONE

13.42 Square and circular tolerance zones.
A Toleranced location dimensions give a square tolerance zone for the axis of the hole.
B Untoleranced basic dimensions (in frames) locate the true position about which a circular tolerance zone of 0.8 mm is specified.

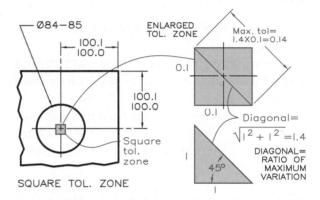

SQUARE TOL. ZONE

13.43 Square tolerancing.
Toleranced coordinates give a square tolerance zone with a diagonal that exceeds the specified tolerance by a factor of 1.4.

specifies the diameter of the circular tolerance zone inside which the hole's center must lie. A circular position zone gives a more precise tolerance of the hole's true position than a square.

Figure 13.43 shows an enlargement of the square tolerance zone resulting from the use of toleranced location dimension to locate a hole's center. The diagonal across the square zone is greater than the specified tolerance by a factor of 1.4. Therefore the **true-position method,** shown enlarged in **Figure 13.44,** can have a larger circular tolerance zone by a factor of 1.4 and still have the same degree of accuracy specified by the 0.1 square zone. If a variation of 0.14 across the diagonal of the square tolerance zone is acceptable in the coordinate

method, a circular tolerance zone of 0.14, which is greater than the 0.1 tolerance permitted by the square zone, should be acceptable in the true-position tolerance method.

The circular tolerance zone specified in the circular view of a hole extends the full depth of the hole. Therefore, the tolerance zone for the centerline of the hole is a cylindrical zone inside which the axis must lie. Because both the size of the hole and its position are toleranced, these two tolerances establish the diameter of a gauge cylinder for checking conformance of hole sizes and their locations against specifications **(Figure 13.44).**

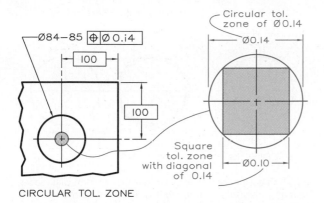

CIRCULAR TOL. ZONE

13.44 True-position tolerancing.
The true-position method of locating holes results in a circular tolerance zone. The circular zone can be 1.4 times greater than the square tolerance zone and still be as accurate.

Concentricity: Concentricity (closely related to a new term, *coaxiality*) is a feature of location because it specifies the location of two cylinders that share the same axis. In **Figure 13.45,** the large cylinder is labeled as datum A to be used as the datum for locating the small cylinder's axis.

Feature control frames of the type shown in **Figure 13.46** are used to specify concentricity and other geometric characteristics throughout the remainder of this chapter.

Symmetry: Symmetry also is a feature of location in which a feature is symmetrical with the same contour and size on opposite sides of a central plane. **Figure 13.47A** shows how to apply a symmetry feature symbol to the notch that is symmetrical about the part's central datum plane B for a zone of 0.6 mm **(Figure 13.47B).**

13.16 Form Tolerancing

Flatness: A surface is flat when all its elements are in one plane. A feature control frame specifies flatness within a 0.4 mm tolerance zone in **Figure 13.48** where no point on the surface may vary more than 0.40 from the highest to the lowest point.

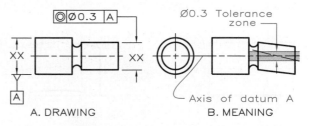

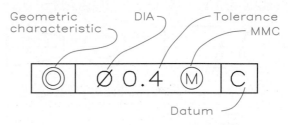

13.45 Concentricity (related to coaxiality) is a tolerance of location. This feature control frame specifies that the axis of the small cylinder be concentric to datum cylinder A, within a tolerance of a 0.3 mm diameter.

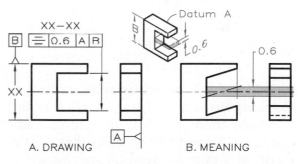

13.46 This typical feature control frame indicates that a surface is concenteric to datum C within a cylindrical diameter of 0.4 mm at MMC.

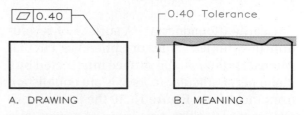

13.47 Symmetry is a tolerance of location that specifies that a part's features be symmetrical about the center plane between parallel surfaces of the part.

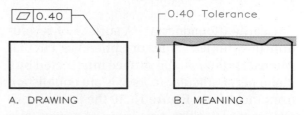

13.48 Flatness is a tolerance of form that specifies a tolerance zone within which a surface must lie.

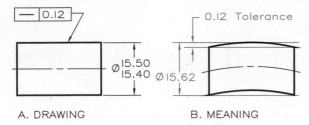

A. DRAWING   B. MEANING

13.49 Straightness is a tolerance of form that indicates that elements of a surface are straight lines. The tolerance frame is applied to the views in which elements appear as lines, not as points.

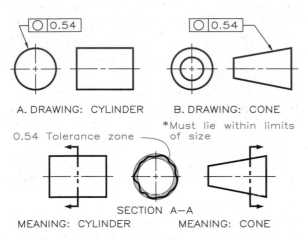

A. DRAWING: CYLINDER   B. DRAWING: CONE

MEANING: CYLINDER   MEANING: CONE

13.50 Circularity (roundness) is a tolerance of form. It indicates that a cross section through a surface of revolution is round and lies within two concentric circles.

Straightness:
A surface is straight if all its elements are straight lines within a specified tolerance zone. The feature control frame shown in **Figure 13.49** specifies that the elements of a cylinder must be straight within 0.12 mm. On flat surfaces, straightness is measured in a plane passing through control-line elements, and it may be specified in two directions (usually perpendicular) if desired.

Circularity (Roundness):
A surface of revolution (a cylinder, cone, or sphere) is circular when all points on the surface intersected by a plane perpendicular to its axis are equidistant from the axis. In **Figure 13.50** the feature control frame specifies circularity of a cone and cylinder, permitting a tolerance of 0.54 mm on

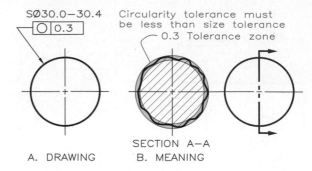

A. DRAWING   B. MEANING

13.51 Circularity of a sphere is a tolerance of form that means that any cross section through it is round within the specified tolerance.

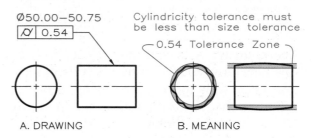

A. DRAWING   B. MEANING

13.52 Cylindricity is a tolerance of form that is a combination of roundness and straightness. It indicates that the surface of a cylinder lies within a tolerance zone formed by two concentric cylinders that are 0.54 apart.

the radius. **Figure 13.51** specifies a 0.30 mm tolerance zone for the roundness of a sphere.

Cylindricity:
A surface of revolution is cylindrical when all its elements lie within a cylindrical tolerance zone, which is a combination of tolerances of roundness and straightness (**Figure 13.52**). Here, a cylindricity tolerance zone of 0.54 mm on the radius of the cylinder is specified.

13.17 Profile Tolerancing

Profile tolerancing involves specifying tolerances for a contoured shape formed by arcs or irregular curves, and it can apply to a surface or a single line. The surface with the unilateral profile tolerance shown in **Figure 13.53A** is defined by coordinates. **Figure 13.54B** shows how to specify bilateral and unilateral tolerance zones.

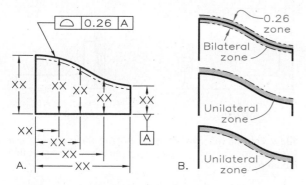

13.53 Profile is a tolerance for irregular curving planes and lines. (A) The curving plane is located by coordinates and is toleranced unidirectionally. (B) The tolerance may be applied by any of these methods.

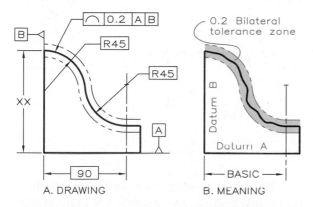

13.54 The profile of a line is a tolerance of profile that specifies the variation allowed from the path of a line. Here, the line is formed by tangent arcs. The tolerance zone may be either bilateral or unilateral, as shown in Figure 13.53.

A profile tolerance for a single line is specified as shown in **Figure 13.54.** The curve is formed by tangent arcs whose radii are given as basic dimensions. The radii are permitted to vary ±0.10 mm from the basic radii.

13.18 Orientation Tolerancing

Tolerances of orientation include **parallelism, perpendicularity,** and **angularity.**

Parallelism: A surface or line is parallel when all its points are equidistant from a datum plane or axis. Two types of parallelism tolerance zones are:

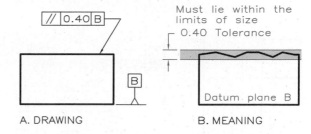

13.55 Parallelism is a tolerance of orientation indicating that a plane is parallel to a datum plane within specified limits. Here, plane B is the datum plane.

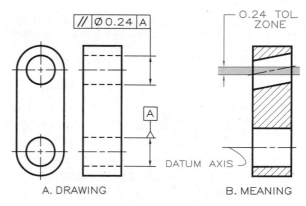

13.56 Parallelism of one centerline to another centerline can be specified by using the diameter of one of the holes as the datum.

1. A planar tolerance zone parallel to a datum plane within which the axis or surface of the feature must lie **(Figure 13.55).** This tolerance of orientation also controls flatness.

2. A cylindrical tolerance zone parallel to a datum feature within which the axis of a feature must lie **(Figure 13.56).**

Figure 13.57 shows the effect of specifying parallelism at MMC, where the modifier *M* is given in the feature control frame. Tolerances of form apply at RFS when not specified. Specifying parallelism at MMC means that the axis of the cylindrical hole must vary no more than 0.20 mm when the holes are at their smallest permissible size.

As the hole approaches its upper limit of 30.30, the tolerance zone increases to a maximum of 0.50 DIA. Therefore a greater variation is given at MMC than at RFS.

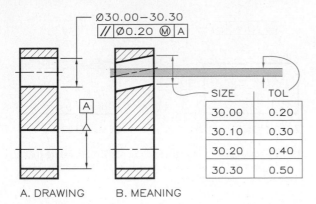

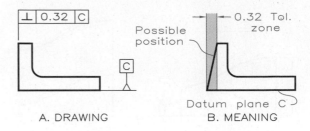

A. DRAWING   B. MEANING

13.58 Perpendicularity is a tolerance of orientation that gives a tolerance zone of 0.32 for a plane perpendicular to a specified datum plane.

13.57 The critical tolerance exists when features are at MMC. (A) The upper hole must be parallel to the hole used as datum A within a 0.20 DIA. (B) As the hole approaches its maximum size of 30.30 mm, the tolerance zone approaches 0.50 mm.

Perpendicularity: The specifications for the perpendicularity of a plane to a datum are shown in **Figure 13.58.** The feature control frame shows that the surface perpendicular to datum plane C has a tolerance of 0.32 mm. In **Figure 13.59** a hole is specified as perpendicular to datum plane B.

Angularity: A surface or line is angular when it is at an angle (other than 90°) from a datum or an axis. The angularity of the surface shown in **Figure 13.60** is dimensioned with a basic angle (exact angle) of 30° and an angularity tolerance zone of 0.20 mm inside of which the plane must lie.

13.19 Runout Tolerancing

Runout tolerancing is a way of controlling multiple features by relating them to a common datum axis. Features so controlled are surfaces of revolution about an axis and surfaces perpendicular to the axis.

The datum axis, such as diameter B in **Figure 13.61,** is established by a circular feature that rotates about the axis. When the part is rotated about this axis, the features of rotation must fall within the prescribed tolerance at full indicator movement (FIM).

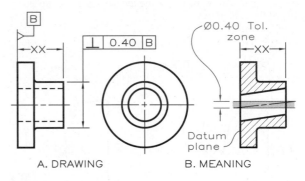

A. DRAWING   B. MEANING

13.59 Perpendicularity is a tolerance of orientation. It can apply to the axis of a feature such as the centerline of a cylinder.

The two types of runout are **circular runout** and **total runout.** One arrow in the feature control frame indicates circular runout and two arrows indicate total runout.

Circular Runout: Rotating an object about its axis 360° determines whether a circular cross section exceeds the permissible runout tolerance at any point **(Figure 13.61).** This same technique is used to measure the amount of wobble in surfaces perpendicular to the axis of rotation.

Total Runout: Used to specify cumulative variations of circularity, straightness, concentricity, angularity, taper, and profile of a surface **(Figure 13.62),** total runout tolerances are measured for all circular and profile positions as the part is rotated 360°. When applied to surfaces perpendicular to the axis, total

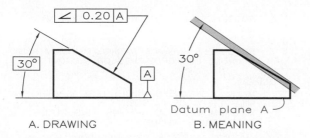

A. DRAWING   B. MEANING

13.60 Angularity is a tolerance of orientation specifying a tolerance zone for an angular surface with respect to a datum plane. Here, the 30° angle is a true (basic) angle from which a tolerance of 0.20 mm is applied.

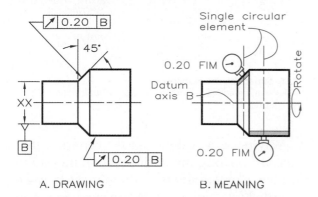

A. DRAWING   B. MEANING

13.61 Circular runout tolerance, a composite of several tolerance of form characteristics, is used to specify concentric cylindrical parts. The part is mounted on the datum axis and is gauged as it is rotated.

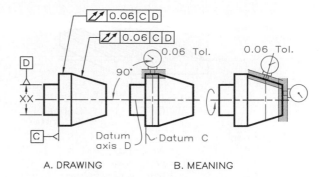

A. DRAWING   B. MEANING

13.62 Total runout tolerance is measured by mounting the object on the primary datum plane C and the secondary datum cylinder D. The cylinder and conical surface are gauged to check their conformity to a tolerance zone of 0.06 mm. The runout at the end of the cone could have been noted.

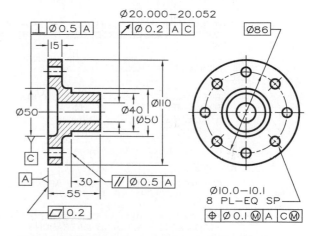

13.63 A combination of notes and symbols describe this part's geometric features.

runout tolerances control variations in perpendicularity and flatness.

The dimensioned part shown in **Figure 13.63** illustrates several of the techniques of geometric tolerancing described in this and previous sections.

13.20 SurfaceTexture

Because the surface texture of a part affects its function, it must be precisely specified instead of giving an unspecified finished mark such as a V. **Figure 13.64** illustrates most of the terms that apply to surface texture (surface control).

Surface texture: The variation in a surface, including roughness, waviness, lay, and flaws.

Roughness: The finest of the irregularities in the surface caused by the manufacturing process used to smooth the surface.

Roughness height: The average deviation from the mean plane of the surface measured in microinches (μin) or micrometers (μm), or millionths of an inch and a meter, respectively.

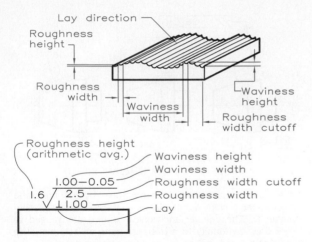

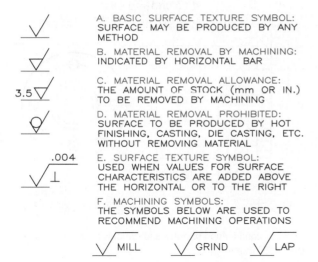

13.64 Surface texture terminology.
These are the definitions and terminology of surface texture for a finished surface.

A. BASIC SURFACE TEXTURE SYMBOL: SURFACE MAY BE PRODUCED BY ANY METHOD

B. MATERIAL REMOVAL BY MACHINING: INDICATED BY HORIZONTAL BAR

3.5 C. MATERIAL REMOVAL ALLOWANCE: THE AMOUNT OF STOCK (mm OR IN.) TO BE REMOVED BY MACHINING

D. MATERIAL REMOVAL PROHIBITED: SURFACE TO BE PRODUCED BY HOT FINISHING, CASTING, DIE CASTING, ETC. WITHOUT REMOVING MATERIAL

.004 E. SURFACE TEXTURE SYMBOL: USED WHEN VALUES FOR SURFACE CHARACTERISTICS ARE ADDED ABOVE THE HORIZONTAL OR TO THE RIGHT

F. MACHINING SYMBOLS: THE SYMBOLS BELOW ARE USED TO RECOMMEND MACHINING OPERATIONS

MILL   GRIND   LAP

13.65 Surface texture symbols.
Use surface texture symbols to specify surface finish are placed on the edge views of finished surfaces.

Roughness width: The width between successive peaks and valleys forming the roughness measured in microinches or micrometers.

Roughness width cutoff: The largest spacing of repetitive irregularities that includes average roughness height (measured in inches or millimeters). When not specified, a value of 0.8 mm (0.030 inch) is assumed.

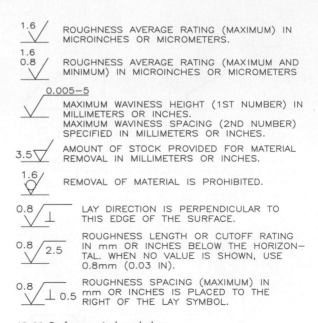

1.6 ROUGHNESS AVERAGE RATING (MAXIMUM) IN MICROINCHES OR MICROMETERS.

1.6
0.8 ROUGHNESS AVERAGE RATING (MAXIMUM AND MINIMUM) IN MICROINCHES OR MICROMETERS

0.005−5 MAXIMUM WAVINESS HEIGHT (1ST NUMBER) IN MILLIMETERS OR INCHES. MAXIMUM WAVINESS SPACING (2ND NUMBER) SPECIFIED IN MILLIMETERS OR INCHES.

3.5 AMOUNT OF STOCK PROVIDED FOR MATERIAL REMOVAL IN MILLIMETERS OR INCHES.

1.6 REMOVAL OF MATERIAL IS PROHIBITED.

0.8 LAY DIRECTION IS PERPENDICULAR TO THIS EDGE OF THE SURFACE.

0.8 2.5 ROUGHNESS LENGTH OR CUTOFF RATING IN mm OR INCHES BELOW THE HORIZON-TAL. WHEN NO VALUE IS SHOWN, USE 0.8mm (0.03 IN).

0.8 0.5 ROUGHNESS SPACING (MAXIMUM) IN mm OR INCHES IS PLACED TO THE RIGHT OF THE LAY SYMBOL.

13.66 Surface control symbols.
Values may be added to surface control symbols for more precise specifications.

Waviness: A widely spaced variation that exceeds the roughness width cutoff measured in inches or millimeters. Roughness may be regarded as a surface variation superimposed on a wavy surface.

Waviness height: The peak-to-valley distance between waves measured in inches or millimeters.

Waviness width: The spacing between wave peaks or wave valleys measured in inches or millimeters.

Lay: The direction of the surface pattern caused by the production method used.

Flaws: Irregularities or defects occurring infrequently or at widely varying intervals on a surface, including cracks, blow holes, checks, ridges, scratches, and the like. The effect of flaws is usually omitted in roughness height measurements.

Contact area: The surface that will make contact with a mating surface.

Symbols for specifying **surface texture** are shown in **Figure 13.65.** The point of the V

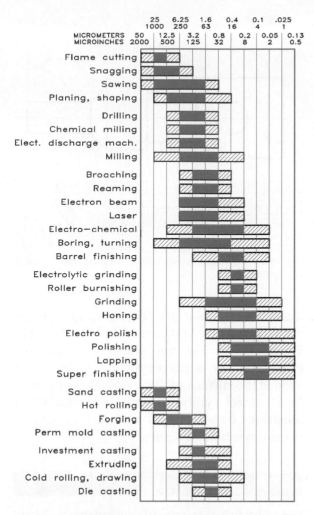

13.67 Various types of production methods result in the surface roughness heights shown in micrometers and microinches (millionths of a meter or an inch, respectively).

MILLIMETERS	0.08	0.25	0.80	2.5	8.0	25
INCHES	.003	.010	.030	.1	.3	1

13.69 Roughness width cutoff values.
This range of roughness width cutoff values is recommended in the ANSI Y14.36 standards. When unspecified, assume a value of 0.80 or 0.03 inches.

MAXIMUM WAVINESS HEIGHT VALUES

mm	in.	mm	in.
0.0005	.00002	0.025	.001
0.0008	.00003	0.05	.002
0.0012	.00005	0.08	.003
0.0020	.00008	0.12	.005
0.0025	.0001	0.20	.008
0.005	.0002	0.25	.010
0.008	.0003	0.38	.015
0.012	.0005	0.50	.020
0.020	.0008	0.80	.030

13.70 Waviness height values.
This range of maximum waviness height values is recommended in the ANSI Y14.36 standards.

PREFERRED ROUGHNESS AVERAGE VALUES

Micrometers μm	Microinches μin.	Micrometers μm	Microinches μin.
0.025	1	1.6	63
0.050	2	3.2	125
0.10	4	6.3	250
0.20	8	12.5	500
0.40	16	25	1000
0.80	32	Micrometers=0.001 mm	

13.68 Roughness heights.
This range of roughness heights is recommended by the *ANSI Y14.36* standards for metric and English units.

must touch the edge view of the surface, an extension line from it, or a leader pointing to the surface. **Figure 13.66** shows how to specify values as a part of surface texture symbols. Roughness height values are related to the processes used to finish surfaces and may be taken from the table in **Figure 13.67.** The preferred values of roughness height are listed in **Figure 13.68.**

The preferred roughness width cutoff values in **Figure 13.69** are for specifying the sampling width used to measure roughness height. A value of 0.80 mm is assumed if no value is given. When required, maximum waviness height values may be selected from the recommended values shown in **Figure 13.70.**

Lay symbols indicating the direction of texture (markings made by the machining

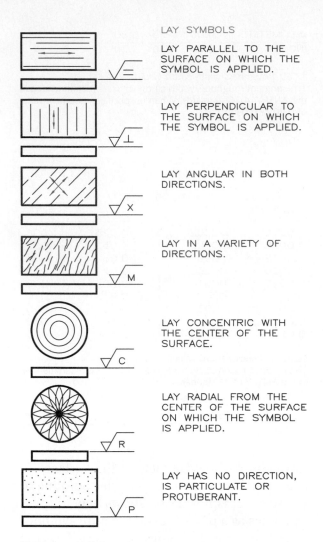

LAY SYMBOLS

LAY PARALLEL TO THE SURFACE ON WHICH THE SYMBOL IS APPLIED.

LAY PERPENDICULAR TO THE SURFACE ON WHICH THE SYMBOL IS APPLIED.

LAY ANGULAR IN BOTH DIRECTIONS.

LAY IN A VARIETY OF DIRECTIONS.

LAY CONCENTRIC WITH THE CENTER OF THE SURFACE.

LAY RADIAL FROM THE CENTER OF THE SURFACE ON WHICH THE SYMBOL IS APPLIED.

LAY HAS NO DIRECTION, IS PARTICULATE OR PROTUBERANT.

13.71 Lay symbols.
These symbols are used to indicate the direction of lay with respect to the surface where the control symbol is placed.

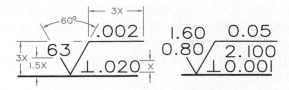

13.72 Surface texture symbols.
These are examples and proportions of typical, fully specified surface texture symbols.

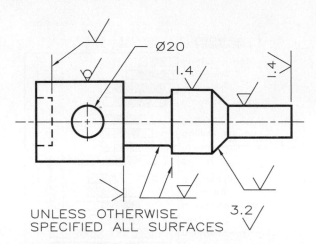

UNLESS OTHERWISE SPECIFIED ALL SURFACES

13.73 Application of surface texture symbols.
Various techniques of applying surface texture symbols to a part are illustrated here.

operation) on a surface (**Figure 13.71**) may be added to **surface texture symbols** as shown in **Figure 13.72.** The perpendicular sign indicates that lay is perpendicular to the edge view of the surface in this view (where the surface control symbol appears). **Figure 13.73** illustrates how to apply a variety of surface texture symbols to a part.

Problems

Solve the following problems on size A sheets laid out on a grid of 0.20 in. or 5 mm.

Cylindrical Fits

1. (Sheet 1) Draw the shaft and hole shown (it need not be to scale), give the limits for each diameter, and complete the table of values. Basic diameter of 1.00 in. (25 mm) and a class RC 1 fit or H8/f 7.

2. Repeat Problem 1: Basic diameter of 1.75 in. (45 mm) and a class RC 9 fit or H11/c11.

3. Repeat Problem 1: Basic diameter of 2.00 in. (51 mm) and a class RC 5 fit or H9/d9.

4. Repeat Problem 1: Basic diameter of 12.00 in. (305 mm) and a class LC 11 fit or H7/h6.

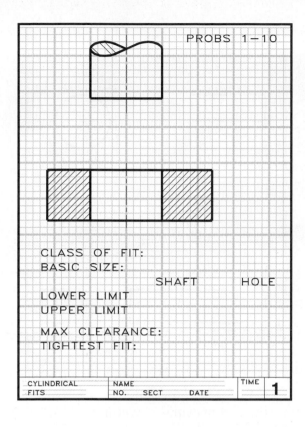

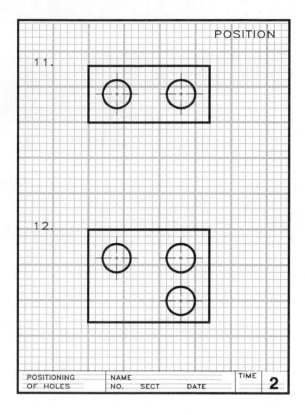

5. Repeat Problem 1: Basic diameter of 3.00 in. (76 mm) and a class LC 1 fit or H7/h6.

6. Repeat Problem 1: Basic diameter of 8.00 in. (203 mm) and a class LC 1 fit or H7/k6.

7. Repeat Problem 1: Basic diameter of 102 in. (2591 mm) and a class LN 3 fit or H7/n6.

8. Repeat Problem 1: Basic diameter of 11.00 in. (279 mm) and a class LN 2 fit or H7/p6.

9. Repeat Problem 1: Basic diameter of 6.00 in. (152 mm) and a class FN 5 fit or H7/s6.

10. Repeat Problem 1: Basic diameter of 2.60 in. (66 mm) and a class FN 1 fit or H7/u6.

Tolerances of Location

11A. (Sheet 2) Make an instrument drawing of the part and locate the holes with a size tolerance of 1.00 mm and a position toler-ance of 0.50 DIA. Give the symbols and dimensions.

11B. (Sheet 2) Using position tolerances lo-cate the holes and note them to provide a size tolerance of 1.00 mm and a locational toler-ance of 0.40 DIA.

12. (Sheet 3) Using position tolerances, locate the holes and properly note them to provide a size tolerance of 1.50 mm and a locational tolerance of 0.60 DIA.

13A. (Sheet 4) Using a feature control symbol and the necessary dimensions, indicate that the notch is symmetrical at the left-hand end of the part within 0.60 mm.

13B. (Sheet 4) Using a feature control symbol and the necessary dimensions, indicate that the small cylinder is concentric with the large

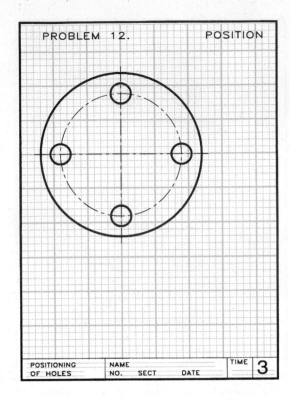

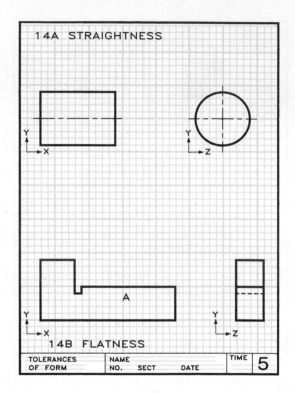

one (the datum cylinder) within a tolerance of 0.80 mm.

Tolerances of Form

14A. (Sheet 5) Using a feature control symbol and the necessary dimensions, indicate that the elements of the cylinder are straight within a tolerance of 0.20 mm.

14B. (Sheet 5) Using a feature control symbol and the necessary dimensions, indicate that surface A of the object is flat within a tolerance of 0.08 mm.

15A–C. (Sheet 6) Using feature control symbols and the necessary dimensions, indicate that the cross sections of the cylinder, cone, and sphere are round within a tolerance of 0.40 mm.

Tolerances of Profile

16A. (Sheet 7) Using a feature control symbol and the necessary dimensions, indicate that the profile of the irregular surface of the object

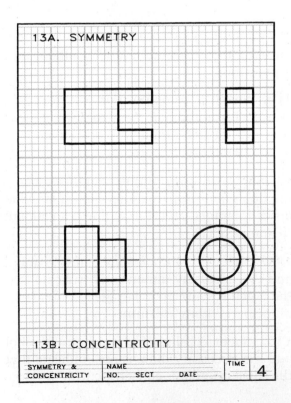

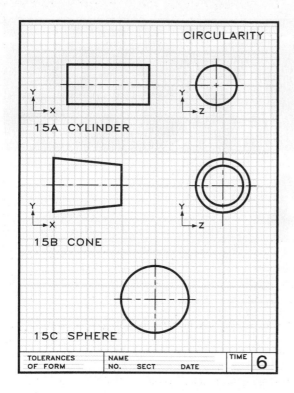

15A CYLINDER

15B CONE

15C SPHERE

CIRCULARITY

TOLERANCES OF FORM	NAME			TIME	6
	NO.	SECT	DATE		

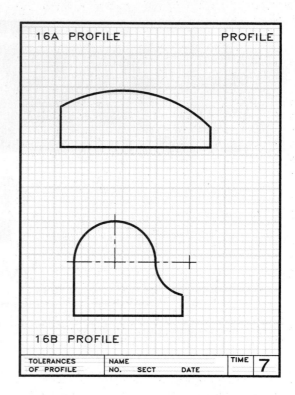

16A PROFILE

16B PROFILE

PROFILE

TOLERANCES OF PROFILE	NAME			TIME	7
	NO.	SECT	DATE		

lies within a bilateral or unilateral tolerance zone of 0.40 mm.

16B. (Sheet 7) Using a feature control symbol and the necessary dimensions, indicate that the profile of the line formed by tangent arcs lies within a bilateral or unilateral tolerance zone of 0.40 mm.

Tolerances of Orientation

17A. (Sheet 8) Using a feature control symbol indicate that surface B of the object is parallel to datum A within 0.30 mm.

17B. (Sheet 8) Using a feature control symbol, indicate that the angularity tolerance of the inclined plane is 0.7 mm from the bottom of the object, the datum plane.

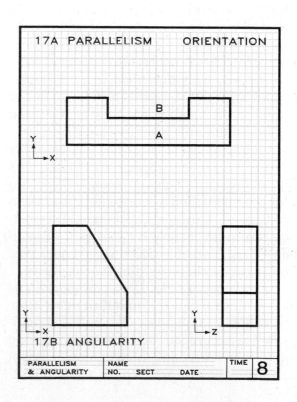

17A PARALLELISM   ORIENTATION

17B ANGULARITY

PARALLELISM & ANGULARITY	NAME			TIME	8
	NO.	SECT	DATE		

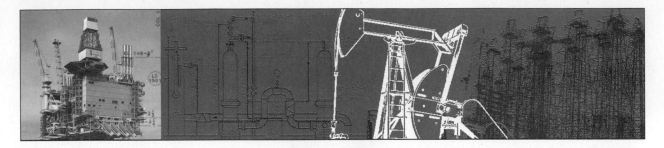

14

Welding

14.1 Introduction

Welding is the process of permanently joining metal by heating a joint to a suitable temperature with or without applying pressure and with or without using filler material. The welding practices described in this chapter comply with the standards developed by the American Welding Society and the American National Standards Institute (ANSI).

Welding is done in shops, on assembly lines, or in the field as shown in **Figure 14.1** where a welder is joining pipes. Welding is a widely used method of fabrication with its own language of notes, specifications, and symbology. You must become familiar with this system of notations in order to make and read drawings containing welding specifications.

Advantages of welding over other methods of fastening include (1) simplified fabrication, (2) economy, (3) increased strength and rigidity, (4) ease of repair, (5) creation of gas- and liquid-tight joints, and (6) reduction in weight and size.

14.1 This welder is joining two pipes in accordance with specifications on a set of drawings. (*Courtesy of Texas Eastern Transmission Corp.;* TE Today; *photo by Bob Thigpen.*)

14.2 Welding Processes

Figure 14.2 shows various types of welding processes. The three main types are **gas welding, arc welding,** and **resistance welding.**

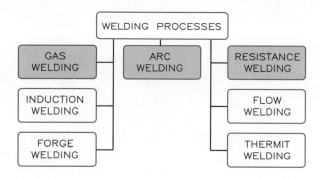

14.2 The three main types of welding processes are gas welding, arc welding, and resistance welding.

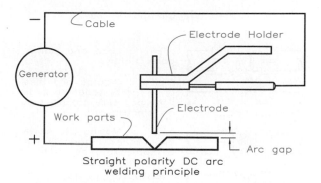

14.4 In arc welding, either AC or DC current is passed through an electrode to heat the joint.

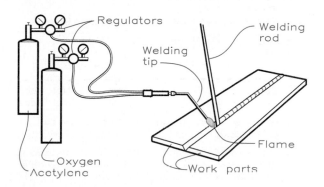

14.3 The gas welding process burns gases such as oxygen and acetylene in a torch to apply heat to a joint. The welding rod supplies the filler material.

Gas welding involves the use of gas flames to melt and fuse metal joints. Gases such as acetylene or hydrogen are mixed in a welding torch and burned with air or oxygen **(Figure 14.3).** The oxyacetylene method is widely used for repair work and field construction.

Most oxyacetylene welding is done manually with a minimum of equipment. Filler material in the form of welding rods is used to deposit metal at the joint as it is heated. Most metals, except for low- and medium-carbon steels, require fluxes to aid the process of melting and fusing the metals.

Arc welding involves the use of an electric arc to heat and fuse joints, with pressure sometimes required in addition to heat **(Figure 14.4).** The filler material is supplied

by a consumable or nonconsumable electrode through which the electric arc is transmitted. Metals well-suited to arc welding are wrought iron, low- and medium-carbon steels, stainless steel, copper, brass, bronze, aluminum, and some nickel alloys. In electric-arc welding, **the flux is a material coated on the electrodes that forms a coating on the metal being welded.** This coating protects the metal from oxidation so that the joint will not be weakened by overheating.

Flash welding is a form of arc welding, but it is similar to resistance welding because both pressure and electric current **(Figure 14.5)** are applied. The pieces to be welded are brought together, and an electric current is passed through them, causing heat to build up between them. As the metal burns, the current is turned off, and the pressure between the pieces is increased to fuse them.

Resistance welding comprises several processes by which metals are fused both by the heat produced from the resistance of the parts to an electric current and by pressure. Fluxes and filler materials normally are not used. All resistance welds are either lap- or butt-type welds.

Resistance spot welding is performed by pressing the parts together, and an electric current fuses them as illustrated in the lap joint weld in **Figure 14.6.** A series of small

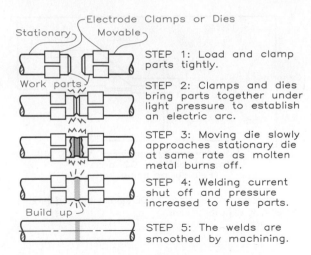

Electrode Clamps or Dies

Stationary Movable

Work parts

Build up

STEP 1: Load and clamp parts tightly.

STEP 2: Clamps and dies bring parts together under light pressure to establish an electric arc.

STEP 3: Moving die slowly approaches stationary die at same rate as molten metal burns off.

STEP 4: Welding current shut off and pressure increased to fuse parts.

STEP 5: The welds are smoothed by machining.

14.5 Flash welding, a type of arc welding, uses a combination of electric current and pressure to fuse two parts.

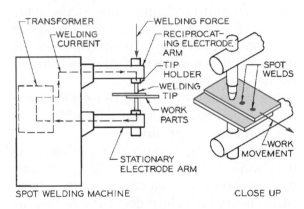

TRANSFORMER

WELDING CURRENT

WELDING FORCE

RECIPROCAT- ING ELECTRODE ARM

TIP HOLDER

WELDING TIP

WORK PARTS

SPOT WELDS

WORK MOVEMENT

STATIONARY ELECTRODE ARM

SPOT WELDING MACHINE   CLOSE UP

14.6 Resistance spot welding may be used to join lap and butt joints.

Material	Spot Welding	Flash Welding
Low—carbon mild steel		
SAE 1010	Rec.	Rec.
SAE 1020	Rec.	Rec.
Medium—carbon steel		
SAE 1030	Rec.	Rec.
SAE 1050	Rec.	Rec.
Wrought alloy steel		
SAE 4130	Rec.	Rec.
SAE 4340	Rec.	Rec.
High—alloy austenitic stainless steel		
SAE 30301—30302	Rec.	Rec.
SAE 30309—30316	Rec.	Rec.
Ferritic and martensistic stainless steel		
SAE 51410—51430	Satis.	Satis.
Wrought heat—resisting alloys		
19—9—DL	Satis.	Satis.
16—25—6	Satis.	Satis.
Cast iron	NA	Not Rec.
Gray iron	NA	Not Rec.
Aluminum & alum. alloys	Rec.	Satis.
Nickel & nickel alloys	Rec.	Satis.

Rec.—Recommended
Not Rec.— Not recommended

Satis.—Satisfactory
NA—Not applicable

14.7 Resistance welding processes for various materials are shown here.

welds spaced at intervals, called **spot welds,** secure the parts. **Figure 14.7** lists the recommended materials and processes to be used for resistance welding.

14.3 Weld Joints and Welds

Figure 14.8 shows the five standard weld joints: **butt joint, corner joint, lap joint, edge joint,** and **tee joint.** The **butt joint** can be joined with the square groove, V-groove, bevel groove, U-groove, and J-groove welds. The **corner joint** can be joined with these welds

and with the fillet weld. The **lap joint** can be joined with the bevel groove, J-groove, fillet, slot, plug, spot, projection, and seam welds. The **edge joint** uses the same welds as the lap joint along with the square groove, V-groove, U-groove, and seam welds. The **tee joint** can be joined by the bevel groove, J-groove, and fillet welds.

Figure 14.9 depicts commonly used welds and their corresponding ideographs (symbols). The fillet weld is a built-up weld at the intersection (usually 90°) of two surfaces. The square, bevel, V-groove, J-groove, and U-groove welds all have grooves, and the weld is made in these grooves. Slot and plug welds have intermittent holes or openings where the parts are welded.

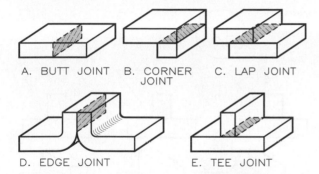

A. BUTT JOINT   B. CORNER JOINT   C. LAP JOINT

D. EDGE JOINT   E. TEE JOINT

14.8 These diagrams depict the five standard weld joints.

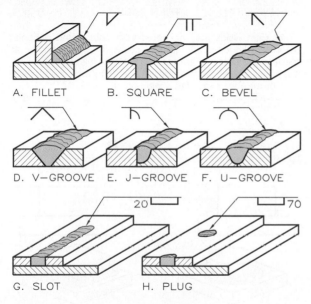

A. FILLET   B. SQUARE   C. BEVEL

D. V-GROOVE   E. J-GROOVE   F. U-GROOVE

20   70

G. SLOT   H. PLUG

14.9 These views illustrate standard welds and their corresponding ideographs.

Holes are unnecessary when resistance welding is used.

14.4 Welding Symbols

If a drawing has a general welding note such as, ALL JOINTS ARE WELDED THROUGH-OUT, the designer has transferred responsibility to the welder. Welding is too important to be left to chance and should be specified more precisely.

Symbols are used to convey welding specifications on a drawing. The complete

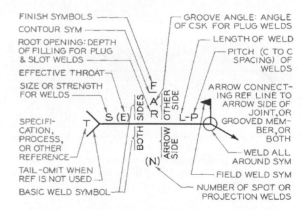

FINISH SYMBOLS
CONTOUR SYM
ROOT OPENING: DEPTH OF FILLING FOR PLUG & SLOT WELDS
EFFECTIVE THROAT
SIZE OR STRENGTH FOR WELDS
SPECIFI-CATION, PROCESS, OR OTHER REFERENCE
TAIL-OMIT WHEN REF IS NOT USED
BASIC WELD SYMBOL

GROOVE ANGLE: ANGLE OF CSK FOR PLUG WELDS
LENGTH OF WELD
PITCH (C TO C SPACING) OF WELDS
ARROW CONNECT-ING REF LINE TO ARROW SIDE OF JOINT, OR GROOVED MEM-BER, OR BOTH
WELD ALL AROUND SYM
FIELD WELD SYM
NUMBER OF SPOT OR PROJECTION WELDS

14.10 The welding symbol. Usually it is modified to a simpler form for use on drawings.

welding symbol is shown in **Figure 14.10,** but it usually appears on a drawing in modified, more general form with less detail. The scale of the welding symbol is based on the letter height used on the drawing, which is the size of the grid on which the symbol is drawn in **Figure 14.11.** The standard height of lettering on a drawing is usually 1/8 inch or 3 mm.

The **ideograph** is the symbol that denotes the type of weld desired, and it generally depicts the cross section representation of the weld. **Figure 14.12** shows the ideographs used most often. They are drawn to scale on the 1/8-inch (3-mm) grid (equal to the letter height), which represents their full size when added to the welding symbol.

14.5 Application of Symbols

Fillet Welds

In **Figure 14.13A,** placement of the fillet weld ideograph below the horizontal line of the symbol indicates that the weld is at the joint on the arrow side—the right side in this case. The vertical leg of the ideograph is always on the left side.

A numeral (either a common fraction or a decimal value) to the right of the ideograph

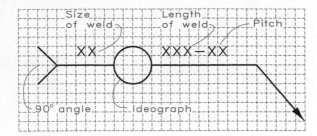

14.11 Welding symbol proportions are based on the letter height used on a drawing, usually 1/8 inch or 3 mm. This grid is equal to the letter height.

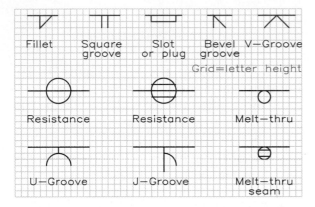

14.12 The sizes of the ideographs shown on the 1/8-inch (3 mm) grid (the letter height) are proportional to the size of the welding symbol (Figure 14.11).

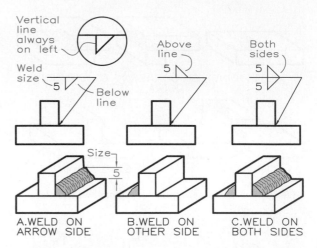

14.13 Fillet welds may be noted with abbreviated symbols. (A) When the ideograph appears below the horizontal line, it specifies a weld on the arrow side. (B) When it is above the line, it specifies a weld on the opposite side. (C) When it is on both sides of the line, it specifies a weld on each side.

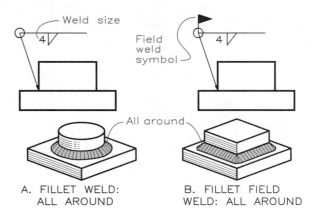

14.14 These symbols indicate fillet welds are to be made all around two types of parts.

indicates the size of the weld. You may omit this number from the symbol if you insert a general note elsewhere on the drawing to specify the fillet size, such as:

ALL FILLET WELDS 1/4 IN.
UNLESS OTHERWISE NOTED.

Placing the ideograph above the horizontal line **Figure 14.13B** indicates that the weld is to be on the other side, that is, the joint on the other side of the part away from the arrow. When the part is to be welded on both sides, use the ideograph shown in **Figure 14.13C.** You may omit the tail and other specifications from the symbol when you provide detailed specifications elsewhere.

A single arrow is often used to specify a weld that is to be made all around two joining parts **(Figure 14.14A).** A circle of 6 mm (twice the letter height) in diameter, drawn at the bend in the leader of the symbol, denotes this type of weld. If the welding is to be done in the field rather than in the shop, a solid black triangular "flag" is added also **(Figure 14.14B).**

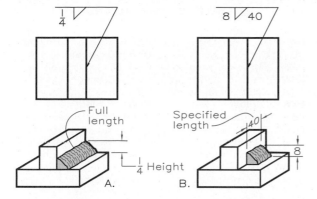

14.15 Fillet weld symbols.
A This symbol indicates full-length fillet welds on the arrow side.
B This symbol indicates fillet welds of specified, but less than full length, of the welded parts.

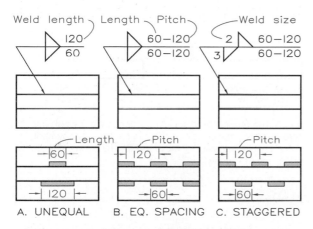

14.16 These symbols specify intermittent welds of varying lengths and alignments.

You may specify a **full-length fillet weld** that is to run the full length of the two parts as in **Figure 14.15A.** The ideograph is on the lower side of the horizontal line, so the weld is on the arrow side. You may specify a fillet weld that is to run shorter than full length as in **Figure 14.15B,** where 40 represents the weld's length in millimeters.

You may specify fillet welds to run different lengths and be positioned on both sides of a part as in **Figure 14.16A.** The dimension on the lower side of the horizontal gives the length of the weld on the arrow side, and the dimension on the upper side of the horizontal gives the length on the opposite side.

Intermittent welds have a specified length and are spaced uniformly, center to center, at an interval called the pitch. In **Figure 14.16B,** the welds are equally spaced on both sides, are 60 mm long, and have pitches of 120 mm, as indicated by the symbol shown. The symbol shown in **Figure 14.16C** specifies intermittent welds that are staggered in alternate positions on opposite sides.

Groove Welds
The standard types of groove welds are: **V-groove, bevel groove, double V-groove, U-groove,** and **J-groove (Figure 14.17).** When you do not give the depth of the grooves, angle of the chamfer, and root openings on a symbol, you must specify them elsewhere on the drawing or in supporting documents. In **Figures 14.17A** and **B,** the angles of the V-joints are labeled 60° and 90° under the ideographs. In **Figure 14.17B,** the depths of the weld (6) and the root opening (2)—the gap between the two parts—are given.

In a bevel groove weld, only one of the parts is beveled. The symbol's leader is bent and pointed toward the beveled part to call attention to it (**Figures 14.17C** and **14.18B**). This practice also applies to J-groove welds, where one side is grooved and the other is not (**Figure 14.18A**).

Notate double V-groove welds by weld size, bevel angle, and root opening (**Figures 14.17D** and **E**). Omit root opening sizes or show a zero on the symbol when parts fit flush. Give the angle and depth of the groove in the symbol for a U-groove weld (**Figure 14.17F**).

Seam Welds
A seam weld joins two lapping parts with either a continuous weld or a series of closely

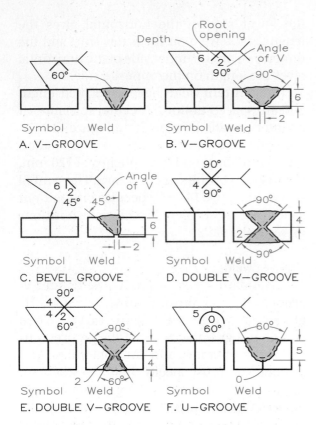

A. V—GROOVE

B. V—GROOVE

C. BEVEL GROOVE

D. DOUBLE V—GROOVE

E. DOUBLE V—GROOVE

F. U—GROOVE

14.17 This drawing shows the various types of groove welds and their general specifications.

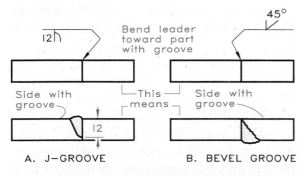

A. J—GROOVE

B. BEVEL GROOVE

14.18 J-groove welds and bevel welds are specified by bent arrows pointing to the side of the joint to be grooved or beveled.

spaced spot welds. The seam weld process to be used is identified by abbreviations in the tail of the weld symbol **(Figure 14.19).** The circular ideograph for a resistance weld is about 12 mm (four times the letter height) in diameter and is

WELDING ABBREVIATIONS

CAW	Carbon—arc w.	IB	Induction brazing
CW	Cold welding	IRB	Infrared brazing
DB	Dip brazing	OAW	Oxyacetylene w.
DFW	Diffusion welding	OHW	Oxyhydrogen w.
EBW	Electric beam w.	PGW	Pressure gas w.
ESW	Electroslag welding	RB	Resist. brazing
EXW	Explosion welding	RPW	Projection weld.
FB	Furnace brazing	RSEW	Resist. seam w.
FOW	Forge welding	RSW	Resist. spot w.
FRW	Friction welding	RW	Resist. welding
FW	Flash welding	TB	Torch brazing
GMAW	Gas metal arc w.	UW	Upset welding
GTAW	Gas tungsten w.		*w.=welding

14.19 These abbreviations represent the various types of welding processes and are used in welding symbols.

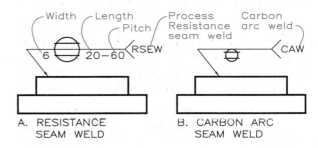

A. RESISTANCE SEAM WELD

B. CARBON ARC SEAM WELD

14.20 The process used for (A) resistance seam welds and (B) carbon-arc seam welds is indicated in the tail of the symbol. For the arc weld the symbol must specify the arrow side or the other side of the piece.

centered over the horizontal line of the symbol **(Figure 14.20A).** The weld's width, length, and pitch are given.

When the seam weld is to be made by arc welding (CAW), the diameter of the ideograph is about 6 mm (twice the letter height) and goes on the upper or lower side of the symbol's horizontal line to indicate whether the seam is to be applied to the arrow side or opposite side **(Figure 14.20B).** When the length of the weld is not shown, the seam weld is understood to extend between abrupt changes in the direction of the seam.

Spot welds are similarly specified with ideographs and specifications by diameter,

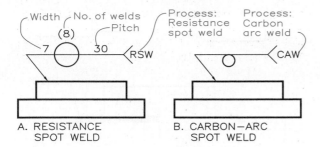

A. RESISTANCE SPOT WELD

B. CARBON–ARC SPOT WELD

14.21 The process to be used for (A) resistance spot welds and (B) arc spot welds is indicated in the tail of the symbol. For the arc weld the symbol must specify the arrow side or the other side of the piece.

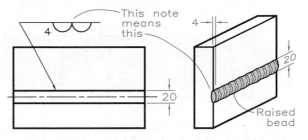

14.22 Use this method to apply a symbol to a built-up weld on a surface.

number of welds, and pitch between the welds. The process of resistance spot welding (RSW) is noted in the tail of the symbol **(Figure 14.21A).** For arc welding, the arrow side or other side must be indicated by a symbol **(Figure 14.21B).**

Built-Up Welds

When the surface of a part is to be enlarged, or built-up, by welding, indicate this process with a symbol as shown in **Figure 14.22.** Dimension the width of the built-up weld in the view. Specify the height of the weld above the surface in the symbol to the left of the ideograph. The radius of the circular segment is 6 mm (twice the letter height).

14.6 Surface Contouring

Contour symbols are used to indicate which of the three types of contours, **flush, concave,** or **convex,** is desired on the surface of the weld. Flush contours are smooth with the surface or flat across the hypotenuse of a fillet weld. Concave contours bulge inward with a curve, and convex contours bulge outward with a curve **(Figure 14.23).**

Finishing the weld by an additional process to obtain the desired contour is often necessary. These processes, which may be indicated by their abbreviations, are **chipping**

14.23 These contour symbols specify the desired surface finish of a weld.

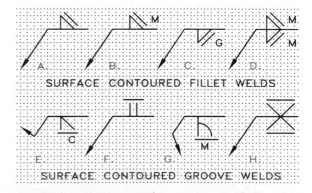

14.24 These examples of contoured weld symbols with letters added indicate the type of finishing to be applied to the weld (M, machining; G, grinding; C, chipping).

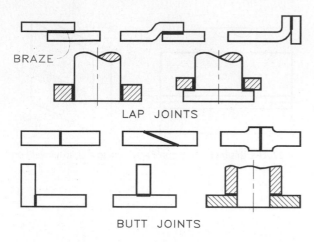

14.25 The two basic types of brazing joints are lap joints and butt joints.

BRAZE

LAP JOINTS

BUTT JOINTS

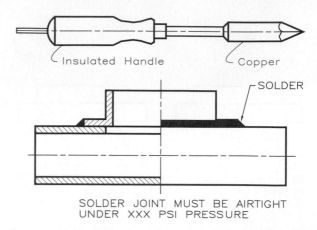

Insulated Handle   Copper

SOLDER

SOLDER JOINT MUST BE AIRTIGHT
UNDER XXX PSI PRESSURE

14.26 This typical hand-held soldering iron is used to soft-solder two parts together. The method of notating a drawing for soldering also is shown.

(C), **grinding** (G), **hammering** (H), **machining** (M), **rolling** (R), and **peening** (P), as shown in **Figure 14.24.**

14.7 Brazing

Brazing is a method much like welding for joining pieces of metal. Brazing entails heating joints to more than 800°F and distributing by capillary action a nonferrous filler material, with a melting point below that of the base materials, between the closely fitting parts.

Before brazing, the parts must be cleaned and the joints fluxed. The brazing filler is added before or just as the joints are heated beyond the filler's melting point. After the filler material has melted, it is allowed to flow between the parts to form the joint. As **Figure 14.25** shows, there are two basic brazing joints: lap joints and butt joints.

Brazing is used to join parts, to provide gas- and liquid-tight joints, to ensure electrical conductivity, and to aid in repair and salvage. Brazed joints withstand more stress, higher temperature, and more vibration than soft-soldered joints.

14.8 Soldering

Soldering is the process of joining two metal parts with a third metal that melts below the temperature of the metals being joined. Solders are alloys of nonferrous metals that melt below 800°F. Widely used in the automotive and electrical industries, soldering is one of the basic techniques of welding and often is done by hand with a soldering iron like the one depicted in **Figure 14.26.** The iron is placed on the joint to heat it and to melt the solder. Basic soldering is noted on the drawing with a leader simply as SOLDER, and other specifications noted as needed, as shown in **Figure 14.26.**

By necessity, the covering in this chapter is introductory in nature, but adequate for a basic understanding of how to specify welding on an engineering drawing. More detailed information on welding is available from the American Welding Society, 550 N.W. LeJeune Road, Miami, FL 33126, 305-443-9353. This society maintains and publishes guidelines and standards for the technology of welding.

Problems

Solve these problems on size A sheets laid out on a grid of 0.20 inch (5 mm).

1–8. **(Figure 14.27)** Give welding notes to include the information specified for each problem. Omit instructional information from the solution.

9. **(Figure 14.28)** The shaft socket has a base of 4 in. × 4 in. Draw a top and front view of it, approximate its dimensions, and show the appropriate welding notes for its fabrication.

10. **(Figure 14.29)** The angle fixture has faces of 9 × 12 and 10 × 12. Redesign the fixture to have ribs that are welded instead of cast and show welding symbols. Estimate the unspecified dimensions.

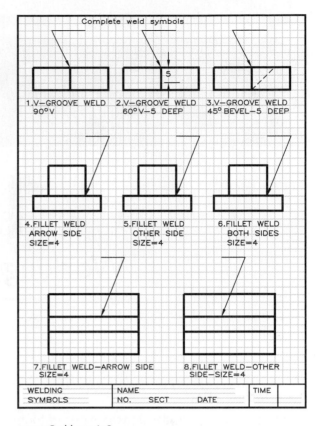

14.27 Problems 1–8.

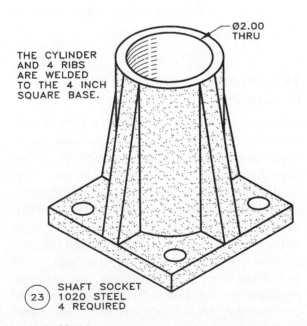

14.28 Problem 9.

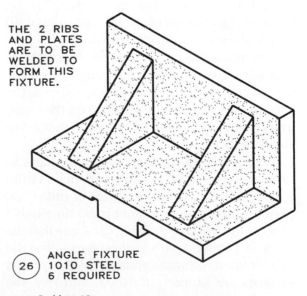

14.29 Problem 10.

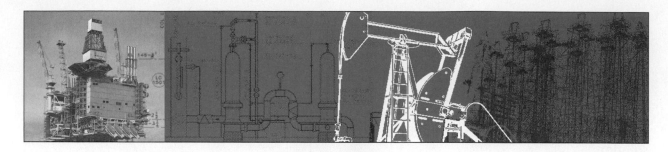

15

Working Drawings

15.1 Introduction

Working drawings are the drawings from which a design is implemented. All principles of orthographic projection and techniques of graphics can be used to communicate the details of a project in working drawings. A detail drawing is a working drawing of a single part (or detail) within the set of working drawings.

Specifications are the written instructions that accompany working drawings. When the design can be represented on a few sheets, the specifications are usually written on the drawings to consolidate the information into a single format.

All parts must interact with other parts to some degree to yield the desired function from a design. Before detail drawings of individual parts are made, the designer must thoroughly analyze the working drawing to ensure that the parts fit properly with mating parts, that the correct tolerances are applied, that the contact surfaces are properly finished, and that the proper motion is possible between the parts.

Much of the work in preparing working drawings is done by the drafter, but the designer, who is usually an engineer, is responsible for their correctness. It is working drawings that bring products and systems into being.

15.2 Working Drawings as Legal Documents

Working drawings are legal contracts that document the design details and specifications as directed by the engineer. Therefore drawings must be as clear, precise, and thorough as possible. Revisions and modifications of a project at the time of production or construction are much more expensive than when done in the preliminary design stages.

Poorly executed working drawings result in wasted time and resources and increase implementation costs. To be economically competitive, drawings must be as error-free as possible.

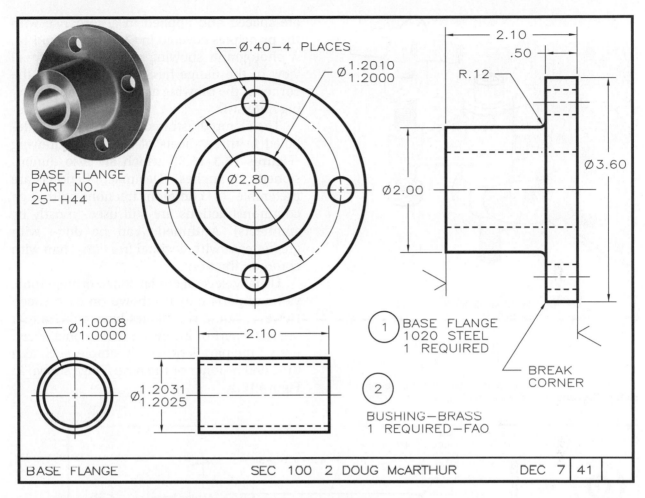

15.1 This revolving clamp assembly holds parts while they are being machined.

Working drawings specify all aspects of the design, reflecting the soundness of engineering and function of the finished product and economy of fabrication. The working drawing is the instrument that is most likely to establish the responsibility for any failure to meet specifications during implementation.

15.3 Dimensions and Units: Inches

English System

The inch is the basic unit of the English system, and virtually all shop drawings made with English units are dimensioned in inches. This practice is followed even when dimensions are several feet in length.

The base flange shown in **Figure 15.1** is an example of a relatively simple working drawing of a part. However, there are many drawings of this simplicity that must be designed, developed, and detailed in order for the overall project to come into being.

The base flange is dimensioned with two-place decimal inches except where four-place decimal inches are used for toleranced dimensions. Inch marks (″) are omitted from dimensions on working drawings because the units are understood to be in inches, and their omission saves drafting time. Finish marks are applied to the surfaces that must be machined smooth. Notice that the dimensions

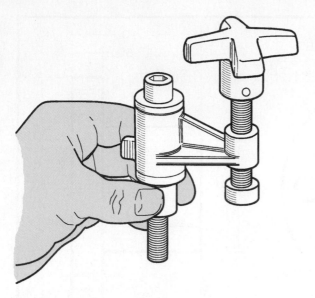

15.2 This revolving clamp assembly holds parts while they are being machined. (*Figures 15.2–15.5 courtesy of Jorgens, Inc.*)

are spaced and applied in accordance with the principles covered in Chapters 13 and 14. A photograph showing a three-dimensional view of the flange has been inserted in the corner of the drawing as a raster image by using AutoCAD.

The clamp illustrated in **Figure 15.2** is detailed in three sheets of a working drawing (**Figures 15.3–15.5**), which are also dimensioned in inches. Decimal fractions are preferable to common fractions, although common fractions are still used (mostly by architects). Arithmetic can be done with greater ease with decimal fractions than with common fractions.

Usually, several dimensioned orthographic views of parts may be shown on each sheet. However, some companies have policies that only one part of a view be drawn on a sheet, even if the part is extremely simple, such as a threaded fastener or the base flange shown in **Figure 15.1.**

15.3 Sheet 1 of 3: This computer-drawn working drawing of parts of the clamp assembly shown in Figure 15.2.

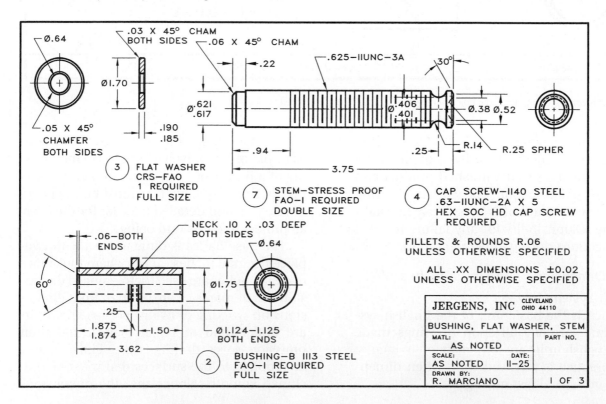

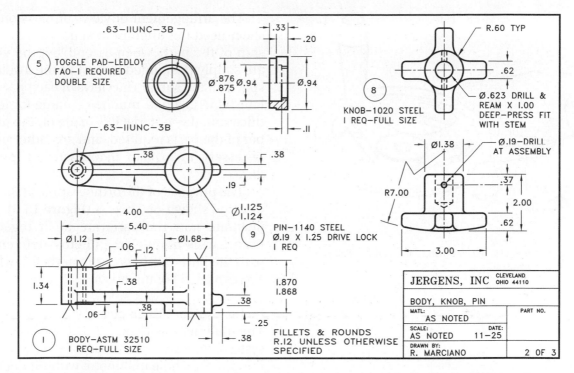

15.4 Sheet 2 of 3 (above): This continuation of Figure 15.3 shows other parts of the clamp assembly.

15.5 Sheet 3 of 3 (below): A pad assembly, an overall assembly drawing, and a parts list are shown.

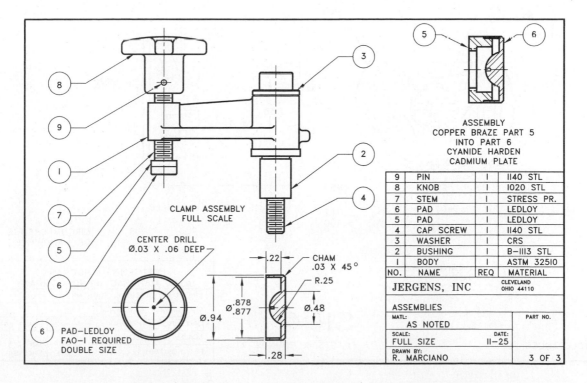

NO.	NAME	REQ	MATERIAL
9	PIN	I	1140 STL
8	KNOB	I	1020 STL
7	STEM	I	STRESS PR.
6	PAD	I	LEDLOY
5	PAD	I	LEDLOY
4	CAP SCREW	I	1140 STL
3	WASHER	I	CRS
2	BUSHING	I	B-1113 STL
I	BODY	I	ASTM 32510

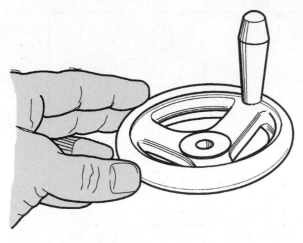

15.6 This photo is of a left-end handcrank that is detailed in the working drawing in Figures 15.7 and 15.8.

15.7 Sheet 1 of 2: This set of working drawings (dimensions in mm) depicts the crank wheel of the left-end handcrank shown in Figure 15.6.

The arrangement of views of parts on the sheet need not attempt to show the relationship of the parts when assembled; the views are simply positioned to best fit the available space on the sheet. The views of each part are labeled with a part number, a name for identification, the material it is made of, the number of the parts required, and any other notes necessary to explain manufacturing procedures.

The purpose of the orthographic assembly drawing shown on sheet 3 (**Figure 15.5**) is to illustrate how the parts are to fit together. Each part is numbered and cross-referenced with the part numbers in the parts list, which serves as a bill of materials.

Metric System: Millimeter

The millimetre is the basic unit of the metric system, and dimensions usually are given to the nearest whole millimetre without decimal

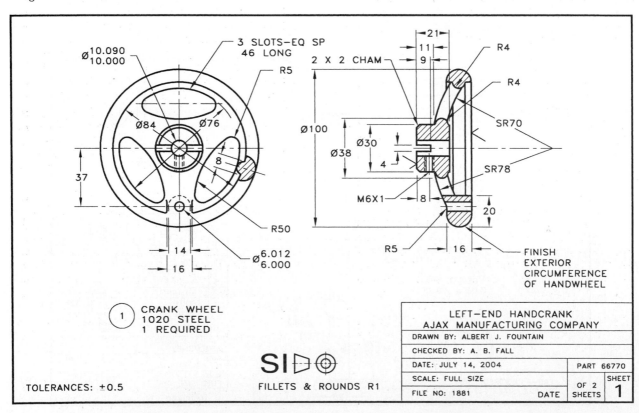

fractions (except to specify tolerances, which may require three-place decimals). Metric abbreviations (mm) after the numerals are omitted from dimensions because the SI symbol near the title block indicates that all units are metric. If you have trouble relating to the length of a millimeter, recall that the fingernail of your index finger is about 10 mm wide.

The left-end handcrank (**Figure 15.6**) is depicted and dimensioned in millimeters in the working drawings shown in **Figures 15.7** and **Figure 15.8** on two size B sheets. Dimensions and notes along with the descriptive views give the information needed to construct the four pieces.

The orthographic, sectioned assembly drawing of the left-end handcrank shown in **Figure 15.8** illustrates how the parts are to be put together. The numbers in the balloons, the part numbers, provide a cross-reference to the parts list placed just above the title block.

Dual Dimensions

Some working drawings carry both inch and millimeter dimensions as shown in **Figure 15.9** where the dimensions in parentheses or brackets are millimeters. The units may also appear as millimeters first and then be converted and shown in brackets as inches. Converting from one unit to the other results in fractional round-off errors. An explanation of the primary unit system for each drawing should be noted in the title block.

Metric Working Drawing Example

Figure 15.10 is a leveling device used to level heavy equipment such as lathes and milling machines. The device raises or lowers the

15.8 Sheet 2 of 2: This continuation of Figure 15.7 includes an assembly drawing and parts list.

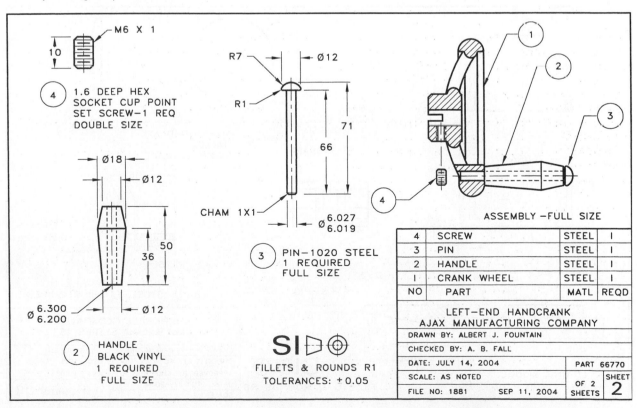

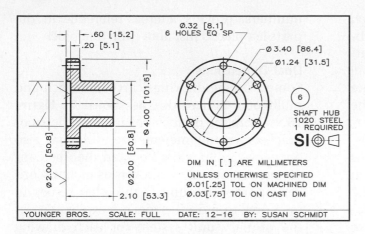

.60 [15.2]
.20 [5.1]
Ø.32 [8.1]
6 HOLES EQ SP
Ø 3.40 [86.4]
Ø1.24 [31.5]

Ø 4.00 [101.6]
Ø 2.00 [50.8]
Ø 2.00 [50.8]
2.00 [50.8]
2.10 [53.3]

⑥ SHAFT HUB
1020 STEEL
1 REQUIRED

SI ◐◯◁

DIM IN [ ] ARE MILLIMETERS
UNLESS OTHERWISE SPECIFIED
Ø.01[.25] TOL ON MACHINED DIM
Ø.03[.75] TOL ON CAST DIM

YOUNGER BROS.   SCALE: FULL   DATE: 12−16   BY: SUSAN SCHMIDT

15.9 In this dual-dimensioned drawing, dimensions are shown in milli-meters; their equivalents in inches given in brackets.

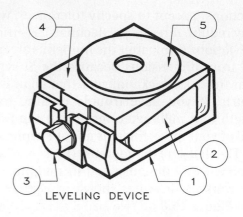

LEVELING DEVICE

15.10 This Lev-L-Line lifting device is used to level heavy machinery.

15.11 Sheet 1 of 2: This working drawing (dimensioned in SI units) is of the lifting device shown in Figure 15.10.

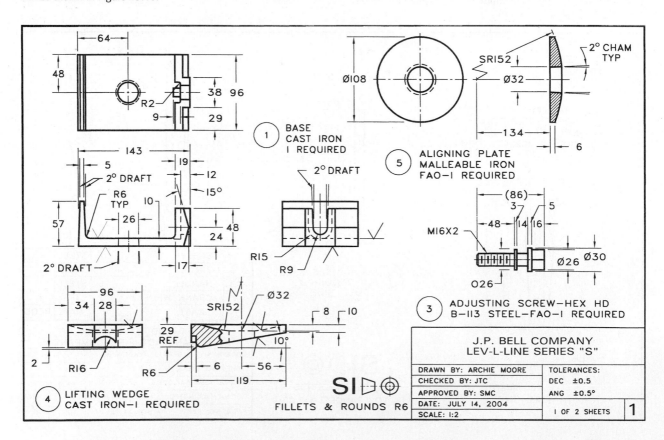

64
48
38 96
R2
9
29
① BASE
CAST IRON
1 REQUIRED

143
5
2° DRAFT
R6 TYP
57
26
10
19
12
15°
48
24
17
2° DRAFT

2° DRAFT
R15
R9

Ø108
SRI52
Ø32
2° CHAM TYP
134
6
⑤ ALIGNING PLATE
MALLEABLE IRON
FAO-1 REQUIRED

M16X2
O26
(86)
3   5
48 14 16
Ø26 Ø30
③ ADJUSTING SCREW-HEX HD
B-113 STEEL-FAO-1 REQUIRED

96
34 28
2
R16
29 REF
R6
SRI52
Ø32
6   56
119
8   10
10°

④ LIFTING WEDGE
CAST IRON-1 REQUIRED

SI ▷◯◎
FILLETS & ROUNDS R6

J.P. BELL COMPANY
LEV-L-LINE SERIES "S"

DRAWN BY: ARCHIE MOORE	TOLERANCES:	
CHECKED BY: JTC	DEC ±0.5	
APPROVED BY: SMC	ANG ±0.5°	
DATE: JULY 14, 2004		
SCALE: 1:2	1 OF 2 SHEETS	1

machinery when the screw is rotated, which slides two wedges together. A two-sheet working drawing that gives the details of the parts of the lifting device is shown in **Figures 15.11** and **15.12.** The SI symbol indicates that the dimensions are in millimeters and the truncated cone indicates that the orthographic views are drawn using third-angle projection. The assembly drawing in **Figure 15.12** illustrates how the parts are to be assembled after they have been made.

15.4 Laying Out a Detail Drawing

When making a drawing with instruments on paper or film, first lay out the views and dimensions on a different sheet of paper. Then overlay the drawing with vellum or film and trace it to obtain the final drawing. You must use guidelines for lettering for each dimension and note. Lightly draw the guidelines or underlay the drawing with a sheet containing guidelines.

Figure 15.13 shows the standard sheet sizes for working drawings. Paper, film, cloth, and reproduction materials are available in these modular sizes; good practice requires that you make drawings in one of these standard sizes. Modular sized drawing can be folded to fit standard-sized envelopes, match the sizes of print paper, and fit in standard size filing cabinets.

15.12 Sheet 2 of 2: This continuation of Figure 15.11 is a further working drawing and assembly drawing of the lifting device.

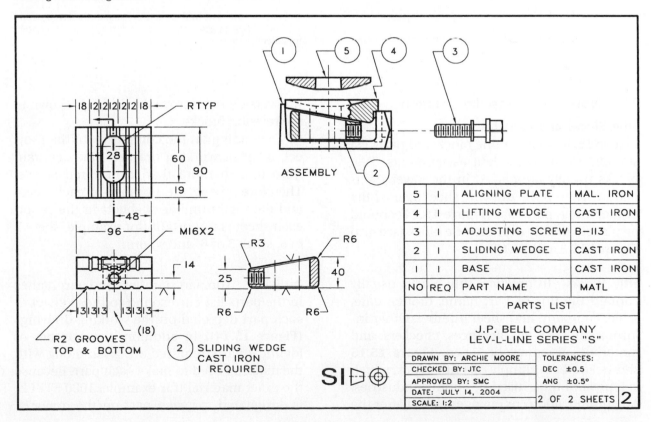

A	11 x 8.5		A4	297 X 210
B	17 X 11		A3	420 X 297
C	22 X 17		A2	594 X 420
D	34 X 22		A1	841 X 594
E	44 X 34		A0	1189 X 841

ENGLISH SIZES   METRIC SIZES

15.13 These are the standard sheet sizes for working drawings dimensioned in inches and millimeters.

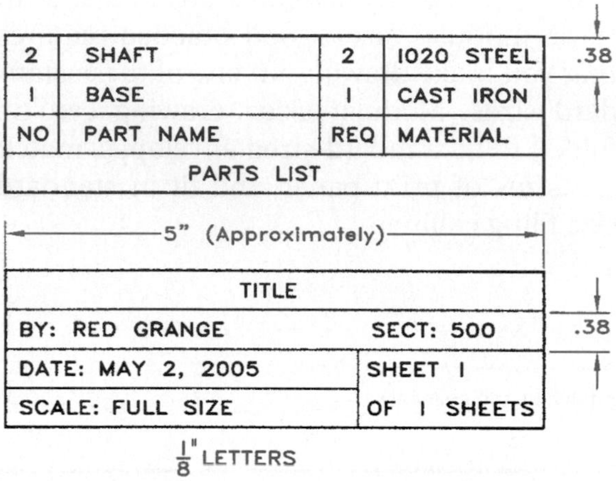

2	SHAFT	2	1020 STEEL	.38
I	BASE	I	CAST IRON	
NO	PART NAME	REQ	MATERIAL	
	PARTS LIST			

←————— 5" (Approximately) —————→

TITLE		
BY: RED GRANGE	SECT: 500	.38
DATE: MAY 2, 2005	SHEET I	
SCALE: FULL SIZE	OF I SHEETS	

$\frac{1}{8}"$ LETTERS

15.14 This typical title block and parts list is suitable for most student assignments.

REVISIONS	COMPANY NAME COMPANY ADDRESS	
CHG. HEIGHT	TITLE: LEFT—END BEARING	
FAO	DRAWN BY: JOHNNY RINGO	
	CHECKED BY: FRED J. DODGE	
	DATE: JULY 14, 2004	
	SCALE HALF SIZE	SHEET 2 OF 3 SHEETS

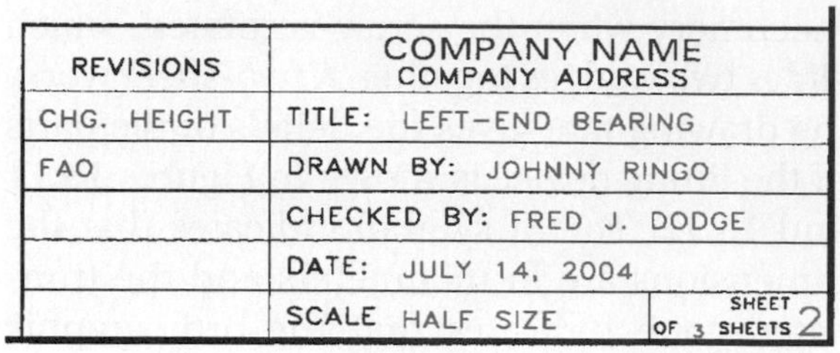

15.15 This title block, which includes a revision block, is typical of those used in industry.

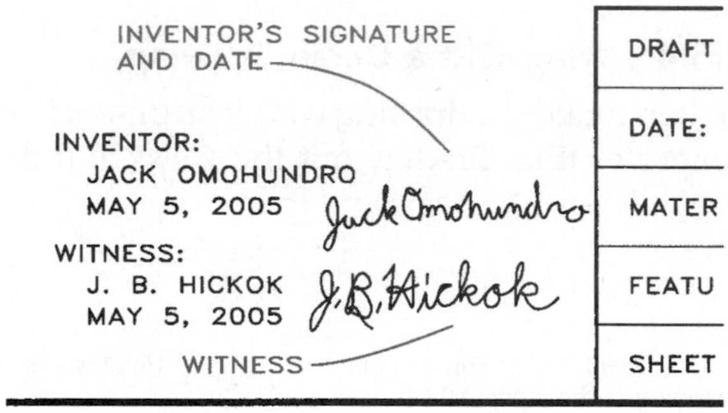

INVENTOR'S SIGNATURE AND DATE

INVENTOR:
JACK OMOHUNDRO
MAY 5, 2005   *Jack Omohundro*

WITNESS:
J. B. HICKOK
MAY 5, 2005   *J. B. Hickok*

WITNESS

DRAFT

DATE:

MATER

FEATU

SHEET

15.16 This note next to the title block names the inventor and is witnessed by an associate to establish ownership of a design for patent purposes.

15.5 Notes and Other Information

Title Blocks and Parts Lists

Figure 15.14 shows a title block and parts list suitable for most student assignments. Title blocks usually are placed in the lower right-hand corner of the drawing sheet against the borders. The parts list **(Figure 15.14)** should be placed directly over the title block (see also **Figures 15.8** and **15.12**).

Title Blocks In practice, title blocks usually contain the title or part name, drafter, date, scale, company, and sheet number. Other information, such as tolerances, checkers, and materials, also may be given. **Figure 15.15** shows another example of a title block, which is typical of those used by various industries. Any modifications or changes added after the

first version to improve the design is shown in the revision blocks.

Depending on the complexity of the project, a set of working drawings may contain from one to more than a hundred sheets. Therefore, giving the number of each sheet and the total number of sheets in the set on each sheet is important (for example, sheet 2 of 6, sheet 3 of 6, and so on).

Parts List The part numbers and part names in the parts list correspond to those given to each part depicted on the working drawings **(Figure 15.14)**. In addition, the number of identical parts required are given along with the material used to make each part. Because the exact material (for example, 1020 STEEL) is designated for each part on the drawing,

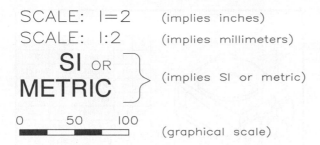

SCALE: 1=2   (implies inches)

SCALE: 1:2   (implies millimeters)

SI OR
METRIC   } (implies SI or metric)

0   50   100
▬▬▬▬   (graphical scale)

15.17 Specify scales in English and SI units on working drawings with these methods.

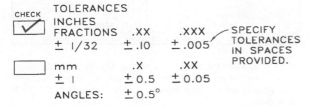

15.18 General tolerance notes on working drawings specify the dimension tolerances permitted.

the material in the parts list may be shortened to STEEL, which requires less space.

Patent Rights Note

A note near the title block that names Jack Omohundro as the inventor of the part or process is used to establish ownership of the design **(Figure 15.16).** An associate, J. B. Hickok, signs and dates the drawing as a witness to the designer's work. This type of note establishes ownership of the ideas and dates of their development to help the inventor obtain a patent. An even better case for design ownership is made if a second witness signs and dates the drawing. As modifications to the design are made, those drawings should receive the same documentation.

Scale Specification

If all working drawings in a set are the same scale, you need to indicate it only once in the title block on each sheet. If several detail drawings on a working drawing are different scales, indicate them on the drawing under each set of views. In this case, write AS SHOWN in the

15.19 Name and number each part on a working drawing for use in the parts list, and indicate the number of parts of this particular part that are needed.

scale area of the title block. When a drawing is not to scale, place the abbreviation NTS (not to scale) in the title block.

Figure 15.17 shows several methods of indicating scales. Use of the colon (for example, 1:2) implies the metric system; use of the equal sign (for example, 1=2) implies the English system—but these are not absolute rules. The SI symbol or metric designation on a drawing specifies that millimeters are the units of measurement.

In some cases, you may want to show a graphical scale with calibrations on a drawing to permit the interpretation of linear measurements by transferring them with dividers from the drawing to the scale.

Tolerances

Recall from Chapter 13 that you may use general notes on working drawings to specify the dimension tolerances. **Figure 15.18** shows a table of values with boxes in which you can make a check mark to indicate whether the units are in inches or millimeters. Position plus-and-minus tolerance values under each common or decimal fraction. For example, this table specifies that each dimension with two-place decimals will have a tolerance of ±0.10 inch. You may also give angular tolerances in general notes (±0.5°, for example).

Part Labeling

Give each part a name and number, using letters and numbers 1/8-inch (3 mm) high **(Figure 15.19).** Place part numbers inside circles, called balloons, having diameters approximately four times the height of the numerals.

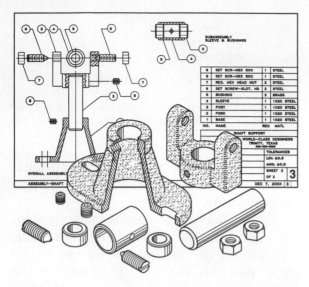

15.20 An assembly drawing explains how the parts of a product are to be assembeled.

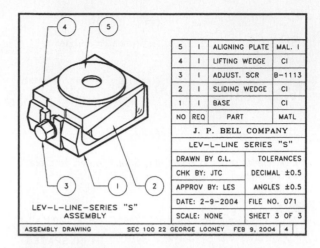

15.21 This isometric assembly drawing depicts the parts of the lifting device shown in Figure 15.10 fully assembled. Dimensions usually are omitted from assembly drawings and a parts list is given.

Place part numbers near the views to which they apply, so their association will be clear. On assembly drawings, balloons are especially important because the same part numbers are used in the parts list. Show the number of parts required near the part name.

15.6 Drafter's Log

In addition to the individual revision records, drafters should keep a log of all changes made during a project. As the project progresses, the drafter should record the changes, dates, and people involved. Such a log allows anyone reviewing the project in the future to understand easily and clearly the process used in arriving at the final design. Calculations often are made during a drawing's preparation, therefore they should be a permanent part of the log to preserve previously expended work.

15.7 Assembly Drawings

After parts have been made according to the specifications of the working drawings, they will be assembled **(Figure 15.20)** in accordance with the directions of an assembly drawing. Two general types of assembly drawings are **orthographic assemblies** and **pictorial assemblies.** Dimensions usually are omitted from assembly drawings.

The lifting device shown in **Figure 15.10** is depicted in an isometric assembly in **Figure 15.21.** Each part is numbered with a balloon and leader to cross-reference them to the parts list where more information about each part is given.

Figure 15.22 shows an orthographic exploded assembly drawing. In many applications, the arrangement of parts may be easier to understand when the parts are shown exploded along their centerlines. These views are shown as regular orthographic views, with some lines shown as hidden lines and others omitted.

Assembly of the same part is shown in **Figure 15.23** in an orthographic assembly drawing, in which the parts are depicted in their assembled positions. The views are sectioned to make them easier to understand.

Figure 15.24 shows a belt tensioner in an exploded pictorial assembly drawing, illustrating how the parts fit together, leaving no doubt as to how the parts relate to each other.

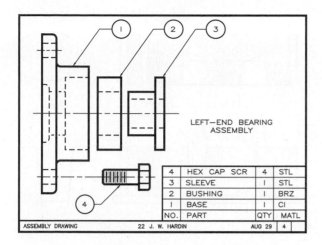

15.22 This exploded orthographic assembly illustrates how the parts shown are to be put together.

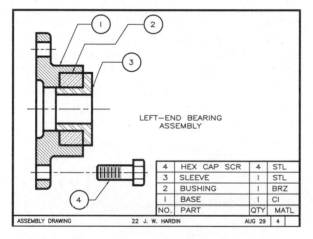

15.23 This sectioned orthographic assembly shows the parts from Figure 15.22 in their assembled positions, except for the exploded bolt.

Part numbers are given in balloons to complete the drawing.

15.8 Freehand Working Drawings

A freehand sketch can serve the same purpose as an instrument drawing, provided that the part is sufficiently simple and that the essential dimensions are shown **(Figure 15.25).** Use the same principles of making working drawings with instruments when making working

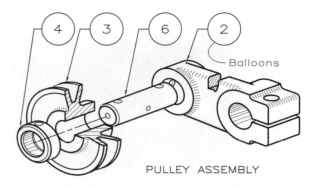

15.24 This exploded pictorial assembly drawing is of a belt tensioner.

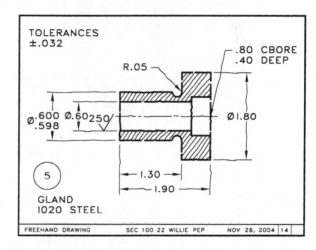

15.25 A freehand working drawing with the essential dimensions can be as adequate as an instrument-drawn detail drawing for simple parts.

drawings freehand. A sketch can be made quickly, and it can be made in the field, fabrication shop, or other locations where drafting-room instruments are not readily available.

15.9 Forged Parts and Castings

Two versions of a part shown in **Figure 15.26** illustrate the difference between a part that has been forged into shape (sometimes called a **blank**) and its final state after the forging has been machined. Recall from Chapter 12 that a forging is a rough form made by hammering (forging) the metal into shape or pressing it

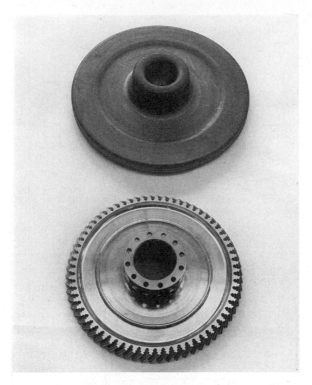

15.26 The upper part is a blank that has been forged. It will look like the part on the bottom after it has been machined. *(Courtesy Lycoming Engines.)*

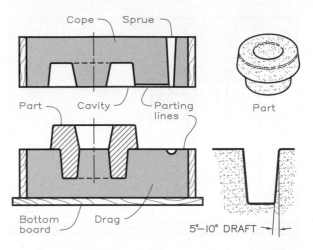

15.27 A two-part sand mold is used to produce a casting. A draft of from 5° to 10° is needed to permit withdrawal of the pattern from the sand. Some machining is usually required to finish various features of the casting within specified tolerances.

between two forms (called dies). The forged part is then machined to its specified finished dimensions and tolerances so that it will function as intended.

A casting **(Figure 15.27),** like a forging, must be machined so that it too will fit and

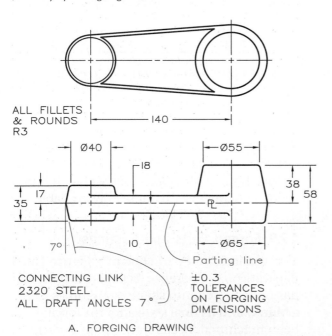

15.28 These separate working drawings, (A) a forging drawing and (B) a machining drawing, give the details of the same part. Often, this information is combined into a single drawing.

function with other parts when assembled; therefore additional material is added to the areas where metal will be removed by the machining processes. As covered in Chapter 11, castings are formed by pouring molten metal into a mold formed by a pattern that is slightly larger than the finished part to compensate for metal shrinkage (**Figure 15.27**). For the pattern to be removable from the sand that forms the mold, its sides have a taper, called a **draft,** of about 5° to 10°.

Some industries require that separate working drawing be made for forged and cast parts (**Figure 15.28A**). More often, however, the parts are detailed on the regular working drawing with the understanding that the features that are to be machined by operations such as grinding or shaping are made oversize by the fabrication shop (**Figure 15.28B**).

Problems

The following problems will provide exercises in the application of the principles required in making working drawings. You should make rapid, freehand sketches of the views before drawing them with instruments or computer to determine the appropriate layout of each sheet. Material covered in all of the previous chapters must be applied in the completion of these assignments.

Working Drawing Practice

Reproduce the drawings shown in **Figures 15.29–15.34** as directed. The purpose of these assignments is to provide experience in laying out working drawings and improving your draftsmanship on the board or at the computer.

15.29 Duplicate the working drawing of the base plate mount on a size B sheet.

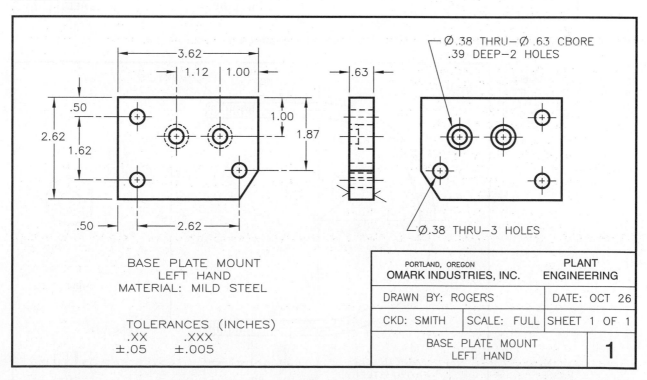

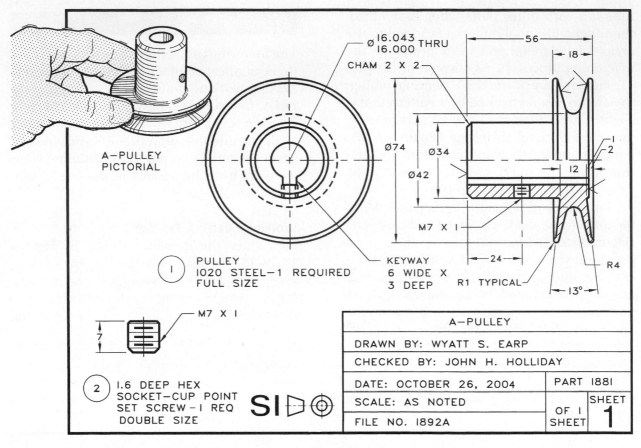

A–PULLEY
PICTORIAL

⌀ 16.043 / 16.000 THRU
CHAM 2 X 2

⌀74
⌀34
⌀42

56
18
12
24
R4
R1 TYPICAL
13°

M7 X 1

1 2

① PULLEY
1020 STEEL–1 REQUIRED
FULL SIZE

KEYWAY
6 WIDE X
3 DEEP

M7 X 1

7

② 1.6 DEEP HEX
SOCKET–CUP POINT
SET SCREW–1 REQ
DOUBLE SIZE

SI ▷◉

A–PULLEY		
DRAWN BY: WYATT S. EARP		
CHECKED BY: JOHN H. HOLLIDAY		
DATE: OCTOBER 26, 2004	PART 1881	
SCALE: AS NOTED	OF 1 SHEET	SHEET 1
FILE NO. 1892A		

15.30 Duplicate this working drawing of the pulley by computer or drafting instruments on a size A sheet.

15.31 Sheet 1 of 3: Duplicate these full-size working drawings and assembly (in mm) of the pipe hanger on size A sheets.

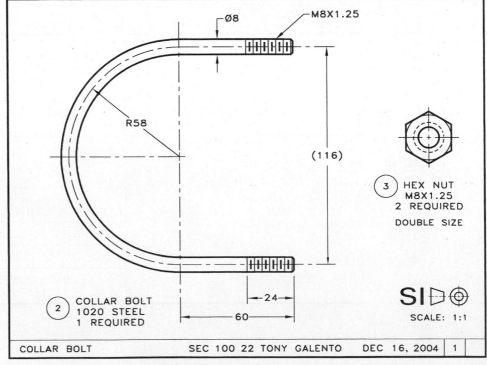

⌀8
M8X1.25

R58

(116)

③ HEX NUT
M8X1.25
2 REQUIRED

DOUBLE SIZE

② COLLAR BOLT
1020 STEEL
1 REQUIRED

24
60

SI ▷◉
SCALE: 1:1

COLLAR BOLT	SEC 100 22 TONY GALENTO	DEC 16, 2004	1

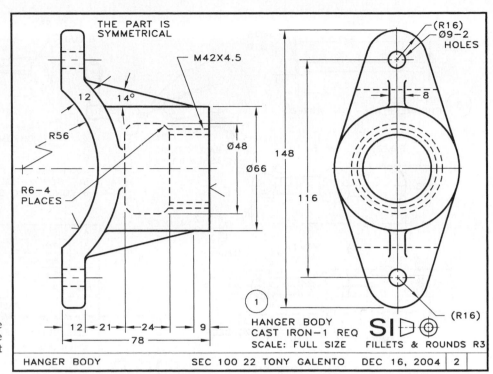

15.32 Sheet 2of 3: Duplicate this second sheet of the pipe hanger working drawing that shows the collar bolt.

THE PART IS SYMMETRICAL

M42X4.5

12   14°

R56

R6-4 PLACES

Ø48

Ø66

148

116

(R16) Ø9-2 HOLES

8

12 ← 21 → ← 24 → ← 9 →

78

① HANGER BODY
CAST IRON-1 REQ
SCALE: FULL SIZE

(R16)

SI ▷⊕

FILLETS & ROUNDS R3

| HANGER BODY | SEC 100 22 TONY GALENTO | DEC 16, 2004 | 2 | |

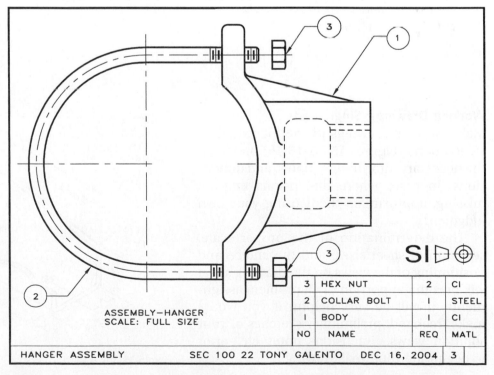

15.33 Duplicate this working drawing of the switch by computer or drafting instruments on a size A sheet. Note that the views are enlarged by a factor of 4.

③

①

③

②

ASSEMBLY-HANGER
SCALE: FULL SIZE

SI ▷⊕

NO	NAME	REQ	MATL
3	HEX NUT	2	CI
2	COLLAR BOLT	I	STEEL
I	BODY	I	CI

| HANGER ASSEMBLY | SEC 100 22 TONY GALENTO | DEC 16, 2004 | 3 | |

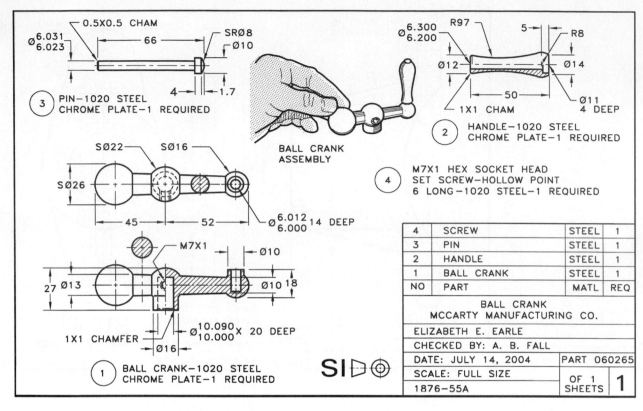

15.34 The ball crank (above) has been detailed in this working drawing. Duplicate the working drawing on a size B sheet. On a second size B sheet, make an assembly drawing of the parts.

Working Drawings: Single parts

Make working drawings of the following assigned parts (**Figures 15.35–15.46**) providing the necessary information, notes, and dimensions. In cases where dimensions may be missing, approximate them using your own judgment.

The determination of the proper scale, selection of sheet sizes, and the choice and positioning of the views on the drawing sheet will be a major portion of all problem assignments. It will be very helpful if you would make freehand, preliminary sketches of your solution (views, dimensions, notes, etc.) prior to beginning your final drawing.

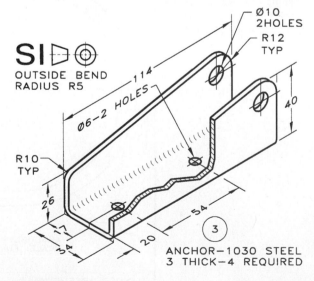

15.35 Size A sheet.

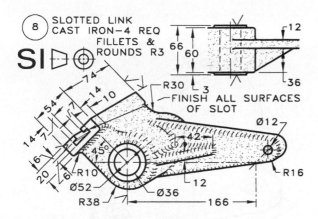

8 SLOTTED LINK
   CAST IRON—4 REQ
   FILLETS &
   ROUNDS R3

SI⊳⊕

66 60
      12
      36

R30 3
—FINISH ALL SURFACES
   OF SLOT

74
54 14 10
14 7
6
20 45°
6 R10
Ø52
R38
42
12
Ø36
166
Ø12
R16

15.36 Size B sheet.

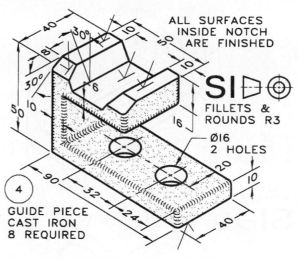

ALL SURFACES
INSIDE NOTCH
ARE FINISHED

40 30° 10
8
30° 6
50
10
50
10
16

SI⊳⊕
FILLETS &
ROUNDS R3

Ø16
2 HOLES

90
32
24
40
20
10

4 GUIDE PIECE
   CAST IRON
   8 REQUIRED

15.39 Size B sheet.

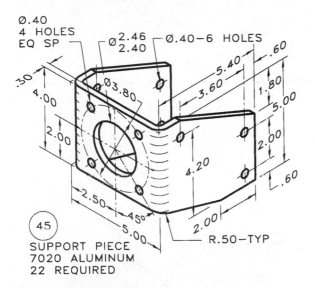

Ø.40
4 HOLES
EQ SP

Ø2.46
2.40
Ø.40—6 HOLES

.30
4.00
Ø3.80
2.00

5.40 .60
3.60 1.80
5.00
2.00
.60

4.20

2.50
5.00
45°
2.00
R.50—TYP

45 SUPPORT PIECE
    7020 ALUMINUM
    22 REQUIRED

15.37 Size B sheet.

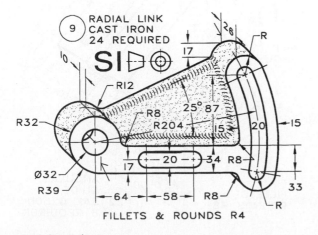

9 RADIAL LINK
   CAST IRON
   24 REQUIRED

SI⊳⊕

28
17
R
10
R12
R32
R8
R204
25° 87
15
20
15
20
Ø32
R39
17
20 34 R8
64 58 R8
33
R

FILLETS & ROUNDS R4

15.38 Size B sheet.

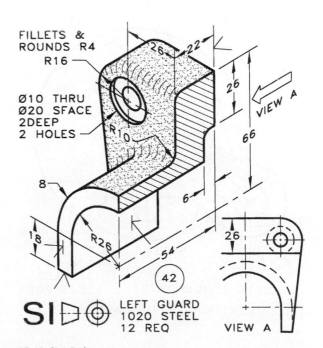

FILLETS &
ROUNDS R4
R16

26 22
26

Ø10 THRU
Ø20 SFACE
2DEEP
2 HOLES

R10
66

VIEW A

8
6

18 R26
54

42

SI⊳⊕
LEFT GUARD
1020 STEEL
12 REQ

26

VIEW A

15.40 Size B sheet.

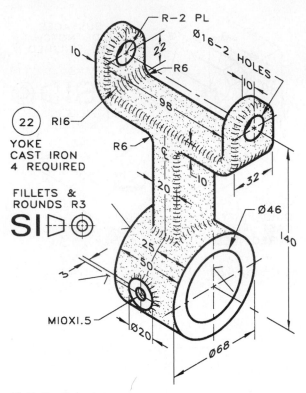

R–2 PL

22

Ø16–2 HOLES

10

R6

98

10

22 RI6

R6

YOKE
CAST IRON
4 REQUIRED

20

10

32

FILLETS &
ROUNDS R3

SI◗⊕

Ø46

25

3

50

140

MI0XI.5

Ø20

Ø68

15.41 Size B sheet.

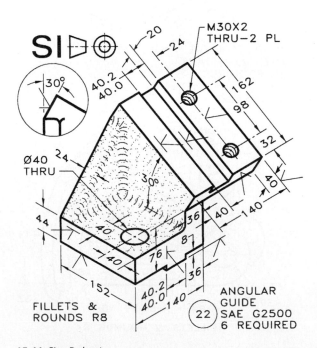

.375–16UNC–2B
2 HOLES

.80

.70

1.30

2.70

1.40

R.50

Ø2.00

1.40

Ø.44
Ø.70 CSK
3 HOLES

3.10

2.40

33

4.50

.30

R.70
3 PL

2.25

2.25

FLANGE
COUPLING
ALUMINUM
25 REQUIRED

FILLETS &
ROUNDS R.10

15.43 Size B sheet.

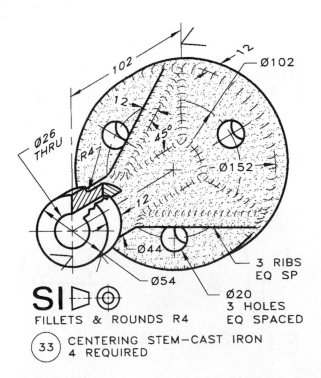

102

12

Ø102

12

Ø26
THRU

R4

45°

Ø152

12

Ø44

3 RIBS
EQ SP

Ø54

Ø20
3 HOLES
EQ SPACED

SI◗⊕
FILLETS & ROUNDS R4

33 CENTERING STEM–CAST IRON
4 REQUIRED

15.42 Size B sheet.

SI◗⊕

30°

20

M30X2
THRU–2 PL

24

40.2
40.0

162

98

Ø40
THRU

24

32

30°

40

44

40

36

40

140

76

8

40

152

40.2
40.0

36

140

FILLETS &
ROUNDS R8

22 ANGULAR
GUIDE
SAE G2500
6 REQUIRED

15.44 Size B sheet.

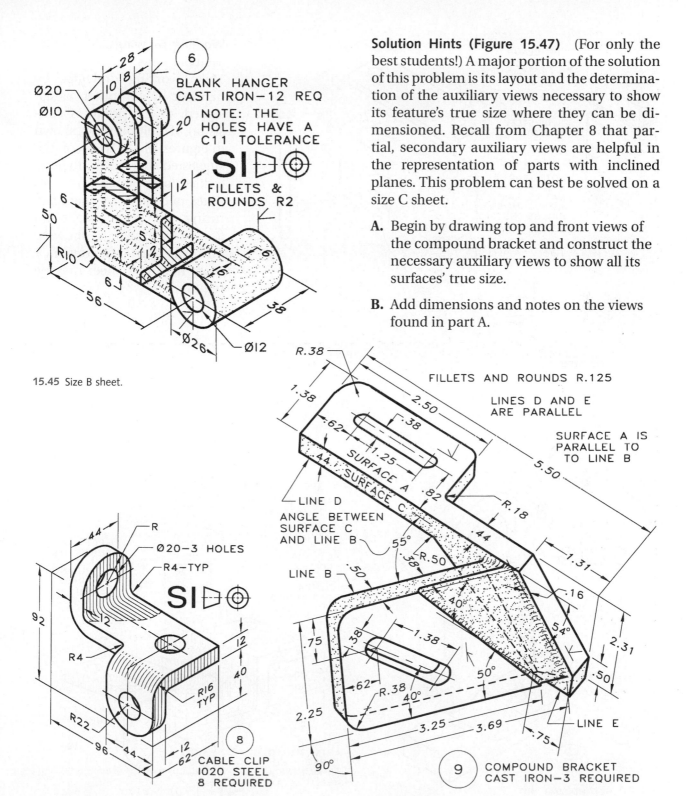

6

BLANK HANGER
CAST IRON—12 REQ

NOTE: THE
HOLES HAVE A
C11 TOLERANCE

SI ▷⊙

FILLETS &
ROUNDS R2

Ø20
Ø10
28
10 8
20
12
6
50
5
12
R10
6
6
56
38
Ø26
Ø12

15.45 Size B sheet.

Solution Hints (Figure 15.47) (For only the best students!) A major portion of the solution of this problem is its layout and the determination of the auxiliary views necessary to show its feature's true size where they can be dimensioned. Recall from Chapter 8 that partial, secondary auxiliary views are helpful in the representation of parts with inclined planes. This problem can best be solved on a size C sheet.

A. Begin by drawing top and front views of the compound bracket and construct the necessary auxiliary views to show all its surfaces' true size.

B. Add dimensions and notes on the views found in part A.

R
Ø20—3 HOLES
R4—TYP

SI ▷⊙

44
92
12
R4
R16
TYP
12
40
R22
96 44
12
62

8

CABLE CLIP
1020 STEEL
8 REQUIRED

15.46 Size B sheet.

R.38
FILLETS AND ROUNDS R.125
LINES D AND E
ARE PARALLEL
1.38
2.50
.62
.38
SURFACE A IS
PARALLEL TO
TO LINE B
.44
SURFACE A
5.50
SURFACE C
1.25
.82
LINE D
R.18
ANGLE BETWEEN
SURFACE C
AND LINE B
55°
.44
LINE B
.50
.38 R.50
1.31
.16
40°
.38
.75
.38 1.38
54°
2.31
.62
50°
.50
R.38
40°
2.25
3.25 3.69
LINE E
90°
.75

9 COMPOUND BRACKET
CAST IRON—3 REQUIRED

15.47 Size B sheet.

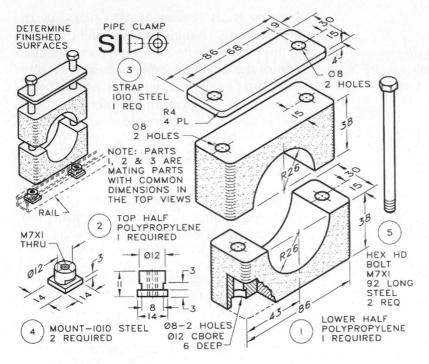

Working Drawings: Multiple Parts

Make working drawings by hand or by computer, as assigned, of the products consisting of multiple parts, shown in **Figures 15.48–15.71,** on the suggested sheet sizes. Include a title block, dimensions, and notes necessary for manufacturing the parts. Make an assembly drawings showing how the parts fit together. More than one sheet may be required.

15.48 Size B sheet.

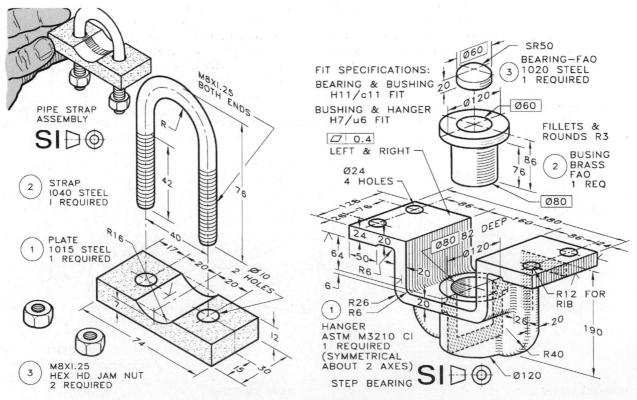

15.49 Size B sheet.

15.50 Size B sheet.

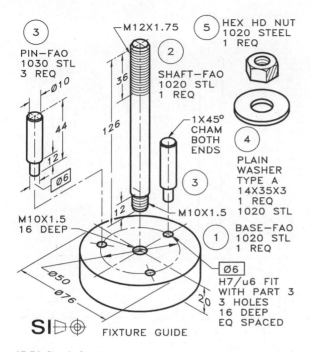

3 PIN–FAO 1030 STL 3 REQ

Ø10

44

12

Ø6

M10X1.5 16 DEEP

—M12X1.75

2 SHAFT–FAO 1020 STL 1 REQ

36

126

12

M10X1.5

5 HEX HD NUT 1020 STEEL 1 REQ

1X45° CHAM BOTH ENDS

4 PLAIN WASHER TYPE A 14X35X3 1 REQ 1020 STL

1 BASE–FAO 1020 STL 1 REQ

Ø6 H7/u6 FIT WITH PART 3 3 HOLES 16 DEEP EQ SPACED

Ø50

Ø76

20

SI◁⊕

FIXTURE GUIDE

15.51 Size B sheets.

Important Points

1. Fill out the title block completely on all sheets of a working drawing: drafter, checker, company (with address), sheet number, and so forth. This is the beginning step of preparing a working drawing that will be a legal document.

2. Prepare the details of each part in a very precise, accurate manner as if you were preparing a legal contract, because that is exactly what working drawings are.

3. Provide adequate space between multiple-view drawings to avoid the crowding of dimensions and notes.

4. The best means of checking your drawing for completeness is by drawing each part using your notes and dimensions.

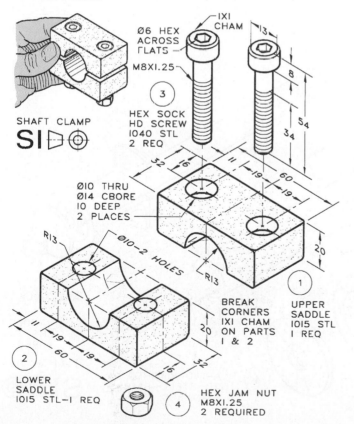

SHAFT CLAMP

SI◁⊕

Ø6 HEX ACROSS FLATS

M8X1.25

3 HEX SOCK HD SCREW 1040 STL 2 REQ

IXI CHAM

13

8

54

34

Ø10 THRU Ø14 CBORE 10 DEEP 2 PLACES

32 16

11 60

19 19

20

1 UPPER SADDLE 1015 STL 1 REQ

R13

Ø10–2 HOLES

R13

BREAK CORNERS IXI CHAM ON PARTS 1 & 2

2 LOWER SADDLE 1015 STL–1 REQ

11 19 60 19

20

16 32

4 HEX JAM NUT M8X1.25 2 REQUIRED

15.52 Size B sheets.

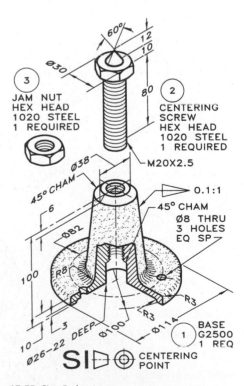

3 JAM NUT HEX HEAD 1020 STEEL 1 REQUIRED

60°

Ø30

12 10

80

2 CENTERING SCREW HEX HEAD 1020 STEEL 1 REQUIRED

M20X2.5

Ø38

45° CHAM

6

0.1:1

45° CHAM Ø8 THRU 3 HOLES EQ SP

Ø82

100

R8

3

Ø100

R3

R3

Ø114

R3

1 BASE G2500 1 REQ

10

Ø26–22 DEEP

SI◁⊕ CENTERING POINT

15.53 Size B sheets.

15.54 Make a working drawing with an assembly of this adjustable milling stop on size B sheets. The column (part 3) that fits into the base (part 1) must be held in position with a set screw. Select an appropriate type and size set screw that passes through the end of the base and bear against the shaft.

15.55 Make a working drawing with an assembly of the flange jig on size B sheets. Give a parts list on the assembly sheet using good drawing and dimensioning practices.

15.56 Make a working drawing with an assembly of the rocker tool post on size B sheets. Give a parts list on the assembly sheet using good drawing and dimensioning practices.

15.57 Make a working drawing with an assembly of the hoist ring on size B sheets. Give a parts list on the assembly sheet using good drawing and dimensioning practices.

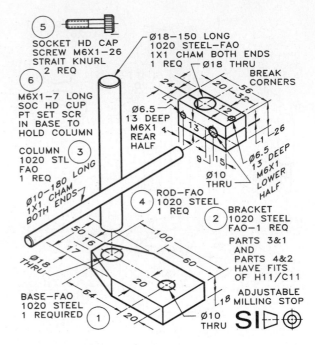

15.54 Size B sheet.

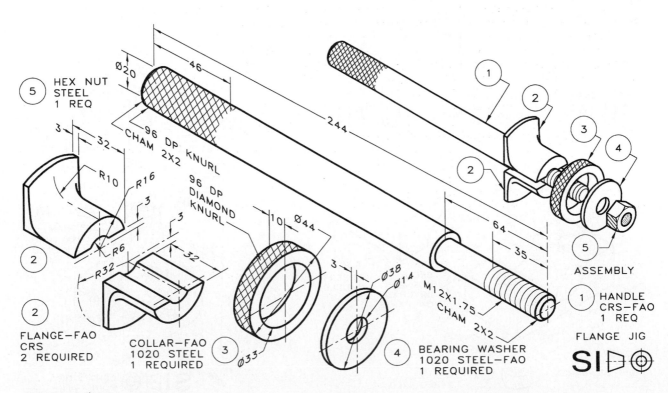

15.55 Size B sheet.

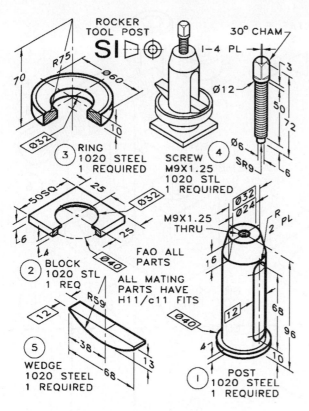

15.56 Size B sheet.

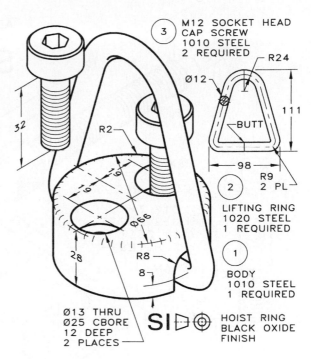

15.57 Size B sheet.

Thought Questions

1. What is "knurling" and what is its purposes?

2. What does the note "FAO" mean?

3. Which orthographic view of a cylindrical hole is best for giving its diameter with a leader and a note? Why?

4. Which orthographic view of a cylindrical hole is best for giving location dimension to show its location? Why?

5. What does it mean when a dimension on a working drawing is enclosed in a box?

6. What does it mean when two dimensions are given on a single dimension line?

7. What term describes a hole with a flat bottom that is made part way inside a smaller hole?

8. What general information should be provided that names and identifies each part shown on a working drawing?

9. What is the variation in size of a part in a working drawing called?

10. How many types of set screws are there? Name them.

11. Who is ulitimately responsible for the correctness of a working drawing?

12. How many dimensions are necessary on an assembly drawing?

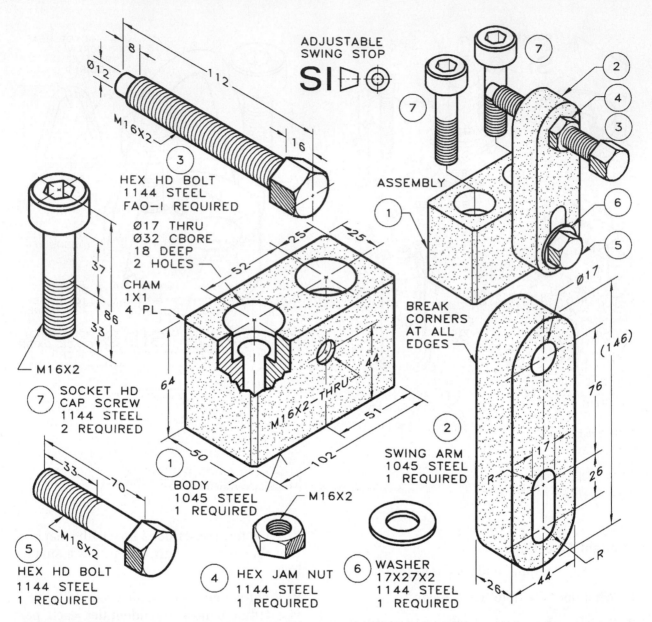

ADJUSTABLE
SWING STOP

SI ▷⊕

Ø12

8

112

16

M16X2

③

HEX HD BOLT
1144 STEEL
FAO-1 REQUIRED

Ø17 THRU
Ø32 CBORE
18 DEEP
2 HOLES

CHAM
1X1
4 PL

37

86

33

M16X2

⑦ SOCKET HD
CAP SCREW
1144 STEEL
2 REQUIRED

33

70

M16X2

⑤
HEX HD BOLT
1144 STEEL
1 REQUIRED

ASSEMBLY

①

②

⑦

⑦

④
③

⑥
⑤

25   25

52

64

44

M16X2-THRU

51

50

102

① BODY
1045 STEEL
1 REQUIRED

M16X2

BREAK
CORNERS
AT ALL
EDGES

Ø17

(146)

76

26

17

R

R

② SWING ARM
1045 STEEL
1 REQUIRED

26   44   R

④ HEX JAM NUT
1144 STEEL
1 REQUIRED

⑥ WASHER
17X27X2
1144 STEEL
1 REQUIRED

15.58 Make working drawings with an assembly drawing of this adjustable swing stop on size B sheets.

Thought Questions

1. What is the purpose of the washer (part 6) in this assembly?

2. Why is there a slot in the swing arm (part 2) instead of the circular hole?

3. Why did the designer of this assembly use both socket-head cap screws and hex-head cap screws instead of using one or the other?

4. When assembled, what is the range of adjustment from the upper surface of the body (part 1) and the centerline of the hex-head bolt (part 3)?

5. What is the maximum distance that the 12 DIA end of hex-head bolt (part 3) can extend beyond the face of swing arm (part 2)?

6. Write a paragraph to verbally give a description of either part 1 or part 2 in the absence of a drawing.

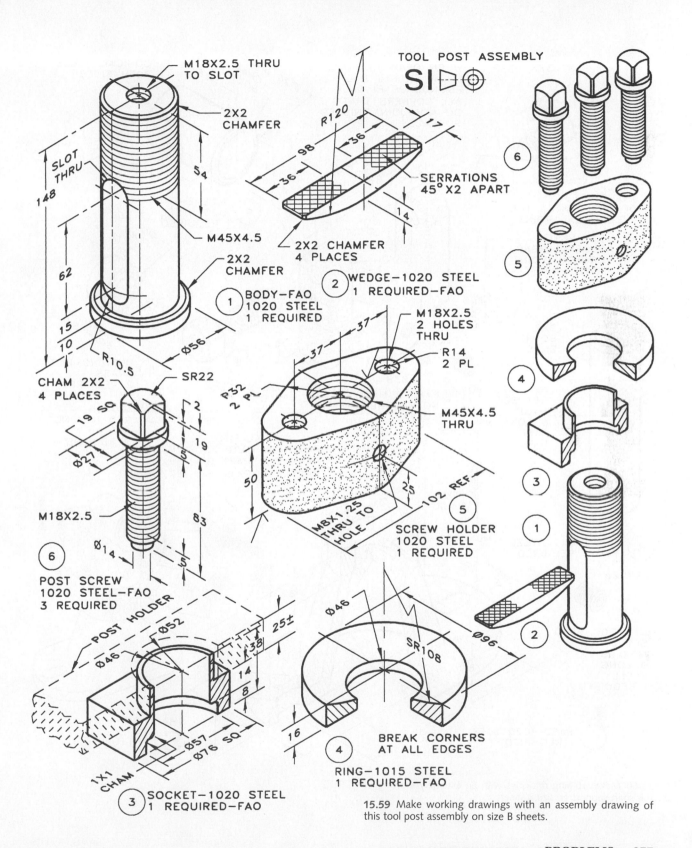

M18X2.5 THRU TO SLOT

2X2 CHAMFER

SLOT THRU

148

54

M45X4.5

2X2 CHAMFER

62

15

10

R10.5

Ø56

① BODY—FAO 1020 STEEL 1 REQUIRED

TOOL POST ASSEMBLY

SI

R120

98

36

36

17

SERRATIONS 45°X2 APART

14

2X2 CHAMFER 4 PLACES

② WEDGE—1020 STEEL 1 REQUIRED—FAO

CHAM 2X2 4 PLACES

SR22

19 SQ

2

19

5

Ø27

M18X2.5

83

5

Ø14

5

⑥ POST SCREW 1020 STEEL—FAO 3 REQUIRED

37

37

M18X2.5 2 HOLES THRU

R14 2 PL

P32 2 PL

M45X4.5 THRU

50

25

102 REF

M8X1.25 THRU TO HOLE

⑤ SCREW HOLDER 1020 STEEL 1 REQUIRED

POST HOLDER

Ø52

Ø46

25±

38

14

8

Ø57

Ø76 SQ

1X1 CHAM

③ SOCKET—1020 STEEL 1 REQUIRED—FAO

Ø46

SR108

16

④ RING—1015 STEEL 1 REQUIRED—FAO

BREAK CORNERS AT ALL EDGES

⑥

⑤

④

③

①

②

Ø96

15.59 Make working drawings with an assembly drawing of this tool post assembly on size B sheets.

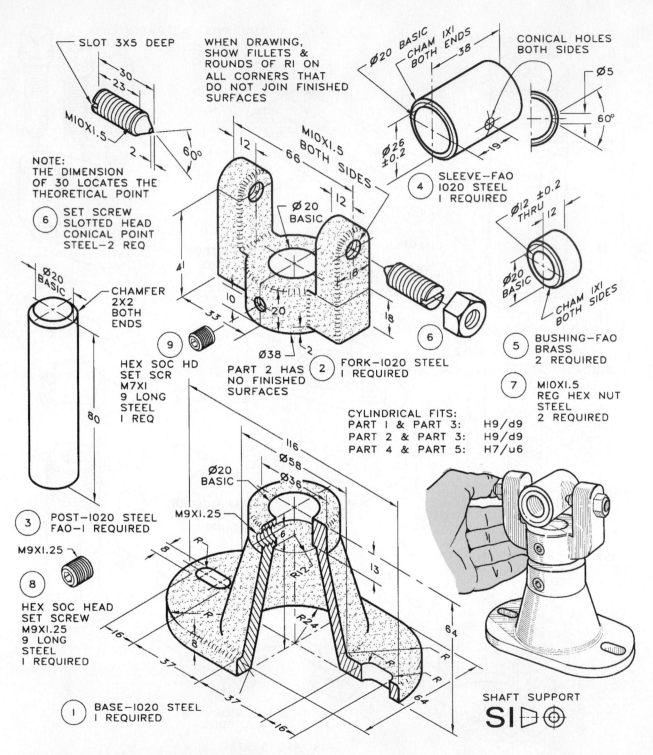

SLOT 3X5 DEEP
30
23
M10X1.5
2
60°

WHEN DRAWING, SHOW FILLETS & ROUNDS OF RI ON ALL CORNERS THAT DO NOT JOIN FINISHED SURFACES

Ø20 BASIC
CHAM IXI BOTH ENDS
38
CONICAL HOLES BOTH SIDES
Ø5
60°
Ø26 ±0.2
19
④ SLEEVE–FAO 1020 STEEL 1 REQUIRED

NOTE: THE DIMENSION OF 30 LOCATES THE THEORETICAL POINT

⑥ SET SCREW SLOTTED HEAD CONICAL POINT STEEL–2 REQ

12
66
M10X1.5 BOTH SIDES
12
Ø20 BASIC
41
18
10
33
20
Ø38
2
② FORK–1020 STEEL 1 REQUIRED
PART 2 HAS NO FINISHED SURFACES
18

Ø12 ±0.2 THRU
12
Ø20 BASIC
CHAM IXI BOTH SIDES
⑤ BUSHING–FAO BRASS 2 REQUIRED

⑥

⑦ M10X1.5 REG HEX NUT STEEL 2 REQUIRED

Ø20 BASIC
CHAMFER 2X2 BOTH ENDS

⑨ HEX SOC HD SET SCR M7X1 9 LONG STEEL 1 REQ

80

③ POST–1020 STEEL FAO–1 REQUIRED

M9X1.25
⑧ HEX SOC HEAD SET SCREW M9X1.25 9 LONG STEEL 1 REQUIRED

CYLINDRICAL FITS:
PART 1 & PART 3:   H9/d9
PART 2 & PART 3:   H9/d9
PART 4 & PART 5:   H7/u6

116
Ø58
Ø36
Ø20 BASIC
M9X1.25
8
6
R12
R24
R
R
13
64
64
16
37
37
16
8
R

① BASE–1020 STEEL 1 REQUIRED

SHAFT SUPPORT
SI ⫼ ⊕

15.60 Make working drawings with an assembly drawing of this shaft support on size B sheets.

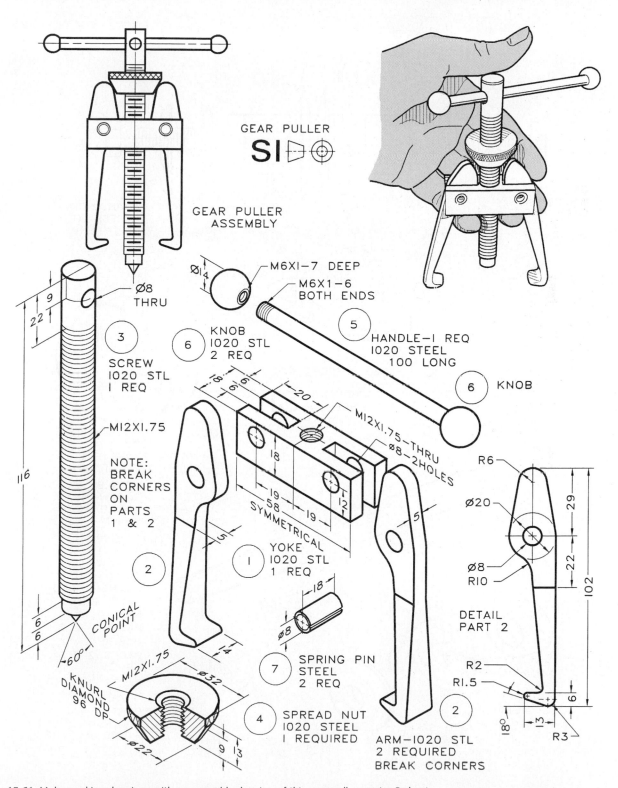

GEAR PULLER

SI ⊏⊙⊙

GEAR PULLER
ASSEMBLY

Ø8
THRU

③
SCREW
1020 STL
1 REQ

Ø14

M6X1−7 DEEP

M6X1−6
BOTH ENDS

⑥
KNOB
1020 STL
2 REQ

⑤
HANDLE−1 REQ
1020 STEEL
100 LONG

⑥ KNOB

M12X1.75

NOTE:
BREAK
CORNERS
ON
PARTS
1 & 2

9
22

116

6
6

60°

CONICAL
POINT

②

18
6
6

20

M12X1.75−THRU
ø8−2HOLES

18

19
58
19

12

SYMMETRICAL

①
YOKE
1020 STL
1 REQ

5

R6

Ø20

Ø8
R10

29
22
102

DETAIL
PART 2

②

18

ø8

⑦
SPRING PIN
STEEL
2 REQ

5

4

R2
R1.5

18°

13

6

R3

M12X1.75

KNURL
DIAMOND
96 DP

ø32

④
SPREAD NUT
1020 STEEL
1 REQUIRED

ø22

9 13

ARM−1020 STL
2 REQUIRED
BREAK CORNERS

15.61 Make working drawings with an assembly drawing of this gear puller on size B sheets.

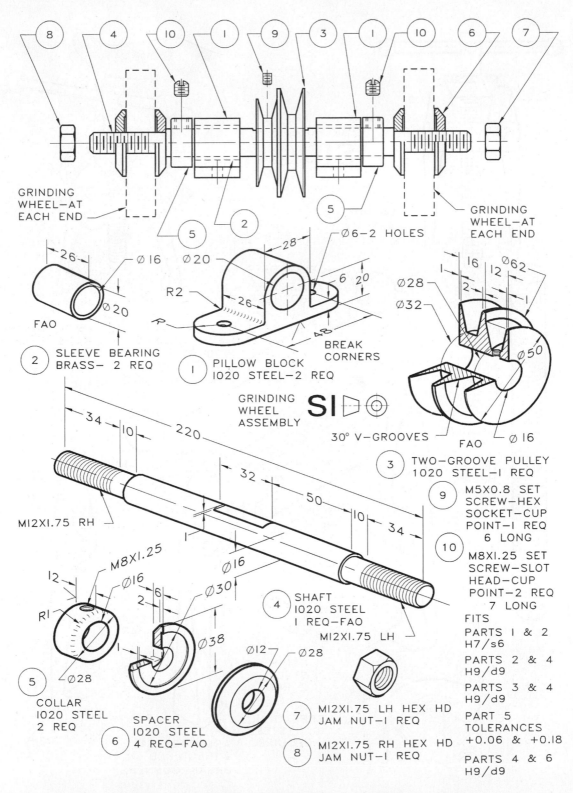

8 GRINDING WHEEL—AT EACH END

4 **10** **1** **9** **3** **1** **10** **6** **7** GRINDING WHEEL—AT EACH END

5 **2**

5

26 Ø16 Ø20

Ø20

FAO

2 SLEEVE BEARING BRASS— 2 REQ

Ø6—2 HOLES

28

R2

26

R

BREAK CORNERS

6 20

48

1 PILLOW BLOCK 1020 STEEL—2 REQ

GRINDING WHEEL ASSEMBLY

SI

30° V-GROOVES

16 12 Ø62

2 1

Ø28

Ø32

Ø50

FAO

Ø16

3 TWO—GROOVE PULLEY 1020 STEEL—1 REQ

9 M5X0.8 SET SCREW—HEX SOCKET—CUP POINT—1 REQ 6 LONG

10 M8X1.25 SET SCREW—SLOT HEAD—CUP POINT—2 REQ 7 LONG

34 10 220

32

50

10 34

MI2XI.75 RH

M8XI.25

12 Ø16

RI

Ø16

2 6

Ø30

Ø16

SHAFT 1020 STEEL 1 REQ—FAO

MI2XI.75 LH

Ø12 Ø28

MI2XI.75 LH HEX HD JAM NUT—1 REQ

MI2XI.75 RH HEX HD JAM NUT—1 REQ

Ø28

5 COLLAR 1020 STEEL 2 REQ

Ø38

6 SPACER 1020 STEEL 4 REQ—FAO

7

8

4

FITS

PARTS 1 & 2 H7/s6

PARTS 2 & 4 H9/d9

PARTS 3 & 4 H9/d9

PART 5 TOLERANCES +0.06 & +0.18

PARTS 4 & 6 H9/d9

15.62 Make working drawings with an assembly drawing of this grinding wheel assembly on size B sheets.

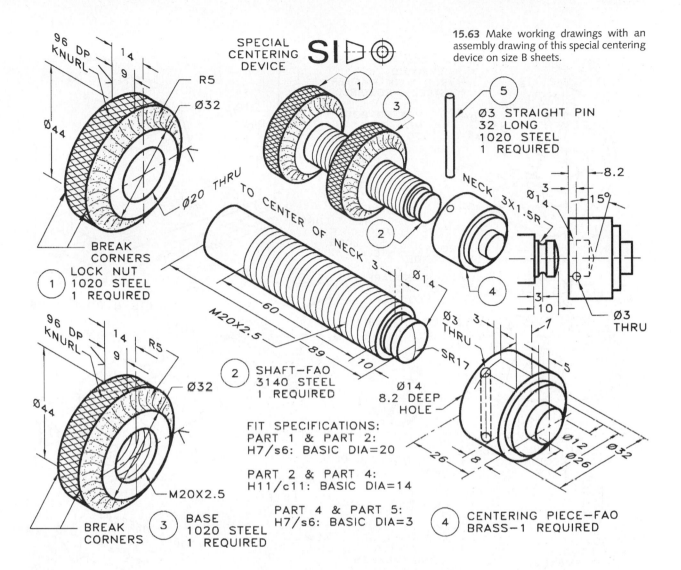

15.63 Make working drawings with an assembly drawing of this special centering device on size B sheets.

SPECIAL CENTERING DEVICE   SI ▷ ◎

① LOCK NUT
1020 STEEL
1 REQUIRED

② SHAFT—FAO
3140 STEEL
1 REQUIRED

③ BASE
1020 STEEL
1 REQUIRED

④ CENTERING PIECE—FAO
BRASS—1 REQUIRED

⑤ Ø3 STRAIGHT PIN
32 LONG
1020 STEEL
1 REQUIRED

FIT SPECIFICATIONS:
PART 1 & PART 2:
H7/s6: BASIC DIA=20

PART 2 & PART 4:
H11/c11: BASIC DIA=14

PART 4 & PART 5:
H7/s6: BASIC DIA=3

Thought Questions

1. What conversion factor would you use to convert metric dimensions to English units?

2. Why is knurling given on parts 1 and 3? Why was knurling not given on part 4?

3. Why was part 1 and part 2 not designed with threads for attachment to each other?

4. When assembled, will part 4 rotate about the end of part 2? Explain and determine why it attaches as it does.

5. Why was the fit between part 4 and 5 selected to be H7/s6 instead of H11/c11?

6. What would be the approximate weight of the total assembly if all materials were assumed to weight 490 lbs per cu ft?

7. Can you explain why parts 1 and 3 were designed with bosses as shown?

8. Which of the parts can be specified on a working drawing by a note without a drawing?

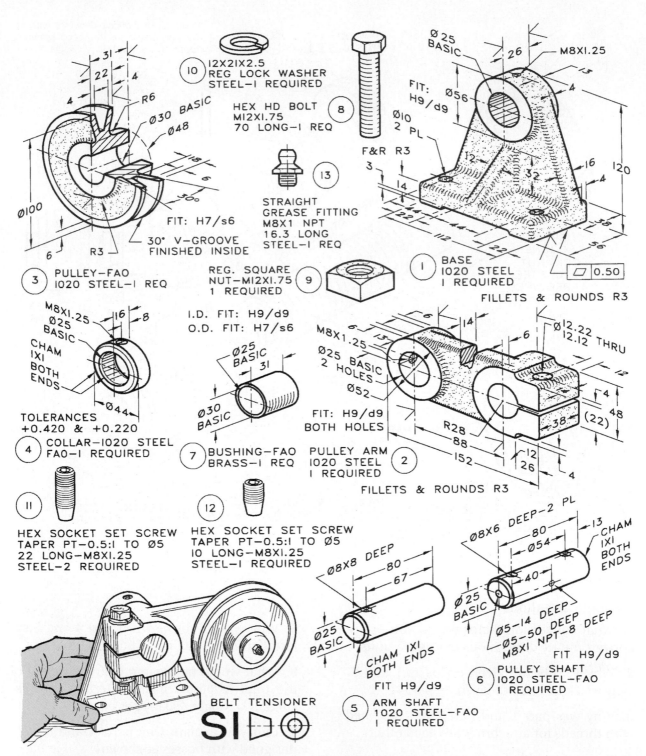

10 12X21X2.5
REG LOCK WASHER
STEEL-1 REQUIRED

8 HEX HD BOLT
M12X1.75
70 LONG-1 REQ

13 STRAIGHT
GREASE FITTING
M8X1 NPT
16.3 LONG
STEEL-1 REQ

9 REG. SQUARE
NUT-M12X1.75
1 REQUIRED

31
22
4
4
R6
Ø30 BASIC
Ø48
Ø100
18
6
30°
6
R3
FIT: H7/s6
30° V-GROOVE
FINISHED INSIDE

3 PULLEY-FAO
1020 STEEL-1 REQ

Ø25
BASIC
26
M8X1.25
13
FIT:
H9/d9
Ø56
Ø10
2 PL
F&R R3
3
14
12
32
16
4
120
44
22
44
112
22
38
56

1 BASE
1020 STEEL
1 REQUIRED

☐ 0.50

FILLETS & ROUNDS R3

M8X1.25
Ø25
BASIC
16 8
CHAM
1X1
BOTH
ENDS
Ø44

TOLERANCES
+0.420 & +0.220

4 COLLAR-1020 STEEL
FAO-1 REQUIRED

I.D. FIT: H9/d9
O.D. FIT: H7/s6

Ø25
BASIC
31
Ø30
BASIC

7 BUSHING-FAO
BRASS-1 REQ

M8X1.25
Ø25 BASIC
2 HOLES
Ø52
6
14
6
Ø12.22 THRU
12.12
12
4
48
38
(22)
R28
88
152
12
26
4
FIT: H9/d9
BOTH HOLES

PULLEY ARM
1020 STEEL
1 REQUIRED

2

FILLETS & ROUNDS R3

11 HEX SOCKET SET SCREW
TAPER PT-0.5:1 TO Ø5
22 LONG-M8X1.25
STEEL-2 REQUIRED

12 HEX SOCKET SET SCREW
TAPER PT-0.5:1 TO Ø5
10 LONG-M8X1.25
STEEL-1 REQUIRED

Ø8X6 DEEP-2 PL
80
13
CHAM
1X1
BOTH
ENDS
Ø54
40
Ø25
BASIC
Ø5-14 DEEP
Ø5-50 DEEP
M8X1 NPT-8 DEEP
FIT H9/d9

6 PULLEY SHAFT
1020 STEEL-FAO
1 REQUIRED

Ø8X8 DEEP
80
67
Ø25
BASIC
CHAM 1X1
BOTH ENDS
FIT H9/d9

5 ARM SHAFT
1020 STEEL-FAO
1 REQUIRED

BELT TENSIONER

SI ◁ ◉

15.64 Make working drawings with an assembly drawing of
this belt tensioner device on size B sheets.

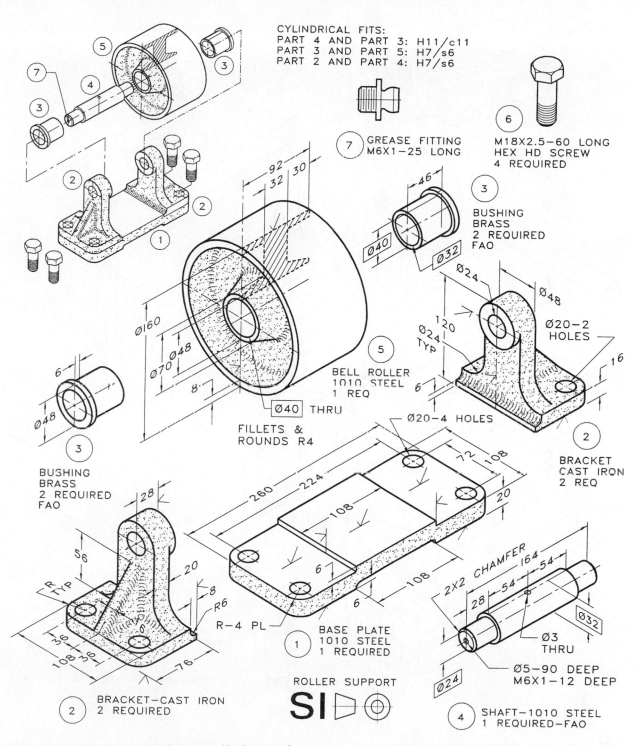

CYLINDRICAL FITS:
PART 4 AND PART 3: H11/c11
PART 3 AND PART 5: H7/s6
PART 2 AND PART 4: H7/s6

⑦ GREASE FITTING
M6X1-25 LONG

⑥ M18X2.5-60 LONG
HEX HD SCREW
4 REQUIRED

92
32 30

③ BUSHING
BRASS
2 REQUIRED
FAO

46
Ø40
Ø32

Ø160
Ø70 Ø48
8

⑤ BELL ROLLER
1010 STEEL
1 REQ

Ø40 THRU

FILLETS &
ROUNDS R4

Ø24
Ø48
120
Ø24
TYP
6
Ø20-2
HOLES
16

② BRACKET
CAST IRON
2 REQ

6
Ø48

③ BUSHING
BRASS
2 REQUIRED
FAO

Ø20-4 HOLES
72 108
260 224
108
108
6
6
20

① BASE PLATE
1010 STEEL
1 REQUIRED

28
56
20
R
TYP
8
R6
36
108 36
76
R-4 PL

② BRACKET—CAST IRON
2 REQUIRED

ROLLER SUPPORT

SI

2X2 CHAMFER
164
28 54 54
Ø3
THRU
Ø32
Ø24
Ø5-90 DEEP
M6X1-12 DEEP

④ SHAFT-1010 STEEL
1 REQUIRED-FAO

15.65 Make working drawings with an assembly drawing of
this roller support fixture on size B sheets.

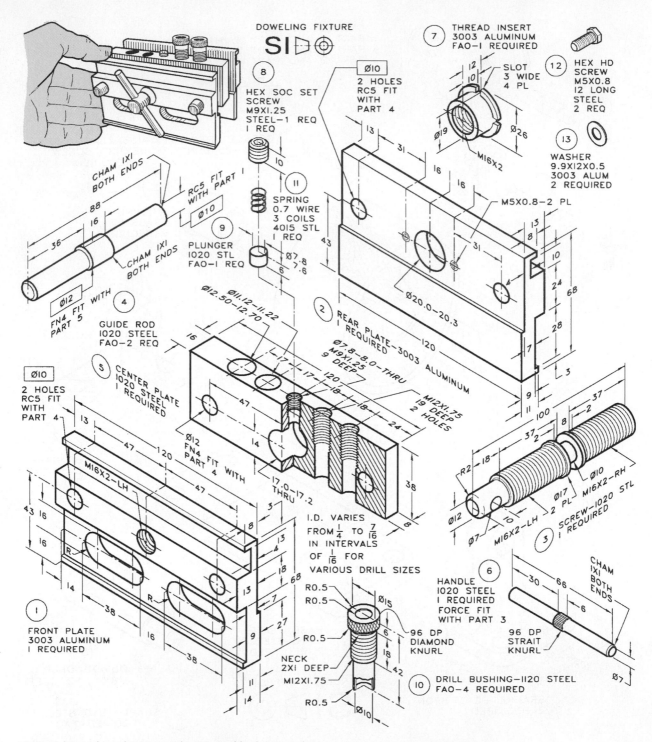

DOWELING FIXTURE

SI ⊳⊙

⑧ HEX SOC SET SCREW M9X1.25 STEEL—1 REQ I REQ

⑪ SPRING 0.7 WIRE 3 COILS 4015 STL I REQ

⑨ PLUNGER 1020 STL FAO-1 REQ

Ø10 2 HOLES RC5 FIT WITH PART 4

⑦ THREAD INSERT 3003 ALUMINUM FAO-1 REQUIRED

SLOT 3 WIDE 4 PL

M16X2

⑫ HEX HD SCREW M5X0.8 12 LONG STEEL 2 REQ

⑬ WASHER 9.9X12X0.5 3003 ALUM 2 REQUIRED

M5X0.8—2 PL

② REAR PLATE—3003 ALUMINUM I REQUIRED

Ø20.0—20.3

CHAM IXI BOTH ENDS

RC5 FIT WITH PART I

Ø10

CHAM IXI BOTH ENDS

Ø12 FN4 FIT WITH PART 5

④ GUIDE ROD 1020 STEEL FAO-2 REQ

⑤ CENTER PLATE 1020 STEEL I REQUIRED

Ø12 FN4 FIT WITH PART 4

Ø11.12—11.22 Ø12.50—12.70

Ø7.8—8.0—THRU M9X1.25 9 DEEP

M12X1.75 19 DEEP 2 HOLES

17.0—17.2 THRU

Ø10 2 HOLES RC5 FIT WITH PART 4

M16X2-LH

① FRONT PLATE 3003 ALUMINUM I REQUIRED

R

R

I.D. VARIES FROM 1/4 TO 7/16 IN INTERVALS OF 1/16 FOR VARIOUS DRILL SIZES

R0.5
R0.5
R0.5
Ø15

96 DP DIAMOND KNURL

NECK 2X1 DEEP
M12X1.75
R0.5
Ø10

⑥ HANDLE 1020 STEEL I REQUIRED FORCE FIT WITH PART 3

96 DP STRAIT KNURL

M16X2-LH — 2 PL
Ø17
Ø10
M16X2-RH

③ SCREW-1020 STL I REQUIRED

Ø12
Ø7

CHAM IXI BOTH ENDS

Ø7

⑩ DRILL BUSHING—1120 STEEL FAO-4 REQUIRED

15.66 Make working drawings with an assembly drawing of this doweling fixture on size B sheets.

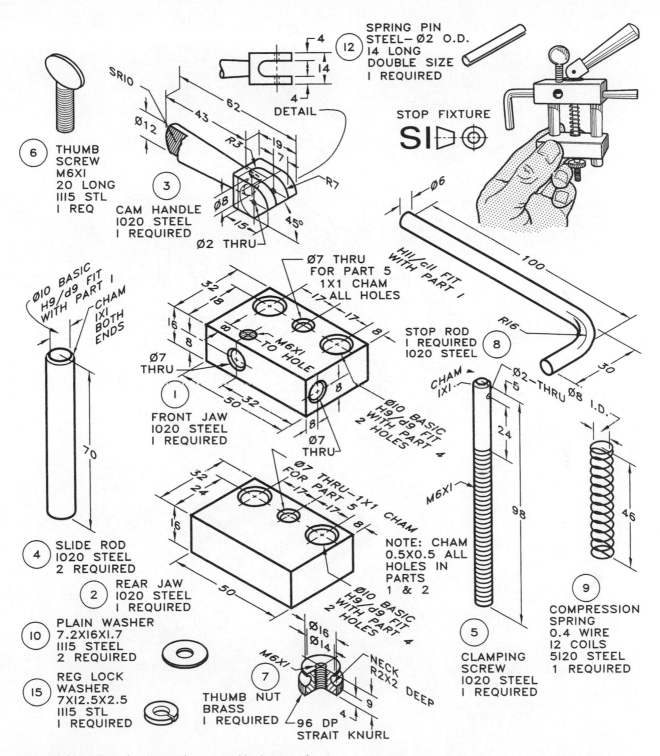

SPRING PIN
STEEL—Ø2 O.D.
14 LONG
DOUBLE SIZE
1 REQUIRED
⑫

DETAIL
4
14
4

STOP FIXTURE
SI⊳⊕

⑥ **THUMB SCREW** M6X1 20 LONG 1115 STL 1 REQ

SR10
Ø12
43
62
R3
19
7
Ø8
15
45°
Ø2 THRU

③ **CAM HANDLE** 1020 STEEL 1 REQUIRED

R7

Ø6

H11/c11 FIT WITH PART 1

100

R16

STOP ROD 1 REQUIRED 1020 STEEL ⑧

30

Ø2-THRU Ø8 I.D.

Ø10 BASIC H9/d9 FIT WITH PART 1

CHAM 1X1 BOTH ENDS

70

Ø7 THRU FOR PART 5 1X1 CHAM ALL HOLES
32
18
16
8
8
Ø7 THRU
M6X1 TO HOLE
① **FRONT JAW** 1020 STEEL 1 REQUIRED
50
32
8
Ø7 THRU
Ø10 BASIC H9/d9 FIT WITH PART 4 2 HOLES
8

CHAM 1X1
Ø2 THRU
5
24
M6X1
98

④ **SLIDE ROD** 1020 STEEL 2 REQUIRED

Ø7 THRU—1X1 CHAM FOR PART 5
32
24
17
17
8
16
50
Ø10 BASIC H9/d9 FIT WITH PART 4 2 HOLES

② **REAR JAW** 1020 STEEL 1 REQUIRED

NOTE: CHAM 0.5X0.5 ALL HOLES IN PARTS 1 & 2

⑤ **CLAMPING SCREW** 1020 STEEL 1 REQUIRED

⑩ **PLAIN WASHER** 7.2X16X1.7 1115 STEEL 2 REQUIRED

⑮ **REG LOCK WASHER** 7X12.5X2.5 1115 STL 1 REQUIRED

M6X1
⑦
THUMB NUT BRASS 1 REQUIRED
Ø16
Ø14
NECK R2X2 DEEP
9
4
96 DP STRAIT KNURL

⑨ **COMPRESSION SPRING** 0.4 WIRE 12 COILS 5120 STEEL 1 REQUIRED
46

15.67 Make working drawings with an assembly drawing of this stop fixture on size B sheets.

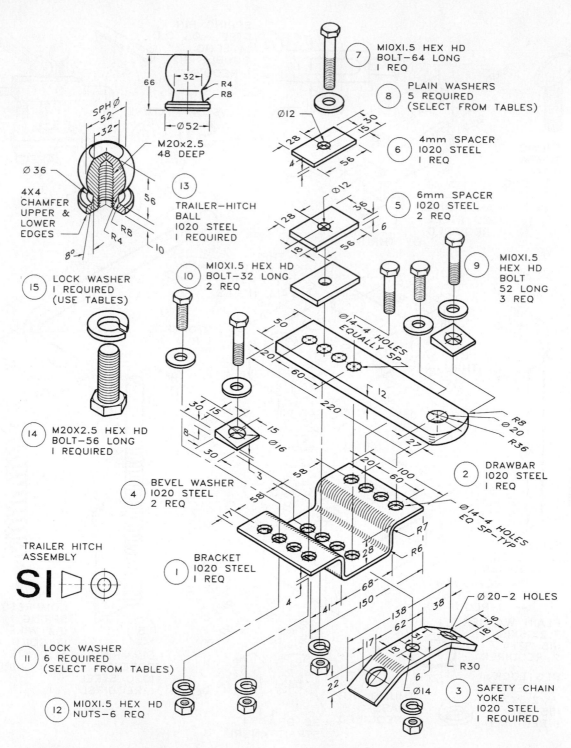

66

32

R4
R8

Ø52

SPHØ
52
32

M20x2.5
48 DEEP

Ø 36

56

4X4
CHAMFER
UPPER &
LOWER
EDGES

R8
R4

8°

10

⑬ TRAILER—HITCH
BALL
1020 STEEL
1 REQUIRED

⑦ M10X1.5 HEX HD
BOLT—64 LONG
1 REQ

⑧ PLAIN WASHERS
5 REQUIRED
(SELECT FROM TABLES)

Ø12

30
15

28

4

56

⑥ 4mm SPACER
1020 STEEL
1 REQ

Ø12

30

28

6

18

56

⑤ 6mm SPACER
1020 STEEL
2 REQ

⑮ LOCK WASHER
1 REQUIRED
(USE TABLES)

⑩ M10X1.5 HEX HD
BOLT—32 LONG
2 REQ

⑨ M10X1.5
HEX HD
BOLT
52 LONG
3 REQ

Ø14—4 HOLES
EQUALLY SP

50

20

60

220

12

R8
Ø 20
R36

27

20

100

60

② DRAWBAR
1020 STEEL
1 REQ

Ø14—4 HOLES
EO SP—TYP

30
15

8

30

15
Ø16

3

58

④ BEVEL WASHER
1020 STEEL
2 REQ

58

17

R7

28

R6

⑭ M20X2.5 HEX HD
BOLT—56 LONG
1 REQUIRED

TRAILER HITCH
ASSEMBLY

SI ▷ ⊙

① BRACKET
1020 STEEL
1 REQ

4

41

68

150

138

62

Ø 20—2 HOLES

30
18

17

8

R30

6

Ø14

③ SAFETY CHAIN
YOKE
1020 STEEL
1 REQUIRED

22

⑪ LOCK WASHER
6 REQUIRED
(SELECT FROM TABLES)

⑫ M10X1.5 HEX HD
NUTS—6 REQ

15.68 Make working drawings with an assembly drawing of
this trailer hitch on size B sheets.

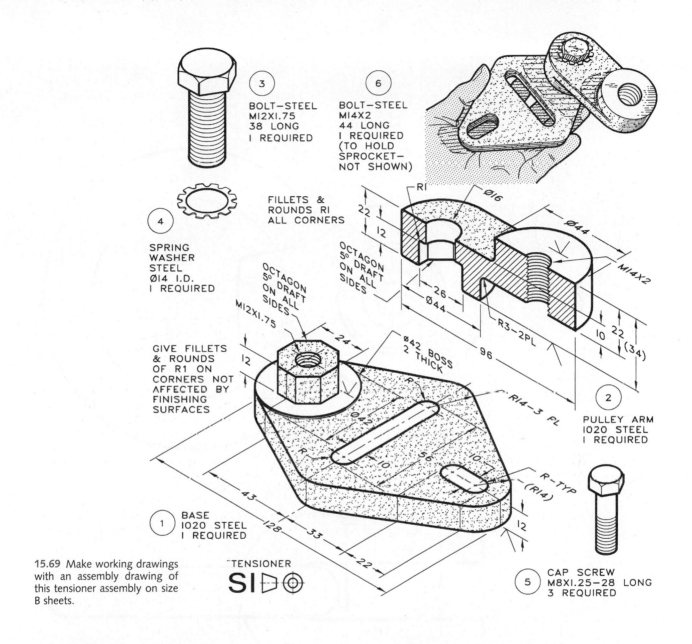

3 BOLT—STEEL
M12X1.75
38 LONG
1 REQUIRED

6 BOLT—STEEL
M14X2
44 LONG
1 REQUIRED
(TO HOLD
SPROCKET—
NOT SHOWN)

4 SPRING
WASHER
STEEL
Ø14 I.D.
1 REQUIRED

FILLETS &
ROUNDS R1
ALL CORNERS

R1   Ø16   Ø44   M14X2

22   12

OCTAGON
5° DRAFT
ON ALL
SIDES

26   Ø44   R3—2PL   10   22   (34)

96

2 PULLEY ARM
1020 STEEL
1 REQUIRED

OCTAGON
5° DRAFT
ON ALL
SIDES

M12X1.75

GIVE FILLETS
& ROUNDS
OF R1 ON
CORNERS NOT
AFFECTED BY
FINISHING
SURFACES

12   24   Ø42 BOSS
2 THICK

1 BASE
1020 STEEL
1 REQUIRED

43   128   33   22

R14—3 PL

R—TYP
(R14)

12

5 CAP SCREW
M8X1.25—28 LONG
3 REQUIRED

15.69 Make working drawings with an assembly drawing of this tensioner assembly on size B sheets.

TENSIONER
S1 ⊳ ⊕

Working Drawings: Multiple Parts with Design Applications

Make dimensioned working drawings of the multiple parts shown in **Figures 15.70–15.71** on a sheet size of your choice with the necessary dimensions and notes to fabricate the parts. Each part is given in a general format, which requires some design on your part. You must consider the addition of fillets and rounds, the application of finish marks, and the modification of features of the parts to make them functional and practical. Apply tolerances to the parts in limit form by using the tables of cylindrical fits in the Appendix. Make an assembly drawing and parts list to show how the parts are to be put together.

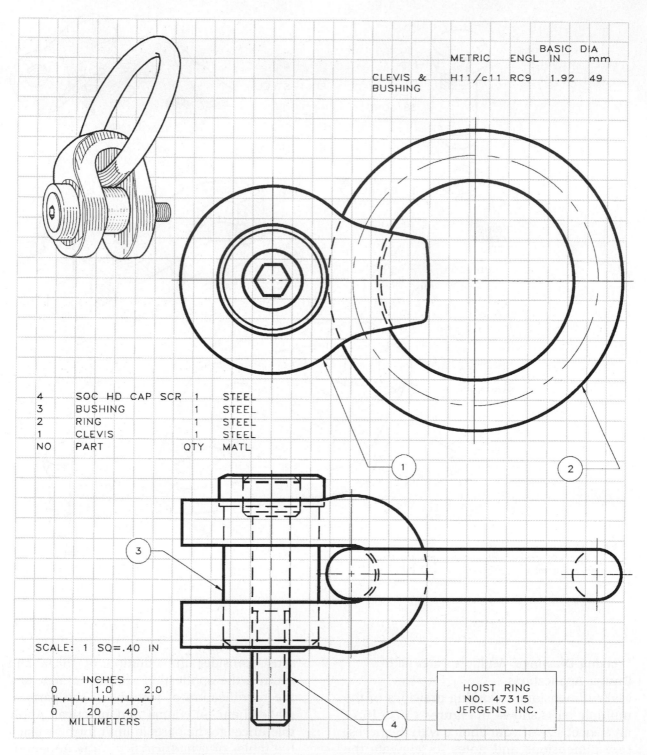

			BASIC DIA
	METRIC	ENGL IN	mm
CLEVIS & BUSHING	H11/c11	RC9 1.92	49

4	SOC HD CAP SCR	1	STEEL
3	BUSHING	1	STEEL
2	RING	1	STEEL
1	CLEVIS	1	STEEL
NO	PART	QTY	MATL

SCALE: 1 SQ=.40 IN

INCHES
0   1.0   2.0
0   20   40
MILLIMETERS

HOIST RING
NO. 47315
JERGENS INC.

15.70 Make working drawings with an assembly drawing of the hoist ring on size B sheets.

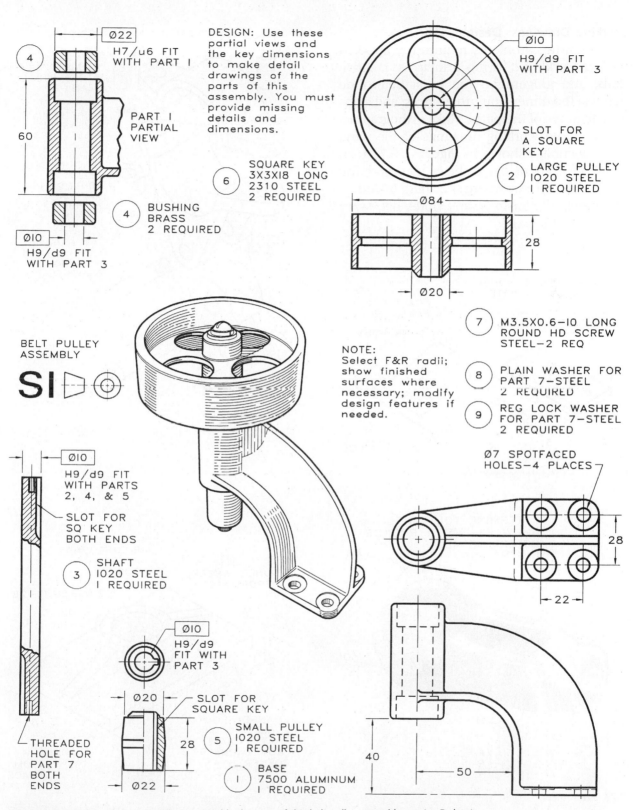

Ø22

④

H7/u6 FIT
WITH PART 1

PART I
PARTIAL
VIEW

60

Ø10

H9/d9 FIT
WITH PART 3

④ BUSHING
BRASS
2 REQUIRED

DESIGN: Use these
partial views and
the key dimensions
to make detail
drawings of the
parts of this
assembly. You must
provide missing
details and
dimensions.

⑥ SQUARE KEY
3X3X18 LONG
2310 STEEL
2 REQUIRED

Ø10

H9/d9 FIT
WITH PART 3

SLOT FOR
A SQUARE
KEY

② LARGE PULLEY
1020 STEEL
1 REQUIRED

Ø84

28

Ø20

⑦ M3.5X0.6-10 LONG
ROUND HD SCREW
STEEL-2 REQ

⑧ PLAIN WASHER FOR
PART 7-STEEL
2 REQUIRED

⑨ REG LOCK WASHER
FOR PART 7-STEEL
2 REQUIRED

BELT PULLEY
ASSEMBLY

SI

NOTE:
Select F&R radii;
show finished
surfaces where
necessary; modify
design features if
needed.

Ø7 SPOTFACED
HOLES-4 PLACES

Ø10

H9/d9 FIT
WITH PARTS
2, 4, & 5

SLOT FOR
SQ KEY
BOTH ENDS

③ SHAFT
1020 STEEL
1 REQUIRED

28

22

THREADED
HOLE FOR
PART 7
BOTH
ENDS

Ø10

H9/d9
FIT WITH
PART 3

Ø20

SLOT FOR
SQUARE KEY

28

Ø22

⑤ SMALL PULLEY
1020 STEEL
1 REQUIRED

① BASE
7500 ALUMINUM
1 REQUIRED

40

50

15.71 Make working drawings with an assembly drawing of this belt pulley assembly on size B sheets.

Working Drawings: Design

The following problems require the application of working drawing principles, creative skills, and judgement. You must determine many of the dimensions, tolerances, and standard features of the parts. Make orthographic, dimensioned working drawings of the parts and assemblies shown in **Figures 15.72** thru **15.86** on size A or size B sheets incorporating the design features where specified. Include a title block, dimensions, and notes necessary for making the part.

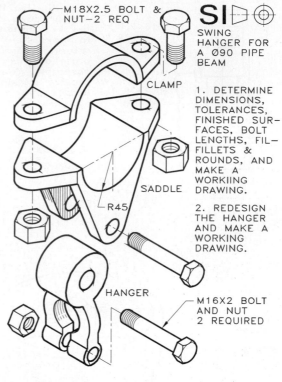

SI

SWING HANGER FOR A Ø90 PIPE BEAM

M18X2.5 BOLT & NUT-2 REQ

CLAMP

SADDLE

R45

HANGER

M16X2 BOLT AND NUT 2 REQUIRED

1. DETERMINE DIMENSIONS, TOLERANCES, FINISHED SURFACES, BOLT LENGTHS, FIL-FILLETS & ROUNDS, AND MAKE A WORKIING DRAWING.

2. REDESIGN THE HANGER AND MAKE A WORKING DRAWING.

15.74 Size B sheets.

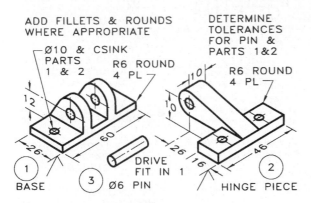

ADD FILLETS & ROUNDS WHERE APPROPRIATE

Ø10 & CSINK PARTS 1 & 2

R6 ROUND 4 PL

12

26

60

1

BASE

DRIVE FIT IN 1 Ø6 PIN

3

DETERMINE TOLERANCES FOR PIN & PARTS 1&2

R6 ROUND 4 PL

10

10

26

16

46

2

HINGE PIECE

15.72 Size A sheet. Hinge assembly.

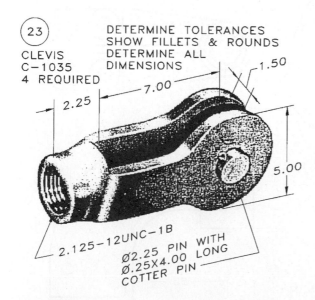

23

CLEVIS C-1035 4 REQUIRED

DETERMINE TOLERANCES SHOW FILLETS & ROUNDS DETERMINE ALL DIMENSIONS

2.25

7.00

1.50

5.00

2.125-12UNC-1B

Ø2.25 PIN WITH Ø.25X4.00 LONG COTTER PIN

15.73 Size A sheet.

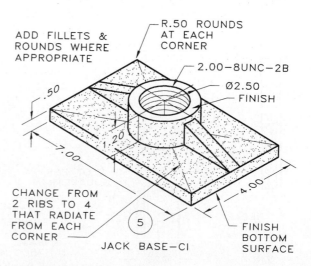

ADD FILLETS & ROUNDS WHERE APPROPRIATE

R.50 ROUNDS AT EACH CORNER

2.00-8UNC-2B

Ø2.50 FINISH

.50

.20

7.00

4.00

CHANGE FROM 2 RIBS TO 4 THAT RADIATE FROM EACH CORNER

5

JACK BASE-CI

FINISH BOTTOM SURFACE

15.75 Size B sheets.

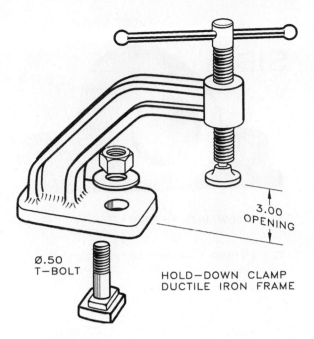

Ø.50
T-BOLT

3.00
OPENING

HOLD-DOWN CLAMP
DUCTILE IRON FRAME

15.76 Size B sheets.

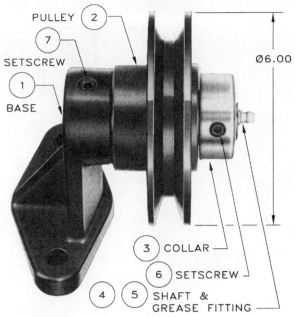

PULLEY ②

⑦

SETSCREW

① BASE

Ø6.00

③ COLLAR

⑥ SETSCREW

④ ⑤ SHAFT &
GREASE FITTING

PULLEY BRACKET ASSEMBLY

15.78 Size B sheets. *(Courtesy of Rockwell Automation.)*

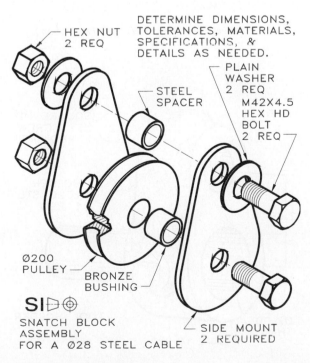

HEX NUT
2 REQ

DETERMINE DIMENSIONS,
TOLERANCES, MATERIALS,
SPECIFICATIONS, &
DETAILS AS NEEDED.

STEEL
SPACER

PLAIN
WASHER
2 REQ
M42X4.5
HEX HD
BOLT
2 REQ

Ø200
PULLEY

BRONZE
BUSHING

SI ▷ ⊕

SNATCH BLOCK
ASSEMBLY
FOR A Ø28 STEEL CABLE

SIDE MOUNT
2 REQUIRED

15.77 Size B sheets.

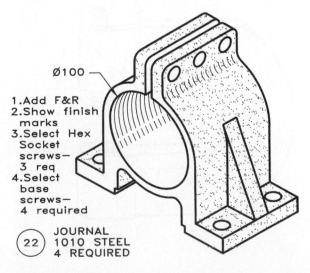

Ø100

1. Add F&R
2. Show finish
   marks
3. Select Hex
   Socket
   screws—
   3 req
4. Select
   base
   screws—
   4 required

㉒ JOURNAL
1010 STEEL
4 REQUIRED

15.79 Size B sheets.

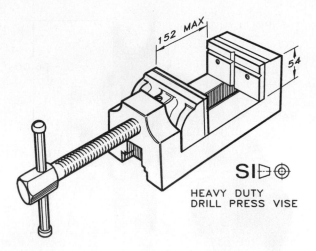

152 MAX

54

SI⊳⊕

HEAVY DUTY
DRILL PRESS VISE

15.80 Size B sheets.

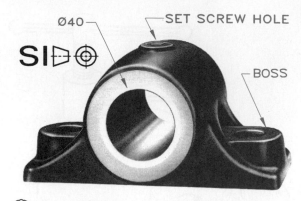

Ø40

SET SCREW HOLE

SI⊳⊕

BOSS

33 PILLOW BLOCK—1020 STEEL

15.82 Size B sheets. *(Courtesy of Rockwell Automation.)*

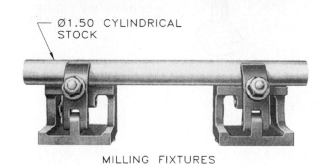

Ø1.50 CYLINDRICAL
STOCK

MILLING FIXTURES

15.83 Size B sheets. *(Courtesy of Walter E. Field & Son.)*

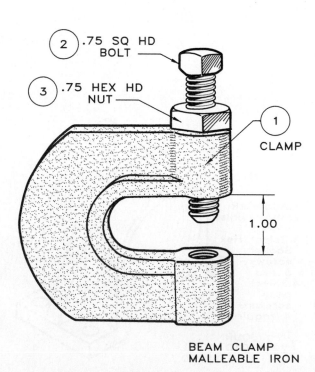

2 .75 SQ HD
BOLT

3 .75 HEX HD
NUT

1

CLAMP

1.00

BEAM CLAMP
MALLEABLE IRON

15.81 Size B sheets. Beam clamp.

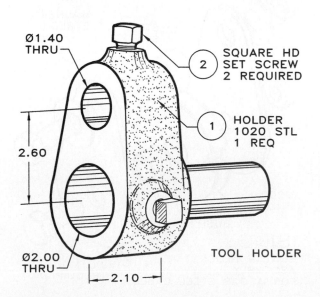

Ø1.40
THRU

2 SQUARE HD
SET SCREW
2 REQUIRED

1 HOLDER
1020 STL
1 REQ

2.60

Ø2.00
THRU

2.10

TOOL HOLDER

15.84 Size B sheets. .

15.85 Size A sheets. Handwheel.

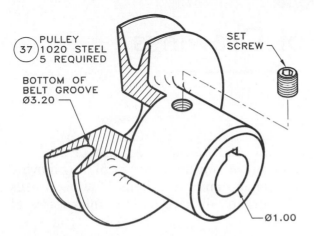

PULLEY
(37) 1020 STEEL
5 REQUIRED

BOTTOM OF
BELT GROOVE
Ø3.20

SET
SCREW

Ø1.00

15.86 Make working drawings of the pulley, shaft, key, and set screw based on the general dimensions given. Select the proper set screw and key for holding the pulley in position. Show finish marks and tolerances where appropriate. Use size A sheets.

Product Design (Figure 15.87 and 15.88)

Photographs of a nut cracker and can crusher are given to illustrate products currently on the market. Although not complicated, these products fill a specific need which makes them marketable and profitable to the inventor and manufacturer. Place yourself in the mode of the designer of the devices by applying the following steps.

1. Sketch orthographic details of the nut cracker or the can crusher (as assigned) as the preliminary step toward making a working drawing. Determine the dimensions by using your judgement.

2. Convert you freehand sketches into a finished working drawing that includes dimensions, specifications and an assembly.

3. As an alternate approach, design a product of your own creation that will serve the same purpose as those given in the examples. Begin with freehand sketches and conclude with a final working drawing.

15.88 Size A sheets. Can crusher.

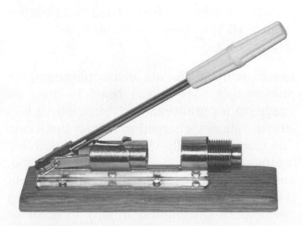

15.87 Size A sheets. Nut cracker.

16

Reproduction of Drawings

16.1 Introduction

So far we have discussed the preparation of drawings and specifications through the working-drawing stage where detailed drawings are completed on tracing film or paper. Now the drawings must be reproduced, folded, and prepared for transmittal to those who will use them to prepare bids or to fabricate the parts.

Several methods of reproduction are available to engineers and technologists for making copies of their drawings. However, most reproduction methods require strong, well-executed line work on the originals in order to produce good copies.

16.2 Computer Reproduction

Three major types of computer reproduction are (A) **pen plotting,** (B) **ink jet printing,** and (C) **laser printing.**

Pen plotting is done by plotter with a single or a multiple ink pen holder with a fiber point that "draws" on the paper or film by moving the pen in x and y directions. Multiple strokes of the pen will give various thicknesses of lines.

Ink jet printing is the process of spraying ink from tiny holes in a flat, disposable printhead onto the drawing surface as it passes through the printer. Prints can be obtained in color in addtion to black and white. Ink jet printers vary in size from 8-1/2 × 11 output **(Figure 16.1)** to large engineering print sizes **(Figure 16.2).**

Laser printing is an electrophotographic process that uses a laser beam to draw an image on a photosensitive drum where it is electrostatically charged to attract the toner. The electrostatically charged paper is rolled against the drum, the image is transferred, and the toner fused to the paper by heat **(Figure 16.3).** Laser printers make sharp drawings of the highest quality in color or black and white.

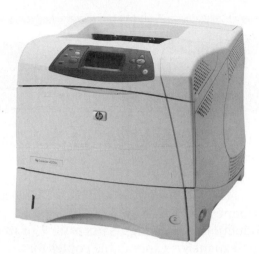

16.1 The DesignJet 995ck printer provides quiet high-speed operation and high print quality. Its letter- and legal-size format produces excellent color plots of text and graphics. *(Courtesy of Hewlett-Packard Company).*

16.3 This new workgroup laser printer, the HP LaserJet 4200n, has a speed of 35 pages per minute at 1200 dpi to accommodate more users and higher print volumes. *(Courtesy of Hewlett-Packard Company.)*

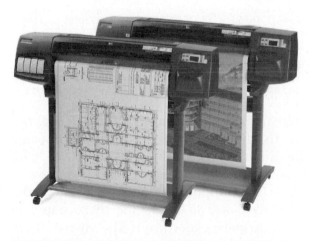

16.2 The DesignJet 1050c printer provides quiet high-speed operation and high print quality. This large-format color plotter can print a D-size color line drawing in less than one minute. *(Courtesy of Hewlett-Packard Company.)*

16.4 This HP LaserJet 1300n has a speed of 20 pages per minute and a print quality of 1200 dots per inch for black and white prints. *(Courtesy of Hewlett-Packard Company.)*

Figure 16.4 shows the LaserJet 1300, which is a favorite of offices whose needs do not exceed A-size sheets for both text and graphics. It prints with the highest laser quality of 1200 dots per inch.

16.3 Types of Reproduction

Drawings made by a drafter are of little use in their original form. If original drawings were handled by checkers and by workers in the field or shop, they would quickly be soiled and

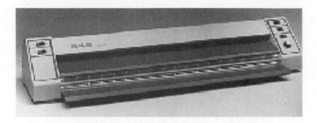

16.5 This typical whiteprinter operates on the diazo process. *(Courtesy of Bidwell Industrial Group, Blu-Ray Division.)*

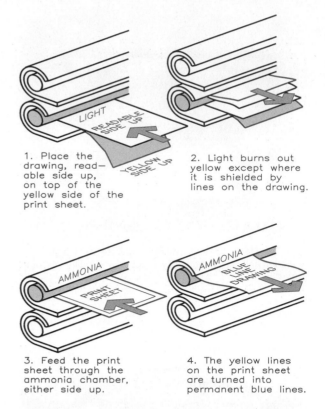

1. Place the drawing, read-able side up, on top of the yellow side of the print sheet.

2. Light burns out yellow except where it is shielded by lines on the drawing.

3. Feed the print sheet through the ammonia chamber, either side up.

4. The yellow lines on the print sheet are turned into permanent blue lines.

16.6 Diazo (blue-line) prints are made by placing the original readable side up and on top of the yellow side of the diazo paper and feeding them under the light as shown in the steps above.

damaged, and no copy would be available as a permanent record of the job. Therefore the reproduction of drawings is necessary for making inexpensive, expendable copies for use by the people who need to use them.

The most often used processes of reproducing engineering drawings are (1) **diazo printing,** (2) **microfilming,** (3) **xerography,** and (4) **photostating.**

Diazo Printing

The **diazo print** could be more correctly called a **whiteprint** or **blue-line print** rather than a **blueprint** because it has a white background and blue lines. Other colors of lines are available, depending on the type of diazo paper used. (Blueprinting, which creates a print with white lines and a blue background, is a wet process that is almost obsolete at the present.) **Figure 16.5** shows an example of a typical diazo printer.

Diazo printing requires that original drawings be made on semitransparent tracing paper, cloth, or film that light can pass through except where lines have been drawn. The diazo paper on which the blue-line print is copied is chemically treated giving it a yellow tint on one side. Diazo paper must be stored away from heat and light to prevent spoilage.

The sequential steps of making a diazo print are shown in **Figure 16.6.** The drawing is placed face up on the yellow side of the diazo paper and then fed through the diazo-process machine, which exposes the drawing to a built-in light. Light rays pass through the tracing paper and burn away the yellow tint on the diazo paper except where the drawing lines have shielded the paper from the light, similar to how a photographic negative is used. (It is important that your lines be adequately dense to shield the diazo paper in order to make a good print.) The exposed diazo paper becomes a duplicate of the original drawing except that the lines are light yellow and are not permanent.

When the diazo paper is passed through the developing unit of the diazo machine, ammonia fumes develop the yellow lines on it into permanent blue lines. The speed at which the drawing passes under the light determines the darkness of the blue-line copy; the faster the speed, the darker the print. A slow speed burns out more of the yellow and produces a clear white background, but some of the lighter lines of the drawing may be lost. Most diazo copies are made at a speed fast enough to give a light tint of blue in the background to obtain the darkest lines on the copy. Ink drawings, whether made by hand or by computer, give the best reproductions.

Diazo printing has been enhanced by the advent to the computer since computer drawings are made in ink. Thus the print quality is much better than pencil drawings. Also, drawing made by different drafters are more uniform in line weight, lettering, and technique than drawings made by hand.

Microfilming

Microfilming is a photographic process that converts large drawings into film copies— either aperture cards or roll film. Drawings are placed on a copy table and photographed on either 16-mm or 35-mm film.

The roll film or aperture cards are placed in a microfilm enlarger-printer, where the individual drawings can be viewed on a built-in screen. The selected drawings can be printed from the film in standard sizes. Microfilm copies are usually made smaller than the original drawings to save paper and make the drawings easier to use.

Microfilming eliminates the need for large, bulky files of drawings because hundreds of drawings can be stored in permanent archives in miniature on a small amount of film. This is the same process used to preserve newspapers and other large materials by libraries and archives.

Xerography

Xerography is an electrostatic process of duplicating drawings on ordinary, unsensitized paper. Originally developed for business and clerical uses, xerography more recently is currently used for the reproduction of engineering drawings. The xerographic process can also be used to reduce the sizes of the drawings being copied to more convenient and easier to use sizes. The Xerox 2080 can reduce a 24 × 36-inch drawing to 8 × 10 inches.

Photostating

Photostating is a method of enlarging or reducing drawings photographically. The drawing is placed under the glass of the exposure table, which is lit by built-in lamps. The image appears on a glass plate inside the darkroom where it is exposed on photographically sensitive paper. The exposed negative paper is placed in contact with receiver paper, and the two are fed through the developing solution to obtain a photostatic copy. Photostating also can be used to make reproductions on transparent films and for reproducing halftones (photographs with tones of gray).

16.4 Assembling Drawing Sets

After the original drawings have been copied, they should be stored flat and unfolded in a flat file for future use and updating. Prints made from the originals, however, usually are folded or rolled for ease of transmittal from office to office. The methods of folding size B, C, D, and E sheets so that the image will appear on the outside of the fold are shown in **Figure 16.7.**

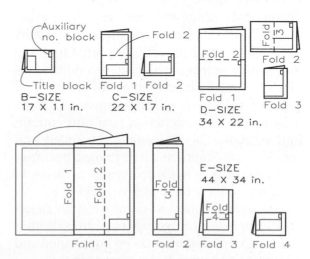

16.7 All standard drawing sheets can be folded to 8-1/2″ × 11″ size for filing and storage.

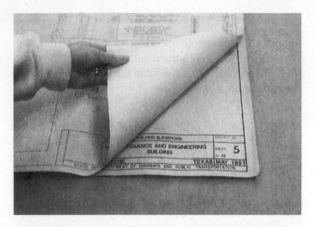

16.8 The title block should appear at the right, usually in the lower right-hand corner of the sheet.

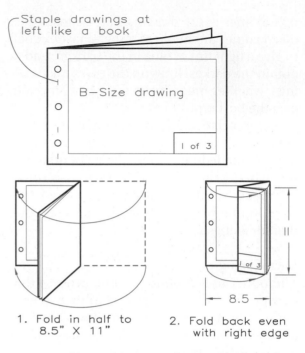

Staple drawings at left like a book

B—Size drawing

1 of 3

1. Fold in half to 8.5" X 11"

2. Fold back even with right edge

8.5

16.9 A set of size B drawings can be assembled by stapling, punching, and folding, as shown here, for safekeeping in a three-ring notebook with the title block visible on top.

Drawings should be folded to show the title block always on the outside at the right, usually in the lower right-hand corner of the page **(Figure 16.8).** The final size after folding is 8 1/2 × 11 inches (or 9 × 12 inches).

An alternative method of folding and stapling size B sheets is used often for student assignments so that they can be kept in a three-ring notebook **(Figure 16.9).** The basic rules of assembling drawings are listed in **Figure 16.10.**

16.5 Transmittal of Drawings

Prints of drawings are delivered to contractors, manufacturers, fabricators, and others who must use the drawings for implementing the project. Prints usually are placed in standard 9 × 12-inch envelopes for delivery by hand or mail. Sets of large drawings, which may be 30 × 40 inches in size and contain four or more sheets, usually are rolled and sent in a mailing tube when folding becomes impractical. It is not uncommon for a set of drawings to have forty or fifty sheets.

An advanced method of transmitting drawings is by use of large fax machines. Within minutes, large documents can be scanned and transmitted to their destination sites.

Computer drawings can be transmitted on disk by mailing them to their destination, where hard copies can be plotted and reproduced. This procedure offers substantial savings in shipping charges.

Computer drawings can also be transmitted over the Internet in the form of data that is downloaded at its destination. The downloaded data is then printed in the form of a drawing and is maintained in the database of the computer. In the future, more drawings, documents, and photographs will be sent electronically as data and as scanned images over telephone wires, making them available instantaneously at the desired location.

Only a few years ago, transmission of information and data across the state or nation was time consuming, with the risk of loss. Today, any document can be transmitted overnight with certainty of delivery, and most can be transmitted to the receiver within minutes.

WORKING DRAWING CHECKLIST

1. Staple along left edge, like a book. Use several staples, never just one.

2. Fold with drawing on outside.

3. Fold drawings as a set, not one at a time separately.

4. Fold to an 8.5"X 11" modular size.

5. The title block must be visible after folding.

6. Sheets of a set should be uniform in size.

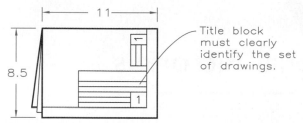

16.10 Follow these basic rules for assembling sets of working drawing prints.

16.11 Hewlett Packard's OmniShare conferencer enables people in two locations to "meet" and collaborate on the same document, at the same time, over a single phone line. *(Courtesy of Hewlett-Packard Company.)*

The OmniShare conferencer **(Figure 16.11)** lets people in two locations collaborate on the same document at the same time over a single phone line.

Hewlett Packard's LaserJet printers have accessories available for fax, copy, file, and read capabilities. Today, the communication of engineering data can be done instantaneously and easily, contributing to an increased productivity **(Figure 16.12)**.

Numerically controlled manufacturing systems can be actuated directly from engineering data once the designs have been digitized. Such systems can be controlled from remote sites to produce products that previously required a high intensity of work hours by individuals. The future holds many unique innovations in the manner in which business, manufacturing, and construction is done. The transfer of voice communication, hard-copy communications, working drawings, and specifications will be instantaneous.

16.12 Fast, high-quality output and paper-handling flexibility required of today's business user can be found in the HP LaserJet 4100 printer. In addition, users can add copy, fax, file, and read capabilities by adding the optional LaserJet Companion printer accessory. *(Courtesy of Hewlett-Packard Company.)*

17

Three-Dimensional Pictorials

17.1 Introduction

A three-dimensional pictorial is a drawing that shows an object's three principal planes, much as they would be captured by a camera. This type of pictorial is an effective means of illustrating a part that is difficult to visualize when only orthographic views are given. Pictorials are especially helpful when a design is complex and when the reader of the drawings is unfamiliar with orthographic drawings.

Sometimes called **technical illustrations,** pictorials are widely used to describe products in catalogs, parts manuals, and maintenance publications **(Figure 17.1).** The ability to sketch pictorials rapidly to explain a detail to an associate in the field is an important communication skill.

The four commonly used types of pictorials are **(1) obliques, (2) isometrics, (3) axonometrics,** and **(4) perspectives (Figure 17.2).**

Oblique pictorials: Three-dimensional drawings made by projecting from the object

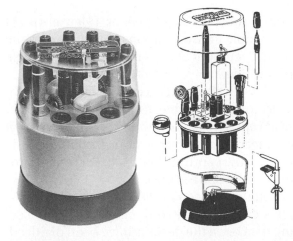

17.1 Many objects cannot be seen as well in real life as they can in a drawing, as shown in this pen set. *(Courtesy of Koh-I-Noor, Inc.)*

with parallel projectors that are oblique to the picture plane **(Figure 17.2A).**

Isometric and **axonometric pictorials:** Three-dimensional drawings made by projecting from the object with parallel projectors that

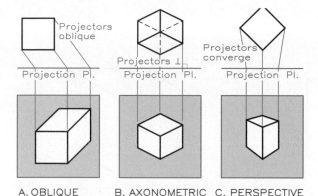

A. OBLIQUE   B. AXONOMETRIC   C. PERSPECTIVE

17.2 The three pictorial projection systems are: (A) oblique pictorials, with parallel projectors oblique to the projection plane; (B) axonometric (including isometric) pictorials, with parallel projectors perpendicular to the projection plane; and (C) perspectives, with converging projectors that make varying angles with the projection plane.

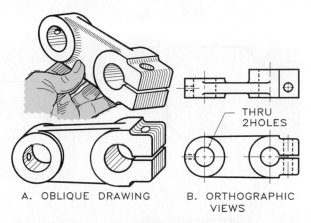

A. OBLIQUE DRAWING   B. ORTHOGRAPHIC VIEWS

17.3 The oblique drawing of this part makes it easier to visualize it than do its orthographic views.

are perpendicular to the picture plane (**Figure 17.2B**).

Perspective-pictorials: Three-dimensional drawings made with projectors that converge at the viewer's eye and make varying angles with the picture plane (**Figure 17.2C**).

17.2 Oblique Drawings

The pulley arm shown in **Figure 17.3** is illustrated by orthographic views and an oblique pictorial. Because most parts are drawn before they are made, photographs cannot be taken; therefore the next best option is to draw a three-dimensional pictorial of the part. Details can usually be drawn with more clarity than can be shown in a photograph.

Oblique pictorials are easy to draw. If you can draw an orthographic view of a part, you are but one step away from drawing an oblique. For example, **Figure 17.4,** shows that drawing a front view of a box twice and connecting its corners yields an oblique drawing.

Thus an oblique is no more than an orthographic view with a receding axis, drawn at

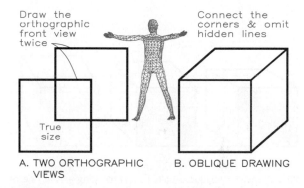

A. TWO ORTHOGRAPHIC VIEWS   B. OBLIQUE DRAWING

17.4 Draw two true-size surfaces of the box, connect them at the corners, and you have an oblique drawing.

an angle to show the depth of the object. An oblique is a pictorial that does not exist in reality (a camera cannot give an oblique). This type of pictorial is called an oblique because its parallel projectors from the object are oblique to the picture plane.

Types of Obliques
The three basic types of oblique drawings are: (1) cavalier, (2) cabinet, and (3) general (**Figure 17.5**). For each type, the angle of the receding axis with the horizontal can be at any angle between 0° and 90°. Measurements along the receding axes of the cavalier

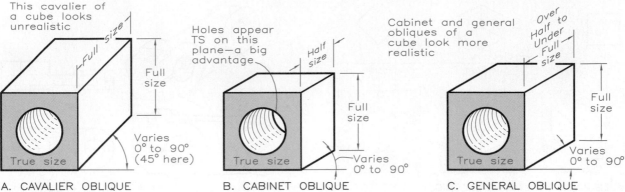

A. CAVALIER OBLIQUE

B. CABINET OBLIQUE

C. GENERAL OBLIQUE

17.5 The three types of obliques:
A The cavalier oblique has a receding axis at any angle and true-length measurements on the receding axis.

B The cabinet oblique has a receding axis at any angle and half-size measurements along the receding axis.

C The general oblique has a receding axis at any angle and measurements along the receding axis larger than half size and less than full size.

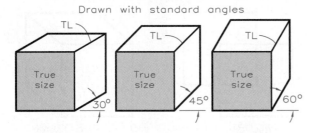

17.6 A cavalier oblique usually has its receding axis at one of the standard angles of drafting triangles. Each gives a different view of a cube.

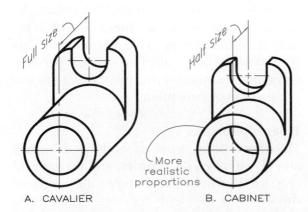

A. CAVALIER   B. CABINET

17.7 Measurements along the receding axis of a cavalier oblique are full size and those in a cabinet oblique are half size.

oblique are laid off true length, and measurements along the receding axes of the cabinet oblique are laid off half size. The general oblique has measurements along the receding axes that are greater than half size and less than full size.

Figure 17.6 shows three examples of cavalier obliques of a cube. The receding axes for each is drawn at a different angle, but the receding axes are drawn true length. **Figure 17.7** compares cavalier with cabinet obliques.

Constructing Obliques
You can easily begin a cavalier oblique by drawing a box using the overall dimensions of height, width, and depth with light construction lines. As demonstrated in **Figure 17.8,** first draw the front view as a true-size orthographic view. True measurements must be made parallel to the three axes and transferred from the orthographic views with your dividers. Then remove the notch from the blocked-in construction box to complete the oblique.

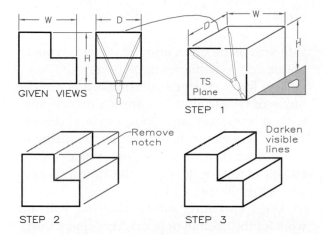

GIVEN VIEWS

STEP 1

TS Plane

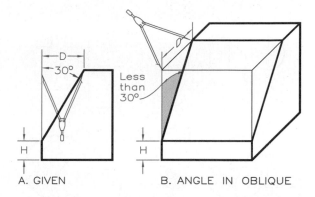

17.10 Angles that do not lie in a true-size plane of an oblique must be located with coordinates.

STEP 2

Remove notch

STEP 3

Darken visible lines

17.8 Constructing a cavalier oblique:
Step 1 Draw the front surface of the object as a true-size plane. Draw the receding axis at a convenient angle and transfer the true distance D from the side view to it using your dividers.
Step 2 Draw the notch on the front plane and project it to the rear plane.
Step 3 Darken the lines to complete the drawing.

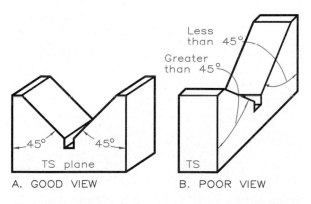

17.9 Objects with angular features should be drawn in oblique so that the angles appear true size. This results in a better pictorial and one that is easier to draw.

Angles

Angular measurements can be made on the true-size plane of an oblique, but not on the other two planes. Note in **Figure 17.9A** that a true angle can be measured on a true-size surface, but in **Figure 17.9B** angles along receding planes are either smaller or larger than their true sizes. A better, easier-to-draw ob-

lique is obtained when angles are drawn to appear true size.

To construct an angle in an oblique on one of the receding planes, you must use coordinates, as shown in **Figure 17.10.** To find the surface that slopes 30° from the front surface, locate the vertex of the angle, H distance from the bottom. To find the upper end of the sloping plane, measure the distance D along the receding axis. Transfer H and D to the oblique using your dividers. The angle in the oblique is not equal to the 30° angle in the orthographic view.

Cylinders

The **major advantage** of an oblique is that **circular features can be drawn as true circles** on its frontal plane **(Figure 17.11).** Draw the centerlines of the circular end at A and construct the receding axis at the desired angle. Locate the end at B by measuring along the axis, draw circles at each end at centers A and B, and draw tangents to both circles.

These same principles apply to construction of the object having semicircular features shown in **Figure 17.12.** Position the oblique so that the semicircular features are true size. Locate centers A, B, and C and the two semicircles. Then complete the cavalier oblique.

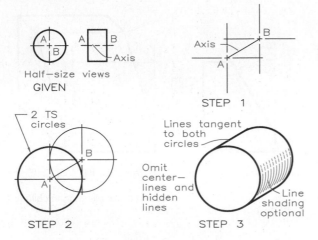

17.11 Drawing a cylinder in oblique:
Step 1 Draw axis AB and locate the centers of the circular ends of the cylinder at A and B. Because the axis is true length, this is a cavalier oblique.
Step 2 Draw a true-size circle with its center at A by using a compass or computer-graphics techniques.
Step 3 Draw the other circular end with its center at B and connect the circles with tangent lines parallel to axis AB.

Circles

Circular features drawn as true circles on a true-size plane of an oblique pictorial appear on the receding planes as ellipses.

The four-center ellipse method is a technique of constructing an approximate ellipse with a compass and four centers (**Figure 17.13**). The ellipse is tangent to the inside of a rhombus drawn with sides equal to the circle's diameter. Drawing the four arcs produces the ellipse.

The four-center ellipse method will not work for the cabinet or general oblique; coordinates must be used. **Figure 17.14** illustrates the method of locating coordinates on the planes of cavalier and cabinet obliques. For the cabinet oblique, the coordinates along the receding axis are half size, and the coordinates along the horizontal axis (true-size axis) are full size. Draw the ellipse with an irregular

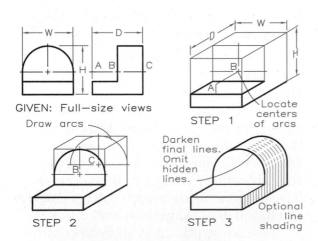

17.12 Drawing semicircular features in oblique:
Step 1 Block in the overall dimensions of the cavalier oblique with light construction lines, ignoring the semicircular feature.
Step 2 Locate centers B and C and draw arcs with a compass or by computer tangent to the sides of the construction boxes.
Step 3 Connect the arcs with lines tangent to each arc and parallel to axis BC and darken the lines.

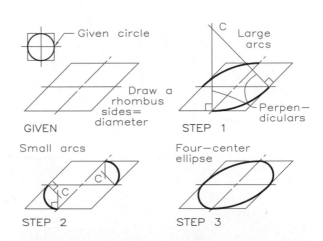

17.13 Constructing a four-center ellipse in oblique:
Given: Block in the circle to be drawn in oblique with a square tangent to the circle. This square becomes a rhombus on the oblique plane.
Step 1 Draw construction lines perpendicular at the points of tangency to locate the centers for drawing two segments of the ellipse.
Step 2 Locate the centers for the two remaining arcs with perpendiculars drawn from adjacent tangent points.
Step 3 Draw the four arcs, which yield an approximate ellipse.

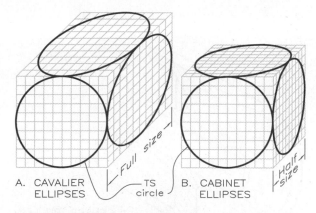

17.14 Circular features on the faces of cavalier and cabinet obliques are compared here. Ellipses on the receding planes of cabinet obliques must be plotted by coordinates. The lengths of the coordinates along the receding axis of cabinet obliques are half size.

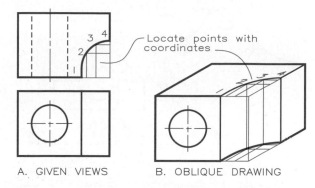

17.16 Coordinates are used to find points along irregular curves in oblique. Projecting the points downward a distance equal to the height of the object yields the lower curve.

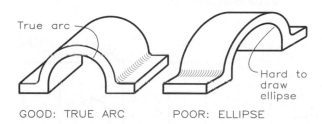

17.15 An obliques should be positioned so that circular and curving features can be drawn most easily.

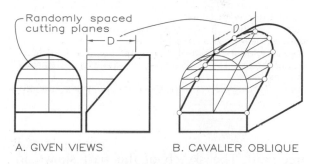

17.17 Construction of an elliptical feature on an inclined surface in oblique requires the use of three-dimensional coordinates to locate points on the curve.

curve or an ellipse template that approximates the plotted points.

Whenever possible, oblique drawings of objects with circular features should be positioned so circles can be drawn as true circles instead of ellipses. The left view in **Figure 17.15** is better than the one at the right because it gives a more descriptive view of the part and is easier to draw.

Curves

Irregular curves in oblique pictorials must be plotted point by point with coordinates (**Figure 17.16**). Transfer the coordinates from the orthographic to the oblique view and draw the

curve through the plotted points with an irregular curve. If the object has a uniform thickness, plot the points for the lower curve by projecting vertically downward from the upper points a distance equal to the object's height.

To obtain the elliptical feature on the inclined surface in **Figure 17.17,** use a series of coordinates to locate points along its curve. Connect the plotted points by using an irregular curve or ellipse template.

Sketching

Understanding the principles of oblique construction is essential for sketching obliques

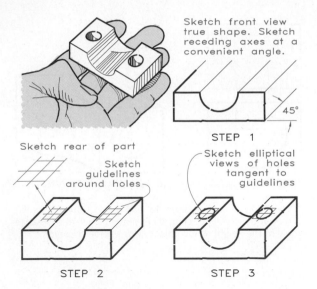

17.18 Sketching obliques:
Step 1 Sketch the front of the object as true-size surface and draw a receding axis from each corner.
Step 2 Lay off the depth, along the receding axes to locate the rear of the part. Lightly sketch pictorial boxes as guidelines for drawing the holes.
Step 3 Sketch the holes inside the boxes and darken all lines.

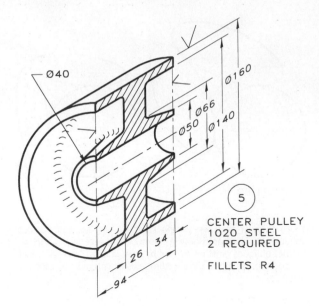

17.19 Oblique pictorials can be drawn as dimensioned sections to serve as working drawings.

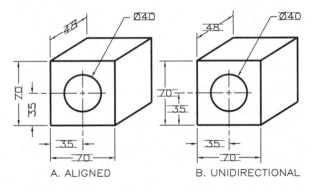

17.20 Either of these methods of lettering, aligned or unidirectional, is acceptable for dimensioning obliques.

freehand. The sketch of the part shown in **Figure 17.18** is based on the principles discussed, but its proportions were determined by eye instead of with scales and dividers.

Lightly drawn guidelines need not be erased when you darken the final lines. When sketching on tracing vellum, you can place a printed grid under the sheet to provide guidelines. Refer to Chapter 6 to review sketching techniques if needed.

Dimensioned Obliques

Dimensioned sectional views of obliques provide excellent, easily understood depictions of objects (**Figure 17.19**). Apply numerals and lettering in oblique pictorials by using either the **aligned** method (with numerals aligned with the dimension lines) or the **unidirectional** method (with numerals positioned horizontally regardless of the direction of the dimension lines), as shown in **Figure 17.20.**

Notes connected with leaders are positioned horizontally in both methods.

17.3 Isometric Pictorials

In **Figure 17.21** the pulley arm is drawn in orthographic views and as a three-dimensional pictorial drawing. The pictorial is an isometric drawing in which the three planes of the object are equally foreshortened, representing the object more realistically than an oblique drawing can.

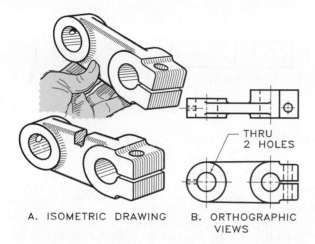

A. ISOMETRIC DRAWING   B. ORTHOGRAPHIC VIEWS

THRU 2 HOLES

17.21 An isometric drawing gives a more realistic view of a part than an oblique drawing.

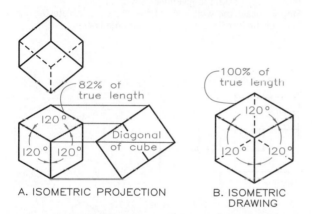

82% of true length

Diagonal of cube

A. ISOMETRIC PROJECTION

100% of true length

B. ISOMETRIC DRAWING

17.22 Projection vs. drawing.
A A true isometric projection is found by constructing a view in which the diagonal of a cube appears as a point and the axes are foreshortened.
B An isometric drawing is not a true projection because the dimensions are true size rather than foreshortened.

With more realism comes more difficulty of construction. In particular, circles and curves do not appear true shape on any of the three isometric planes.

Isometric Projection versus Drawing

In isometric projection, parallel projectors are perpendicular to the imaginary projection (picture) plane in which the diagonal of a cube appears as a point **(Figure 17.22).** An isometric pictorial constructed by projection is

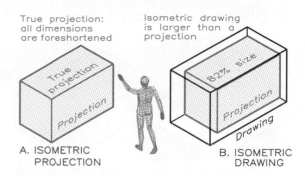

True projection: all dimensions are foreshortened

Isometric drawing is larger than a projection

A. ISOMETRIC PROJECTION

B. ISOMETRIC DRAWING

17.23 The true isometric projection is foreshortened to 82% of full size. The isometric drawing is drawn full size for convenience.

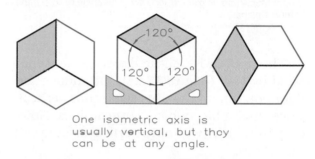

One isometric axis is usually vertical, but they can be at any angle.

17.24 Isometric axes are spaced 120° apart, but they can be revolved into any position. Usually, one axis is vertical, but it can be at any angle with axis spacing remaining the same.

called an **isometric projection,** with the three axes foreshortened to 82% of their true lengths and 120° apart. The name *isometric,* which means equal measurement, aptly describes this type of projection because the planes are equally foreshortened.

An **isometric drawing** is a convenient approximate isometric pictorial in which the measurements are shown full size along the three axes rather than at 82% as in **isometric projection (Figure 17.23).** Thus the isometric drawing method allows you to measure true dimensions with standard scales and lay them off with dividers along the three axes. The only difference between the two is the larger size of the drawing. Consequently, isometric drawings are used much more often than isometric projections.

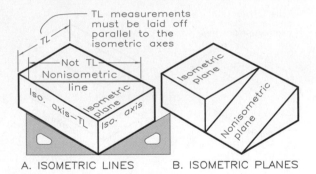

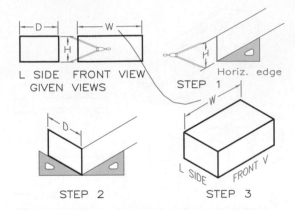

A. ISOMETRIC LINES

B. ISOMETRIC PLANES

17.25 Isometric lines and planes:
A Isometric lines (parallel to the three axes) give true measurements, but nonisometric lines do not.
B Here, the three isometric planes are equally foreshortened, and the nonisometric plane is inclined at an angle to one of the isometric planes.

The axes of isometric drawings are separated by 120° **(Figure 17.24),** but more often than not, one of the axes selected is vertical, since most objects have vertical lines. However, isometrics without a vertical axis are still isometrics.

17.4 Isometric Drawings

An isometric drawing is begun by drawing three axes 120° apart. Lines parallel to these axes are called **isometric lines (Figure 17.25A).** You can make true measurements along isometric lines but not along nonisometric lines. The three surfaces of a cube in an isometric drawing are called **isometric planes (Figure 17.25B).** Planes parallel to those planes also are isometric planes.

To draw an isometric pictorial, you need a scale, dividers, and a 30°–60° triangle **(Figure 17.26).** Begin by selecting the three axes and then constructing a plane of the isometric from the dimensions of height, H, and depth, D. Add the third dimension of width, W, and complete the isometric drawing.

Use light construction lines to block in all isometric drawings **(Figure 17.27)** and the overall dimensions W, D, and H. Take other dimensions from the given views with dividers and

17.26 Drawing an isometric of a box:
Step 1 Use a 30°–60° triangle and a horizontal straight edge to construct a vertical line equal to the height, H, and draw two isometric lines through each end.
Step 2 Draw two 30° lines and locate the depth, D, by transferring depth from the given views with dividers.
Step 3 Locate the width, W, of the object, complete the surfaces of the isometric box, and darken the lines.

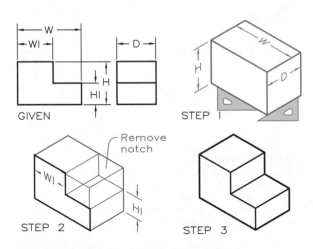

17.27 Constructing an isometric of a simple part:
Step 1 Construct an isometric drawing of a box with the overall dimensions W, D, and H from the given views.
Step 2 Locate the notch by transferring dimensions W1 and H1 from the given views with your dividers.
Step 3 Darken the lines to complete the drawing.

measure along their isometric lines to locate notches in the blocked-in drawing.

Figure 17.28 shows an isometric drawing of a slightly more complex object, with two

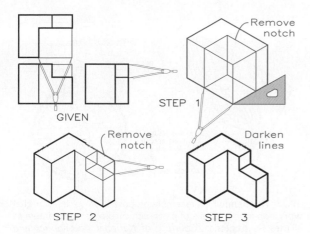

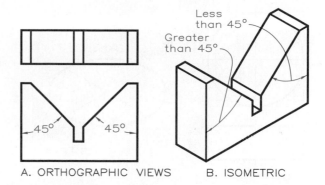

17.30 Angles in isometric may appear larger or smaller than they actually are.

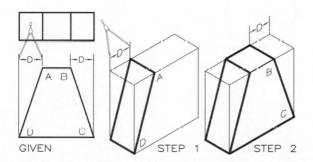

17.28 Laying out an isometric drawing:
Step 1 Use the overall dimensions given to block in the object with light lines and remove the large notch.
Step 2 Remove the small notch.
Step 3 Darken the lines to complete the drawing.

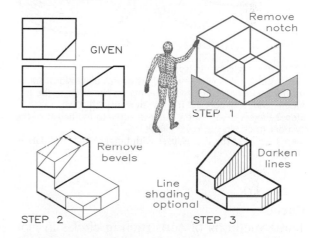

17.29 Use coordinates measured along the isometric axes to obtain inclined surfaces. Angular lines are not true length in isometric.

17.31 Drawing inclined planes in isometric:
Step 1 Block in the object with light lines, using the overall dimensions, and remove the notch.
Step 2 Locate the ends of the inclined planes by using measurements parallel to the isometric axes.
Step 3 Darken the lines to complete the drawing.

notches. The object was blocked in by using the H, W, and D dimensions. Remove the notches in the block to complete the drawing.

Angles

You cannot measure an angle's true size in an isometric drawing because the surfaces of an isometric are not true size. Instead, you must locate angles with isometric coordinates measured parallel to the axes **(Figure 17.29)**. Lines AD and BC are equal in length in the orthographic view, but they are shorter and

longer than true length in the isometric drawing. **Figure 17.30** shows a similar situation, where two angles drawn in isometric are less than and greater than their true dimensions in the orthographic view.

Figure 17.31 shows how to construct an isometric drawing of an object with inclined surfaces. Blocking in the object with its overall dimensions with light construction lines is followed by removal of the inclined portions.

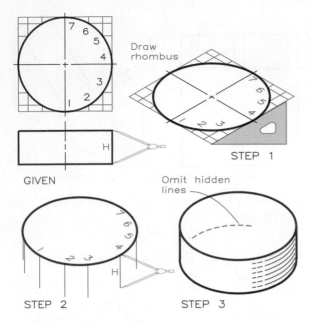

GIVEN

STEP 1

Draw rhombus

Omit hidden lines

STEP 2

STEP 3

17.32 Plotting circles in isometric:
Step 1 Block in the circle by using its overall dimensions. Transfer the coordinates that locate points on the circle to the isometric plane and connect them with a smooth curve.
Step 2 Project each point a distance equal to the height of the cylinder to obtain the lower ellipse.
Step 3 Connect the two ellipses with tangent lines and darken all lines.

Circles

Three methods of constructing circles in isometric drawings are (1) **point plotting,** (2) **four-center ellipse construction,** and (3) **ellipse template usage.**

Point plotting is a method of using a series of *x* and *y* coordinates to locate points on a circle in the given orthographic views **(Figure 17.32).** The coordinates are then transferred with dividers to the isometric drawing to locate the points on the ellipse one at a time.

Block in the cylinder with light construction lines and show the centerlines. Draw coordinates on the upper plane and use the height dimension to locate the points on the lower plane. Draw the ellipses with an irregular curve or an ellipse template.

A plotted ellipse is a true ellipse and is equivalent to a 35° ellipse drawn on an iso-

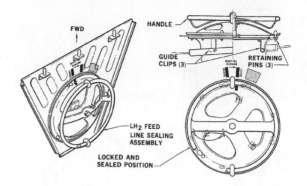

17.33 This handwheel assembly proposed for use in an orbital workshop is an example of parts with circular features drawn as ellipses in isometric. *(Courtesy of National Aeronautics and Space Administration.)*

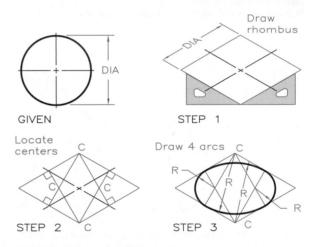

GIVEN

STEP 1

Draw rhombus

Locate centers

STEP 2

Draw 4 arcs

STEP 3

17.34 The four-center ellipse method:
Step 1 Use the diameter of the given circle to draw an isometric rhombus and the centerlines.
Step 2 Draw light construction lines perpendicularly from the midpoints of each side to locate four centers.
Step 3 Draw four arcs from the centers to represent an ellipse tangent to the rhombus.

metric plane. An example of a design composed of circular features drawn in isometric is the handwheel shown in **Figure 17.33.**

Four-center ellipse construction is the method of producing an approximate ellipse **(Figure 17.34)** by using four arcs drawn with a compass. Draw an isometric rhombus with its sides equal to the diameter of the circle

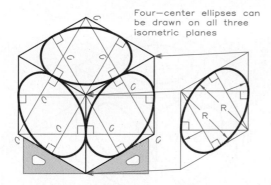

Four—center ellipses can be drawn on all three isometric planes

17.35 Four-center ellipses may be drawn on all three surfaces of an isometric drawing.

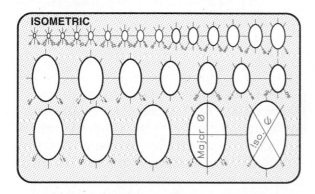

ISOMETRIC

17.36 The isometric template (a 35° ellipse angle) is designed for drawing elliptical features in isometric. The isometric diameters of the ellipses are not major diameters of the ellipses but are diameters that are parallel to the isometric axes.

to be represented. Find the four centers by constructing perpendiculars to the sides of the rhombus at the midpoints of each side, and draw the four arcs to complete the ellipse. You may draw four-center ellipses on all three isometric planes because each plane is equally foreshortened **(Figure 17.35)**. Although it is only an approximate ellipse, the four-center ellipse technique is acceptable for drawing large ellipses and as a way to draw ellipses when an ellipse template is unavailable.

Isometric ellipse templates are specially designed for drawing ellipses in isometric **(Figure 17.36).** The numerals on the templates

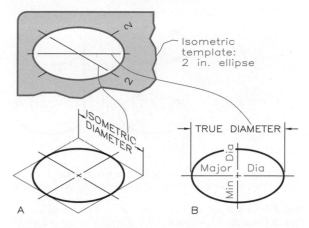

Isometric template: 2 in. ellipse

17.37 Ellipse terminology.
A Measure the diameter of a circle along the isometric axes. The major diameter of an isometric ellipse thus is larger than the measured diameter.
B The minor diameter is perpendicular to the major diameter.

represent the isometric diameters of the ellipses because diameters are measured parallel to the isometric axes of an isometric drawing **(Figure 17.37)**. Recall that the maximum diameter across the ellipse is its major diameter, which is a true diameter. Thus the size of the diameter marked on the template is less than the ellipse's major diameter. You may use the isometric ellipse template to draw an ellipse by constructing centerlines of the ellipse in isometric and aligning the ellipse template with those isometric lines **(Figure 17.37).**

Cylinders

A cylinder may be drawn in isometric by using the four-center ellipse method **(Figure 17.38).** Use the isometric axes and centerline axis to construct a rhombus at each end of the cylinder. Then draw the ellipses at each end, connect them with tangent lines, and darken the lines to complete the drawing.

An easier way to draw a cylinder is to use an isometric ellipse template **(Figure 17.39).** Draw the axis of the cylinder and construct

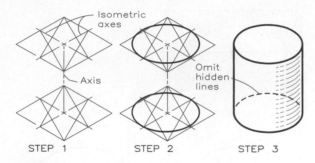

STEP 1   STEP 2   STEP 3

17.38 A cylinder drawn with the four-center method:
Step 1 Draw an isometric rhombus at each end of the cylinder's axis.
Step 2 Draw a four-center ellipse within each rhombus.
Step 3 Draw lines tangent to each rhombus to complete the drawing.

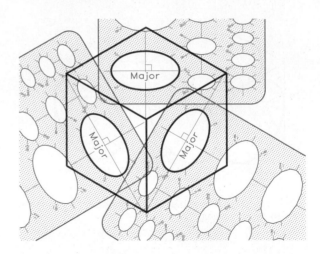

17.40 Position the isometric ellipse template as shown for drawing ellipses of various sizes on the three isometric planes.

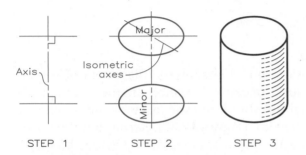

STEP 1   STEP 2   STEP 3

17.39 A cylinder using ellipse template method:
Step 1 Establish the length of the axis of the cylinder and draw perpendicular at each end.
Step 2 Draw the elliptical ends by aligning the major diameter of the ellipse template with the perpendiculars at the ends of the axis. The isometric diameters of the isometric ellipse template will align with two isometric axes.
Step 3 Connect the ellipses with tangent lines to complete the drawing and omit hidden lines.

perpendiculars at each end. Because the axis of a right cylinder is perpendicular to the major diameter of its elliptical ends, position the ellipse template with its major diameter perpendicular to the axis. Draw the ellipses at each end, connect them with tangent lines, and darken the visible lines to complete the drawing.

The isometric ellipses template (**Figure 17.36**) can be used to draw ellipses on all three

planes of an isometric drawing as shown in **Figure 17.40.** On each plane, the major diameter is perpendicular to the isometric axis of the adjacent perpendicular plane. The isometric diameters marked on the template align with the isometric axes. Rounded corners (fillets or rounds) can be drawn with an ellipse template by using only a quarter of the ellipse.

Inclined Planes
Inclined planes in isometric may be located by coordinates, but they cannot be measured with a protractor because they do not appear true size. **Figure 17.41** illustrates the coordinate method. Use horizontal and vertical coordinates (in the x and y directions) to locate key points on the orthographic views. Transfer these coordinates to the isometric drawing with dividers to show the features of the inclined surface.

Curves
Irregular curves in isometric must be plotted point by point, with coordinates locating each point. Locate points A through F in the

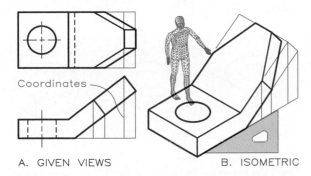

A. GIVEN VIEWS   B. ISOMETRIC

17.41 Inclined surfaces in isometric must be located with three-dimensional coordinates parallel to the isometric axes. True angles cannot be measured in isometric drawings.

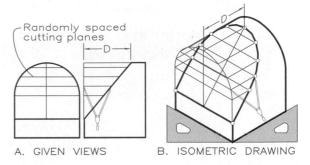

A. GIVEN VIEWS   B. ISOMETRIC DRAWING

Randomly spaced cutting planes

17.43 To construct ellipses on inclined planes, draw coordinates to locate points in the orthographic views. Then transfer the three-dimensional coordinates to the isometric drawing and connect them with a smooth curve.

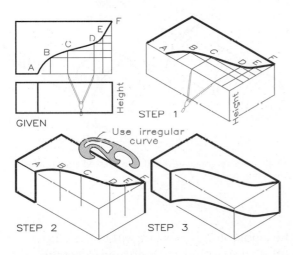

GIVEN

Use irregular curve

STEP 1

STEP 2   STEP 3

17.42 Plotting irregular curves:
Step 1 Block in the shape by using the overall dimensions. Locate points on the irregular curve with coordinates transferred from the orthographic views.
Step 2 Project these points downward the distance H (height) from the upper points to obtain the lower curve.
Step 3 Connect the points and darken the lines.

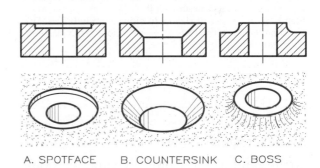

A. SPOTFACE   B. COUNTERSINK   C. BOSS

17.44 These examples of circular features in isometric may be drawn by using ellipse templates.

orthographic view with coordinates of width and depth **(Figure 17.42).** Then transfer them to the isometric view of the blocked-in part and connect them with an irregular curve.

Project points on the upper curve downward a distance of H, the height of the part, to locate points on the lower curve. Connect

these points with an irregular curve and darken the lines to complete the isometric.

Ellipses on Nonisometric Planes

Ellipses on nonisometric planes in an isometric drawing, such as the one shown in **Figure 17.43,** must be found by locating a series of points on the curve. Locate three-dimensional coordinates in the orthographic views and then transfer them to the isometric with your dividers. Connect the plotted points with an irregular curve or an ellipse template selected to approximate the plotted points. The more points you select, the more accurate the final ellipse will be. It will not be drawn by an isometric ellipse template, but by one that fits the plotted points.

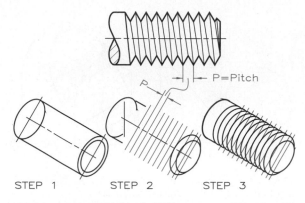

STEP 1   STEP 2   STEP 3

17.45 Threads in isometric:
Step 1 Using an ellipse template, draw the cylinder to be threaded.
Step 2 Lay off perpendiculars, spacing them apart at a distance equal to the pitch of the thread, P.
Step 3 Draw a series of ellipses to represent the threads. Draw the chamfered end by using an ellipse whose major diameter is equal to the root diameter of the threads.

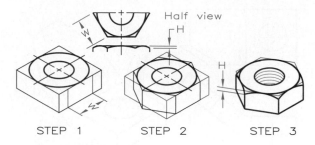

STEP 1   STEP 2   STEP 3

17.46 Constructing a nut:
Step 1 Use the overall dimensions of the nut to block in the nut.
Step 2 Construct the hexagonal sides at the top and bottom.
Step 3 Draw the chamfer with an irregular curve. Draw the threads to complete the drawing.

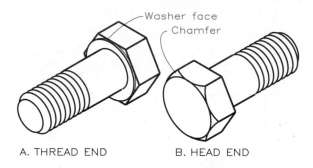

A. THREAD END   B. HEAD END

17.47 Isometric drawings of the lower and upper sides of a hexagon-head bolt.

17.5 Technical Illustration

Orthographic and isometric views of a **spot-face, countersink,** and **boss** are shown in **Figure 17.44.** These features may be drawn in isometric by point-by-point plotting of the circular features, the four-center, or ellipse template method (the easiest method of the three).

A threaded shaft may be drawn in isometric as shown in **Figure 17.45.** First draw the cylinder in isometric. Draw the major diameters of the crest lines equally separated by distance P, the pitch of the thread. Then draw ellipses by aligning the major diameter of the ellipse template with the perpendiculars to the cylinder's axis. Use a smaller ellipse at the end for the 45° chamfered end.

Figure 17.46 shows how to draw a hexagon-head nut with an ellipse template. Block in the nut and draw an ellipse tangent to the rhombus. Construct the hexagon by locating distance W across a flat parallel to the isometric axes. To find the other sides of the

hexagon, draw lines tangent to the ellipse. Lay off distance H at each corner to establish the chamfers.

Figure 17.47 depicts a hexagon-head bolt in two positions. The washer face is on the lower side of the head and the chamfer is on the upper side.

A portion of a sphere is drawn to represent a round-head screw in **Figure 17.48.** Construct a hemisphere and locate the centerline of the slot along one of the isometric planes. Measure the head's thickness, E, from the highest point on the sphere.

Sections

A full section drawn in isometric can clarify internal details that might otherwise be over-

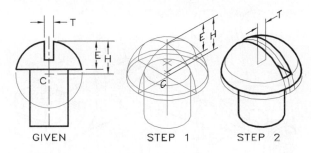

17.48 Drawing spherical features:
Step 1 Use an isometric ellipse template to draw the elliptical features of a round-head screw.
Step 2 Draw the slot in the head and darken the lines to complete the drawing.

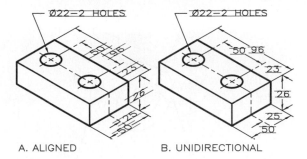

A. ALIGNED   B. UNIDIRECTIONAL

17.50 Either of the techniques shown—aligned or unidirectional—is acceptable for placing dimensions on isometric drawings. Guidelines should always be used for lettering.

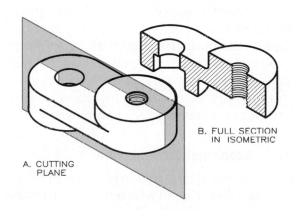

B. FULL SECTION
IN ISOMETRIC

A. CUTTING
PLANE

17.49 Isometric sections can be used to clarify the internal features of a part.

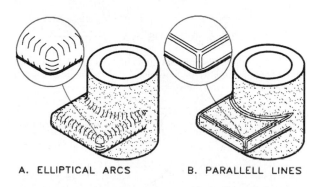

A. ELLIPTICAL ARCS   B. PARALLELL LINES

17.51 Either of these two methods can be used to represent fillets and rounds on the pictorial view of a part.

looked **(Figure 17.49).** Half sections also may be used advantageously.

Dimensioned Isometrics

When you dimension isometric drawings, place numerals on the dimension lines, using either **aligned** or **unidirectional** numerals **(Figure 17.50).** In both cases, notes connected with leaders usually are positioned horizontally, but drawing them to lie in an isometric plane is permissible. Always use guidelines for your lettering and numerals.

Fillets and Rounds

Fillets and rounds in isometric may be represented by either of the techniques shown in **Figure 17.51** for added realism. The enlarged detail in the balloon shows how to draw fillets and rounds with elliptical segments (A) or with parallel lines (B). Elliptical segments are best drawn by using an ellipse template. The stipple shading can be applied by an overlay film or a computer hatch pattern.

When fillets and rounds are shown in a three-dimensional drawing of a part they are much more readily understood than when they are depicted in two-dimensional orthographic views **(Figure 17.52).**

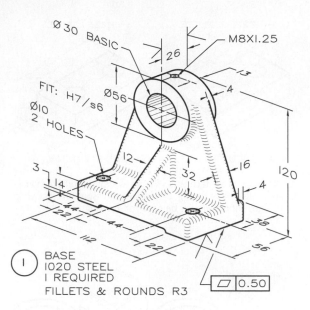

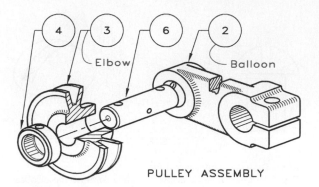

17.54 This exploded isometric assembly shows how parts are to be put together.

17.52 This three-dimensional drawing has been drawn to show fillet and rounds, dimensions, and notes in order for it to be used as a working drawing.

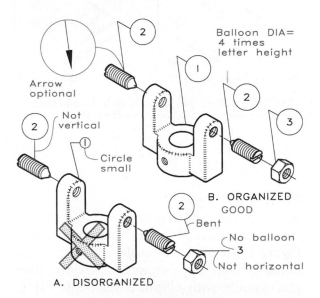

17.53 This drawing shows (A) common mistakes in applying leaders and part numbers in balloons of an assembly, and (B) acceptable techniques of applying leaders and part numbers to an assembly.

Assemblies

Assembly drawings illustrate how to put parts together. **Figure 17.53A** shows common mistakes in applying leaders and balloons to an

assembly, and **Figure 17.53B** shows the correct method of applying them. The numbers in the balloons correspond to the part numbers in the parts list. **Figure 17.54** shows an exploded assembly that illustrates the relationship of four mating parts. Illustrations of this type are excellent for inclusion in parts catalogs and maintenance manuals.

17.6 Axonometric Projection

An axonometric projection is a type of orthographic projection in which the pictorial view is projected perpendicularly onto the picture plane with parallel projectors. The object is positioned at an angle to the picture plane so that its pictorial projection will be a three-dimensional view. The three types of axonometric projections are: (1) **isometric,** (2) **dimetric,** and (3) **trimetric (Figure 17.55).**

Recall that the **isometric projection** is the type of pictorial in which the diagonal of a cube is seen as a point, the three axes and planes of the cube are equally foreshortened, and the axes are equally spaced 120° apart. Measurements along the three axes will be equal but less than true length because the isometric projection is true projection.

A **dimetric projection** is a pictorial in which two planes are equally foreshortened and two of the axes are separated by equal

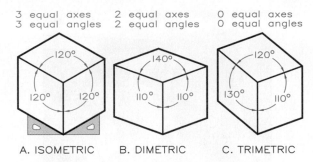

17.55 This drawing illustrates the three types of axonometric projection.

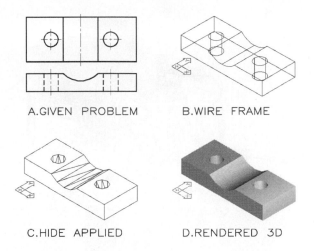

17.56 Modeling a simple part:
A Two orthographic views (a top and front) of the part are given.
B A 3-D wire frame drawing of the part is made .
C The hidden lines are suppressed to give a three-dimensional model.
D The model is rendered to give it a realistic look.

angles. Measurements along two axes of the cube are equal.

A **trimetric projection** is a pictorial in which all three planes are unequally foreshortened. The lengths of the axes are unequal, and the angles between them are different.

17.7 Three-Dimensional Modeling

Objects drawn with AutoCAD as true three-dimensional solids can be rotated and viewed from any angle as if they were held in your hand. The object in **Figure 17.56** is an example of a simple object represented by two orthographic views, a wire frame drawing, a hidden-line wire frame drawing, and a rendered solid. The capability to depict objects as rendered solids is a powerful design and communications tool.

Another example of a three-dimensional part that would be difficult to draw by hand is the pulley shown in **Figure 17.57.** A typical section through the pulley and its axis are drawn, the section is revolved about the axis, and the wire frame diagram is rendered. In addition to being able to select various views of the pulley, different lighting combinations and materials can be applied to it in infinite combinations of effects.

An example of an industrial application is given in **Figure 17.58,** which shows an apparatus of a higher degree of complexity that would be a rigorous assignment if drawn by

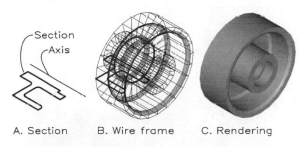

17.57 A model by revolution:
A A typical section of the pulley and its axis are drawn.
B The section is rotated about the axis to obtain a wire frame drawing.
C The wire frame is rendered to obtain a realistic view of the pulley.

hand. Although it is no easy chore to draw it as a series of solids by computer, the computer drawing enables you to obtain many different views of the parts, and to replicate drawings in combination. For example, the apparatus in **Figure 17.58** is applied repetitively in the subsea production-equipment assembly in **Figure 17.59.** The savings in time and effort becomes highly significant, and the final rendering greatly improves the understanding of the unit as a whole.

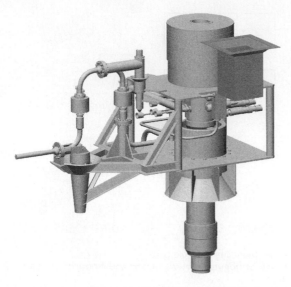

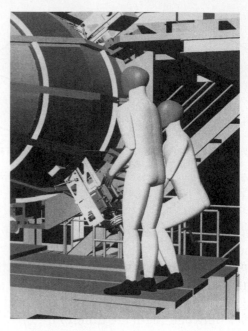

17.58 This apparatus is an example of a rendered three-dimensional model of a moderately complex application. *(Courtesy of Cameron Division, Cooper Cameron Corporation.)*

17.60 This computer-drawn scene illustrates the interaction between people and equipment. *(Courtesy McDonnell Douglas Space & Defense System—Kennedy Space Center, NASA.)*

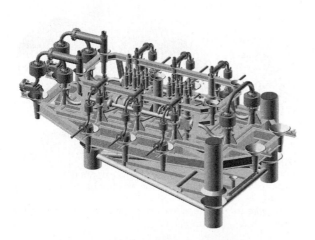

17.59 The apparatus in Figure 17.58 is replicated a number of times in this equipment assembly used in subsea production. *(Courtesy of Cameron Division, Cooper Cameron Corporation.)*

In addition to determining the interactions between parts, assemblies, and equipment, it is equally important to study the relationship of personnel to their working environment. An example of this type of application in **Figure 17.60** shows worker performing maintenance on a spacecraft.

Problems

Draw your solutions to the following problems **(Figure 17.61)** on size A or B sheets, as assigned. Select an appropriate scale to take advantage of the space available on each sheet. By letting each square represent 0.20 inch (5 mm), you can draw two solutions on each size A sheet. By setting each square to 0.40 inch (10 mm), you can draw one solution on each size A sheet.

Oblique Pictorials

1–24. Construct cavalier, cabinet, or general obliques of the parts assigned.

Isometric Pictorials

1–24. Construct isometrics of the parts assigned.

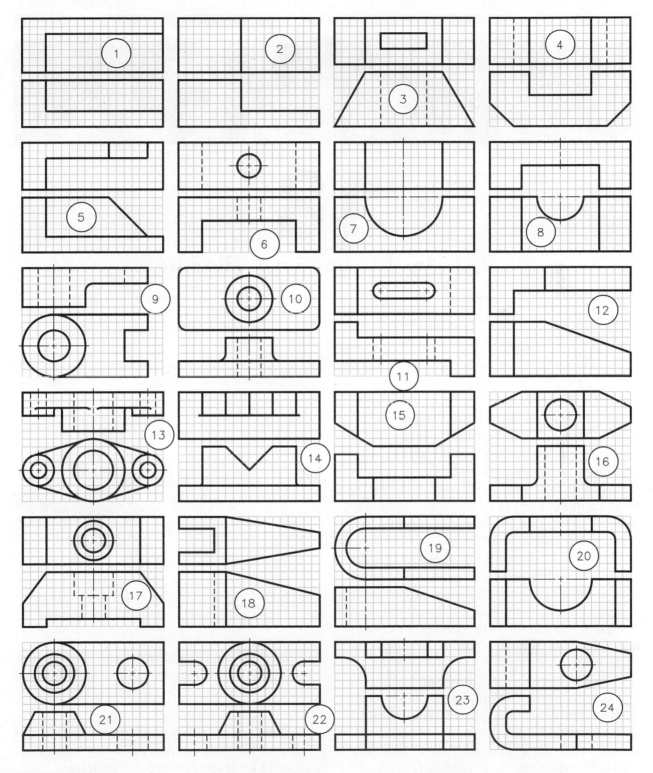

17.61 Problems 1–24.

18

Points, Lines, and Planes

18.1 Introduction

Points, lines, and planes are the basic geometric elements used in three-dimensional (3D) spatial geometry, called **descriptive geometry.** You need to understand how to locate and manipulate these elements in their simplest form because they will be applied to 3D spatial problems in Chapters 18 through 23.

The huge antenna in **Figure 18.1** is composed of many points, lines, and planes that represent its structural members and shapes. Its geometry had to be established one point at a time with great precision in order for it to function and to be properly supported.

The labeling of points, lines, and planes is an essential part of 3D projection because it is your means of analyzing their spatial relationships. **Figure 18.2** illustrates the fundamental requirements for properly labeling these elements in a drawing:

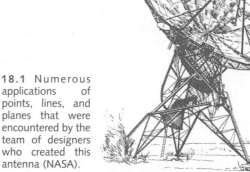

18.1 Numerous applications of points, lines, and planes that were encountered by the team of designers who created this antenna (NASA).

Lettering: use 1/8-inch letters with guidelines for labels; label lines at each end and planes at each corner with either letters or numbers.

Points: mark with two short, perpendicular dashes forming a cross, not a dot; each dash should be approximately 1/8 inch long.

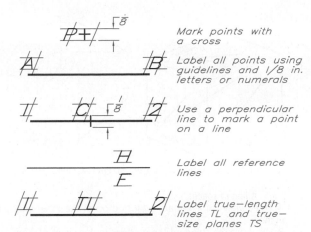

Mark points with a cross

Label all points using guidelines and 1/8 in. letters or numerals

Use a perpendicular line to mark a point on a line

Label all reference lines

Label true–length lines TL and true-size planes TS

18.2 These are standard practices for labeling points, lines, and planes.

Points on lines: mark with a short, perpendicular dash crossing the line, not a dot.

Reference lines: label these thin, dark lines as described in Chapter 7.

Object lines: draw these lines used to represent points, lines, and planes twice as thick as hidden lines with an F or HB pencil; draw hidden lines twice as thick as reference lines.

True-length lines: label true length or TL.

True-size planes: label true size or TS.

Projection lines: draw precisely with a 2H or 4H pencil as thin lines, just dark enough to be visible so they need not be erased.

18.2 Projection of Points

A point is a theoretical location in space having no dimensions other than its location. However, a series of points establishes lengths, areas, and volumes of complex shapes.

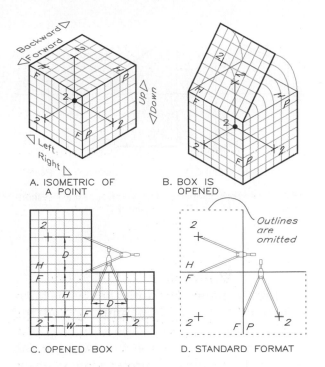

A. ISOMETRIC OF A POINT

B. BOX IS OPENED

C. OPENED BOX

D. STANDARD FORMAT

Outlines are omitted

18.3 Three views of a point:
A The point is projected to three projection planes.
B The projection planes are opened into a single plane.
C In the opened box, point 2 is 5 units to the left of the profile, 5 units below the horizontal, and 4 units behind the frontal plane.
D The outlines of the projection planes are omitted in orthographic projection.

A point must be located in at least two adjacent orthographic views to establish its position in 3D space **(Figure 18.3).** When the planes of the projection box **(Figure 18.3A)** are opened onto the plane of the drawing surface **(Figure 18.3C),** the projectors from each view of point 2 are perpendicular to the reference lines between the views. Letters **H, F,** and **P** represent the **horizontal, frontal,** and **profile planes,** the three principal projection planes.

A point may be located from verbal descriptions with respect to the principal planes. For example, point 2 in **Figure 18.3** may be described as being (A) 5 units left of

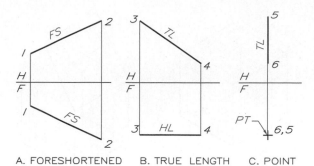

A. FORESHORTENED   B. TRUE LENGTH   C. POINT

18.4 A line in orthographic projection can appear as foreshortened (FS), true length (TL), or a point (PT).

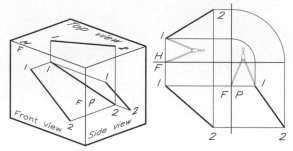

A. THREE—DIMENSIONAL VIEW   B. ORTHOGRAPHIC VIEWS

18.5 A line in space:
A Three views of a line are projected on to the three principal planes.
B These are the standard three orthographic views of a line.

the profile plane, (B) 5 units below the horizontal plane, and (C) 4 units behind the frontal plane.

When you look at the front view of the box, the horizontal and profile planes appear as edges. In the top view, the frontal and profile planes appear as edges. In the side view, the frontal and horizontal planes appear as edges.

18.3 Lines

A **line** is the straight path between two points in 3D space. A line may appear as (1) **foreshortened,** (2) **true-length,** or (3) a **point** (**Figure 18.4**). Oblique lines are neither parallel nor perpendicular to a principal projection plane (**Figure 18.5**). When line 1-2 is projected onto the horizontal, frontal, and profile planes, it appears foreshortened in each view.

Principal lines are parallel to at least one of the principal projection planes. A principal line is true length in the view where the principal plane to which it is parallel appears true size. The three types of principal lines are **horizontal, frontal,** and **profile lines.**

Figure 18.6A shows a **horizontal line** (HL), which appears true length in the horizontal (top) view. Any line shown in the top view will appear true length as long as it is parallel to the horizontal plane.

When looking at the top view, you cannot tell whether the line is horizontal. You must look at the front or side views to do so. In those views, a horizontal line (HL) will be parallel to the edge view of the horizontal, the HF fold line (**Figure 18.7**). A line that projects as a point in the front view is a combination horizontal and profile line.

A **frontal line** (FL) is parallel to the frontal projection plane. It appears true length in the front view because your line of sight is perpendicular to it in this view. In **Figure 18.6B** line 3-4 is an FL because it is parallel to the edge of the frontal plane in the top and side views.

A **profile line** (PL) is parallel to the profile projection planes and appears true length in the side (profile) views. To tell whether a line is a PL, you must look at a view adjacent to the profile view: the top or front view. In **Figure 18.6C,** line 5-6 is parallel to the edge view of the profile plane in both the top and side views.

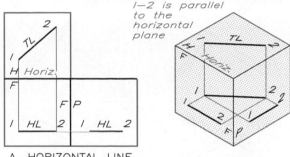

1–2 is parallel to the horizontal plane

A. HORIZONTAL LINE

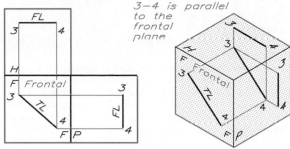

3–4 is parallel to the frontal plane

B. FRONTAL LINE

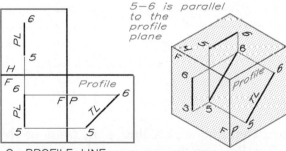

5–6 is parallel to the profile plane

C. PROFILE LINE

18.6 Principal lines:
A A horizontal line is true length in the horizontal (top) view. It is parallel to the edge view of the horizontal plane in the front and side views.
B The frontal line is true length in the front view. It is parallel to the edge view of the frontal plane in the top and side views.
C The profile line is true length in the profile (side) view. It is parallel to the edge view of the profile plane in the top and front views.

Locating a Point on a Line

Figure 18.8 shows the top and front views of line 1-2 with point O located at its midpoint. To find the front view of the point, recall that

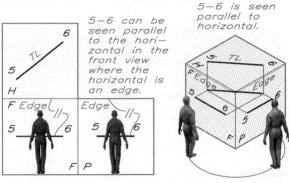

5–6 can be seen parallel to the horizontal in the front view where the horizontal is an edge.

5–6 is seen parallel to horizontal.

A. ORTHOGRAPHIC VIEWS   B. PICTORIAL VIEW

18.7 In order to determine that a line is horizontal, you must look at the front or side views in which the horizontal plane is an edge. Line 5-6 is seen parallel to the horizontal edge and is a horizontal line, too.

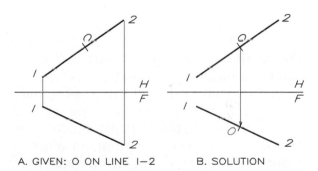

A. GIVEN: O ON LINE 1–2   B. SOLUTION

18.8 Point O on the top view of line 1-2 can be found in the front view by projection. The projector is perpendicular to the HF reference line between the views.

in orthographic projection the projector between the views is perpendicular to the HF fold line. Use that projector to project point O to line 1-2 in the front view. A point located at a line's midpoint will be at the line's midpoint in all orthographic views of the line.

Intersecting and Nonintersecting Lines

Lines that intersect have a common point of intersection lying on both lines. Point O in

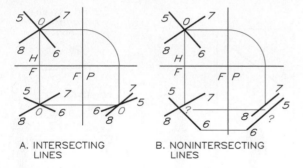

A. INTERSECTING
LINES

B. NONINTERSECTING
LINES

18.9 Crossing lines:
A These lines intersect because O, the point of intersection, projects as a common point of intersection in all views.
B The lines cross in the top and front views, but they do not intersect because there is no common point of intersection in all views.

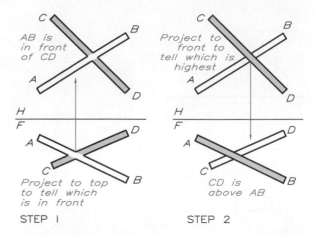

STEP 1

STEP 2

18.10 Visibility of lines:
Step 1 Project the crossing point from the front to the top view. This projector strikes line AB before it strikes line CD, indicating that line AB is in front and thus is visible in the front view.
Step 2 Project the crossing point from the top view to the front view. This projector strikes line CD before it strikes line AB, indicating that line CD is above line AB and thus is visible in the top view.

Figure 18.9A is a point of intersection because it projects to a common crossing point in all three views. However, the crossing point of the lines in **Figure 18.9B** in the top and front views is not a point of intersection. Point O does not project to a common crossing point in the top and front views, so the lines do not intersect; they simply cross, as shown in the profile view.

18.4 Visibility

Crossing Lines

In **Figure 18.10** nonintersecting lines AB and CD cross in certain views. Therefore portions of the lines are visible or hidden at the crossing points (here, line thickness is exaggerated for purposes of illustration). Determining which line is above or in front of the other is referred to as finding a line's visibility, a requirement of many 3D problems.

You have to determine line visibility by analysis. For example, select a crossing point in the front view and project it to the top view to determine which line is in front of the other. Because the projector contacts line AB first, you know that line AB is in front of CD and is visible in the front view.

Repeat this process by projecting downward from the intersection in the top view to find that line CD is above line AB and is visible in the top view. If only one view were available, visibility would be impossible to determine.

A Line and a Plane

The principles of visibility analysis also apply to determining visibility for a line and a plane (**Figure 18.11**). First, project the intersections of line AB with lines 4-5 and 5-6 to the top view to determine that the lines of the plane (4-5 and 5-6) lie in front of line AB in the front

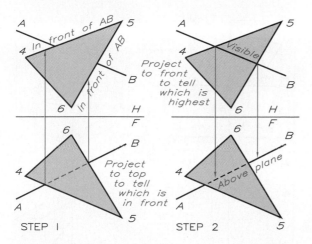

STEP 1   STEP 2

18.11 Visibility of a line and a plane:
Step 1 Project the points where line AB crosses the plane from the front view to the top view. These projectors intersect lines 4-5 and 5-6 of the plane first, indicating that the plane is in front of the line and making line AB hidden in the front view.
Step 2 Project the points where line AB crosses the plane in the top view to the front view. These projectors encounter line AB first, indicating that line AB is higher than the plane, thus the line is visible in the top view.

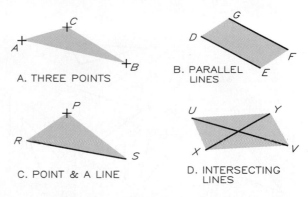

A. THREE POINTS   B. PARALLEL LINES

C. POINT & A LINE   D. INTERSECTING LINES

18.12 A plane can be represented as (A) three points not on a straight line, (B) two parallel lines, (C) a line and a point not on the line or its extension, and (D) two intersecting lines.

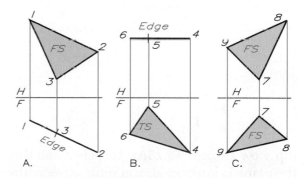

A.   B.   C.

18.13 Views of a plane:
A plane in orthographic projection can appear as (A) an edge, (B) true size (TS), or (C) foreshortened (FS). A plane that is foreshortened in all principal views is an oblique plane.

view. Therefore line AB is a hidden line in the front view.

Similarly, project the two intersections of line AB in the top view to the front view, where line AB is found to lie above lines 4-5 and 5-6 of the plane. Because line AB is above the plane, it is a visible line in the top view.

18.5 Planes

A plane may be represented in orthographic projection by any of the four combinations shown in **Figure 18.12**. In orthographic projection, a plane may appear as (1) an edge, (2) a true-size plane, or (3) a foreshortened plane **(Figure 18.13)**.

Oblique planes (the general case) are not parallel to principal projection planes in any view **(Figure 18.14)**. **Principal planes** are parallel to principal projection planes **(Figure 18.15)**. The three types of principal planes are horizontal, frontal, and profile planes.

A **horizontal plane** is parallel to the horizontal projection plane and is true size in the

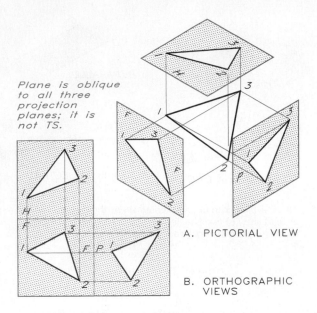

Plane is oblique to all three projection planes; it is not TS.

A. PICTORIAL VIEW

B. ORTHOGRAPHIC VIEWS

18.14 Oblique plane:
An oblique plane is neither parallel nor perpendicular to a projection plane. It is the general-case plane.

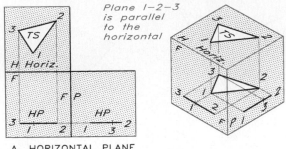

Plane 1–2–3 is parallel to the horizontal

A. HORIZONTAL PLANE

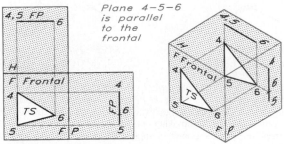

Plane 4–5–6 is parallel to the frontal

B. FRONTAL PLANE

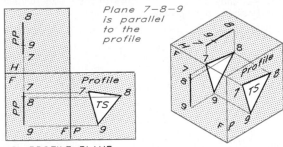

Plane 7–8–9 is parallel to the profile

C. PROFILE PLANE

18.15 Principal planes:
A The horizontal plane is true size in the horizontal (top) view. It is parallel to the edge view of the horizontal plane in the front and profile views.
B The frontal plane is true size in the front view. It is parallel to the edge view of the frontal plane in the top and profile views.
C The profile plane is true size in the profile view. It is parallel to the edge view of the profile plane in the top and front views.

top view **(Figure 18.15A).** To determine that the plane is horizontal, you must observe the front or profile views, where you can see its parallelism to the edge view of the horizontal plane.

A **frontal plane** is parallel to the frontal projection plane and appears true size in the front view **(Figure 18.15B).** To determine that the plane is frontal, you must look at the top or profile views, where you can see its parallelism to the edge view of the frontal plane.

A **profile plane** is parallel to the profile projection plane and is true size in the side view **(Figure 18.15C).** To determine that the plane is profile, you must observe the top or front views, where you can see its parallelism to the edge view of the profile plane.

A Point on a Plane
Point O on the front view of plane 4-5-6 in **Figure 18.16** is to be located on the plane in the top view. First, draw a line in any direction

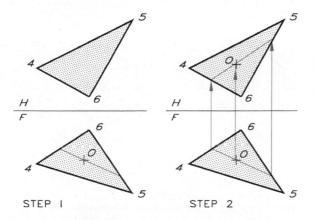

STEP 1   STEP 2

18.16 A point on a plane:
Step 1 In the front view, draw a line through point O in any convenient direction except vertical.
Step 2 Project the ends of the line to the top view and draw the line. Project point O to this line.

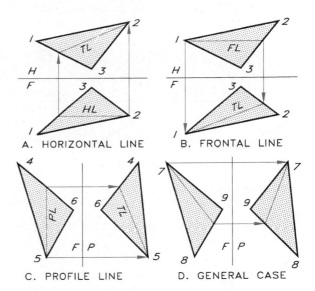

A. HORIZONTAL LINE   B. FRONTAL LINE

C. PROFILE LINE   D. GENERAL CASE

18.17 Principal lines on a plane:
A First, draw a horizontal line in the front view parallel to the edge view of the horizontal plane. Then, project it to the top view, where it is true length.
B First, draw a frontal line in the top view parallel to the edge view of the frontal plane. Then, project it to the front view, where it is true length.
C First, draw a profile line in the front view parallel to the edge view of the profile plane. Then project it to the profile view, where it is true length.
D A general-case line is not parallel to the frontal, horizontal, or profile planes and is not true length in any principal view.

(except vertical) through the point to establish a line on the plane. Then project this line to the top view and project point O from the front view to the top view of the line.

Principal Lines on a Plane

Principal lines may be found in any view of a plane when at least two orthographic views of the plane are given. Any number of principal lines can be drawn on any plane.

Figure 18.17A shows a horizontal line parallel to the edge view of the horizontal projection plane in the front view. When projected to the top view, this line is true length.

Figure 18.17B shows a frontal line parallel to the edge view of the frontal projection plane in the top view. When projected to the front view, this line is true length.

Figure 18.17C shows a profile line parallel to the edge view of the profile projection plane in the front. When projected to the profile view, this line is true length.

In the general case (oblique), a line is not parallel to the edge view of any principal projection plane **(Figure 18.17D).** Therefore it is not true length in any principal view.

18.6 Parallelism

Lines

Two parallel lines appear parallel in all views, except in views where both appear as points. Parallelism of lines in 3D space cannot be determined without at least two adjacent orthographic views. In **Figure 18.18,** line AB was

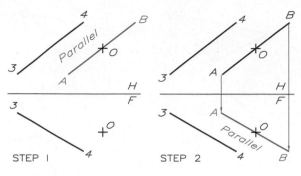

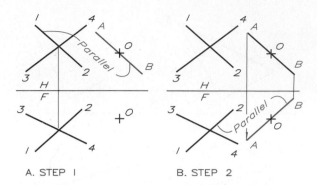

18.18 A line parallel to a line:
Step 1 Draw line AB parallel to the top view of line 3-4, with its midpoint at O.
Step 2 Draw the front view of line AB parallel to the front view of 3-4 through point O.

18.20 A line parallel to plane:
Step 1 Draw line AB parallel to line 1-2 through point O.
Step 2 Draw line AB parallel to the same line, line 1-2, in the front view, which makes line AB parallel to the plane.

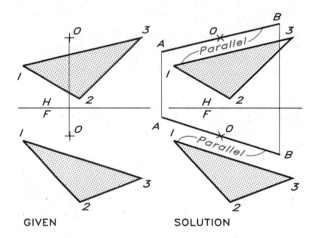

GIVEN · SOLUTION

18.19 A line parallel to plane:
A line may be drawn through point O parallel to plane 1-2-3 if the line is parallel to any line in the plane. Draw line AB parallel to line 1-3 of the plane in the front and top views, making it parallel to the plane.

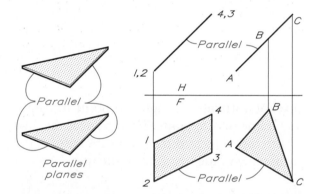

18.21 Parallel planes:
Definition. Two planes are parallel when intersecting lines in one are parallel to intersecting lines in the other. When parallel planes appear as edges, their edges are parallel.

drawn parallel to the horizontal view of line 3-4 and through point O, which is the midpoint. Projecting points A and B to the front view with projectors perpendicular to the HF reference plane yields the length of line AB, which is parallel to line 3-4 through point O.

A Line and a Plane

A line is parallel to a plane when it is parallel to any line in the plane. In **Figure 18.19** a line with its midpoint at point O is to be drawn parallel to plane 1-2-3. In this case line AB was drawn parallel to line 1-3 in the plane in the top and front views. The line could have been drawn parallel to any line in the plane, making infinite solutions possible.

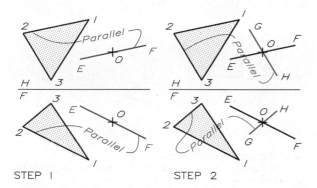

18.22 A plane through a point parallel to a plane:
Step 1 Draw EF parallel to any line in plane (1-2 in this case). Show the line in both views.
Step 2 Draw a second line parallel to 2-3 in the top and front views. These intersecting lines passing through O represent a plane parallel to 1-2-3.

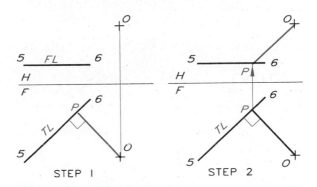

18.24 A line perpendicular to a principal line:
Step 1 Line 5-6 is a frontal line and is true length in the front view, so a perpendicular from point O makes a true 90° angle with it in the front view.
Step 2 Project point P to the top view and connect it to point O. As neither line is true length in the top view, they do not intersect at 90° in this view.

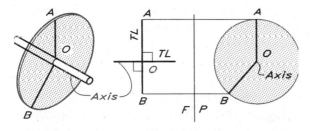

18.23 Perpendicular lines:
Perpendicular lines have a true angle of 90° between them in a view where one or both of them appear true length.

In **Figure 18.22,** a plane is to be drawn through point O parallel to plane 1-2-3. First, draw line EF through point O parallel to line 1-2 in the top and front views. Then draw a second line through point O parallel to line 2-3 of the plane in the front and top views. These two intersecting lines form a plane parallel to plane 1-2-3, as intersecting lines on one plane are parallel to intersecting lines on the other.

18.7 Perpendicularity

Lines

When two lines are perpendicular, draw them with a true 90° angle of intersection in views where one or both of them appear true length (**Figure 18.23**). In a view where neither of two perpendicular lines is true length, the angle between is not a true 90° angle.

In **Figure 18.23,** the axis is true length in the front view; therefore any spoke of the circular wheel is perpendicular to the axis in the front view. Spokes OA and OB are examples of true-length foreshortened axes, respectively, in the front view.

Figure 18.20 shows a similar example. Here, a line parallel to the plane with its midpoint at O was drawn. In this case, the plane is represented by two intersecting lines instead of an outlined area.

Planes

Two planes are parallel when intersecting lines in one plane are parallel to intersecting lines in the other (**Figure 18.21**). Determining whether planes are parallel is easy when both appear as edges in a view.

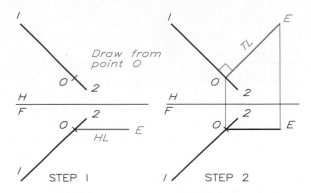

18.25 A line perpendicular to an oblique line:
Step 1 Draw a horizontal line (OE) from O in the front view.
Step 2 Horizontal line OE is true length in the top view, so draw it perpendicular to line 1-2 in this view.

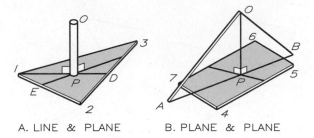

A. LINE & PLANE   B. PLANE & PLANE

18.26 Perpendicularity of line and plane:
A A line is perpendicular to a plane when it is perpendicular to two intersecting lines on the plane.
B A plane is perpendicular to another plane if it contains a line that is perpendicular to the other plane.

A Line Perpendicular to a Principal Line

In **Figure 18.24** a line is to be constructed through point O perpendicular to frontal line 5-6, which is true length in the front view. First, draw OP perpendicular to frontal line 5-6 because it is true length. Then, project point P to the top view of line 5-6. In the top view, line OP is not perpendicular to line 5-6 because neither of these lines is true length in this view.

A Line Perpendicular to an Oblique Line

In **Figure 18.25** a line is to be constructed from point O that is perpendicular to oblique line 1-2. First, draw a horizontal line from O to some convenient length in the front view, to E for example.

Locate point O in the top view by projection. OE will be true length in the top view regardless of the direction in which it is drawn because it is a horizontal line. Since it is true length in the top view, draw line OE to make a 90° angle with the top view of 1-2.

Planes

A line is perpendicular to a plane when it is perpendicular to any two intersecting lines in the plane (**Figure 18.26A**). A plane is perpendicular to another plane when a line in one plane is perpendicular to the other plane (**Figure 18.26B**).

A Line Perpendicular to a Plane

In **Figure 18.27** a line is to be drawn perpendicular to the plane from point O on the plane. First, draw a frontal line on the plane in the top view through O. Project the line to the front view, where it is true length. Draw line OP at a convenient length perpendicular to the true-length line.

Next, draw a horizontal line through point O in the front view and project it to the top view of the plane, draw line OP perpendicular to the true-length line. This construction results in a line perpendicular to the plane because the line is perpendicular to two intersecting lines, a horizontal and a frontal line, in the plane.

Plane Perpendicular to a Plane

A plane is perpendicular to another plane if a line on the plane is perpendicular to the second plane.

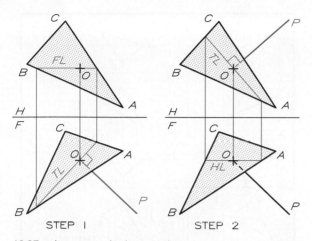

STEP 1   STEP 2

18.27 A line perpendicular to a plane:
Step 1 Draw a frontal line on the plane through O in the top view. This line is true length in the front view, so draw line OP perpendicular to this true-length line.
Step 2 Construct a horizontal line through point O in the front view. This line is true length in the top view, so draw line OP perpendicular to it.

Problems

Use size A sheets for the following problems and lay out your solutions with instruments. Each square on the grid is equal to 0.20 in. or 5 mm. Use either grid paper or plain paper. Label all reference planes and points in each problem with 1/8-in. letters and numbers using guidelines.

1. (Sheet 1) Points and Lines
(A–D) Draw three views (top, front, and right-side views) of the given points.
(E–F) Draw the three views of the points and connect them to form lines.

2. (Sheet 2) Lines
(A–B) Draw right-side view of line 1-2 and plane 3-4-5.

(C–E) Draw the missing views of the planes so that 6-7-8 is a frontal plane, 1-2-3 is a horizontal plane, and 4-5-6 is a profile plane.

3. (Sheet 3) Lines and Planes
(A) Draw the side view of plane 1-2-3 and locate point A on it in all views.
(B) Draw the right-side view of the plane and draw two horizontal lines on it in all views.
(C) Draw the right-side view of the plane and draw two frontal lines on it in all views.
(D) Draw the right-side view of the plane and draw two profile lines on it in all views.
(E) Draw three views of the given line and line through B that is parallel to the plane.
(F) Draw three views of a plane formed by intersecting lines with A at the endpoint of one of the lines.

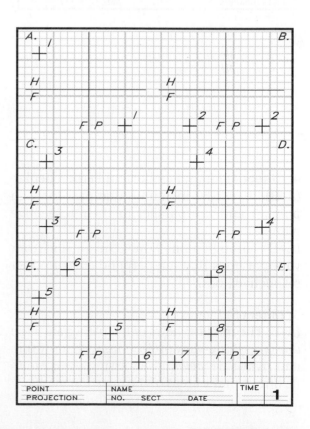

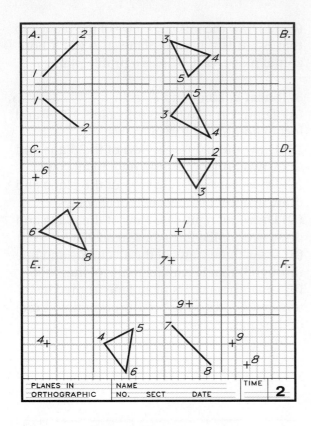

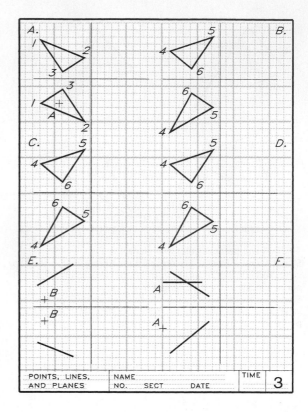

4. (Sheet 4) Lines: Parallel and Perpendicular
(A–B) Draw 1.50-inch lines that pass through point 0 and are parallel to their respective planes.
(C–D) Through point 0 draw the top and front views of lines that are perpendicular to their respective lines.

5. (Sheet 5) Planes
(A–E) The missing views of the planes are to be drawn as three-view projections with top, front, and right-side views. The missing views are to be drawn in the areas of the question marks.

Thought Questions

1. How many principal lines are there, and what are their names?

2. What are the different ways in which a line can appear on a drawing?

3. The three orthographic dimensions are height, width, and depth. Which of these dimensions are necessary to locate a line in the side view? The front view? The top view?

4. What are the three different ways in which a plane can appear on a drawing?

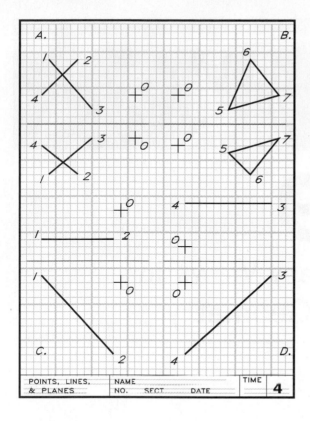

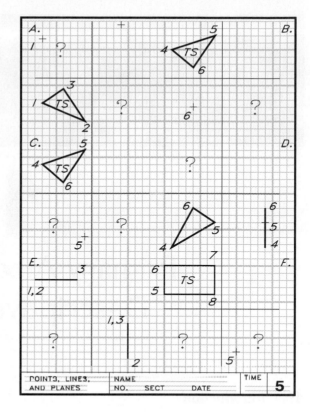

5. If a line is parallel to another line, what can be said about the lines in any view in which they appear in a drawing?

6. When is a line parallel to a plane?

7. When can two perpendicular lines be drawn at a true perpendicular angle on a drawing?

8. If a plane appears as an edge in the front view, what can be said about its top view?

9. A line that is vertical appears how in the top view? The side view?

10. A line that is true length in the top view is what type of principal line? How will it appear in the front view?

11. A line that is perpendicular to a plane in the top view makes what angle with a front line on the plane?

12. How many principal lines can be drawn on any view of a plane?

13. A plane that is vertical is true size in which views? Explain.

14. When three views of a plane are given, describe the conditions that must exist for the plane to be true size in the top view.

15. How many orthographic views are necessary to establish the three-dimensionality of a line or plane?

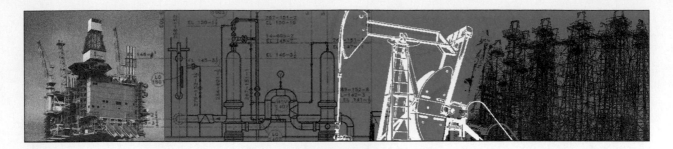

19

Primary Auxiliary Views in Descriptive Geometry

19.1 Introduction

Descriptive geometry is the projection of three-dimensional (3D) orthographic views onto a two-dimensional (2D) plane of paper to allow graphical determination of lengths, angles, shapes, and other geometric information. Orthographic projection is the basis for laying out and solving problems by descriptive geometry.

The primary auxiliary view, which permits analysis of 3D geometry, is essential to descriptive geometry. For example, the design of the helicopter frame shown in **Figure 19.1** contains many complex geometric elements (lines, angles, and surfaces) that were analyzed by descriptive geometry prior to its fabrication.

19.2 True-Length Lines by Primary Auxiliary Views

Figure 19.2 shows the top and front views of line 1-2 pictorially and orthographically. Line 1-2 is not a principal line, so it is not true length

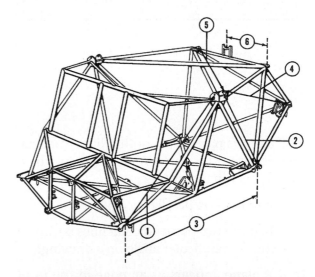

19.1 This helicopter frame was designed using the principles of descriptive geometry to determine lengths, angles, and areas. *(Courtesy of Bell Helicopter Textron Company.)*

in a principal view. Therefore a primary auxiliary view is required to find its true-length view. In **Figure 19.2A** the line of sight is perpendicular to the front view of the line and reference line F1 is parallel to the line's frontal

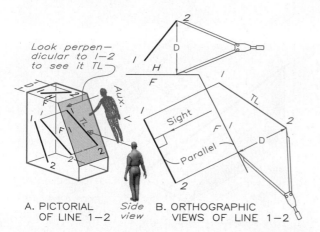

A. PICTORIAL   *Side*   B. ORTHOGRAPHIC
OF LINE 1—2   *view*   VIEWS OF LINE 1—2

19.2 True-length line by auxiliary view:
A A pictorial of line 1-2 is shown inside a projection box where an auxiliary plane is parallel to the line and perpendicular to the frontal plane.
B The auxiliary view is projected from the front orthographic view to find 1-2 true length.

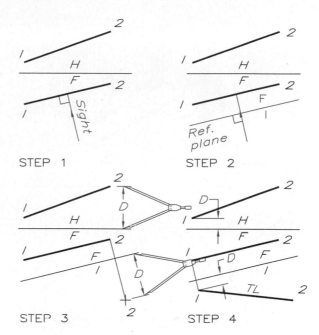

STEP 1   STEP 2

STEP 3   STEP 4

19.3 The true length of a line:
Step 1 To find the true length of line 1-2, the line of sight must be perpendicular to one of its views, the front view here.
Step 2 Draw the F1 reference line parallel to the line and perpendicular to the line of sight.
Step 3 Project point 2 perpendicularly from the front view. Transfer distance D from the top view to locate point 2 in the auxiliary view.
Step 4 Locate point 1 in the same manner to find line 1-2 true length in the auxiliary view.

view. The auxiliary plane is parallel to the line and perpendicular to the frontal plane, accounting for its label, F1, where F and 1 are abbreviations for frontal and primary planes, respectively.

Projecting parallel to the line of sight and perpendicular to the F1 reference line yields the auxiliary view **(Figure 19.2B).** Transferring distance D with dividers to the auxiliary view locates point 2 because the frontal plane appears as an edge in both the top and auxiliary views. Point 1 is located in the same manner, and the points are connected to find the true-length view of the line.

Figure 19.3 summarizes the steps of finding the true-length view of an oblique line. Letter all reference planes using the notation suggested in Chapter 18 and as shown in the examples throughout this chapter, with the

exception of explanatory dimensions such as D. Use your dividers to transfer dimensions.

True Length by Analytical Geometry

The method of finding the true length of frontal line 3-4 mathematically is shown in **Figure 19.4.** The Pythagorean theorem states that the hypotenuse of a right triangle is equal to the square root of the sum of the squares of the other two sides. Because the line is true length in the front view, measuring that

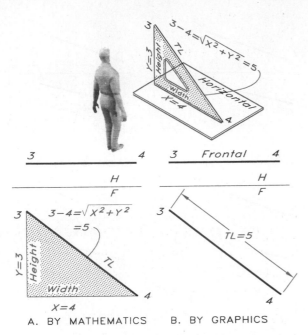

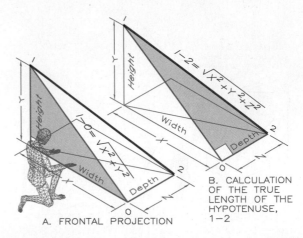

19.5 To calculate the true length of a 3D line that is not true length in principal view, find (A) the frontal projection, line 1-0, by using the x and y distances, and (B) the hypotenuse of the right triangle 1-0-2 by using the length of line 1-0 and the z distance. Then apply the Pythagorean theorem to find its length of 5.

19.4 Apply the Pythagorean theorem to calculate the length of a line that appears true length in a view, the front view here. Because line 3-4 is true length in the front view, it can be measured to find its length.

length provides a check on the mathematical solution.

The true length of a line shown pictorially in **Figure 19.5A** (line 1-2) is determined by analytical geometry from its length in the front view where the *x* and *y* distances form a right triangle. **Figure 19.5B** shows a second right triangle, 1-0-2, whose hypotenuse is the true length of line 1-2. Thus the true length of an oblique line is the square root of the sum of the squares of the *x*, *y*, and *z* distances that correspond to the width, height, and depth of the triangles.

True-Length Diagram

A true-length diagram is two perpendicular lines used to find a line true length **(Figure 19.6).**

The two measurements laid out on the true-length diagram may be transferred from any two adjacent orthographic views. One measurement is the distance between the endpoints in one of the views. The other measurement, from the adjacent view, is the distance between the endpoints perpendicular to the reference line between the two views. Here, these dimensions are vertical, V, and horizontal, H, between points 1 and 2. This method does not give the line's direction, only its true length.

19.3 Angles Between Lines and Principal Planes

To measure the angle between a line and a plane, the line must appear true length and the plane as an edge in the same view **(Figure 19.7).** A principal plane appears as an

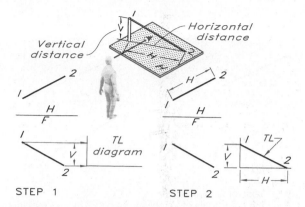

STEP 1   STEP 2

19.6 Using a true-length diagram:
Step 1 Transfer the vertical distance between the ends of line 1-2 to the vertical leg of the TL diagram.
Step 2 Transfer the horizontal length of the line in the top view to the horizontal leg of the TL diagram. The diagonal is the true length of line 1-2.

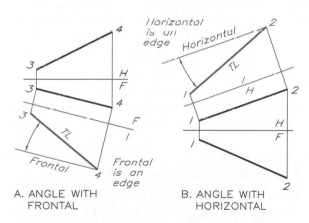

A. ANGLE WITH FRONTAL   B. ANGLE WITH HORIZONTAL

19.7 Angles between lines and principal planes:
A An auxiliary view projected from the front view shows the line true length and the frontal plane as an edge. Its angle with the frontal plane can be measured here.
B An auxiliary view projected from the top shows the line true length and the horizontal plane as an edge. Its angle with the horizontal can be measured here.

edge in a primary auxiliary view projected from it, so the angle a line makes with this principal plane can be measured if the line is true length in this auxiliary view.

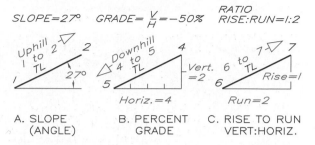

A. SLOPE (ANGLE)   B. PERCENT GRADE   C. RISE TO RUN VERT:HORIZ.

19.8 Slope terminology:
The inclination of a line with the horizontal may be measured and expressed as: (A) slope angle, (B) percent grade, and (C) slope ratio.

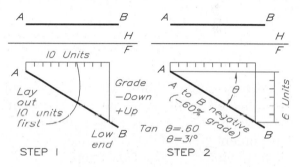

STEP 1   STEP 2

19.9 Percent grade of a line:
Step 1 The percent grade of a line can be measured in the view where the horizontal appears as an edge and the line is true length (here, the front view). Lay off ten units parallel to the horizontal from the end of the line.
Step 2 A vertical distance from the end of the 10 units to the line measures 6 units. The percent grade is 6 divided by 10, or 60%. This is a negative grade from A to B because the line slopes downward from A. The tangent of this slope angle is 6/10, or 0.60, which can be used to verify the slope of 31° from trigonometric tables.

19.4 Sloping Lines

Slope is the angle that a line makes with the horizontal plane when the line is true length and the plane is an edge. **Figure 19.8** shows the three methods for specifying slope: **slope angle, percent grade,** and **slope ratio.**

The **slope angle** of line AB in **Figure 19.9** is 31°. It can be measured in the front view where the line is true length.

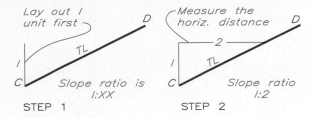

19.10 Slope ratio:

Step 1 Slope ratio always begins with 1, so lay out a vertical distance of 1 from end C.

Step 2 Lay off a horizontal distance from the end of the vertical line and measure it. It is 2, so the slope ratio (always expressed as 1:XX) of this line is 1:2.

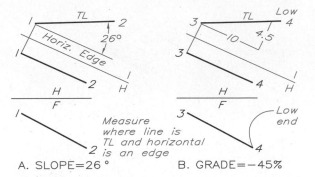

A. SLOPE=26 °   B. GRADE=−45%

19.11 Slope of an oblique line:

A Find the slope angle of an oblique line: (26° in this case) in a view where the horizontal appears as an edge and the line is true length.

B Find the percent grade in an auxiliary view projected from the top view where line 3-4 is true length (−45% from 3 to 4, the low end, in this case).

Percent grade is the ratio of the vertical (rise) divided by the horizontal (run) between the ends of a line, expressed as a percentage. The percent grade of line AB is determined in the front view of **Figure 19.9** where the line is true length and the horizontal plane is an edge. Line AB has a −60% grade from A to B because the line slopes downward; it would be positive (upward) from B to A. Trigonometric tables verify that an angle whose tangent is 0.60 (6/10) is 31°.

The slope ratio is the ratio of a rise of 1 to the run. The rise is always written as 1, followed by a colon and the run (for example, 1:10, 1:200). **Figure 19.10** illustrates the graphical method of finding the slope ratio. The rise of 1 unit is laid off on the true-length view of CD. The corresponding run measures 2 units, for a slope ratio of 1:2.

The slope of oblique lines are found true length in an auxiliary view projected from the top view so that the horizontal reference plane will appear as an edge (**Figure 19.11A**). The slope is expressed as an angle, or 26°.

To find the percent grade of an oblique line (**Figure 19.11B**) lay off 10 units horizontally,

parallel to the H1 reference line. The corresponding vertical distance measures 4.5 units, for a −45% grade from point 3 to point 4.

The principles of true-line length and angles between lines and planes are useful in applications such as the design of aggregate conveyors (**Figure 19.12**) where slope is crucial to optimal operation of the equipment.

19.5 Bearings and Azimuths

Two types of bearings of a line's direction are compass bearings and azimuths. Compass bearings are angular measurements from north or south. The line in **Figure 19.13A** that makes a 30° angle with north toward the west has a bearing of N 30° W. The line making a 60° angle with south toward the east has a bearing of south 60° east, or S 60° E. Because a compass can be read only when held level, bearings of a line must be found in the top, or horizontal, view.

Azimuths are measured clockwise from north through 360° (**Figure 19.13B**). Azimuth bearings are written N 120°, N 210°, and so on, indicating they are measured from north.

19.12 The design of these aggregate conveyors required the application of sloping-line principles in order to obtain their optimal slopes. *(Courtesy of FMC Technologies.)*

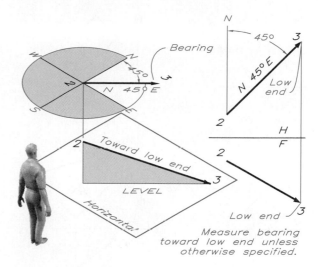

19.14 Compass bearing:
Measure the compass bearing of a line in the top view toward its low end (unless otherwise specified). Line 2-3 has a bearing of N 45° E from 2 to 3 toward the low end at point 3.

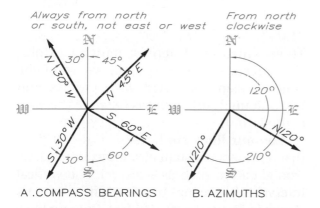

A. COMPASS BEARINGS   B. AZIMUTHS

19.13 Compass directions:
A Compass bearings are measured with respect to north and south, in an east or west direction.
B Azimuths are measured clockwise from north up to 360°.

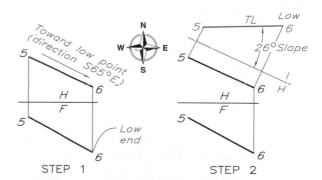

19.15 The slope and bearing of a line:
Step 1 Measure the bearing in the top view toward its low end, or S 65° E in this case.
Step 2 Measure the slope angle of 26° from the H1 reference line in an auxiliary view projected from the top view where the line is true length.

The bearing of a line is toward the low end of the line unless otherwise specified. For example, line 2-3 in **Figure 19.14** has a bearing of N 45° E because the line's low end is point 3 in the front view.

Figure 19.15 shows how to find the bearing and slope of a line. This information may be used verbally to describe the line as having a bearing of S 65° E and a slope of 26° from point 5 to point 6. This information and the location of one point in the top and front views is sufficient to complete a 3D drawing of a line as illustrated in **Figure 19.16.**

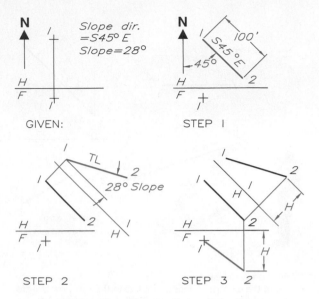

GIVEN:

STEP 1

STEP 2

STEP 3

19.16 A line from slope specifications:
Required: Draw a line through point 1 that bears S 45° E for 100 ft horizontally and slopes 28°.
Step 1 Draw the bearing and the horizontal distance in the top view.
Step 2 Project an auxiliary view from the top view and draw the line at a slope of 28°.
Step 3 Find the front view of line 1-2 by locating point 2 in the front view.

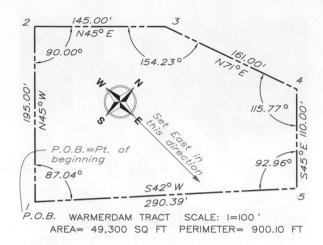

19.17 Plot plan:
This typical plot plan shows the lengths and bearings of each side of a tract of land, the interior angles, and the north arrow by using AutoCAD's *Surveyor's Units* option.

19.6 Application: Plot Plans

A typical plot plan for a tract of land is shown in **Figure 19.17.** The boundary lines of the tract, their bearings, and the interior angles are used to legally define the property. Auto-CAD provides an option called *Surveyor's Units* that is an excellent means of drawing and labeling a plot plan. The lengths and directions of lines are measured in a clockwise direction from the point of beginning (P.O.B.) and labeled as shown.

19.7 Contour Maps and Profiles

A contour map depicts variations in elevation of the earth in two dimensions **(Figure 19.18).**

Three-dimensional representations involve (1) conventional orthographic views of the contour map combined with profiles and (2) contoured surface views (often in the form of models).

Contour lines are horizontal (level) lines that represent constant elevations from a horizontal datum such as sea level. The vertical interval of spacing between the contours shown in **Figure 19.18** is 10 feet. Contour lines may be thought of as the intersection of horizontal planes with the surface of the earth.

Contour maps contain contour lines that connect points of equal elevation on the earth's surface and therefore are continuous **(Figure 19.18).** The closer the contour lines are to each other, the steeper the terrain is.

Profiles are vertical sections through a contour map that show the earth's surface at any desired location **(Figure 19.18).** Contour lines represent edge views of equally spaced horizontal planes in profiles. True representation of a profile involves the use of a vertical scale equal to the scale of the contour

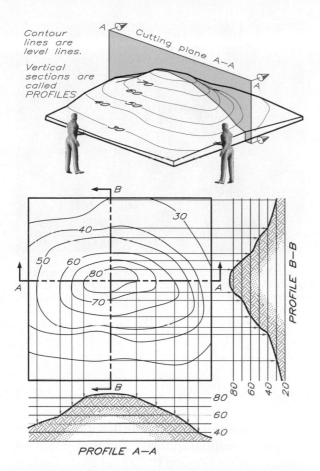

Contour lines are level lines.

Vertical sections are called PROFILES

Cutting plane A-A

PROFILE B-B

PROFILE A—A

19.18 Plan-profile:
A contour map shows variations in elevation on a surface. A profile is a vertical section through the contour map. To construct a profile, draw elevation lines parallel to the cutting planes, spacing them equally to show the difference in elevations of the contours (10 ft in this case). Then project crossing points of contours and the cutting plane to their respective elevations in the profile and connect them.

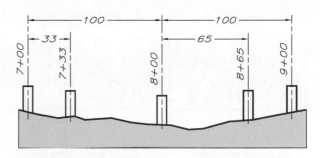

19.19 Station numbers:
Primary station points are located 100 ft apart: for example, station 7 is 700 ft from station 0 (not shown). A point 33 ft beyond station 7 is labeled station 7 + 33. A point 865 ft from the origin is labeled station 8 + 65.

mobile bodies, ship hulls, and household appliances—this technique of showing contours is called **lofting.**

Station numbers identify distances on a contour map. Surveyors use a chain (metal tape) 100 feet long, so primary stations are located 100 feet apart **(Figure 19.19).** For example, station 7 is 700 feet from the beginning point, station 0; a point 33 feet beyond station 7 is station 7 + 33.

A **plan-profile** combines a section of a contour map (a plan view) and a vertical section (a profile view). Engineers use plan-profile drawings extensively for construction projects such as pipelines, roadways, and waterways.

Application: Vertical Sections
In **Figure 19.20,** a vertical section passed through the top view of an underground pipe gives a **profile view.** The pipe is known to have elevations of 80 feet at point 1 and 60 feet at point 2. Project an auxiliary view perpendicularly from the top view, locate contour lines, and draw the top of the earth over the pipe in profile. To measure the true lengths and angles

map; however, the vertical scale usually is drawn larger to emphasize changes in elevation that often are slight compared to horizontal dimensions.

Contoured surfaces also are depicted in drawings with contour lines **(Figure 19.18)** or on models. When applied to objects other than the earth's surface—such as airfoils, auto-

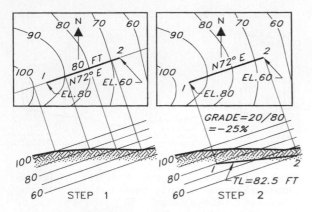

19.20 Drawing profiles:
Step 1 An underground pipe has elevations of 80 ft and 60 ft at its ends. Project an auxiliary view perpendicularly from the top view and draw contours at 10-ft intervals corresponding to their elevations in the plan view. Locate the ground surface by projecting from the contour lines in the plan view.
Step 2 Locate points 1 and 2 at elevations of 80 ft and 60 ft in the profile. Line 1-2 is TL in the section, so measure its slope (percent grade) here and label its bearing and slope in the top view.

19.21 Pipeline construction applies the principles of descriptive geometry, true-length lines, and slopes of 3D lines to develop construction plans. *(Courtesy of Panhandle Energy, Trunkline Gas Co.)*

of slope in the profile, use the same scale for both the contour map and the profile.

Pipeline installation **(Figure 19.21)** requires major outlays for engineering design and construction. The use of profiles, found graphically, is the best way to make cost estimations for constructing ditches for laying underground pipe.

19.8 Plan–Profile Drawings

A plan–profile drawing shows an underground drainage system from manhole 1 to manhole 3 in **Figures 19.22** and **19.23.** The profile has a larger vertical scale to emphasize variations in the earth's surface and the grade of the pipe, although the vertical scale may be drawn at the same scale as the plan if desired.

The location of manhole 1 is projected to the profile orthographically, but the remaining points are not **(Figure 19.22).** Instead, transfer the distances where the contour lines cross the top view of the pipe to their respective elevations in the profile with your dividers to show the surface of the ground over the pipe.

Figure 19.23 shows the manholes, their elevations, and the bottom line of the pipe. Find the drop from manhole 1 to manhole 2 (4.40 ft) by multiplying the horizontal distance of 220.00 ft by a −2.00% grade. The pipes intersect at manhole 2 at an angle, so the flow of the drainage is disrupted at the turn. A drop of 0.20 ft (2.4 in.) across the bottom of the manhole compensates for the loss of pressure (head) through the manhole.

The true lengths of the pipes in the profile view cannot be measured when the vertical scale is different from the horizontal scale. Instead, you must use trigonometry to calculate them.

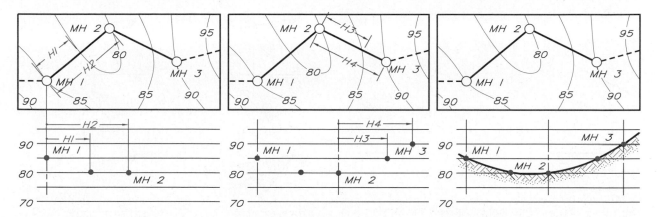

19.22 Plan–profile: Vertical section
Required Find the profile of the earth's surface over the pipeline.
Step 1 Transfer distances H1 and H2 from MH 1 in the plan to their respective elevations in the profile view.

Step 2 Measure distances H3 and H4 from manhole 2 in plan and transfer them to their respective elevations in profile. These points represent elevations of points on the earth above the pipe.

Step 3 Connect these points with a freehand line and crosshatch the drawing to represent the earth's surface. Draw centerlines to show the locations of the three manholes.

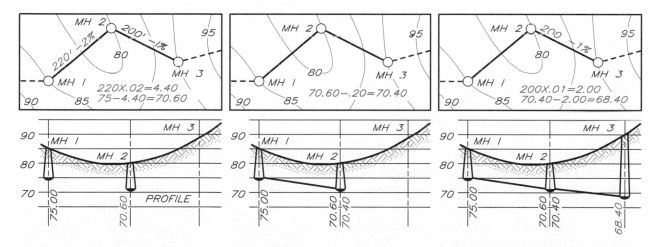

19.23 Plan–profile: Manhole location
Step 1 Multiply the horizontal distance from MH1 to MH2 by −2%. Find the elevation of the bottom of MH2 by subtracting the amount of fall from the elevation of MH1 (70.60′).

Step 2 The lower side of MH2 is 0.20 ft lower than the inlet side to compensate for loss of head (pressure) because of the turn in the pipeline. Find the elevation on the lower side (70.40′) and label it.

Step 3 Calculate the elevation of MH3 (200 ft) at a −1% grade from MH2 (68.40′). Draw the flow line of the pipeline from manhole to manhole and label the elevations at each manhole.

19.9 Edge Views of Planes

The edge view of a plane appears in a view where any line on the plane appears as a point. Recall that you can find a line as a point by projecting from its true-length view (**Figure 19.24**). You may obtain a true-length line on any plane by drawing a line parallel to one of the principal planes and projecting it to the

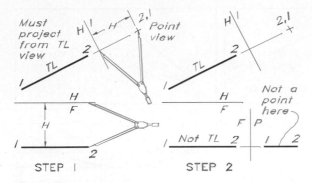

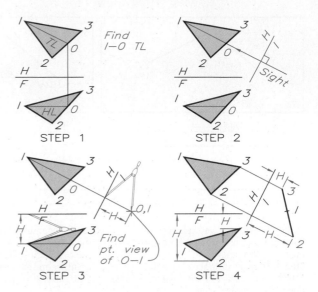

19.24 Point view of a line:
Step 1 Line 1-2 is horizontal in the front view and therefore is true length in the top view.
Step 2 Find the point view of line 1-2 by projecting parallel to its true length to the auxiliary view.

19.25 Edge view of a plane:
Step 1 Draw horizontal line 1-0 on the front view of the plane 1-2-3 and project it to the top view, where it is true length.
Step 2 Draw a line of sight parallel to the true-length line 1-0. Draw H1 perpendicular to the line of sight.
Step 3 Find the point view of 1-0 in the auxiliary view by transferring height (H) from the front view.
Step 4 Locate points 2 and 3 in the same manner to find the edge view of the plane.

adjacent view **(Figure 19.25).** You then get the edge view of the plane in an auxiliary view by finding the point view of line 1-0 on its surface.

Dihedral Angle

The angle between two planes, a **dihedral angle,** is found in a view where the line of intersection between two planes appears as a point. In this view, both planes appear as edges and the angle between them is true size. The line of intersection, line 1-2, between the two planes shown in **Figure 19.26** is true length in the top view. Project an auxiliary view from the top view to find the point view of line 1-2 and the edge views of both planes.

19.10 Planes and Lines

Piercing Points

By Projection Finding the piercing point of line 1-2 passing through the plane by projection is shown in **Figure 19.27.** Pass cutting

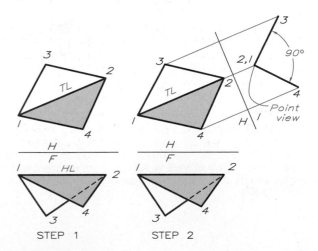

19.26 Dihedral angle:
Step 1 The line of intersection between the planes, line 1-2, is true length in the top view.
Step 2 The angle between the planes (the dihedral angle) is found in the auxiliary view where the line of intersection appears as a point and both planes are edges.

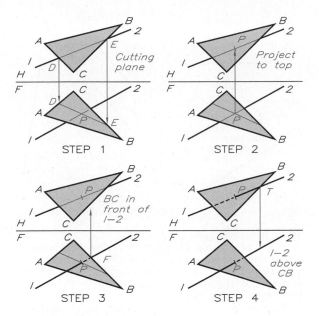

19.27 Piercing point by projection:
Step 1 Pass a vertical cutting plane through the top view of line 1-2, cutting the plane along line DE. Project line DE to the front view to locate piercing point P.
Step 2 Project point P to the top view of line 1-2.
Step 3 Determine visibility in the front view by projecting the crossing point of lines CB and 1-2 to the top view. Because CB is encountered first, it is in front of 1-2, making segment PF hidden in the front view.
Step 4 Determine visibility in the top view by projecting the crossing point of lines CB and 1-2 to the front view. Because line 1-2 is encountered first, it is above line CB, making 2P visible in the top view.

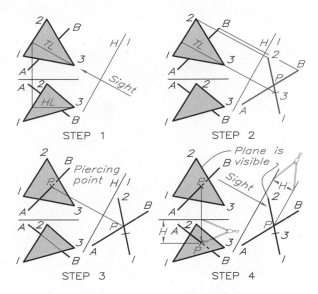

19.28 Piercing point by auxiliary view:
Step 1 Draw a horizontal line on the plane in the front view and project it to the top view where it is true length on the plane.
Step 2 Find the edge view of the plane in an auxiliary view and project AB to this view. P is the piercing point.
Step 3 Project point P to line AB in the top view. Line AP is nearest the H1 reference line, so it is the highest end of the line and is visible in the top view.
Step 4 Project P to line AB in the front view. AP is visible in the front view because line AP is in front of 1-2.

planes through the line and plane in the top view. Then project the trace of this cutting plane, line DE, to the front view to find piercing point P. Locate the top view of P and determine the visibility of the line.

By Auxiliary View You may also find the piercing point of a line and a plane by auxiliary view in which the plane is an edge (**Figure 19.28).** The location of piercing point P in step 2 is where line AB crosses the edge view of the plane. Project point P to AB in the top view from the auxiliary view, and then to the front view. To verify the location of point P in the front view, transfer dimension H from the auxiliary view with dividers.

You can easily determine visibility for the top view because you see in the auxiliary view that line AP is higher than the plane and therefore is visible in the top view. Similarly, the top view shows that endpoint A is the forward-most point and line AP therefore is visible in the front view.

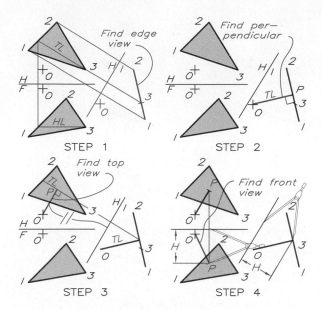

STEP 1

STEP 2

STEP 3

STEP 4

19.29 Line perpendicular to a plane:
Step 1 Find the edge view of the plane by finding the point view of a line on it in an auxiliary view. Project point O to this view, also.
Step 2 Draw line OP perpendicular to the edge view of the plane, which is true length in this view.
Step 3 Because line OP is true length in the auxiliary view, it must be parallel to the H1 reference line in the preceding view. Line OP is visible in the top view because it appears above the plane in the auxiliary view.
Step 4 Project point P to the front view and locate it by transferring height H from the auxiliary view with dividers.

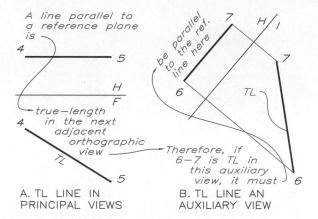

A. TL LINE IN PRINCIPAL VIEWS

B. TL LINE AN AUXILIARY VIEW

19.30 TL line principles:
In the top view of step 3 of Figure 19.29, line OP is parallel to the H1 reference line because it is true length in the auxiliary view. Here, lines 4-5 and 6-7 are examples of this principle: Both are true length in one view and parallel to the reference line in the preceding view.

Perpendicular to a Plane

A perpendicular to a plane where the plane appears as an edge. In **Figure 19.29,** a line is to be drawn from point O perpendicular to the plane. Obtain an edge view of the plane and draw the true-length perpendicular to locate piercing point P. Locate point P in the top view by drawing line OP parallel to the H1 reference line. (This principle is reviewed in **Figure 19.30.**) Line OP also is perpendicular to a true-length line in the top view of the plane. Obtain the front view of point P, along with its visibility, by projection.

Intersection Between Planes

To find the intersection between planes, find the edge view of one of the planes (Figure 19.31). Then project piercing points L and M from the auxiliary view to their respective lines, 5-6 and 4-6, in the top view. Plane 4-5-L-M is visible in the top view because sight line 1 has an unobstructed view of the 4-5-L-M portion of the plane in the auxiliary view. Plane 4-5-L-M is visible in the front view because sight line 2 has an unobstructed view of the top view of this portion of the plane.

19.11 Sloping Planes

Slope and Direction of Slope

The slope of a plane is described using the following definitions:

Angle of Slope: the angle that the plane's edge view makes with the edge of the horizontal plane.

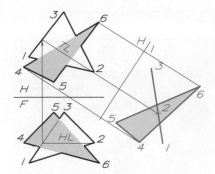
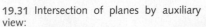

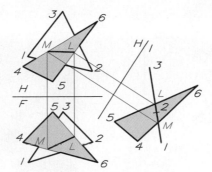

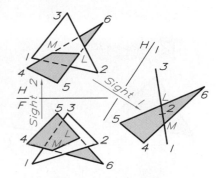

19.31 Intersection of planes by auxiliary view:
Step 1 Locate the edge view of one of the planes in an auxiliary view and project the other plane to this view.

Step 2 Piercing points L and M are found on the edge view of the plane in the auxiliary view. Project the intersection line, LM, back to the top and front views.

Step 3 The line of sight from the top view strikes L-5 first in the auxiliary view, indicating that L-5 is visible in the top view. Line 4-5 is farthest forward in the top view and is visible in the front.

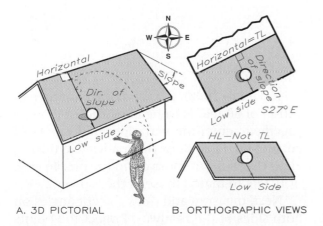

A. 3D PICTORIAL   B. ORTHOGRAPHIC VIEWS

19.32 Slope definition:
A The direction of slope of a plane is the compass bearing of the direction in which a ball on the plane will roll.
B Slope direction is measured in the top view toward the low side of the plane and perpendicular to a horizontal line on the plane.

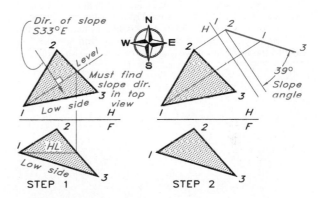

19.33 The slope and bearing of a plane:
Step 1 Slope direction is perpendicular to a true-length, level line in the top view toward the low side of the plane, or S 33° E in this case.
Step 2 Find the slope in an auxiliary view where the horizontal is an edge and the plane is an edge, or 39° in this case.

Direction of Slope: the compass bearing of a line perpendicular to a true-length line in the top view of a plane toward its low side (the direction in which a ball would roll on the plane).

As **Figure 19.32** shows, a ball would roll perpendicular to all horizontal lines on the

roof toward the low side. This direction is the slope direction and it can be measured in the top view as a compass bearing.

The steps of determining the slope and direction of slope of a plane is shown in **Figure 19.33.** Conversely, a plane can be drawn in three-dimensional space by working from

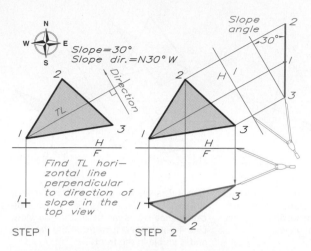

19.34 Plane from slope specifications:
Step 1 If the top view of a plane, the front view of point 1, and slope specifications are given, you can complete the front view. Draw the direction of slope in the top view and a true-length horizontal line on the plane perpendicular to the slope direction.
Step 2 Find a point view of the TL line in the auxiliary view to locate point 1. Draw the edge view of the plane through point 1 at a slope of 30°, according to the specifications. Find the front view by transferring height dimensions from the auxiliary view to the front view.

19.35 This dam was built by applying the principles of cut and fill. *(Courtesy of the Bureau of Reclamation, U.S. Department of the Interior.)*

slope and direction specifications as shown in **Figure 19.34.** Draw the direction of slope in the top view to locate a perpendicular true-length line on the plane. Find the edge view of the plane by locating point 1 and constructing a slope of 30° through it in the auxiliary view. Transfer points 3 and 2 to the front view from the auxiliary view.

Application: Cut and Fill

A level roadway through irregular terrain and the embankment for an earthen dam **(Figure 19.35)** involves the principles of cut and fill. Cut and fill is the process of cutting away high ground and filling low areas, generally of equal volumes.

In **Figure 19.36,** a level roadway at an elevation of 60 feet is to be constructed along a specified centerline with specified angles of cut and fill. First, draw the roadway in the top view. Use contour intervals in the profile view of 10 feet to match those in the top view.

Next, measure and draw the cut angles on both sides of the roadway. Project the points where the cut angles cross each elevation line to the respective contour lines in the plan view to find the limits of cut.

Then, measure and draw the fill angles in the profile view. Project the points where the fill angles cross each elevation line to the respective contour lines in the plan view to find the limits of fill. Finally, draw new contour lines inside the areas of cut parallel to the centerline.

Some of the terminology associated with the design of a dam are **(1) crest,** or the top of the dam; **(2) water level;** and **(3) freeboard,** or the height of the crest above the water level. These terms are illustrated in **Figure 19.37.**

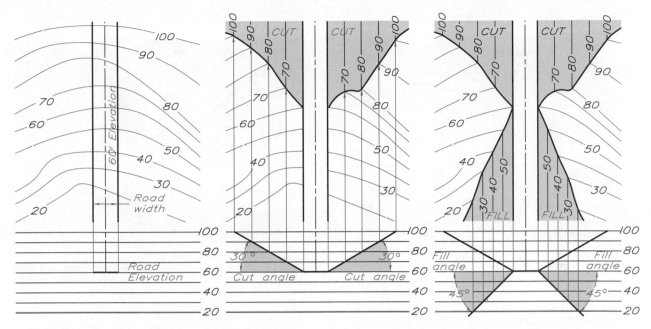

19.36 Cut and fill: Level roadway.
Step 1 Draw and label a series of elevation planes in the front view at the same scale as the contour map. Draw the width and elevation (60 ft. in this case) of the roadway in the top and front views.

Step 2 Draw the cut angles on the higher sides of the road in the front view. Project the points of intersection between the cut angles and the contour planes in the front view to their respective contour lines in the top view to determine the limits of cut.

Step 3 Draw the fill angles on the lower sides of the road in the front. Project the points in the front where the fill angles cross the contour lines to these respective contour lines in the top view. Draw new contours parallel to the centerline in the cut-and-fill areas.

Strike and Dip

Strike and **dip** are terms used in geology and mining engineering to describe the location of strata of ore under the surface of the earth:

Strike: the compass bearings (two are possible) of a level line in the top view of a plane.

Dip: the angle that the edge view of a plane makes with the horizontal and its general compass direction, such as 30° NW or 30° SW.

The dip angle lies in the primary auxiliary view projected from the top view. The dip direction is perpendicular to the strike and toward its low side.

Figure 19.38 demonstrates how to find the strike and dip of a plane. Here, the true-length line in the top view of the plane has a strike of

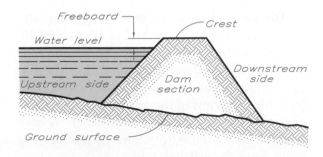

19.37 These terms and symbols are used in the design of a dam.

N 66° W or S 66° E. The dip angle appears in an auxiliary view projected from the top view that shows the horizontal (H1) and the sloping plane as edges, 45° SW in this example.

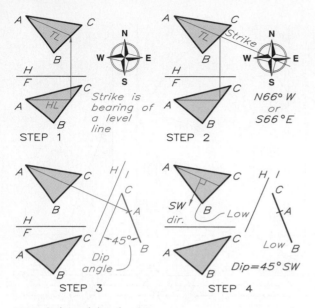

STEP 1

STEP 2

STEP 3

STEP 4

19.38 Strike and dip of a plane:
Step 1 Draw a horizontal line on the plane in the front view and project it to the top view, where it is true length.
Step 2 Strike is the compass direction of a level line on the plane in the top view, either N 66° W or S 66° E.
Step 3 Find the edge view of the plane in the auxiliary view. The dip angle of 45° is the angle between the H1 reference line and the edge view of the plane.
Step 4 The general compass direction of dip is toward the low side and perpendicular to a strike in the top view, SW in this case. Dip and direction is written as 45° SW.

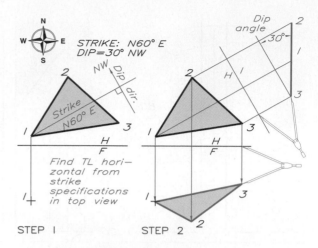

STEP 1

STEP 2

19.39 Strike and dip specifications:
Step 1 Draw the strike in the top view of the plane as a true-length horizontal line. Draw the direction of dip perpendicular to the strike toward the NW as specified.
Step 2 Find the point view of strike in the auxiliary view to locate point 1 where the edge view of the plane passes through it at a 30° dip, as specified. Complete the front view by transferring height (H) dimensions from the auxiliary to the front view.

You can construct a plane from strike and dip specifications as shown in **Figure 19.39.** First, draw the strike as a true-length horizontal line on the plane's top view and the dip direction perpendicular to the strike. Then find the edge view of the plane in the auxiliary view through point 1 at a dip of 30°. Locate points 2 and 3 in the front view by transferring them from the auxiliary view.

19.12 Ore-Vein Applications

The principles of descriptive geometry can be applied to find the distance from a point to a plane. Techniques of finding such distances often are used to solve mining and geological problems. For example, test wells are drilled into coal seams to learn more about them **(Figure 19.40).**

Application: Underground Ore Veins Geologists and mining engineers usually assume that strata of ore veins have upper and lower planes that are parallel. In **Figure 19.41** point O is on the upper surface of the earth and plane 1-2-3 is an underground ore vein. Point 4 is on the lower plane of the vein.

Find the edge view of plane 1-2-3 by projecting from the top view and then draw the lower plane through point 4 parallel to the upper plane. Draw the horizontal distance from point O to the plane parallel to the H1

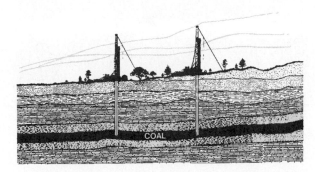

19.40 Test wells are drilled into coal zones to determine the elevations of coal seams that may contribute to the exploration for gas. *(Courtesy of Texas Eastern Products Pipeline Co., LLC.)*

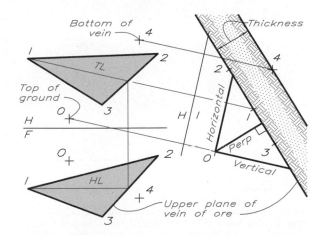

19.41 Distances to an ore vein:
To find the vertical, horizontal, and perpendicular distances from a point to an ore vein, project an auxiliary view from the top view, where the vein appears as an edge. The thickness of an ore vein is perpendicular to the upper and lower planes of the vein.

reference line and the vertical distance perpendicular to line H1. The shortest distance is perpendicular to the ore vein. These three lines from point O are true length in the auxiliary view where the ore vein appears as an edge.

Application: Ore-Vein Outcrop The same assumption is made in **Figure 19.42** that underground ore veins are defined by parallel planes that will outcrop on the earth's surface if they

are inclined (nonhorizontal). When ore veins outcrop, open-pit mining can be employed to reduce costs. The outcrop of an ore vein can be found on the contour map by using the

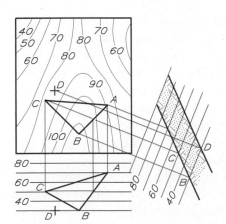

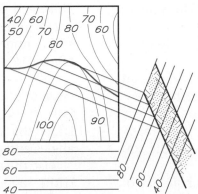

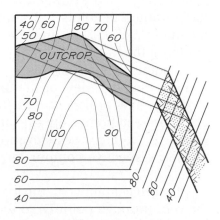

19.42 Ore vein outcrop:
Step 1 Use points A, B, and C on the upper surface of the ore vein to find its edge view by projecting an auxiliary from the top view. Draw the lower surface of the vein parallel to the upper plane through point D.

Step 2 Project points of intersection between the upper plane of the vein and the contour lines in the auxiliary view to their respective contours in the top view to find a line of the outcrop.

Step 3 Project points from the lower plane in the auxiliary view to their respective contours in the top view to find the second line of outcrop. Crosshatch the area between the lines to depict the outcrop of the vein.

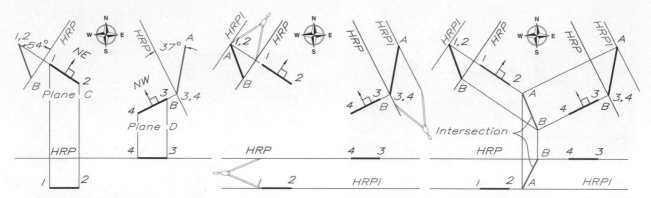

19.43 Intersection of planes: Strike and dip method.

Step 1 Lines 1-2 and 3-4 are strike lines, true length in the top view. Use reference planes, HRP, to find the point view of each strike line by auxiliary views. Find the edge views by drawing the dip angles with the HRP line through the point views.

Step 2 Draw a supplementary horizontal plane, HRP1, at a convenient location in the front view. This plane, shown in both auxiliary views, is located H distance from HRP. The HRP1 cuts through each edge in both auxiliary views, locating A and B.

Step 3 Project points A from both auxiliary views on HRP1 to the top view of their intersection at A. Project points B on HRP to their intersection in the top view. Project points A and B to their respective planes in the front view. AB is the line of intersection.

data found from the exploratory drillings. Use the elevations of points A, B, and C located on the upper plane of the ore vein, and point D found on the lower plane of the vein to draw these points in the front view.

Find the edge view of the ore vein in an auxiliary view projected from the top view. Then project points on the upper surface where the vein crosses elevation lines back to their respective contour lines in the top view.

Also project points on the lower surface of the vein (through point D) to the top view. If the ore vein extends uniformly at its angle of inclination to the earth's surface, the area between these two lines will be the outcrop of the vein. Open-pit mining can be done in this area in lieu of tunneling underground.

19.13 Intersections Between Planes

Strike and Dip Method The method of locating the intersection of two planes located with strike and dip specifications is shown in **Figure 19.43.** The given strike lines are true-length level lines in the top view, so the edge view of the planes appear in the auxiliary views where the strikes appear as points. Draw the edge views using the given dip angles and directions.

Use the additional horizontal datum plane HRP1 to find lines on each plane at equal elevations that intersect when projected to the top view from their auxiliary views. Connect points A and B as the line of intersection between the two planes in the top view and project it to the front view to establish line AB in three dimensions.

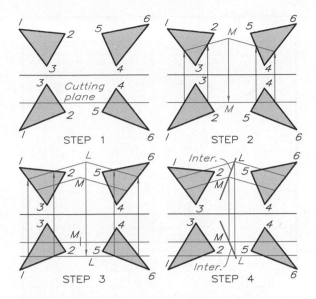

19.44 Intersection of planes: Cutting plane method.
Step 1 Draw a cutting plane that passes through both planes in the front view in any convenient direction.
Step 2 Project the intersections of the cutting plane to the top views of the planes. Find intersection point M in the top view and project it to the front view.
Step 3 Draw a second cutting plane through the front view of the planes and project it to the top view. Find point L in the top view and project it to the front view.
Step 4 Connect L and M in the top and front views to represent the line of intersection of the extended planes.

Cutting-Plane Method In **Figure 19.44,** the top and front views of two planes are given. It is required that the line of intersection between the planes be found if they were extended. (Infinite planes will intersect unless they are parallel planes.) Draw cutting planes through both planes in both views at any angle and project them to the top view. Find points L and M in the top view to establish the line of intersection. Find the front view of line LM, the line of intersection, by projecting its endpoints from the top view to their respective cutting planes in the front view.

Use size A sheets for the following problems (lay out the solutions using instruments). Each square on the grid is equal to 0.20 inch or 5 mm. Use either grid or plain paper. Label all reference planes and points with 1/8-inch or 3 mm letters or numbers with guidelines.

1. (Sheet 1) True-length lines.
(A–D) Find the true-length views of the lines by auxiliary view as indicated by the given lines of sight. *Alternative method:* Find the true-length of the lines by using the Pythagorean theorem.

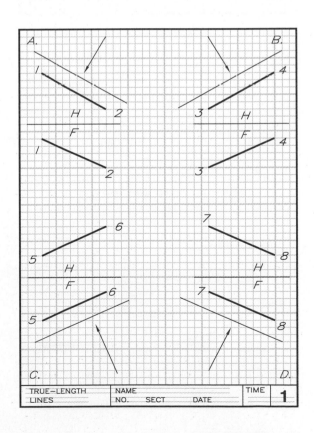

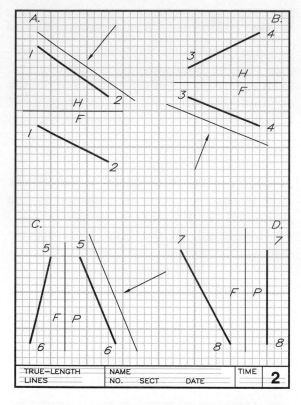

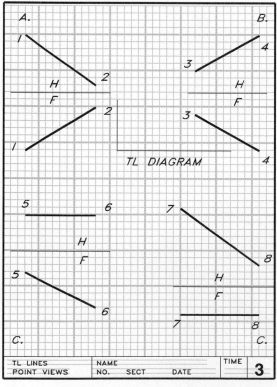

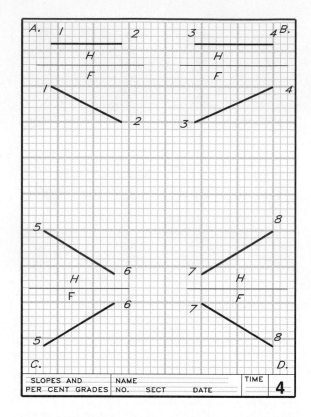

2. **(Sheet 2)** True-length lines.
(A–D) Find the true-length views of the lines by auxiliary view as indicated by the lines of sight. *Alternative method:* Find the true length of the lines by the Pythagorean theorem.

3. **(Sheet 3)** True-length diagram.
(A–B) Find the true length of the lines by using a true-length diagram.
(C–D) Find the point views of the lines.

4. **(Sheet 4)** Sloping lines.
(A–D) Find the slope angle, tangent of the slope angle, and the percent grade of the four lines.

5. **(Sheet 5)** Edge views of planes.
(A–B) Find the edge views of the planes.

6. **(Sheet 6)** Intersections: lines and planes.
(A) Find the angle between the planes.
(B) By projection, find the point of intersection between the line and plane, and show visibility.

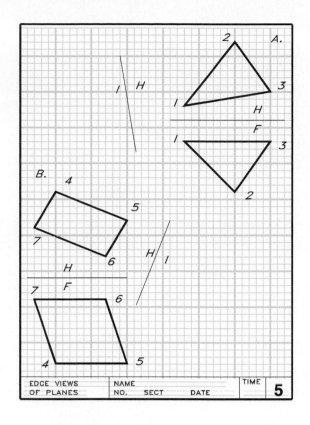

| EDGE VIEWS OF PLANES | NAME | | | TIME | **5** |
| | NO. | SECT | DATE | | |

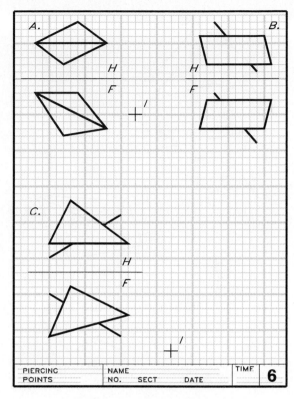

| PIERCING POINTS | NAME | | | TIME | **6** |
| | NO. | SECT | DATE | | |

(C) By the auxiliary view method, find the point of intersection between the line and plane and show visibility.

7. (Sheet 7) Perpendiculars to planes.
(A) Construct a 1 inch line perpendicular from point O on the plane and show it in all views.
(B) Draw a line perpendicular to the plane from O, find the piercing point, and show visibility.

8. (Sheet 8) Dihedral angles.
By using auxiliary views, find:
(A) The line of intersection between the planes.
(B) The angle between the planes.

9. (Sheet 9) Slope and direction of slope.
(A–B) Find the direction of slope and the slope angle of the planes. *Alternative solution:* Find the strike and dip of the planes.

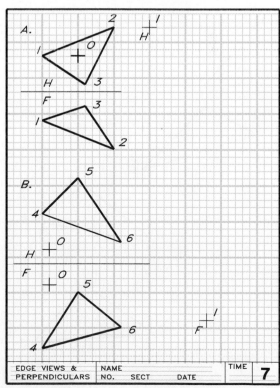

| EDGE VIEWS & PERPENDICULARS | NAME | | | TIME | **7** |
| | NO. | SECT | DATE | | |

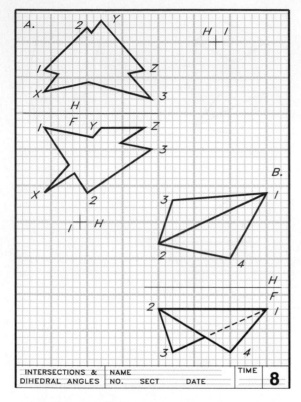

A.

B.

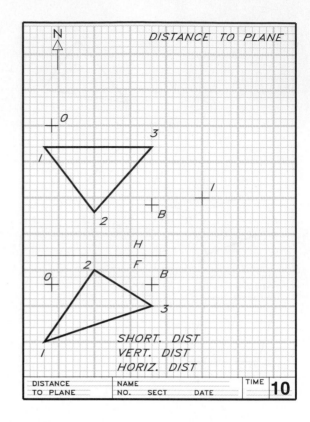

DISTANCE TO PLANE

SHORT. DIST
VERT. DIST
HORIZ. DIST

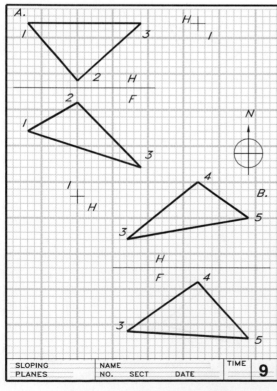

A.

B.

10. (Sheet 10) Distances to a plane.
Find the shortest distance, the horizontal distance, and the vertical distance from point O to the underground ore vein represented by plane 1-2-3. Point B is on the lower plane of the vein. Find the thickness of the vein. Label your solutions. Scale: 1 inch = 10 feet.

11. (Sheet 11) Intersecting planes.
(A) Find the line of intersection between the two planes by the cutting plane method.
(B) Find the line of intersection between the two planes indicated by strike lines 1-2 and 3-4. The plane with strike line 1-2 has a dip of 30°, and the plane with strike line 3-4 has a dip of 55°.

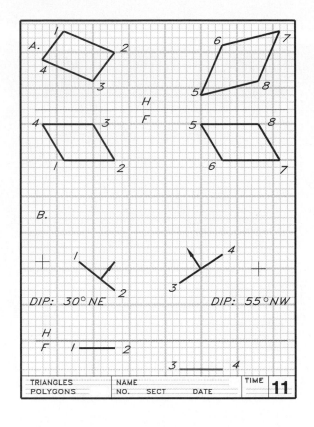

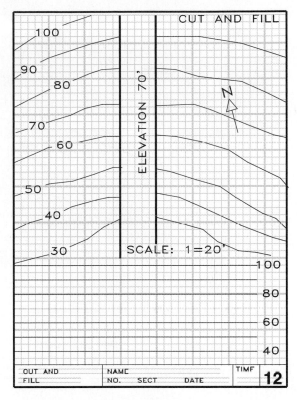

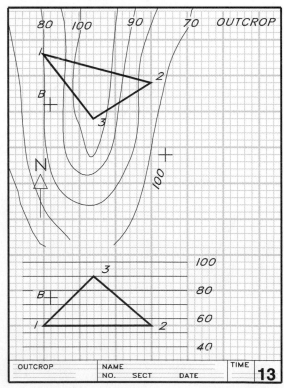

12. (Sheet 12) Cut and Fill.
Find the limits of cut and fill in the plan view of the roadway. Use a cut angle of 35° and fill angle of 40°.

13. (Sheet 13) Outcrop.
Find the outcrop of the ore vein represented by plane 1-2-3 on its upper surface. Point B is on the lower surface. Scale: 1 inch = 20 feet.

14. (Sheet 14) Plan–profile.
Complete the plan–profile drawing of the drainage system from manhole 1 through manhole 2 to manhole 3, using the grades indicated. Allow a drop of 0.20 ft across each manhole to compensate for loss of pressure.

15. (Sheet 15) Contour map.
Draw the contour map with its contour lines. Give the lengths of the sides, their compass

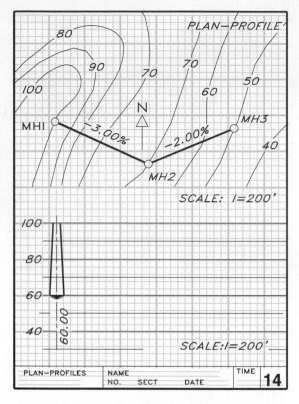

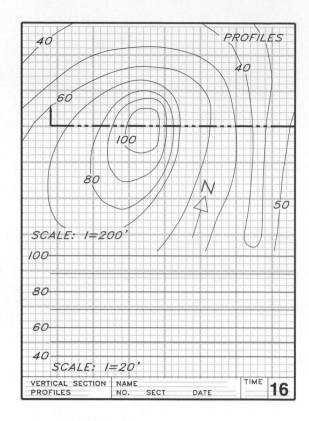

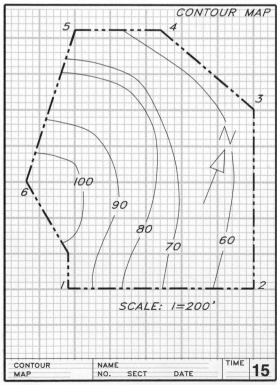

directions, interior angles, scale, and north arrow.

16. (Sheet 16) Vertical section.
Draw the contour map and construct the profile (vertical section) as indicated by the cutting plane line in the plan view. Note that the profile scale is different from the plan scale.

Design Problems
Lay out the necessary orthographic views in order to solve the following problems on size A or size B sheets. Add notes and details to explain your solution. (These problems have been adapted from applications encountered at the Boeing Company.)

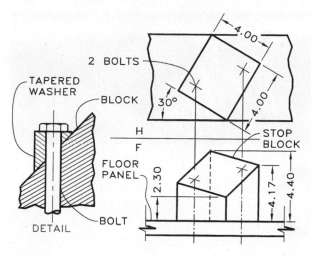

Design 1

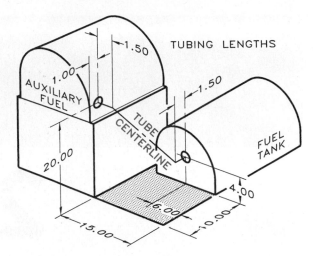

Design 3

Design 1: Bolt Support

A. Use the dimensions given in Design 1 and lay out the views necesary to find the angle the tapered circular washer makes with the inclined block that supports an aircraft seat.

B. Make a working diagram with dimensions to describe the tapered circular washer.

Design 3: Fuel-Tank Tube

Lay out the orthographic views on a size-B sheet that are necessary to find the length of the tube centerline from the stubs given at the fuel and the auxiliary fuel tank. Determine the angles between the stubs at both ends of the tube.

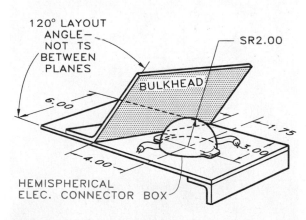

Design 2

Design 2: Bulkhead clearance

Lay out the orthographic views of Design 2 on a size-B sheet that are necessary to find the clearance between the bulkhead and the electrical connector box. If necessary, relocate the connector so it will have a 0.50 clearance with the bulkhead.

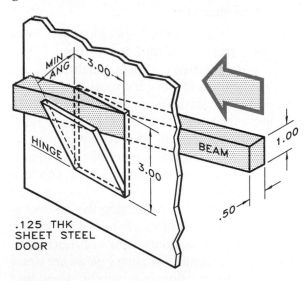

Design 4

Design 4: Door opening.

The beam must pass through an opening installed in an airplane. The gauge of the door and structure is 0.125 inch. Determine the minimum angle of the door opening to permit the beam to pass through it.

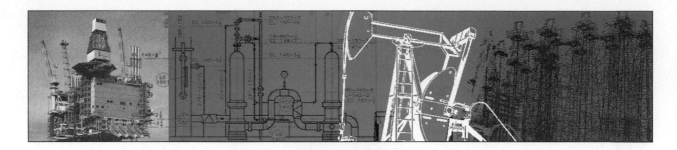

20

Successive Auxiliary Views

20.1 Introduction

A detailed drawing and specifications for a design cannot be completed without determining its geometry, which usually requires the application of descriptive geometry. The structural supports for the lunar lander shown in **Figure 20.1** are examples of complex spatial geometry problems in which lengths must be determined, angles between lines and planes calculated, and three-dimensional connectors designed.

The process of determining the 3D geometry of a design requires the use of secondary and successive auxiliary views of descriptive geometry. Secondary auxiliary views are views projected from primary auxiliary views, and successive auxiliary views are views projected from secondary auxiliary views.

20.1 This structural support of this lunar lander was designed and fabricated through the application of descriptive geometry. (Courtesy of NASA.)

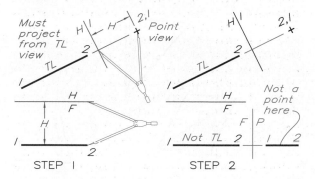

STEP 1   STEP 2

20.2 To find the point view of a line, project an auxiliary view from the true-length view of the line.

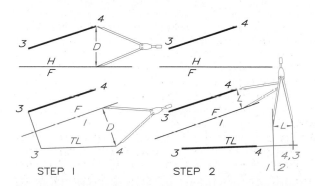

STEP 1   STEP 2

20.3 A point view of an oblique line:
Step 1 Draw a line of sight perpendicular to one of the views, the front view in this case. Line 3-4 is found true length in an auxiliary view projected perpendicularly from the front view.
Step 2 Draw a secondary reference line, 1-2, perpendicular to the true-length view of line 3-4. Find the point view by transferring dimension L from the front view to the secondary auxiliary view.

20.2 Point View of a Line

Recall that when a line appears true length you can find its point view in a primary auxiliary view projected parallel from it. In **Figure 20.2,** line 1-2 is true length in the top view because it is horizontal in the front view. To find its point view in the primary auxiliary view, first construct reference line H1 perpendicular to

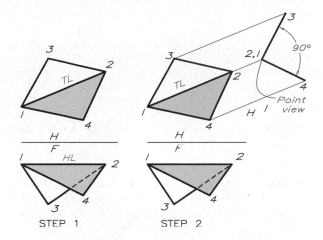

STEP 1   STEP 2

20.4 The angle between planes (the dihedral angle) appears in the view where their line of intersection projects as a point. The line of intersection, 1-2, is true length in the top view, so it can be found as a point in a view projected from the top view.

the true-length line. Transfer the height dimension, H, to the auxiliary view to locate the point view of 1-2.

Line 3-4 in **Figure 20.3** is not true length in either view. Finding the line's true length by a primary auxiliary view enables you to find its point view. To obtain a true-length view of line 3-4, project an auxiliary view from the front view (or from the top view). Projecting parallel from the true-length primary auxiliary view to a secondary auxiliary view gives the point view of line 3-4. Label the point view of 4-3 because you see point 4 first in the secondary auxiliary view. Label the reference line between the primary and secondary planes 1-2 to represent the primary (1) and secondary (2) planes.

20.3 Dihedral Angles

Recall that the angle between two planes is called a **dihedral angle** and can be found in a view where the line of intersection appears as

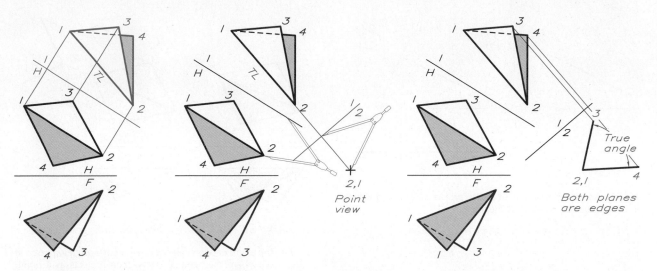

20.5 Angle between two planes:
Step 1 The angle between two planes is found in a view where the line of intersection (1-2) appears as a point. Find the TL view of the intersection in an auxiliary view.

Step 2 Obtain the point view of the line of intersection in the secondary auxiliary view by projecting parallel to the true-length view of line 1-2 in the primary auxiliary view.

Step 3 Complete the edge views of the planes in the secondary auxiliary view by locating points 3 and 4. Measure the angle between the planes (the dihedral angle) in this view.

a point. The line of intersection lies on both planes, so both appear as edges when the point view of the line of intersection is found.

The planes shown in **Figure 20.4** represent a special case because their line of intersection, line 1-2, is true length in the top view. This condition permits you to find the line's point view in a primary auxiliary view and measure the true angle between the planes.

Figure 20.5 presents a more typical case. Here, the line of intersection between the two planes is not true length in either view. The line of intersection, line 1-2, is true length in a primary auxiliary view, and the point view of the line appears in the secondary auxiliary view, where you measure the dihedral angle.

This principle was applied to determine the angles between structural planes of the

Gimbal Rig shown in **Figure 20.6.** That allowed the corner braces to be designed and the structure to be assembled correctly.

20.4 True Size of a Plane

A plane can be found true size in a view projected perpendicularly from an edge view of a plane. The front view of plane 1-2-3 in **Figure 20.7** appears as an edge in the front view as a special case. The plane's true size is in a primary auxiliary view projected perpendicularly from the edge view.

Figure 20.8 depicts a general case in which you can find the true-size view of plane 1-2-3 by finding the edge view of the plane and constructing a secondary auxiliary view projected perpendicularly from the edge view to find the plane true size.

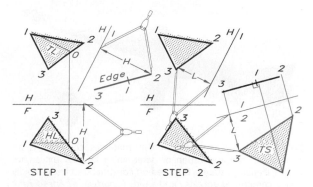

20.8 The true size of a plane: General case.
Step 1 Find the edge view of plane 1-2-3 by obtaining the point view of true-length 1-O in an auxiliary view.
Step 2 Find a true-size view by projecting a secondary auxiliary view perpendicularly from the edge view of the plane found in the primary auxiliary view.

20.6 This Gimbal Rig, used to train astronauts, illustrates the need to determine dihedral angles in order to design connectors and structural members.*(Courtesy of NASA.)*

20.9 NASA's engineering team encountered numerous problems of geometry that were solved in the design of this Mars Exploration Rover. *(Courtesy of NASA.)*

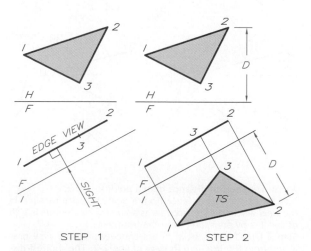

20.7 The true size of a plane (special case):
Step 1 Because plane 1-2-3 appears as an edge in the front view, it is a special case. Draw the sight line perpendicular to its edge and the F1 parallel to the edge.
Step 2 Find the true size of plane 1-2-3 in the primary auxiliary view by locating the vertex points with the depth D dimensions.

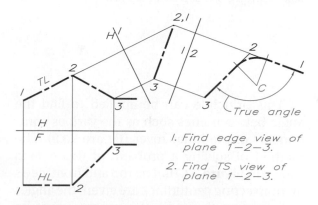

1. Find edge view of plane 1–2–3.

2. Find TS view of plane 1–2–3.

20.10 The angle between two lines is obtained by finding a true-size view of the plane formed by lines 1-2 and 2-3.

20.11 The support frame of this materials conveyor is an example of the application of determining lengths and angles during its design stages. It is used on construction sites to rapidly move cement, aggregate, and sand. *(Courtesy of Speed King Manufacturing Co.)*

auxiliary view and true size in the secondary view, where it can be measured and the radius of curvature drawn. The support frame of the materials conveyor (**Figure 20.11**) is an example of a design involving planes and lines in geometric combinations that was designed with applications of auxiliary views.

20.5 Shortest Distance from a Point to a Line: Line Method

The shortest distance from a point to a line can be measured in the view where the line appears as a point. The shortest distance from

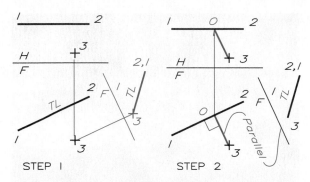

20.12 The shortest distance from a point to a line:
Step 1 The shortest distance from a point to a line is true length where the line (1-2) appears as a point. The true-length view of the connecting line appears in the primary auxiliary view.
Step 2 When the connecting line is projected back to the front view, it must be parallel to the F1 reference line in the front view. Project line 3-O back to the top view.

This principle can be applied to find the angle between lines such as the various components of the lunar rover (**Figure 20.9**). The method of solving a problem of this type is shown in **Figure 20.10.** The top and front views of intersecting centerlines are given, the angles of bend and the radii of curvature must be found. Angle 1-2-3 is an edge in the primary

20.13 The shortest distance from a point to a line:
Step 1 The shortest distance from a point to a line is found in the view where the line appears as a point. Find the true length of line 1-2 by projecting from the front view.
Step 2 Line 1-2 is a point in a secondary auxiliary view projected from the true-length view of line 1-2. The shortest distance to it is true length in this view.
Step 3 Since 3-O is true length in the secondary auxiliary view, it is parallel to the 1-2 reference line in the primary auxiliary view and perpendicular to the line.
Step 4 Find the front and top views of 3-O by projecting from the primary auxiliary view in sequence.

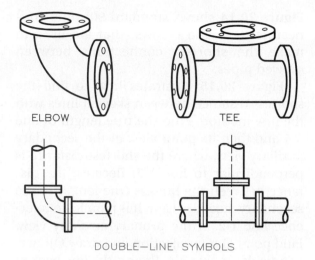

DOUBLE—LINE SYMBOLS

20.14 The shortest distance between two lines, the perpendicular distance, is the most economical connector between them. Perpendicularity also permits the use of standard fittings, 90° tees and elbows.

point 3 to line 1-2 that appears in a primary auxiliary view in **Figure 20.12** (step 1) is a special case. The distance from point 3 to the line is true length in the auxiliary view where the

line is a point, so it is parallel to reference line Fl in the front view.

Figure 20.13 shows how to solve a general-case problem of this type, where neither line appears true length in the given views. Line 1-2 is true length in the primary auxiliary projected perpendicularly from the front view. The point view of line 1-2 lies in the secondary auxiliary view, where the distance from point 3 is true length. Because line O-3 is true length in this view, it will be parallel to reference line 1-2 in the preceding view, the primary auxiliary view. It is also perpendicular to the true-length view of line 1-2 in the primary auxiliary view.

20.6 Shortest Distance Between Skewed Lines: Line Method

Randomly positioned (nonparallel) lines are called **skewed** lines. The shortest distance between two skewed lines is found in the view where one of the lines appears as a point.

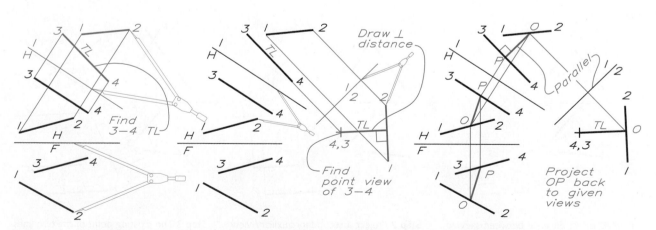

20.15 Shortest distance between skewed lines: Line method
Step 1 The shortest distance between skewed lines appears where one of the lines is a point. Find the 3-4 TL by projecting from the top view.

Step 2 Find the point view of line 3-4 in a secondary auxiliary view projected from the true-length view of line 3-4. The shortest distance between the lines is perpendicular to line 1-2.

Step 3 The shortest distance is true length in the secondary auxiliary view, so it must be parallel to the 1-2 reference line in the preceding view. Project line OP back to the given views.

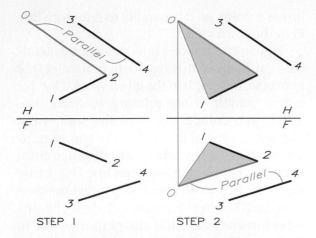

STEP 1   STEP 2

20.16 A plane through a line parallel to another line:
Step 1 Draw line O-2 parallel to line 3-4 to a convenient length.
Step 2 Draw the front view of line O-2 parallel to the front view of line 3-4. Find the length of line O-2 in the front view by projecting from the top view of O. Plane 1-2-O is parallel to line 3-4 because it contains a line that is parallel to line 3-4.

The shortest distance between two lines is a line perpendicular to both lines. The location of the shortest distance between lines is both functional and economical.

Figure 20.14 shows standard 90° pipe connectors (tees and elbows) that are used to make the shortest connections between skewed pipes.

Figure 20.15 illustrates how to find the shortest distance between skewed lines with the line method. Find the true length of line 3-4 and then its point view in the secondary auxiliary view, where the shortest distance is perpendicular to line 1-2. Because the distance between the lines is true length in the secondary auxiliary view, it is parallel to reference line 1-2 in the primary auxiliary view. Find point O by projection and draw OP perpendicular to line 3-4. Project the line back to the given principal views.

20.7 Shortest Distance Between Skewed Lines: Plane Method

You may also determine the shortest distance between skewed lines by the plane method, which requires construction of a plane through one of the lines parallel to the other (**Figure 20.16**). The top and front views of line O-2 are

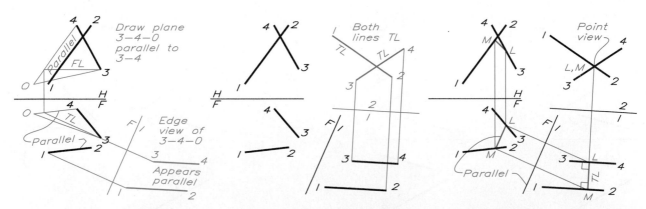

20.17 Shortest distance between skewed lines: Plane method.
Step 1 Construct a plane through 3-4 parallel to 1-2. Find plane 3-4-O as an edge by projecting it from the front view. The lines will appear parallel.

Step 2 Project a secondary auxiliary view perpendicularly from parallel lines 1-2 and 3-4 in the primary auxiliary view to find them true length. The shortest distance between them will be perpendicular to both lines in the primary auxiliary view.

Step 3 The crossing point of the two lines is the point view of the perpendicular distance (LM) between them. Project LM to the primary auxiliary view, where it is true length, and parallel to F1, and back to the given views.

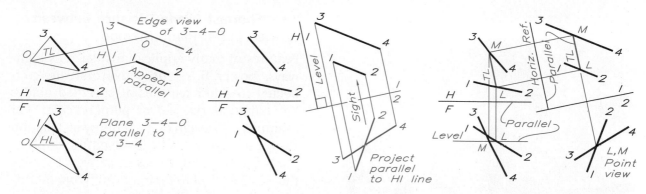

20.18 Shortest level distance between skewed lines: Plane method.

Step 1 Construct plane O-3-4 parallel to 1-2 by drawing O-4 parallel to 1-2. Find the edge view of plane O-3-4 by projecting off the top view; the lines appear parallel. The horizontal plane is an edge when projecting from the top view.

Step 2 Infinitely many horizontal (level) lines may be drawn parallel to reference line H1 in the auxiliary view, but the shortest one appears true length. Construct the secondary auxiliary view by projecting parallel to H1 to find the point view of the shortest level line.

Step 3 The crossing point of the lines in the secondary auxiliary view is the point view of the level connector, LM. Project LM back to the given views. LM is parallel to the horizontal reference plane in the front view, verifying that it is a level line.

parallel to their respective views of line 3-4. Therefore plane 1-2-O is parallel to line 3-4. Both lines will appear parallel in an auxiliary view in which plane 1-2-O appears as an edge.

Figure 20.17 demonstrates this principle. First construct plane 3-4-O. When its edge view is found in a primary auxiliary view, the lines appear parallel. To find the secondary auxiliary view where both lines are true length and cross, project a secondary auxiliary view perpendicularly from these parallel lines. The crossing point is the point view of the shortest distance between the lines. That distance is true length and perpendicular to both lines when projected to the primary auxiliary view as line LM. Project line LM back to the given top and front views to complete the solution.

20.8 Shortest Level Distance Between Skewed Lines: Plane Method

The shortest level (horizontal) distance between two skewed lines can be found by the plane method but not by the line method. In **Figure 20.18** plane 3-4-O is constructed parallel to line 1-2, and its edge view is found in the primary auxiliary view. Lines 1-2 and 3-4 appear parallel in this view, and the horizontal reference plane, H1, appears as an edge.

A line of sight parallel to H1 is used, and the secondary reference line, 1-2, is drawn perpendicular to H1. The crossing point of the lines in the secondary auxiliary view locates the point view of the shortest horizontal distance between the lines. This line, LM, is

20.19 This massive complex with numerous conveyors involved many applications of skewed lines at specified percent grades. *(Courtesy of FMC Corporation)*

20.9 Shortest Grade Distance Between Skewed Lines: Plane Method

Features of many applications (such as highways, power lines, or conveyors) are connected to other features at specified grades other than horizontal or perpendicular. For example, the design of the refinery installation shown in **Figure 20.19** involved the application of slopes and grades of conveyor chutes that were critical to their optimum operation.

If you need to find a 40% grade connector between two lines **(Figure 20.20),** use the plane method. To obtain an edge view of the horizontal plane from which the 40% grade is constructed, you must project the primary auxiliary view from the top view. Construct a view in which the lines appear parallel and draw a 40% grade line from the edge view of the horizontal (H1) by laying off rise and run units of 4 and 10, respectively. The grade line may be constructed in two directions from the H1 reference line, but the shortest distance is the direction most nearly perpendicular to both lines.

true length in the primary auxiliary view and parallel to the H1 plane. Line LM is projected back to the given views. As a check on construction, LM must be parallel to the HF line.

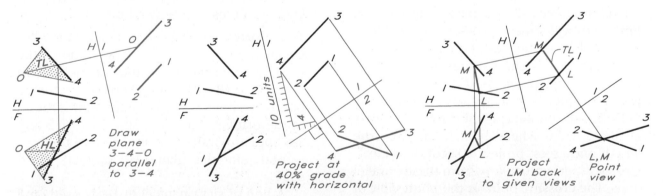

20.20 Grade distance between skewed lines:
Step 1 To find a level line or a line on a grade between two skewed lines, the primary auxiliary must be projected from the top view. Construct plane 3-4-O parallel to 1-2. Find the edge view of the plane; the lines appear parallel.

Step 2 Construct a 40% grade line from the edge view of the H1 reference line in the primary auxiliary view that is most nearly perpendicular to the lines. Project the secondary auxiliary view parallel to the grade line. The shortest grade distance appears true length in the primary auxiliary.

Step 3 The point of crossing of the two lines in the secondary auxiliary view establishes the point view of the 40% grade line, LM. Project LM back to the primary auxiliary view to find its true length. Project LM back to the top and front views to complete the problem.

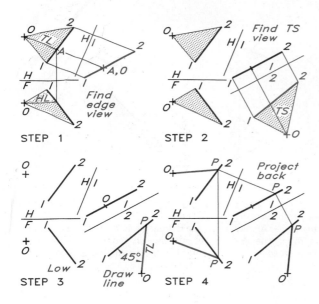

STEP 1   STEP 2

STEP 3   STEP 4

20.21 Line through a point at a given angle to a line:

Step 1 Connect O to each end of the line to form plane 1-2-O in both views. Draw a horizontal line in the front view of the plane and project it to the top view where it is TL. Find the point view of AO and the edge view of the plane.

Step 2 Find the true size of plane 1-2-O in the primary auxiliary view by projecting perpendicularly from its edge view. Omit the outline of the plane in this view and only show line 1-2 and point O.

Step 3 Construct line OP at an angle of 45° with line 1-2. If you draw the angle toward point 2 (the low end), the line slopes downward; if toward point 1. it slopes upward.

Step 4 Project line OP back to the previous views in sequence to complete the solution.

Project the secondary auxiliary view parallel to this 40% grade line to find the crossing point of the lines to locate the shortest connector, LM. Project line LM back to all views; it is true length in the primary auxiliary view, where the given lines appear parallel.

The shortest distances between skewed lines—perpendicular, horizontal, and perpendicular—are true length in the view where the lines appear parallel. The plane method is the general-case method that can be used to find the shortest connector between two lines.

20.10 Angular Distance to a Line

Standard connectors used to connect pipes and structural members are available in standard angles of 90° and 45° (**Figure 20.14**). Specifying these standard connectors in a design is far more economical than calling for the fabrication of specially made connectors.

In **Figure 20.21,** a line from point O that makes an angle of 45° with line 1-2 is to be found. Connect point O with the line's endpoints, 1 and 2, to find plane 1-2-O in the top and front views, and create its edge view in a primary auxiliary view. Find the true-size view of plane 1-2-O by projecting perpendicularly from its edge view. Measure the angle of the line from point O in this view, where the plane of the line and point is true size.

Draw the 45° connector from point O toward point 2 (the low point) if it slopes downward or toward point 1 if it slopes upward. Determine the upper and lower ends of line 1-2 by referring to the front view, where the height is easily seen. Project the 45° line, OP, back to the given views.

20.11 Angle Between a Line and a Plane: Plane Method

The angle between a line and a plane can be measured in the view where the plane appears as an edge and the line appears true length. In **Figure 20.22,** the edge view of plane 1-2-3 lies in a primary auxiliary view

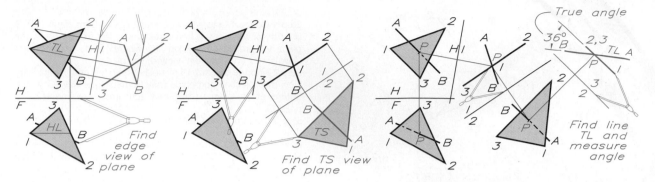

20.22 Angle between a line and plane: Plane method.

Step 1 The angle between a line and a plane is found in the view where the plane is an edge and the line is TL. Find the plane as an edge by projecting it from the top view.

Step 2 To find the plane's true size, project a secondary auxiliary view perpendicularly from the edge view of the plane. A view projected in any direction from a true-size plane will show the plane as an edge.

Step 3 Project a third successive auxiliary perpendicularly from AB. The line is TL, the plane is an edge, and the angle is TS. Project AB back in sequence to the given views; find the piercing points and visibility in the views.

projected from the top view and is true size (step 2) where the line appears foreshortened. Line AB is true length in a third successive auxiliary view projected perpendicularly from the secondary auxiliary view of line AB. The line appears true length and the plane appears as an edge in the third successive auxiliary view. Therefore, the angle between the line and plane can be measured here.

20.12 Angle Between a Line and a Plane: Line Method

An alternative method (not illustrated here) of finding the angle between a line and a plane is the **line method.** In that method, the line, rather than the plane, is the primary geometric element that is projected. The line is found as true length in a primary auxiliary view; it is

found as a point in the secondary auxiliary view. The plane is found as an edge in the third successive auxiliary view.

Because this last view was projected from a point view of the line, the line appears true length in this view where the plane appears as an edge. Project the piercing point back to the secondary, primary, top, and front views in sequence to complete the problem.

Problems

Use size A sheets for the following problems and lay out your solutions with instruments on grid or plain paper. Each square on the grid is equal to 0.20 inch (5 mm). Label all reference planes and points in each problem with 1/8-inch letters or numbers, using guidelines.

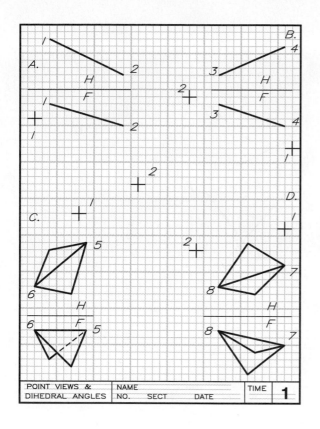

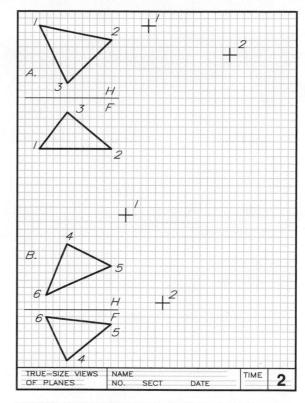

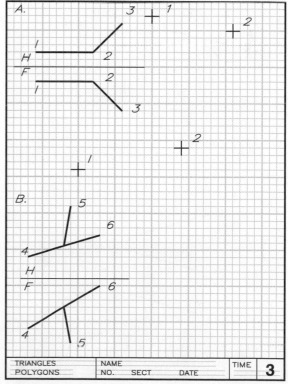

POINT VIEWS & DIHEDRAL ANGLES	NAME		TIME	**1**
	NO. SECT DATE			

TRUE–SIZE VIEWS OF PLANES	NAME		TIME	**2**
	NO. SECT DATE			

TRIANGLES POLYGONS	NAME		TIME	**3**
	NO. SECT DATE			

Use the crosses marked "1" and "2" for positioning the primary and secondary reference lines. Primary reference lines should pass through "1" and secondary reference lines through "2."

1. (Sheet 1) Lines and planes.
(A–B) Find the point views of the lines.
(C–D) Find the angles between the planes.

2. (Sheet 2) True-size planes.
(A–B) Find the true-size views of the planes.

3. (Sheet 3) Angles between lines.
(A–B) Find the angles between the lines.

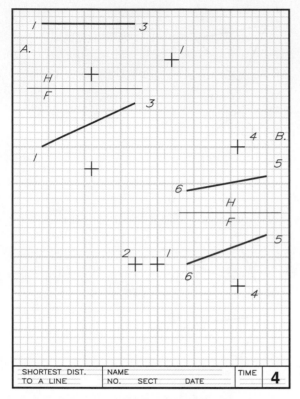

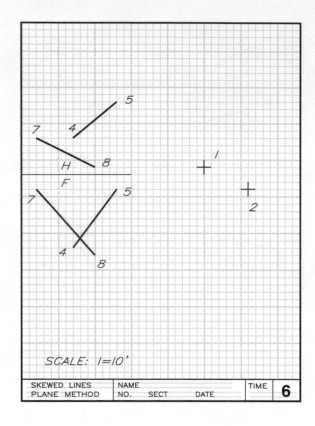

SCALE: 1=10'

| SKEWED LINES | NAME | | TIME | **6** |
| PLANE METHOD | NO. SECT DATE | | | |

| SHORTEST DIST. | NAME | | TIME | **4** |
| TO A LINE | NO. SECT DATE | | | |

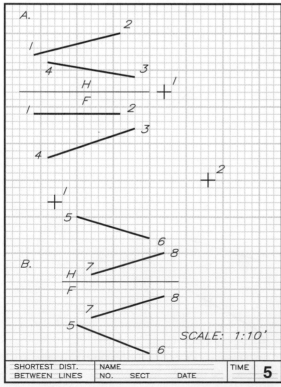

SCALE: 1:10'

| SHORTEST DIST. | NAME | | TIME | **5** |
| BETWEEN LINES | NO. SECT DATE | | | |

4. (Sheet 4) Distance to lines.
(A–B) Find the shortest distances from the points to the lines. Show these distances in all views.

5. (Sheet 5) Distances between lines.
(A–B) Find the shortest distances between the lines by the line method. Show these distances in all views.

6. (Sheet 6) Skewed line problems.
Find the shortest distance between the lines by the plane method. Show the shortest line in all views.

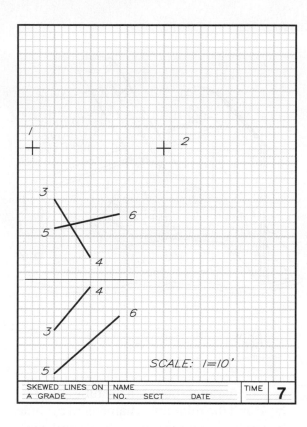

SCALE: 1=10'

SKEWED LINES ON A GRADE	NAME			TIME	7
	NO.	SECT	DATE		

Alternative problem: Find the shortest horizontal distance between the two lines and show the distance in all views.

7. (Sheet 7) Grade distance between lines. Find the shortest 20% grade distance between the two lines. Show this distance in all views.

Design Problems

The following problems must be laid out with the necessary orthographic views in order to determine the specified design information before working drawings can be made. Problems can be drawn on either A size or B size sheets.

(These problems have been adapted from applications encountered at the Boeing Company.)

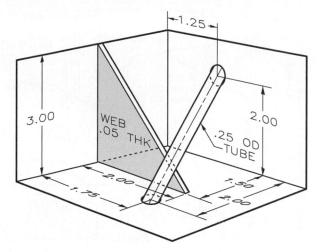

Design 1

Design 1

Lay out the necessary views in order to determine the clearance (if any) between the web and the aircraft tube. If there is interference, modify the web accordingly.

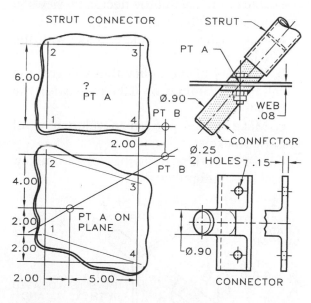

Design 2

Design 2

Design a connector to support a structural strut that intersects the plane of the aircraft. Show your geometric construction and prepare detail drawings to explain your design. Use one or more B-size sheets.

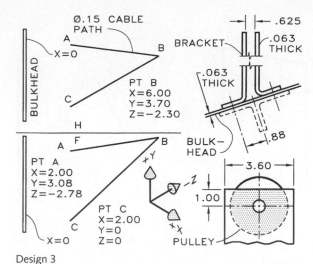

Design 3

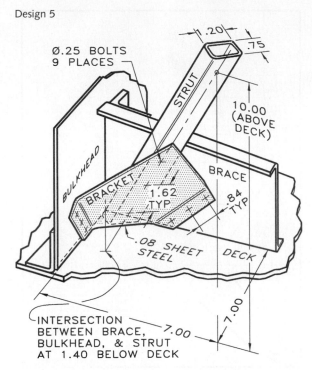

Design 5

Design 3

Points A, B, and C define points on a control cable. Two brackets are to be designed to attach to the bulkhead (with 4–0.25 DIA bolts) and support a pulley that lies in the plane of the cable path. Lay out the necessary views to determine the following:

a. The angles the brackets make with the bulkhead.

b. The size of the pulley that will give the cable a clearance of 0.25 inch with the bulkhead.

c. Make a flat-pattern development of both brackets.

d. Make working drawings of the pulley, brackets, and pulley axle.

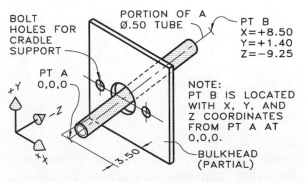

Design 4

Design 4 The 0.50 OD overflow tube extends through an access hole in a bulkhead.

a. Find the angle between the tube and bulkhead and the minimum diameter of the hole.

b. Design a support to cradle the tube and to be bolted to the bulkhead. Make a working drawing of your design.

Design 5

(For only the best students!) Design a bracket so the strut clears the deck, bulkhead, and brace by 0.10 inch. The lower edge of the bracket flanges should clear the deck by 0.10. Draw a dimensioned flat pattern of your bracket design. Suggested scale: 1 = 5 in.

a. Make a working drawing of the bracket.

b. Make a flat-pattern of the bracket.

Design 6

The bracket is made of bent flat stock. The true angle of bend at the left is 60° but it cannot be measured as 60° in the front view.

a. Draw the front and top views of the bracket.

b. Draw a developed flat view of the bracket.

c. Calculate the weight of the bracket. (Steel weighs 490 lbs per cubic foot.)

Design 6

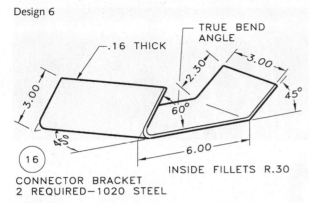

.16 THICK

TRUE BEND ANGLE

3.00

2.30

3.00

60°

45°

3.00

45°

16

6.00

INSIDE FILLETS R.30

CONNECTOR BRACKET
2 REQUIRED—1020 STEEL

Design 7

Determine the angle in the tube at C and make a working drawing of the tube. What is the tube's length?

Design 8

Use single-line representation to show the proposed duct for an aircraft's nose pressurization system and solve for the following:

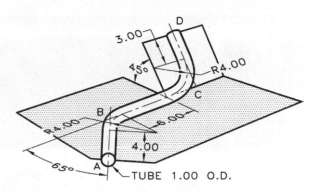

D

3.00

45°

R4.00

B

R4.00

C

8.00

4.00

65°

A

TUBE 1.00 O.D.

Design 7

a. Find the lengths from point A to point F.
b. Find the angles at points C and E.

Alternate solution: Use double-line symbols to represent the ducts; draw the components in the views where the ducts appear true length.

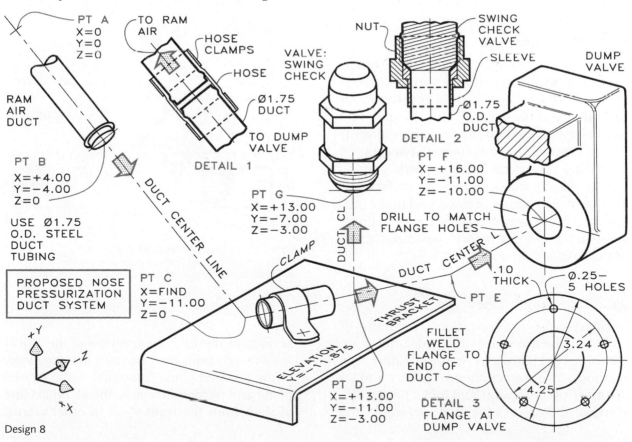

PT A
X=0
Y=0
Z=0

TO RAM AIR

HOSE CLAMPS

HOSE

Ø1.75 DUCT

TO DUMP VALVE

DETAIL 1

VALVE: SWING CHECK

NUT

SWING CHECK VALVE

SLEEVE

DUMP VALVE

Ø1.75 O.D. DUCT

DETAIL 2

PT F
X=+16.00
Y=−11.00
Z=−10.00

DRILL TO MATCH FLANGE HOLES

RAM AIR DUCT

PT B
X=+4.00
Y=−4.00
Z=0

USE Ø1.75 O.D. STEEL DUCT TUBING

DUCT CENTER LINE

PT G
X=+13.00
Y=−7.00
Z=−3.00

DUCT CL

DUCT CENTER L

.10 THICK

PT E

Ø.25— 5 HOLES

3.24

4.25

PROPOSED NOSE PRESSURIZATION DUCT SYSTEM

PT C
X=FIND
Y=−11.00
Z=0

CLAMP

THRUST BRACKET

FILLET WELD FLANGE TO END OF DUCT

DETAIL 3
FLANGE AT DUMP VALVE

+Y

−Z

+X

ELEVATION Y=−11.875

PT D
X=+13.00
Y=−11.00
Z=−3.00

Design 8

21

Revolution

21.1 Introduction

Many products and systems involve revolution, such as an automobile's front suspension, which was designed to revolve about several axes at each wheel. This design is just one of many based on the principles of revolution. Revolution is an alternative technique of solving spatial problems in orthographic views to yield a true-size view of a surface, the angle between planes or lines, and many others. Revolution was used to solve descriptive geometry problems before the introduction of the auxiliary-view method.

21.2 True-Length Lines

Auxiliary-view and revolution methods of obtaining the true size of inclined surfaces are compared in **Figure 21.1.** In the auxiliary view method, the observer changes position to an auxiliary vantage point and looks perpendicularly at the object's inclined surface. In the

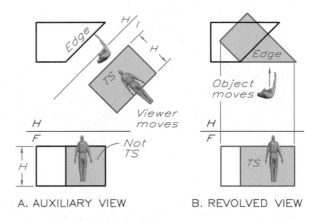

A. AUXILIARY VIEW   B. REVOLVED VIEW

21.1 Auxiliary views vs. revolved views:
A The surface is found true size in an auxiliary view.
B The surface is revolved to be seen true size in the front view.

revolution method, the top view of the object is revolved about the axis until the edge view of the inclined plane is parallel to the frontal plane and perpendicular to the standard line of sight from the front view. In other words,

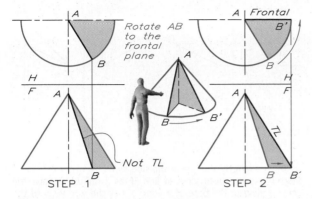

21.2 True length in the front view:
Step 1 Use the top view of line AB as a radius to draw the base of a cone with point A as the apex. Draw the front view of the cone with a horizontal base through point B.
Step 2 Revolve the top view of line AB to be parallel to the frontal plane. When projected to the front view, frontal line AB' is the outside element of the cone and is TL.

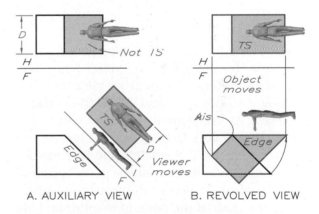

21.3 Principle of a line true length by revolution:
A Line 1-2 does not appear true length in the front view because the observer's line of sight is not perpendicular to it.
B When the triangle is revolved into the frontal plane, his line of sight is perpendicular to it and line 1-2' is seen true length.

the observer's line of sight does not change, but the object is revolved until the plane appears true size in the observer's normal line of sight.

In the Front View

To find a true-length line in the front view by revolution **(Figure 21.2),** revolve line AB into the frontal plane. The top view represents the circular base of a right cone, and the front view is the triangular view of a cone. Line AB' is the outside element of the cone's frontal line and is true length in the front view.

    Figure 21.3 illustrates the technique of finding line 1-2 true length in the front view. The observer's line of sight is not perpendicular to the triangle containing line 1-2 in its first position, and line 1-2 is not seen true length. When the triangle is revolved into the frontal plane, the observer's line of sight is perpendicular to the plane, and line 1-2 is seen true length.

    21.4 Auxiliary views vs. revolved views:
A By looking perpendicular to the edge view of the plane, it appears true size in the primary auxiliary view.
B When the edge view of the plane is revolved to become horizontal, it appears true size in the top view.

In the Top View

A surface that appears as an edge in the front view may be found true size in the top view by a primary auxiliary view or by a single revolution **(Figure 21.4).** The axis of revolution is a

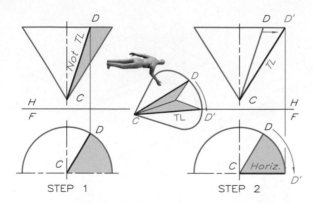

STEP 1   STEP 2

21.5 True length of a line in the top view:
Step 1 Use the front view of line CD as a radius to draw the base of a cone with C as the apex. Draw the top view of the cone with the base as a frontal plane.
Step 2 Revolve the front view of line CD into a horizontal position, CD'. When projected to the top view, CD' is the outside element of the cone and is true length.

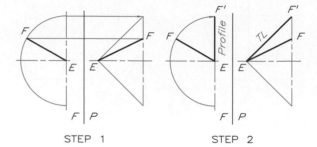

STEP 1   STEP 2

21.6 True length of a line in the side view:
Step 1 Use the front view of line EF as a radius to draw the circular view of the base of a cone. Draw the side view of the cone with its base through F.
Step 2 Revolve EF in the frontal view to position EF' where it is a profile line, the outside element of the cone, and true length in the side view.

point in the front view and true length in the top view. Revolving the edge view of the plane into the horizontal in the front view and projecting it to the top view yields the surface's true size. As in the auxiliary view method, the depth dimension, D, does not change.

In **Figure 21.5,** revolving line CD into the horizontal gives its true length in the top view. The arc of revolution in the front view represents the base of the cone of revolution. Line CD' is true length in the top view because it is horizontal and an outside element of the cone. Note that the depth in the top view does not change.

In the Profile View
In **Figure 21.6,** revolving the front view of line EF into the profile plane gives a true-length view of it. Projecting the circular view of the cone to the side view gives a triangular view of the cone. Because line EF' is a profile line, it is true length in the side view where it is the outside element of the cone.

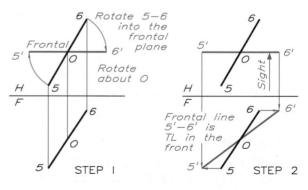

21.7 True-length views of lines may be obtained by revolving them about any point on the lines, not just their endpoints. Line 5-6 is revolved about its midpoint in the top view until it's parallel to the frontal plane and is true length in the front view.

Alternative Points of Revolution
In the preceding examples, each line is revolved about one of its ends. However, a line may be revolved about any point on its length. **Figure 21.7** shows how to find line 5-6 true length by revolving it about point O.

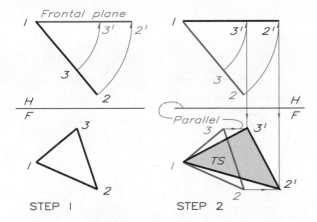

21.8 True size of a plane: Special case.
Step 1 Revolve the edge view of the plane until it is parallel to the frontal plane.
Step 2 Project points 2′ and 3′ to the horizontal projectors from points 2 and 3 in the front view.

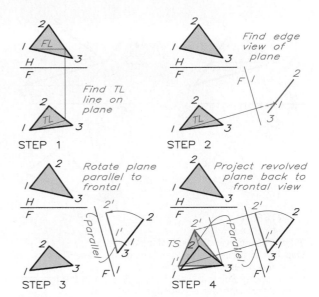

21.9 True size of a plane by revolution: General.
Step 1 To find the edge view of the plane by revolution, draw a frontal line on the plane that is true length in the front view.
Step 2 Find the edge view of the plane by finding the point view of the frontal line.
Step 3 Revolve the edge view of the plane until it is parallel to the F1 reference line.
Step 4 Project the revolved points 1′ and 2′ to the front view to the projectors from points 1 and 2 that are parallel to the F1 reference line.

21.3 True Size of a Plane

When a plane appears as an edge in a principal view (the top view in **Figure 21.8**), it can be revolved to be parallel to the frontal reference plane. The new front view is true size when projected horizontally across from its original front view.

The combination of an auxiliary view and a single revolution finds the plane in **Figure 21.9** true size. After finding the plane as an edge by finding the point of view of a true-length line in the plane, revolve the edge view to be parallel to the F1 reference line. To find the true size of the plane, project the original points (1, 2, and 3) in the front view parallel to the F1 line to intersect the projectors from 1′ and 2′. The true size of the plane also may be found by projecting from the top view to find the edge view.

By Double Revolution

The edge view of a plane can be found by revolution without using auxiliary views

(Figure 21.10). Draw a frontal line on plane 1-2-3, and project it to the front view where it is true length. Revolve the plane until the true-length line is vertical in the front view. The true-length line projects as a point in the top view; therefore the plane appears as an edge in this view. Projectors from points 2 and 3 from the top view are parallel to the HF reference line.

A second revolution, called a double revolution, positions this edge view of the plane parallel to the frontal plane as shown in step 1 of **Figure 21.11**. Projecting the top views of points 1″ and 2″ to the front view gives a true-size plane 1″-2″-3′ (step 2). We could

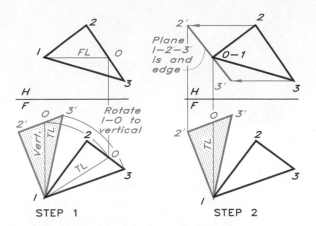

21.10 Finding the edge view of a plane:
Step 1 Draw a true-length frontal line on the plane. Revolve the front view until the TL line is vertical.
Step 2 Locate points 2' and 3' by projection in the top view. Plane 1-2'-3' appears as and edge.

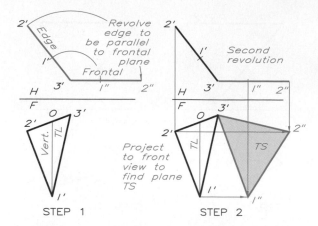

21.11 True size of a plane:
Step 1 The plane in the top view of **Figure 21.10** is revolved to a position parallel to the frontal plane.
Step 2 Project points 1" and 2" to the front view to intersect with the horizontal projectors from the original points 1' and 2'. Plane 1"-2"-3' is true size in this view.

have shown this second revolution in **Figure 21.10,** but it would have resulted in overlapping views, making observation of the separate steps difficult.

Figure 21.12 shows how to use double revolution to find the true size of the oblique plane (1-2-3) of the object. Revolve the true-length line 1-2 on the plane in the top view until it is perpendicular to the frontal plane. Line 1-2 appears as a point in the front view, and the plane appears as an edge. This revolution changes the width and depth, but not the

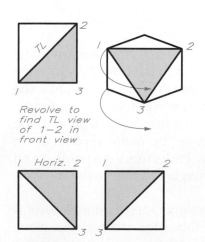

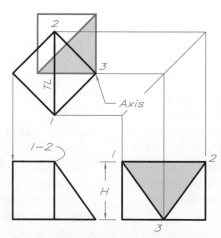

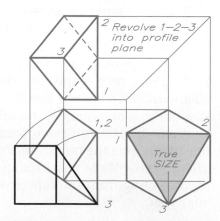

21.12 True size of a plane by double revolution:
Required: Find the true size of the plane by revolution.

Step 1 Line 1-2 is horizontal in the frontal view and true length in the top view. Revolve the top view so that line 1-2 appears as a point in the front view.

Step 2 Plane 1-2-3 is an edge; revolve it into a vertical position in the front view, to find its true size in the side view. The depth does not change.

21.13 The principles of revolution can be used to determine the angles between the planes of the Saturn S-IVB during its design. *(Courtesy of National Aeronautics and Space Administration.)*

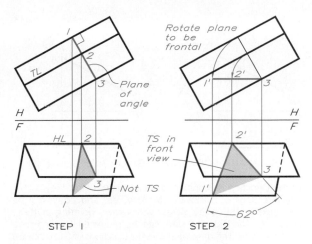

21.14 Angle between planes:
Step 1 Draw the edge of the angle perpendicular to the TL line of intersection between the planes in the top view. Project it to the front view.
Step 2 Revolve the edge view of the plane of the angle to position angle 1'-2'-3 in the top view parallel to the frontal plane. Project this angle to the frontal view, where it is true size.

height. Then, revolve the edge view of the plane into a vertical position parallel to the profile plane. To find the plane in true size, project to the profile view, where the depth remains unchanged but the height is greater.

21.4 Angle Between Planes

The angle between the planes of the nuclear detection satellite (**Figure 21.13**) may be found by revolution instead of by auxiliary views. The application of this geometry was an essential principle in designing the satellite.

In **Figure 21.14,** finding the dihedral angle involves drawing its edge view perpendicular to the line of intersection and projecting the plane of the angle to the front view. When the edge view of the angle is revolved into a frontal plane and projected to the front view, the angle appears true size and can be measured.

Figure 21.15 shows how to solve a similar problem. Here, the line of intersection does not appear true length in the given views; therefore an auxiliary view is needed to find

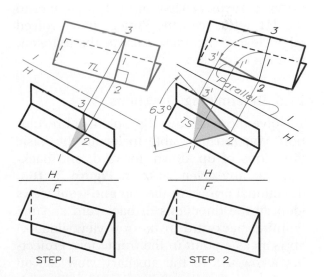

21.15 Angle between oblique planes:
Step 1 Find the true-length view of the line of intersection by projecting perpendicularly from its top view. Draw the edge of the angle perpendicular to the TL of the line of intersection and project it to the top view.
Step 2 Revolve the edge view of the plane of the angle (plane 2-1'-3') until it is parallel to the H1 reference line so that it appears true size in the top view.

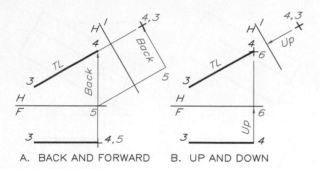

A. BACK AND FORWARD   B. UP AND DOWN

21.16 Direction in a primary auxiliary view:
The directions of backward, forward, up, and down can be identified in the given views with arrows pointing in these directions. Directional arrows can be projected to successive auxiliary views. This drawing shows the directions of (A) backward and (B) up.

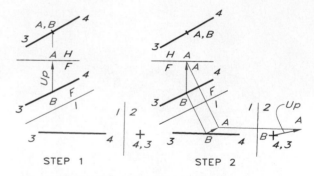

STEP 1   STEP 2

21.17 Direction in a secondary auxiliary view:
Step 1 To find the direction of up in the secondary auxiliary view, draw arrow AB pointing up in the front view. It appears as a point in the top view.
Step 2 Project arrow AB to the primary and secondary auxiliary views to show the direction of up.

its true length. Draw the plane of the dihedral angle as an edge perpendicular to the true-length line of intersection. Project the edge view of plane 1-2-3 to the top view. Then revolve the edge view of plane 1-2-3 in the primary auxiliary view until it is parallel to the H1 reference line. Project the revolved edge view of the angle back to the top view, where it is true size.

21.5 Determining Directions

To solve more advanced problems of revolution, you must be able to locate the basic directions of up, down, forward, and backward in any given view. In **Figure 21.16A,** directional arrows in the top and front views identify the directions of backward and up. Pointing backward in the top view, line 4-5 appears as a point in the front view. Projecting arrow 4-5 to the auxiliary view as you would any other line determines the direction of backward. By drawing the arrow on the other end of the line, you would find the direction of forward.

Locate the direction of up in **Figure 21.16B** by drawing line 4-6 in the direction

of up in the front view and as a point in the top view. Then find the arrow in the primary auxiliary by the usual projection method. The direction of down is in the opposite direction.

You find the directions in secondary auxiliary views in the same way. To determine the direction of up in **Figure 21.17,** begin with an arrow that points up in the front view and appears as a point in the top view. Project the arrow AB from the front view to the primary auxiliary view and then to a secondary auxiliary view to show the direction of up. Identify the other directions in the same way by beginning with the two principal views of a known direction.

21.6 Revolution: Point About an Axis

In **Figure 21.18,** point O is to be revolved about axis 3-4 to its most forward position. Find the axis as a point in the primary auxiliary view and draw the circular path of revolution. Draw the direction of forward and find the new location of point O at O'. Project back through the successive views to find point O' in each view. Note that point O' lies on the line in the

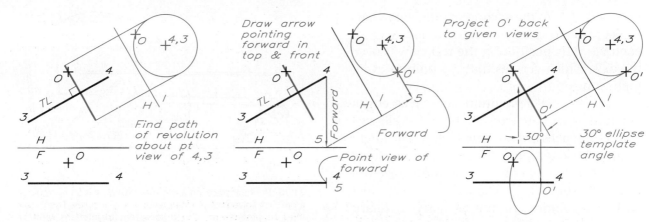

21.18 A point about an axis:
Step 1 To rotate O about axis 3-4 to its most forward position, find the point view of 3-4. The path of revolution is a circle in the auxiliary view and an edge perpendicular to the 3-4 in the top view.

Step 2 Locate the most forward position of point O by drawing an arrow pointing forward in the top view that appears as a point in the front view. Find the arrow, 4-5, in the auxiliary view to locate point O' on the circular path.

Step 3 Project O' back to the given views. The path of revolution is an ellipse in the front view because 3-4 is not true length in this view. Draw a 30° ellipse in the front view, the angle your line of sight makes with the circular path in the front view.

front view, verifying that point O' is in its most forward position.

In **Figure 21.19** an additional auxiliary view is needed in order to rotate a point about an axis because axis 3-4 is not true length in the given views. You must find the true length of the axis before you can find it as a point in the secondary auxiliary view, where the path of revolution appears as a circle. Revolve point O into its highest position, O', and locate the

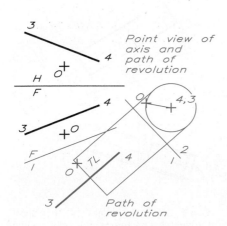

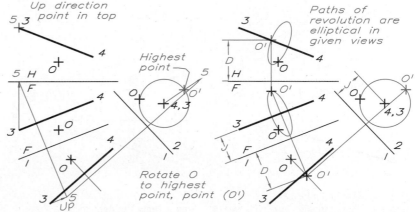

21.19 A point about an oblique axis:
Step 1 To rotate O about axis 3-4 to its highest position, find the point view of 3-4 and draw the circular path. The path of revolution is perpendicular to the 3-4 in the primary auxiliary views.

Step 2 To locate the highest position on the path of revolution, draw arrow 3-5 pointing up in the front and top views and project it to the secondary auxiliary view to find O'.

Step 3 Project point O' back to the given views by transferring the dimensions J and D with your dividers. The highest point lies over the line in the top view. The path of revolution is elliptical where the axis is not true length.

21.6 REVOLUTION: POINT ABOUT AN AXIS • 383

up arrow, 3-5, in the secondary auxiliary view. Project back to the given views to locate O′ in each view. Its position in the top view is over the axis, which verifies that the point is at its highest position.

The paths of revolution appear as edges when their axes are true length and as ellipses when their axes are not true length. The angle of the ellipse template for drawing the ellipse in the front view is the angle the projectors from the front view make with the edge view of the revolution in the primary auxiliary view. To find the ellipse in the top view, project an auxiliary view from the top view to obtain the path of revolution as an edge perpendicular to the true-length axis.

The handcrank of a casement window (**Figure 21.20**) is an example of the application of revolution techniques. The designer must determine the clearances between the sill and the window frame when designing the crank in order for it to operate properly.

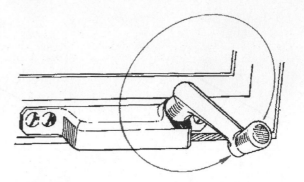

21.20 The handcrank on a casement window is an example of a problem solved by applying revolution principles. The handle must be properly positioned so as not to interfere with the window sill or wall.

A Right Prism

The coal chute shown in **Figure 21.21** conveys coal continuously between two buildings. The sides of the enclosed chute must be vertical and the bottom of the chute's right section

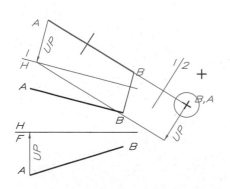

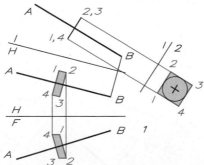

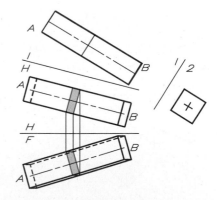

21.22 A prism about its axis:

Step 1 To draw a square chute with two of its sides vertical, find the point view of centerline AB in the secondary auxiliary view. Draw a circle about the axis with a diameter equal to the square section. Draw a vertical arrow in the front and top views; project it to the secondary auxiliary view to show the direction of vertical.

Step 2 Draw the right section, 1-2-3-4, in the secondary auxiliary view with two sides parallel to the vertical directional arrow. Project this section back to the previous views by transferring measurements with dividers. Locate the edge view of the section anywhere along AB in the primary auxiliary view.

Step 3 Draw the edges of the prism through the corners of the right section parallel to AB in all views. Terminate the ends of the prism in the primary auxiliary view where they appear as edges perpendicular to the centerline. Project the corner points of the ends to the top and front views to find the ends.

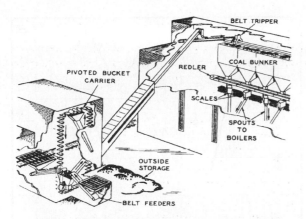

PIVOTED BUCKET CARRIER · REDLER · BELT TRIPPER · COAL BUNKER · SCALES · SPOUTS TO BOILERS · OUTSIDE STORAGE · BELT FEEDERS

21.21 A conveyor chute must be installed so that two edges of its right section are vertical for the conveyors to function properly. *(Courtesy of Goodyear Tire and Rubber Co.)*

must be horizontal. Design of the chute required application of the technique of revolving a prism about its axis.

In **Figure 21.22,** the right section is to be positioned about centerline AB so that two of its sides will be vertical. To do so, find the point view of the axis and project the direction of up to this view. Draw the right section about the axis so that two of its sides are parallel to the up arrow. Find the right section in the other views. Then construct the sides of the chute parallel to the axis. The bottom of the chute's right section will be horizontal and properly positioned for conveying coal.

21.7 A Line at Specified Angles

In **Figure 21.23,** a line is to be drawn through point O at an angle of 35° with the frontal and 44° with the horizontal plane, and slopes forward and down. Draw the cone with elements making 35° with the frontal plane and then a

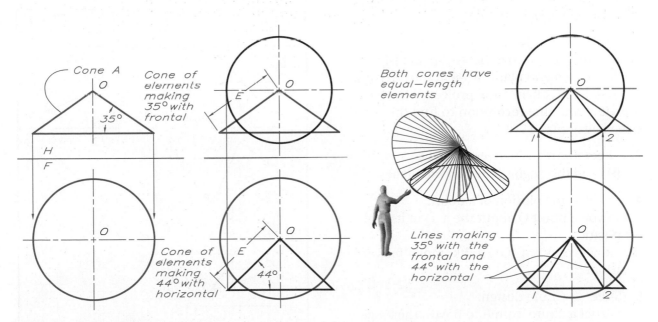

21.23 A line at specified angles:
Step 1 To draw lines at angles of 35° and 44° with the frontal and horizontal, respectively, draw a cone in the top view with outside elements at 35° with the frontal plane. Draw its circular view in the front with O as the apex. All its elements are 35° with the frontal plane.

Step 2 Draw a second cone in the front view with outside elements that make an angle of 44° with the horizontal plane. Draw the elements of this cone equal in length to element E of cone A. All elements of cone B are 44° with the horizontal plane.

Step 3 Because the elements are equal in length, two elements lie on the surface of each cone: lines 0-1 and 0-2. Locate points 1 and 2 where the bases of the cones intersect in both views. These lines slope forward and down at the specified angles.

cone with elements making 44° with the horizontal plane. The length of the elements of both cones must be equal so that the cones will intersect with equal elements. Finally, find lines 0-1 and 0-2, which are elements that lie on both cones and at the specified angles.

Problems

Lay out the problems on size A sheets with instruments. Each grid is equal to 0.20 inch (5mm). Label all reference planes and points in each problem with 1/8-inch letters or numbers.

Primary and secondary reference lines should pass through the crosses marked "1" and "2," repectively. Solve by revolution.

1. (Sheet 1) True-length lines.
(A–B) Find the true-length views of the lines in their front views by revolution.
(C–D) Find the true-length views of the lines in their top views by revolution.

2. (Sheet 2) True-size planes.
(A–B) Find the true-size views of plane 1-2-3 in the front view and plane 4-5-6 in the top view.
(C) Use an auxiliary view projected from the top view, and one revolution to find the true-size view of plane 7-8-9.

3. (Sheet 3) Angles between planes.
(A–B) Find the angles between the planes.

4. (Sheet 4) Revolution of a point.
(A) Revolve point O about the axis to its highest position.
(B) Revolve point O to its most forward position.

5. (Sheet 5) Chute design.
Construct a chute from A to B with the longer sides of its cross section being vertical.

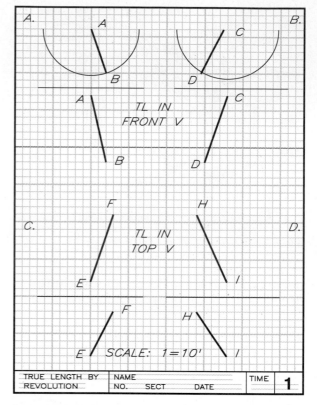

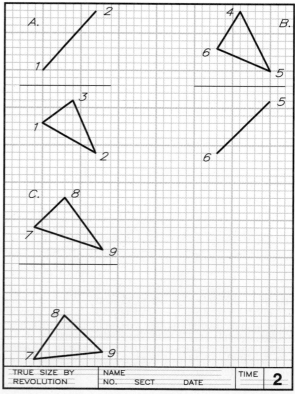

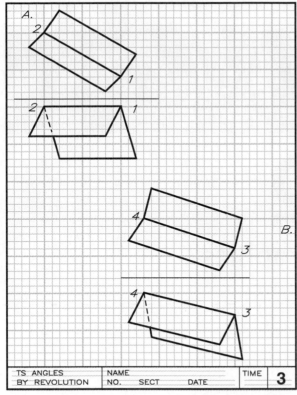

TS ANGLES BY REVOLUTION	NAME		TIME	**3**
	NO.	SECT	DATE	

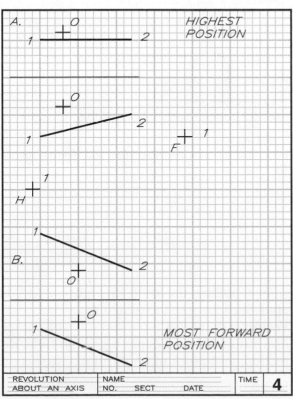

REVOLUTION ABOUT AN AXIS	NAME		TIME	**4**
	NO.	SECT	DATE	

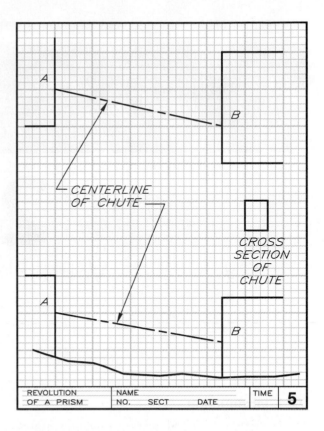

REVOLUTION OF A PRISM	NAME		TIME	**5**
	NO.	SECT	DATE	

Design 1 Aircraft valve. The shut-off valve must be rotated to a position where its handle will clear the bulkhead by 1 inch. Lay out the necessary views to show this revolution. What is the valve's rotation from its lowest position and the distance B that establishes the limit of the handle's path?

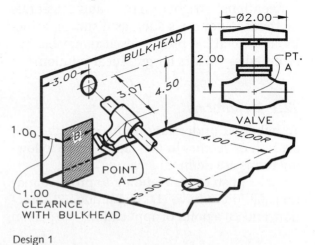

Design 1

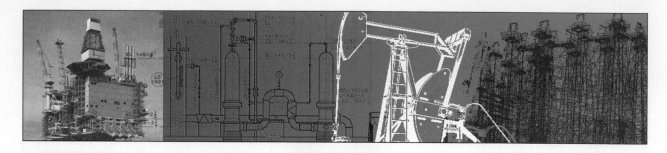

22

Vector Graphics

22.1 Introduction

Design of a structural system requires analysis of each member to determine the loads they must support and whether those loads are in tension or compression. Forces may be represented graphically by vectors and their magnitudes and directions determined in 3D space. Graphical methods are useful in the solution of vector problems as alternatives to conventional trigonometric and algebraic methods. Quantities such as distance, velocity, and electrical properties also may be represented as vectors for graphical solution.

22.2 Definitions

To help you understand more easily the discussion of vectors in this chapter, the following terms are defined:

Force: a push or pull tending to produce motion. All forces have (1) magnitude, (2) direction, and (3) a point of application. The person

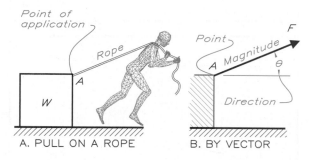

22.1 A force applied to an object (A) may be represented by vectors depicting the magnitude and direction of the force (B).

shown pulling the rope in **Figure 22.1A** is applying a force to the weight W.

Vector: a graphical representation of a force drawn to scale and depicting magnitude, direction, and point of application. The vector

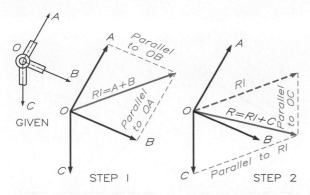

22.2 Resultant by the parallelogram method:
Step 1 Draw a parallelogram with its sides parallel to vectors A and B. The diagonal R1 is the resultant of forces A and B.
Step 2 Draw a parallelogram using vectors R1 and C to find diagonal R, or the overall resultant that can replace forces A, B, and C.

in **Figure 22.1B** represents the force applied through the rope to pull the weight W.

Magnitude: the amount of push or pull represented by the length of the vector line, usually measured in pounds or kilograms.

Direction: the inclination of a force (with respect to a reference coordinate system) indicated by a line with an arrow at one end.

Point of application: the point through which the force is applied on the object or member (point A in **Figure 22.1A**).

Compression: the state created in a member by forces that tend to shorten it. Compression is represented by the letter C or a plus sign (+).

Tension: the state created in a member by pulling forces that tend to stretch it. Tension is represented by the letter T or a minus sign (—).

System of forces: the combination of all forces acting on an object as shown (forces A, B, and C in **Figure 22.2**).

Resultant: a single force that can replace all the forces of a force system and have the same effect (force R in **Figure 22.2**).

Equilibrant: the opposite of a resultant; the single force that can be used to counterbalance all forces of a force system.

Components: separate forces that, if combined, would result in a single force; forces A and B are components of resultant R1 in **Figure 22.2.**

Space diagram: a diagram depicting the physical relationship between structural members, as given in **Figure 22.2.**

Vector diagram: a diagram of vectors representing the forces in a system and used to solve for unknown vectors in the system.

Metric units: standard units of weights and measures. The kilogram (kg) is the unit of mass (load), and one kilogram is approximately 2.2 pounds.

22.3 Coplanar, Concurrent Forces

When several forces, represented by vectors, act through a common point of application, the system is **concurrent.** In **Figure 22.2** vectors A, B, and C act through a single point; therefore this system is concurrent. When all vectors lie in the same plane, the system is **coplanar** and only one view is necessary to show them true length.

The **resultant** is the single vector that can replace all forces acting on the point of application. Resultants may be found graphically by (1) the **parallelogram method** and (2) the **polygon method.**

An **equilibrant** has the same magnitude, orientation, and point of application as the resultant in a system of forces, but in the opposite direction. The resultant of the system of forces shown in **Figure 22.3** is balanced by the equilibrant applied at point O, thereby causing the system to be in equilibrium.

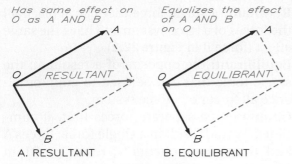

Has same effect on O as A AND B

Equalizes the effect of A AND B on O

A. RESULTANT

B. EQUILIBRANT

22.3 The (A) resultant and (B) equilibrant are equal in all respects except in direction (shown by arrowhead).

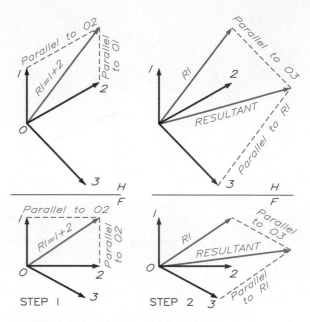

22.5 3D resultant by the parallelogram method:
Step 1 Use vectors 1 and 2 to construct a parallelogram in the top and front views. Diagonal R1 is the resultant of vectors 1 and 2.
Step 2 Use vectors 3 and R1 to construct a second parallelogram to find the overall resultant. The resultant must be found TL in a primary auxiliary view.

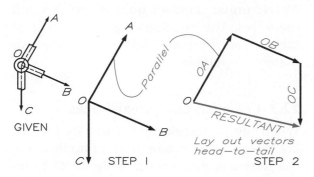

22.4 2D resultant:
The resultant (R) of a coplanar, concurrent system may be determined by the polygon method, in which the vectors are drawn head-to-tail. The vector that closes the polygon is the resultant.

Resultant: Parallelogram Method

In **Figure 22.2,** the vectors lie in the same plane, act through a common point, and are scaled to their known magnitudes. Use of the parallelogram method to determine resultants requires that the vectors be drawn to scale. Vectors A and B form two sides of a parallelogram. Constructing parallels to these vectors completes the parallelogram. Its diagonal, R1, is the resultant of forces A and B; it is the vector sum of vectors A and B.

Replaced by R1, vectors A and B now may be disregarded. Resultant R1 and vector C are two sides of a second parallelogram. Its diagonal, R, is the vector sum of R1 and C and the resultant of the entire system. Resultant R may be thought of as the only force acting on the point, thereby simplifying further analysis.

Resultant: Polygon Method

Figure 22.4 shows the same system of forces, but here the resultant is determined by the polygon method. Again, the vectors are drawn to scale, but in this case head-to-tail, in their true directions to form the polygon. The vectors are laid out in a clockwise sequence beginning with vector A. The polygon does not close, so the system is not in equilibrium but

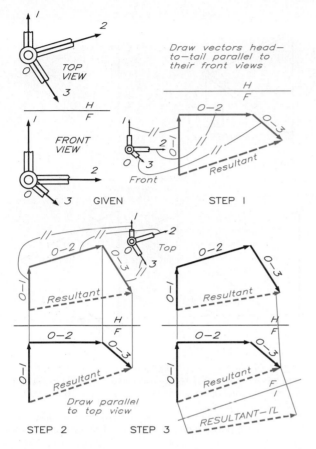

Draw vectors head-to-tail parallel to their front views

22.6 3D Resultant by the polygon method:
Step 1 Lay off each vector head-to-tail parallel to the given view and find the front view of the resultant.
Step 2 Draw the top view of the same vectors head-to-tail projected above the 3D polygon in the front view.
Step 3 The resultant is found true length in an auxiliary view projected from the front view.

The resultant of a system of noncoplanar forces may be obtained by the parallelogram method if their projections are given in two adjacent orthographic views.

3D Resultant: Parallelogram Method

In **Figure 22.5** vectors 1 and 2 are used to construct the top and front views of a parallelogram and its diagonal R1. The front view of R1 must be an orthographic projection of its top view.

Then resultant R1 and vector 3 are resolved to form the overall resultant in both views. The top and front views of the resultant must project orthographically. The overall resultant replaces vectors 1, 2, and 3. However, it is an oblique line, so an auxiliary view (**Figure 22.6**) or revolution must be used to obtain its true length.

3D Resultant: Polygon Method

Figure 22.6 shows the solution of the same system of forces for the resultant by the polygon method. Each vector is laid off head-to-tail clockwise, beginning with vector 1 in the front view.

Then the vectors are projected orthographically from the front view to the top view of the vector polygon. The vector polygon does not close, so the system is not in equilibrium. In both views the resultant (from the tail of vector 1 to the head of vector 3) closes the polygon. However, the resultant is an oblique line, requiring an auxiliary view to obtain its true length.

22.5 Forces in Equilibrium

The manufacturing hoist shown in **Figure 22.7** can be analyzed graphically to determine the loads carried by each member and cable since it is a coplanar, concurrent structure in equilibrium. A structure in equilibrium is one that is static with no motion taking place; the forces balance each other.

tends to be in motion. The resultant (from the tail of vector A to the head of vector C closes the polygon.

22.4 Noncoplanar, Concurrent Forces

When vectors lie in more than one plane of projection, they are noncoplanar, requiring 3D views for analysis of their spatial relationships.

22.7 The loads in the members of this crane can be determined by vector graphics as a coplanar system in equilibrium. *(Courtesy of Pacific Hoist Company.)*

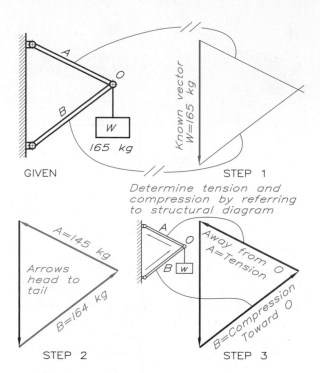

22.8 Coplanar forces in equilibrium:
Step 1 Draw the load of 165 kg as a vector. Draw vectors A and B parallel to their directions from the ends of the 165 kg vector and extend to their intersection.
Step 2 The direction of the 165 kg vector is known. Draw the arrows on the polygon head-to-tail to find the directions of A and B.
Step 3 Vector A points away from point O when transferred to the structural diagram and thus is in tension. Vector B points toward point O and is in compression.

The coplanar, concurrent structure depicted in **Figure 22.8** is designed to support a load of W = 165 kg. The maximum loading of each structural member determines the material and size of the members to be used in the design.

A single view of a vector polygon in equilibrium allows you to find only two unknown values. (Later, we show how to solve for three unknowns by using descriptive geometry.) Lay off the only known force, W = 165 kg, parallel to its given direction (here pointing vertically downward). Then draw the unknown forces A and B parallel to the supports to form the vector polygon and scale (or calculate) the magnitude of these forces.

Analyze vectors A and B to determine whether they are in tension or compression and thus find their direction. Vector B points upward to the right, which is toward point O

when transferred to the structural diagram shown in the small drawing. Vectors that act toward their point of application are in compression. Vector A points away from point O when transferred to the structural diagram and is in tension.

Figure 22.9 is a similar example involving determination of the loads in the structural members caused by the weight of 110 pounds

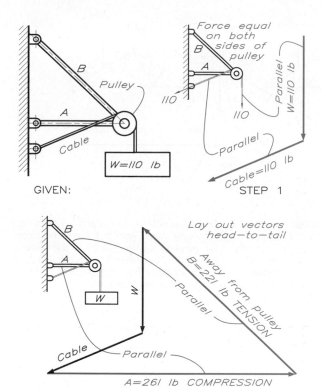

GIVEN:

STEP 1

Force equal on both sides of pulley

Pulley

Cable

W=110 lb

110

110

Parallel W=110 lb

Parallel

Cable=110 lb

STEP 2

Lay out vectors head-to-tail

Away from pulley B=221 lb TENSION

Parallel

Cable

Parallel

W

A=261 lb COMPRESSION
Toward pulley

W

22.9 Forces in equilibrium: Pulley application.
Step 1 The force in the cable is equal to 110 lb on both sides of the pulley. Draw these two forces as vectors head-to-tail and parallel to their directions in the space diagram.
Step 2 Draw A and B head-to-tail to close the polygon. Vector A points toward the point of application and thus is in compression. Vector B points away from the point and is in tension.

acting through a pulley. The only difference between this solution and the previous one is the construction of two equal vectors at the outset to represent the cable loads on both sides of the pulley.

22.6 Coplanar Truss Analysis

Designers use vector polygons to determine the loads in each member of a truss by two graphical methods: (1) joint-by-joint analysis and (2) Maxwell diagrams.

Joint-by-Joint Analysis

In the **Fink truss** shown in **Figure 22.10,** 3000-lb loads are applied at its joints. The exterior forces on the truss are labeled with letters placed between them, and numerals are placed between the interior members. Each vector is referred to by the number on each of its sides, clockwise about its joint. For example, the vertical load at the left is denoted AB, with A at the tail and B at the head of the vector. This method of designating forces is called **Bow's notation.**

First analyze the joint at the left end with a reaction of 4500 lb. When you read clockwise about the joint, the force is EA, where E is the tail and A is the head of the vector. Continuing clockwise, the next forces are A-1 and 1-E, which close the polygon at E, the beginning letter. Place arrowheads in a head-to-tail sequence beginning with the known vector EA.

Determine tension and compression by relating the directions of each vector to the original joint. For example, A-1 points toward the joint and is in compression, whereas 1-E points away and is in tension. The truss is symmetrical and equally loaded, so the loads in the members on the right will be equal to those on the left.

Analyze the other joints in the same way. The directions of the vectors are opposite at each end. For example, vector A-1 is toward the left in step 1 and toward the right in step 2.

Maxwell Diagrams

The **Maxwell diagram** is virtually the same as the joint-by-joint analysis, with the exception that the polygons overlap, with some vectors common to more than one polygon. In **Figure 22.11** (step 1) the exterior loads are laid out

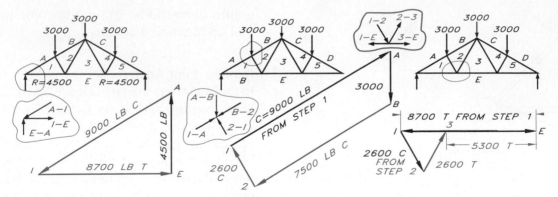

22.10 Truss analysis: By joint.
Step 1 Label the truss, with letters between the exterior loads and numbers between interior members. Analyze the left joint with two unknowns, A-1 and 1-E. Draw vectors A-1 and 1-E parallel to their directions from both ends of EA in a head-to-tail sequence.

Step 2 Use vector 1-A and load AB from step 1 to find B-2 and 2-1. Draw 1-A first, then AB, and draw B-2 and 2-1 to close the polygon, moving clockwise about the joint. A vector pointing toward the point of application is in compression. A vector pointing away from the point of application is in tension.

Step 3 Lay out vectors E-1 and 1-2 from the preceding steps. Vectors 2-3 and 3-E close the polygon and are parallel to their directions in the space diagram. Vectors 2-3 and 3-E point away from the point of application and thus are in tension.

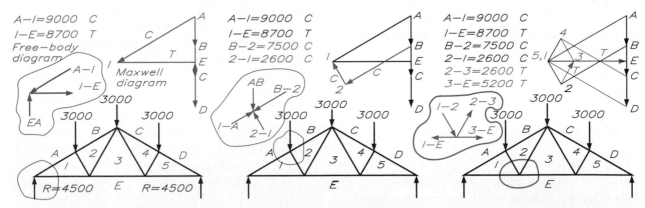

22.11 Truss analysis: Maxwell diagram.
Step 1 Label the outer spaces with letters and the internal spaces with numbers. Draw the loads head-to-tail in a Maxwell diagram; sketch a free-body diagram of the first joint. Use EA, A-1, and 1-E (head-to-tail) to draw a vector diagram. A-1 is in compression (it points toward the joint); 1-E is in tension (it points away).

Step 2 Sketch the next joint to be analyzed. Because AB and A-1 are known, only 2-1 and B-2 are unknown. Draw them parallel to their direction (head-to-tail) in the Maxwell diagram using the previously found vectors. Vectors B-2 and 2-1 are in compression, since each points toward the joint. Vector A-1 becomes vector 1-A when read in a clockwise direction.

Step 3 Sketch a free-body diagram of the next joint to be analyzed, where the unknowns are 2-3 and 3-E. Draw their vectors in the Maxwell diagram parallel to their given members to find point 3. Vectors 2-3 and 3-E are in tension because they point away from the joint. Repeat this process to find the vectors on the opposite side.

head-to-tail in clockwise sequence—AB, BC, CD, DE, and EA—with a letter placed at each end of each vector. The forces are parallel so this force diagram is a vertical line.

Vector analysis begins at the left end where the force EA of 4500 lb is known. A free-body diagram is sketched to isolate this joint. The two unknowns, A-1 and 1-E, are drawn parallel to their directions in the truss, with A-1 beginning at point A, 1-E beginning at point E, and both extended to point 1.

Because resultant EA points upward, A-1 must have its tail at A and its direction toward point 1. The free-body diagram shows that the direction is toward the point of application, which means that A-1 is in compression. Vector 1-E points away from the joint, which means that it is in tension. The vectors are coplanar and may be scaled to determine their magnitudes.

In step 2, where vectors 1-A and AB are known, the unknown vectors, B-2 and 2-1 may be determined. Vector B-2 is drawn parallel to its structural member through point B in the Maxwell diagram, and the line of vector 2-1 is extended from point 1 to intersect with B-2 at point 2. The arrows of each vector are drawn head-to-tail. Vectors B-2 and 2-1 point toward the joint in the free-body diagram; and therefore are in compression.

In step 3, the next joint is analyzed to find the forces in 2-3 and 3-E. The truss and its loading are symmetrical, so the Maxwell diagram will be symmetrical when completed.

If the last force polygon in the series does not close perfectly, an error in construction has occurred. A slight error may be disregarded, as a rounding error may be disregarded in mathematics. Arrowheads are unnecessary and usually are omitted on Maxwell diagrams because each vector will have the opposite direction when applied to a different joint.

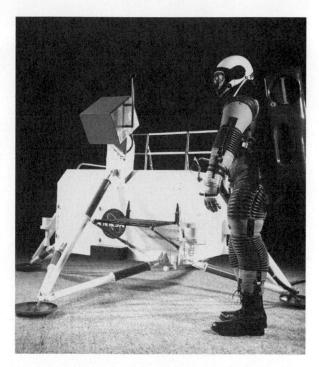

22.12 The structural members of this tripod support for a moon vehicle may be analyzed graphically to determine design load requirement. *(Courtesy of National Aeronautics and Space Administration.)*

22.7 Noncoplanar Vectors: Special Case

The solution of 3D vector systems requires the use of descriptive geometry because the system must be analyzed in 3D space. An example is the manned flying system (MFS) shown in **Figure 22.12,** which was analyzed to determine the loads on its support members. Weight on the moon is 0.165 of earth weight. Thus a tripod that must support 182 lb on earth needs to support only 30 lb on the moon.

In general, only two unknown vectors can be determined in a single view of a vector polygon that is in equilibrium. However, the system

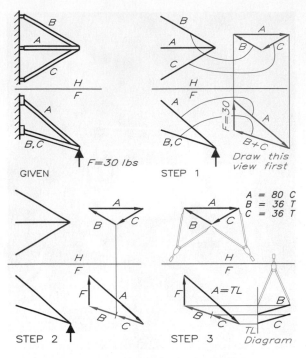

GIVEN

STEP 1 — Draw this view first

F = 30 lbs

A = 80 C
B = 36 T
C = 36 T

STEP 2

STEP 3 — TL Diagram

A = TL

22.13 Noncoplanar structural analysis (special case):
Step 1 Forces B and C coincide in the front view, resulting in only two unknowns. Draw vector F (30 lb) and the two unknown forces parallel to their front view in the front view of the vector polygon. Find the top view of A by projecting from the front. Draw vectors B and C parallel to their top views.
Step 2 Project the point of intersection of vectors B and C in the top view to the front view to separate the head-to-tail vectors.
Step 3 Vectors B and C are in tension because they point away from the point of application in the space diagram. Vector A is in compression because it points toward the point of application. Find the vector true length in the TL diagram.

shown in **Figure 22.13** is a special case because members B and C lie in the same edge view of the plane in the front view. Therefore solving for three unknowns is possible in this case.

Construct a vector polygon in the front view by drawing force F as a vector and using the other vectors as the sides of the polygon. Draw the top view using vectors B and C to

form the polygon that closes at each end of vector A. Then find the front view of vectors B and C.

A true-length diagram gives the lengths of the vectors; measure them to determine their magnitudes. Vector A is in compression because it points toward the point of application. Vectors B and C are in tension because they point away from the point.

General Case

The structural frame shown in **Figure 22.14** is attached to a vertical wall to support a load of W = 1200 lb. There are three unknowns in each of the views, so begin by projecting an auxiliary view from the top view to obtain the edge view of a plane containing vectors A and B, thereby reducing the number of unknowns to two. You no longer need refer to the front view.

Draw a vector polygon with vectors parallel to their members in the auxiliary view. Then draw an adjacent orthographic view of the vector polygon with vectors parallel to their members in the top view. Use a true-length diagram to find the true length of the vectors and measure their magnitudes.

22.8 Resultant of Parallel, Nonconcurrent Forces

The beam in **Figure 22.15** supports the three loads shown. It is necessary to determine the magnitude of supports R1 and R2, the magnitude of the resultant of the loads, and the resultant's location. Begin by labeling the spaces between all vectors clockwise with Bow's notation and draw a vector diagram.

Extend the lines of force in the space diagram and draw the strings from the vector diagram in their respective spaces, parallel to their original directions. For example, string oa is parallel to string oA in space A between forces EA and AB, and string ob is in space B, beginning at the intersection of oa with vector

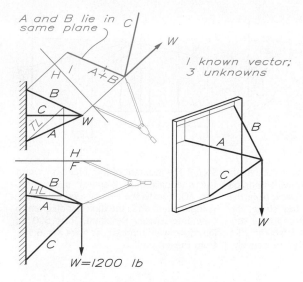

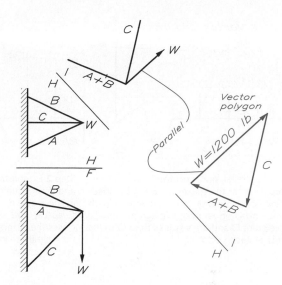

22.14 Noncoplanar structural analysis (general case):
Required: Find the loads in the members.
Step 1 Draw an auxiliary view to show the edge view of the plane containing vectors A and B, which are both unknowns. The load of W = 120 lb is true length and the only known vector in this view.

Step 2 Construct a vector polygon with its vectors parallel to the members found in the auxiliary view in step 1. Beginning with true-length vector W, draw the vector polygon head-to-tail using the two unknown vectors. Refer to the solution in Figure 22.13.

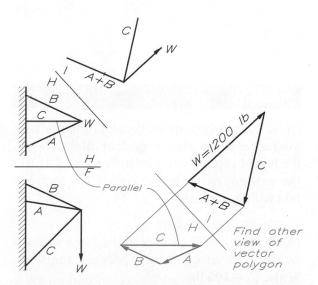

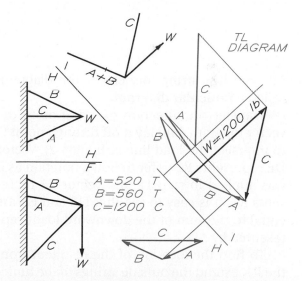

Step 3 Draw the adjacent orthographic view of the vector polygon by projecting perpendicularly to the H1 reference line. Load W appears as a point in this view and vectors A and B are parallel to their corresponding members in the top view of the space diagram.

Step 4 Project the intersection of A and B to the adjacent vector polygon. A and B are in tension since they point toward the application point when transferred to the space diagram. The magnitudes of vectors A, B, and C are found in a true-length diagram and are listed in a table.

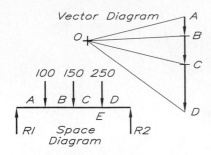

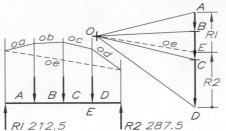

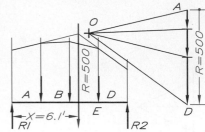

22.15 Parallel, nonconcurrent loads:
Step 1 Letter the spaces between the loads using Bow's notation. Sum the vertical loads head-to-tail in a vector diagram. Locate pole point O and draw strings from O to the ends of each vector.

Step 2 Extend the lines of the vertical loads and draw a funicular diagram with string oa in the A space, ob in the B space, oc in the C space, and so on. The last string, oe, closes the diagram. Transfer oe to the vector diagram to locate E, thus establishing R1 and R2, which are EA and DE, respectively.

Step 3 The resultant of the three downward forces equals their graphical summation, line AD. Locate the resultant by extending strings oa and od in the funicular diagram to their intersection. Resultant, R = 500 lb, acts downward through this point X = 6.1 ft. from the left end.

AB. The last string, oe, closes the diagram, called a **funicular diagram.**

Transfer the direction of string oe to the vector diagram, and lay it off through point O to intersect the load line at E (step 2). Vector DE represents R2 (refer to Bow's notation as it was applied in step 1), and vector EA represents R1. It is easy to see that DE and EA are equal to the sum of the downward loads represented by AD.

To find the location of the resultant from the R1, extend the outside strings of the funicular diagram, oa and od, to their intersection. The resultant will pass downward through this point of intersection. The resultant has a magnitude of 500 lb, a vertical downward direction, and a point of application at X = 6.1 ft. Notice that two scales are used in this problem: one for the vectors in pounds, and one in feet for the space diagram.

Problems

Draw your solutions to these problems with instruments on size A grid or plain sheets. Each grid represents 0.20 inch (5mm). Letter the written matter legibly, using 1/8-inch letters with guidelines.

1. (Sheet 1) Resultants.
(A–B) Find the resultants of the force systems by the parallelogram and polygon methods. Scale: 1″ = 100 lb.

2. (Sheet 2) Resultants.
(A–B) Find the resultants of the force systems by the parallelogram and polygon methods. Scale: 1″ = 100 lb.

3. (Sheet 3) Concurrent, coplanar.
(A–B) Find the forces in the coplanar force systems; label your construction properly.

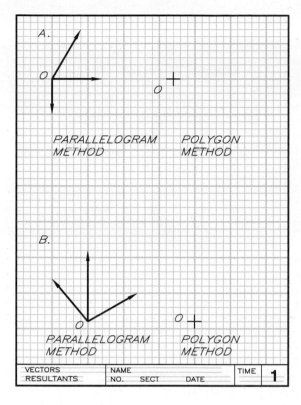

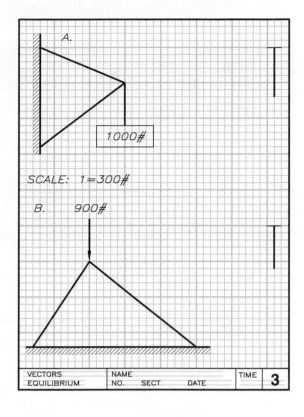

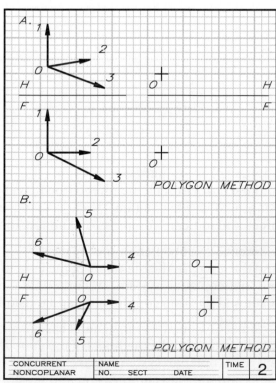

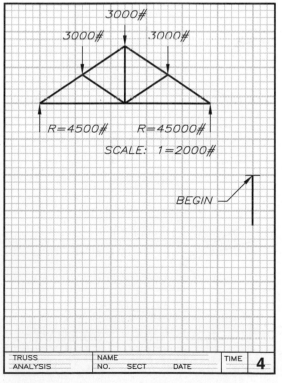

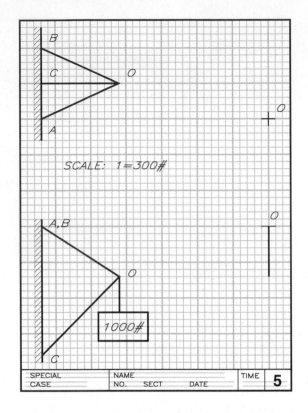

SCALE: 1=300#

SCALE: 1=300#

1000#

SPECIAL CASE / NAME NO. SECT DATE / TIME **5**

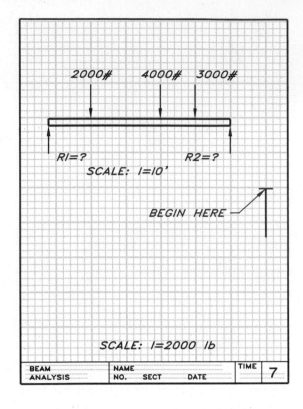

2000#   4000#   3000#

R1=?   R2=?

SCALE: 1=10'

BEGIN HERE

SCALE: 1=2000 lb

BEAM ANALYSIS / NAME NO. SECT DATE / TIME **7**

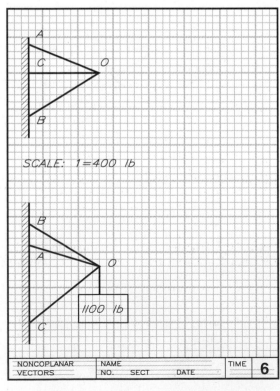

SCALE: 1=400 lb

1100 lb

NONCOPLANAR VECTORS / NAME NO. SECT DATE / TIME **6**

4. (Sheet 4) Truss analysis.
Find the loads in each member of the truss by using a Maxwell diagram. Make a table of forces and indicate compression and tension.

5. (Sheet 5) Noncoplanar, special case.
Find the forces in the members of the concurrent noncoplanar system. Make a table of forces and indicate compression and tension.

6. (Sheet 6) Noncoplanar, general case.
Find the forces in the members of the concurrent noncoplanar system. Make a table of forces and indicate compression and tension.

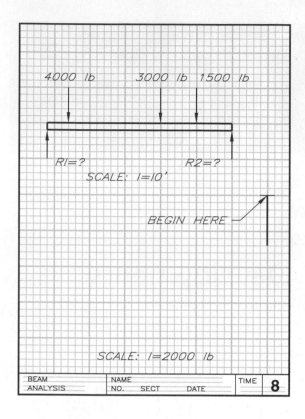

4000 lb　　　3000 lb　1500 lb

R1=?　　　　　R2=?

SCALE: 1=10'

BEGIN HERE

SCALE: 1=2000 lb

BEAM ANALYSIS	NAME			TIME	8
	NO.	SECT	DATE		

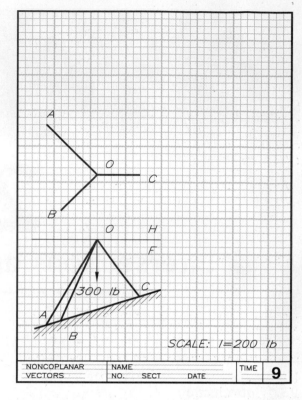

A

O

C

B

O　H

F

300 lb　　C

A

B

SCALE: 1=200 lb

NONCOPLANAR VECTORS	NAME			TIME	9
	NO.	SECT	DATE		

7. (Sheet 7) Beam analysis.
(A) Find forces R1 and R2 necessary to equalize the loads applied to the beam.
(B) Find the value and location of the single support that could replace both R1 and R2.

8. (Sheet 8) Beam analysis.
Repeat Problem 7 for this configuration.

9. (Sheet 9) Concurrent, coplanar.
Find the forces in the coplanar system. Make a table of forces; indicate compression or tension.

10. (Sheet 10) Concurrent, noncoplanar forces. Find the forces in the support members. Make a table of forces; indicate compression or tension.

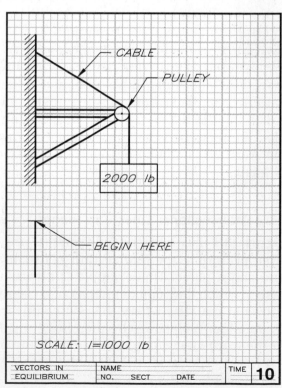

CABLE

PULLEY

2000 lb

BEGIN HERE

SCALE: 1=1000 lb

VECTORS IN EQUILIBRIUM	NAME			TIME	10
	NO.	SECT	DATE		

23

Intersections and Developments

23.1 Introduction

Several methods may be used to find lines of intersection between parts that join. Usually such parts are made of sheet metal, or of plywood if used as forms for concrete. After **intersections** are found, **developments,** or flat patterns, can be laid out on sheet metal and cut to the desired shape. You will see examples of intersections and developments ranging from air-conditioning ducts to massive refineries.

23.2 Intersections: Lines and Planes

Figure 23.1 illustrates the fundamental principle of finding the intersection between a line and a plane. This example is a special case in which the point of intersection clearly shows in the view where the plane appears as an edge. Projecting the piercing point P to the front view completes the visibility of the line.

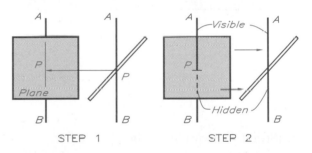

23.1 Intersection: Line and plane.
Step 1 Find the point of intersection in the view where the plane appears as an edge, the side view in this case, and project it to the front view.
Step 2 Determine visibility in the front view by looking from the front view to the right-side view.

This same principle is applied to finding the line of intersection between two planes **(Figure 23.2).** By locating the piercing points of lines AB and DC and connecting these points, the line of intersection is found.

The angular intersection of two planes at a corner gives a line of intersection that bends

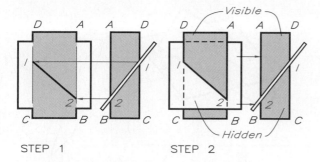

STEP 1   STEP 2

23.2 Intersection: Planes.
Step 1 Find the piercing points of lines AB and DC with the plane where the plane appears as an edge and project them to the front view, points 1 and 2.
Step 2 Line 1-2 is the line of intersection. Determine visibility by looking from the front view to the right-side view.

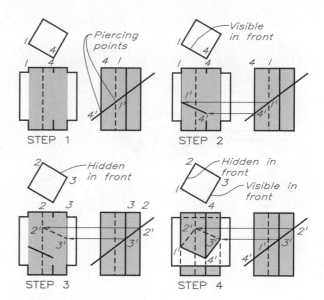

STEP 1   STEP 2

STEP 3   STEP 4

23.4 Intersection: Plane and prism.
Step 1 Vertical corners 1 and 4 intersect the edge view of the plane in the side view at points 1' and 4'.
Step 2 Project points 1' and 4' from the side view to lines 1 and 4 in the front view. Connect them to form a visible line of intersection.
Step 3 Vertical corners 2 and 3 intersect the edge view of the plane at points 2' and 3' in the side view. Project points 2' and 3' to the front view to form a hidden line of intersection.
Step 4 Connect points 1',-2',-3',-and -4' and determine visibility by analyzing the top and side views.

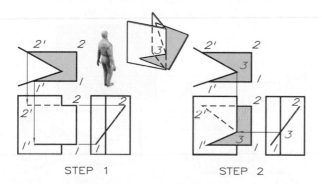

STEP 1   STEP 2

23.3 Intersection: A plane around a corner.
Step 1 The intersecting plane appears as an edge in the side view. Project intersection points 1' and 2' from the top and side views to the front view.
Step 2 The line of intersection from 1' to 2' must bend around the vertical corner at 3 in the top and side views. Project point 3 to the front view to locate line 1'-3-2'.

around the corner **(Figure 23.3).** First, find piercing points 2' and 1'. Then project corner point 3 from the side view where the vertical corner pierces the plane to the front view of the corner. Point 2' is hidden in the front view because it is on the back side.

Figure 23.4 shows how to find the intersection between a plane and prism where the

plane appears as an edge. Obtain the piercing points for each corner line and connect them to form the line of intersection. Show visibility to complete the intersection.

Figure 23.5 depicts a more general case of an intersection between a plane and prism. Passing vertical cutting planes through the planes of the prism in the top view yields traces (cut lines) on the front view of the oblique plane on which the piercing points of the vertical corner lines lie. Connect the points and determine visibility to complete the solution.

In **Figure 23.6,** finding the intersection between a foreshortened plane and an oblique prism involves finding an auxiliary view to

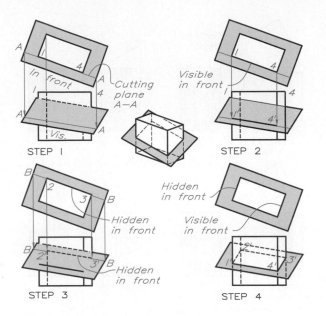

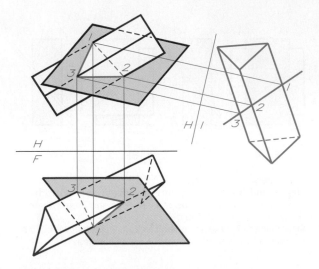

23.6 Intersection: Plane and prism.
To find the intersection between a plane and a prism, construct a view in which the plane appears as an edge. Project piercing points 1, 2, and 3 back to the top and front views.

23.5 Intersection: Oblique plane and prism.
Step 1 Pass vertical cutting plane A-A through corners 1 and 4 in the top view and project endpoints to the front view.
Step 2 Locate piercing points 1′ and 4′ in the front view where line A-A crosses lines 1 and 4.
Step 3 Pass vertical cutting plane B-B through corners 2 and 3 in the top view and project them to the front view to locate piercing points 2′ and 3′.
Step 4 Connect the four piercing points and determine visibility by analysis of the top view.

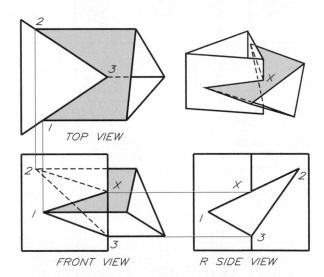

23.7 Intersection: Prisms at corner.
These are three views of intersecting prisms. The points of intersection are best found where intersecting planes appear as edges.

obtain the edge view of the plane and simplify the problem. The piercing points of the corner lines of the prism lie in the auxiliary view and project back to the given views. Points 1, 2, and 3, projected from the auxiliary view to the given views, are shown as examples. Analysis of crossing lines determines visibility to complete the line of intersection in the top and front views.

23.3 Intersections: Prisms

The techniques used to find the intersections between planes and lines also apply to finding the intersections between two prisms **(Figure 23.7).** Project piercing points 1, 2, and 3 from the side and top views to the front

view. Point X lies in the side view where line of intersection 1-2 bends around the vertical corner of the vertical prism. Connect points 1, X, and 2 and determine visibility.

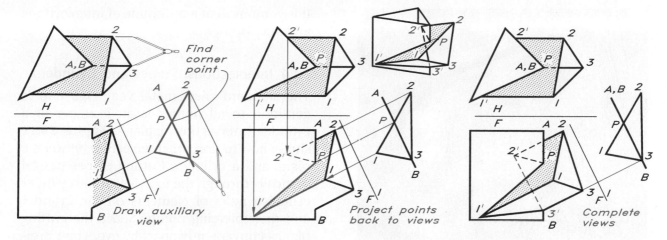

23.8 Intersection: Prisms by auxiliary view. **Step 1** Draw the end view of the inclined prism in an auxiliary view from the front view. Show line AB of the vertical prism in the auxiliary view.

Step 2 Locate piercing points 1' and 2' in the top and front views. Intersection line 1'-2' bends around corner AB at P projected from the auxiliary view.

Step 3 The intersection lines from 2' and 1' to 3' do not bend around the corner, but are straight lines. Line 1'-3' is visible, and line 2'-3' is invisible.

Figure 23.8 illustrates how to find the line of intersection between an inclined prism and a vertical prism. An auxiliary view reveals the end view of the inclined prism where its planes appear as edges. In the auxiliary view, plane 1-2 bends around corner AB at point P. Project points of intersection 1' and 2' from the top and auxiliary to their intersections in the front view. Then draw the line of intersection 1'-P-2' for this portion of the line of intersection. Connect the remaining lines, 1'-3' and 2'-3' to complete the solution.

Figure 23.9 shows an alternative method of solving this type of problem. Piercing points 1' and 2' appear in the front view as projections from the top view. Point 5 is the point where line 1'-5-2' bends around vertical corner AB. To find point 5 in the front view, pass a frontal plane through corner AB in the top view and

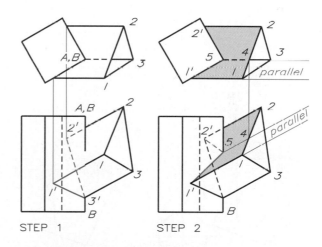

23.9 Intersection: Prisms by projection.
Step 1 Project the piercing points of lines 1, 2, and 3 from the top view to the front view to locate piercing points 1', 2', and 3'.
Step 2 Pass a plane through corner AB in the top view to locate point 5, where intersection line 1'-2' bends around the vertical prism. Find point 5 in the front view and draw line 1'-5'-2'.

project its trace (4-5) to the front view. Draw the lines of intersection, 1'-5-2'.

Some applications of intersections and developments are massive in size as illustrated by the clean-air system shown in **Figure 23.10.**

23.10 The design of this clean-air system used many applications of large-scale intersections and developments of sheet metal components. *(Courtesy of Kirk & Blum, a CECO Environmental Company.)*

It is composed of a multitude of intersections and developments..

23.4 Intersections: Planes and Cylinders

The standard sheet-metal vent pipe that is common to all homes is an example of a cylinder intersecting a plane. **Figure 23.11** shows how to find the intersection between a plane and a cylinder. Cutting planes passed vertically through the top view of the cylinder establish pairs of elements on the cylinder and their piercing points. Space the cutting planes conveniently apart by eye. Then project the piercing points to each view and draw the elliptical line of intersection.

Figure 23.12 shows the solution of a more general problem. Here, the cylinder is vertical and the plane is oblique and does not appear as an edge. Passing vertical cutting planes through the cylinder and the plane in the top view gives elements on the cylinder and their piercing points on the plane. Projecting these points to the front view completes the elliptical

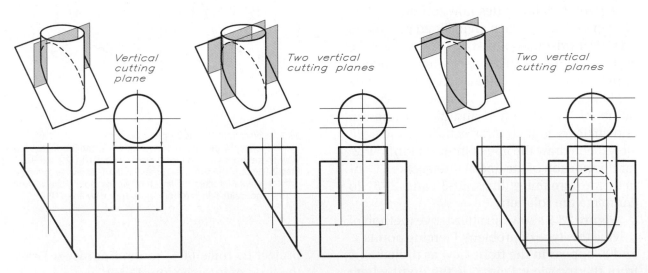

23.11 Intersection: Cylinder and plane.
Step 1 Pass a vertical cutting plane through the cylinder parallel to its axis to find two points of intersection on this plane in the front view.

Step 2 Use two more cutting planes to find four additional points in the top and left side views. Project these points to the front view.

Step 3 Use additional cutting planes to find more points. Determine visibility and connect these points to give an elliptical line of intersection.

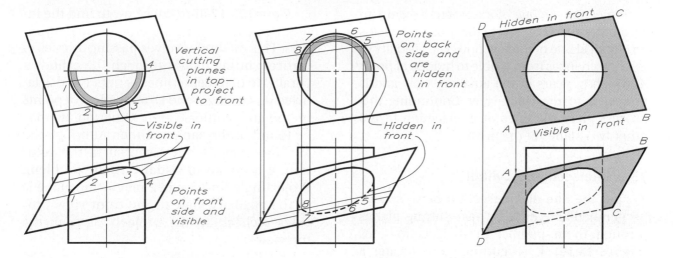

23.12 Intersection: Cylinder and oblique plane.

Step 1 Pass vertical cutting planes through the cylinder in the top view to find elements on it and the plane. Project points 1, 2, 3, and 4 to the front view of their respective lines and connect them with a visible line.

Step 2 Use additional cutting planes to find piercing points—5, 6, 7, and 8; project them to the front view of their respective lines on the plane. The points are on the back side and are hidden; connect them with a hidden line.

Step 3 Determine visibility of the plane and cylinder in the front view. Line AB is visible by inspection of the top view, since it is the farthest out in front. Line CD is the farthest back in the top view and is hidden in the front view.

line of intersection (step 2). The more cutting planes used, the more accurate the line of intersection will be.

Figure 23.13 demonstrates the general case of the intersection between a plane and cylinder, where both the plane and cylinder are oblique in the given views. An auxiliary view is used to show the edge view of the plane. Cutting planes passed through the cylinder parallel to its axis in the auxiliary view establish elements on the cylinder and their piercing points. The points are projected back to the given views and connected to give elliptical line of intersection in the front and side views.

23.5 Intersections: Cylinders and Prisms

An inclined prism intersects a vertical cylinder in **Figure 23.14.** A primary auxiliary view is drawn to show the end view of the inclined prism where its planes appear as edges. A series

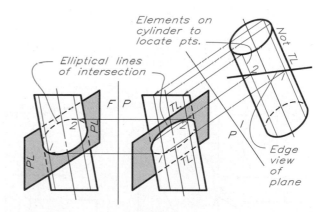

23.13 Intersection: Cylinder and plane by auxiliary.
To find the intersection between an oblique cylinder and an oblique plane, construct a view that shows the plane as an edge. Cutting planes passed through the cylinder locate points on the line of intersection.

of vertical cutting planes in the top view establish lines lying on the surfaces of the cylinder and prism. The cutting planes, also shown in the auxiliary view, are the same distance apart as in the top view.

Projecting the line of intersection from 1 to 3 from the auxiliary view to the front view yields an elliptical line of intersection. The visibility of this line changes from visible to hidden at point X, which appears in the auxiliary view and is projected to the front view. Continuing this process gives the lines of intersection of the other two planes of the prism.

23.6 Intersections: Cylinders

To find the line of intersection between two perpendicular cylinders, pass cutting planes through them parallel to their centerlines (**Figure 23.15**). Each cutting plane locates a pair of elements on both cylinders that intersect at a piercing point. Connecting the points and determining visibility completes the solution. An example of an air-duct system fabricated with intersecting cylinders is shown in **Figure 23.16**. Each cylinder had to be precisely cut to form accurate intersections for tight joints before joining them together.

Figure 23.17 illustrates how to find the intersection between nonperpendicular cylinders. This method involves passing a series of vertical cutting planes through the cylinders parallel to their centerlines. Points 1 and 2, labeled on cutting plane D, are typical of points on the line of intersection. Other points may be found in the same manner. Although the auxiliary view is not essential to the solution, it is an aid in visualizing the problem. Projecting points 1 and 2 on cutting plane D in the auxiliary view to the front view provides a check on the projections from the top view.

23.7 Intersections: Planes and Cones

To find points of intersection on a cone, use cutting planes that are (1) perpendicular to the cone's axis or (2) parallel to the cone's axis. The vertical planes in the top view of **Figure 23.18A** cut radial lines on the cone and establish elements on its surface. The horizontal planes in

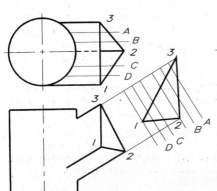

23.14 Intersection: Cylinder and inclined prism.
Step 1 Find the edge views of the planes of the triangular prism in an auxiliary view projected from the front view. Draw frontal cutting planes through the top view and locate them in the auxiliary view with dividers.

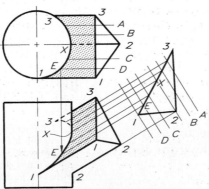

Step 2 Locate points along intersection line 1-3 in the top view and project them to the front view. For example, find point E on cutting plane D in the top and auxiliary views and project it to the front view where the projectors intersect. Visibility changes in the front view at point X.

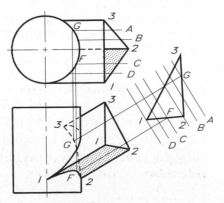

Step 3 Determine the remaining points of intersection by using the other cutting planes. Project point F, shown in the top and auxiliary views, to the front view of line 1-2. Connect the points and determine visibility.

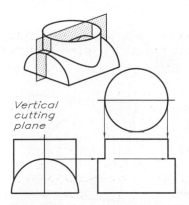

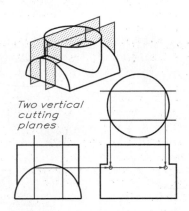

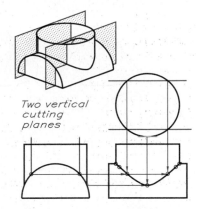

23.15 Intersection: Perpendicular cylinders.
Step 1 Pass a cutting plane through the cylinders parallel to their axes, locating two points of intersection.

Step 2 Use two more cutting planes to find four additional points on the line of intersection.

Step 3 Use two more cutting planes to locate four more points. Connect the points with a smooth curve to complete the line of intersection.

23.16 Intersecting cylindrical ducts of this air-handling system were designed to gather and exhaust contaminants. *(Courtesy of Kirk & Blum, a CECO Environmental Company.)*

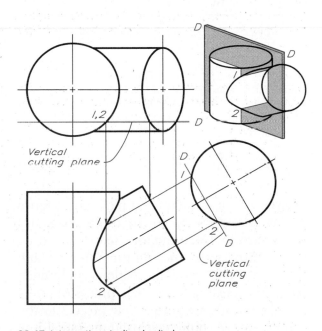

23.17 Intersection: Inclined cylinders.
To find this intersection, find the end view of the inclined cylinder in an auxiliary view. Use a series of vertical cutting planes to find the piercing points of the cylindrical elements and the line of intersection.

Figure 23.18B cut circular sections that appear true size in the top view of a right cone.

A series of radial cutting planes define elements on a cone **(Figure 23.19)**. These elements cross the edge view of the plane in the front view to locate piercing points of each element that, when projected to the top view of the same elements, lie on the line of intersection.

A series of horizontal cutting planes may be used to determine the line of intersection

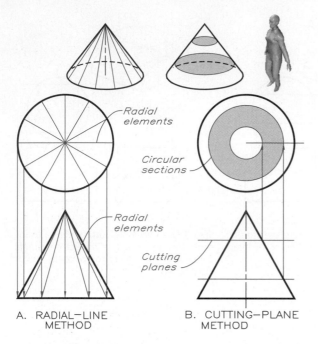

A. RADIAL—LINE
METHOD

B. CUTTING—PLANE
METHOD

23.18 Cutting-plane methods:
To find intersections on conical surfaces, use (A) radial cutting planes that pass through the cone's centerline and are perpendicular to its base, or (B) cutting planes that are parallel to the cone's base.

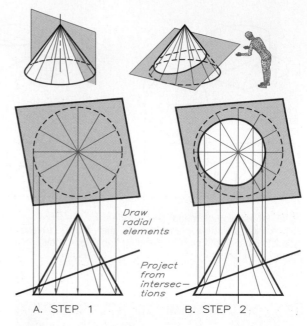

A. STEP 1

B. STEP 2

23.19 Intersection: Plane and a cone:
Step 1 Divide the base evenly in the top view and connect these points with the apex to establish elements on the cone. Project these elements to the front view.
Step 2 Project the piercing point of each element on the edge view of the plane to the top view of the same elements, and connect them to form the line of intersection.

between a cone and an oblique plane **(Figure 23.20).** The sections cut by these imaginary planes are circles in the top view. The cutting planes also locate lines on the oblique plane that intersect the circular sections cut by each respective cutting plane. The points of intersection found in the top view project to the front view. We could have used the radial-line method shown in **Figure 23.19** to obtain the same results.

23.8 Intersections: Cones and Prisms

A primary auxiliary view gives the end view of the inclined prism that intersects the cone in **Figure 23.21.** Cutting planes that radiate from the apex of the cone in the top view locate elements on the cone's surface that intersect the

edge view of the prism in the auxiliary view. These elements are projected to the front view.

Wherever the edge view of plane 1-3 intersects an element in the auxiliary view, the piercing points project to the same element in the front and top views. Passing an extra cutting plane through point 3 in the auxiliary view locates an element that projects to the front and top views. Piercing point 3 projects to this element in sequence from the auxiliary view to the top view. This same procedure yields the piercing points of the other two planes of the prism. All projections of points of intersection originate in the auxiliary view, where the planes of the prism appear as edges.

In **Figure 23.22,** horizontal cutting planes passed through the front view of the cone and

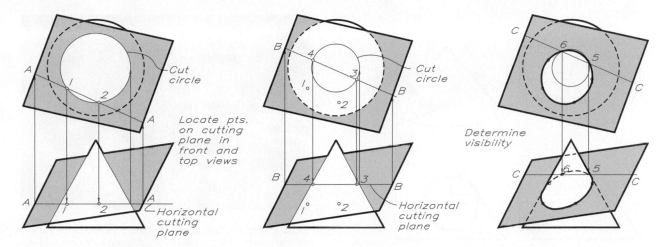

23.20 Intersection: Oblique plane and cone.

Step 1 Pass a horizontal cutting plane through the front view to find a circular section on the cone and a line on the plane in the top view. The piercing points of this line are on the circle. Project points 1 and 2 to the front view.

Step 2 Pass horizontal cutting plane B-B through the front view in the same manner to locate piercing points 3 and 4 in the top view. Project these points to the horizontal plane in the front view from the top view.

Step 3 Use additional horizontal planes to find a sufficient number of points to complete the line of intersection in the same manner as covered in the previous steps. Draw the intersection and determine visibility.

cylinder give a series of circular sections in the top view. Points 1 and 2, shown on cutting plane C in the top view, are typical and project to the front view. The same method produces similar points.

This method is feasible only when the centerline of the cylinder is perpendicular to the axis of the cone, producing circular sections in the top view (rather than elliptical sections, which would be difficult to draw). A series of

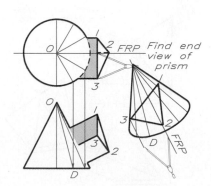

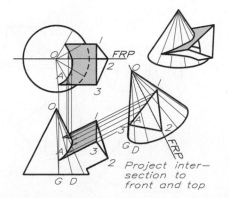

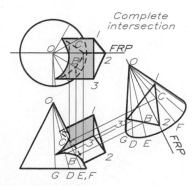

23.21 Intersection: Cone and prism.

Step 1 Draw an auxiliary view to obtain the edge views of the planes of the prism. In the top view, pass vertical cutting planes through the cone through apex O. Project these elements to the front and auxiliary views.

Step 2 Find the piercing points of the cone's elements with the edge view of plane 1-3 in the auxiliary view and project them to the front and top views. For example, point A lies on element OD in the auxiliary view, so project it to the front and top views of OD.

Step 3 Locate the piercing points where the conical elements intersect the edge views of the planes of the prism in the auxiliary view. For example, find point B on OE in the primary auxiliary view and project it to the front and top views of OE.

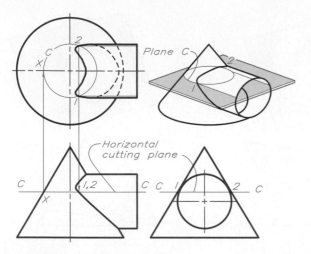

23.23 Examples of intersecting cylinders are shown in this gas turbine power plant in Hong Kong. (*Courtesy of General Electric Company.*)

23.22 Intersection: Horizontal planes.
Horizontal cutting planes are used to find the intersection between the cone and the cylinder. The cutting planes cut circles in the top view. Only one cutting plane is shown here as an example.

intersecting cylinders can be seen in this gas-powered turbine in **Figure 23.23.**

23.9 Intersections: Pyramids

Figure 23.24 shows how to find the intersection of an inclined prism with a pyramid. An auxiliary view shows the end view of the inclined

prism and the pyramid. The radial lines OB and OA drawn through corners 1 and 2 in the auxiliary view project back to the front and top views. Projection locates intersecting points 1 and 2 on lines OB and OA in each view. Point P is the point where line 1-2 bends around corner OC. Finding lines of intersection 1-3 and 2-3 and determining visibility completes the solution.

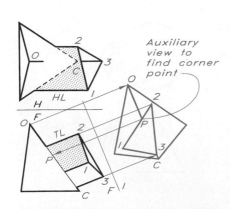

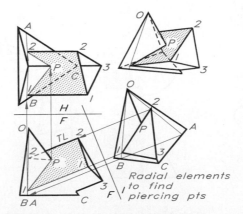

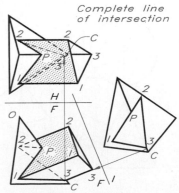

23.24 Intersection: Prism and pyramid.
Step 1 Find the edge views of the planes of the prism in an auxiliary view. Project the pyramid into this view also, showing only the visible surfaces.

Step 2 Pass planes A and B through O and points 1 and 2 in the auxiliary view. Project OA and OB to the front and top views; project 1 and 2 to them. Point P lies on OC. Connect 1, 2, and P for the intersection of this plane.

Step 3 Point 3 lies on OC in the auxiliary view. Project this point to the principal views. Connect 3 to points 1 and 2 to complete the intersections and show visibility. Assume that these shapes are constructed of sheet metal.

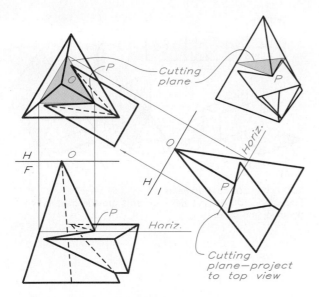

23.25 Intersection: Pyramid and prism.
The intersection of this pyramid and prism is found by obtaining the end view of the prism in an auxiliary view. Horizontal cutting planes are passed through the fold lines of the prism to find the piercing points on the line of intersection. One cutting plane that is used to find corner point P is shown.

23.26 Essentially the entire body of an aircraft, such as the Super Hornet shown in the foreground, is a series of applications of intersections and developments.
(Courtesy of Northrop Grumman Corporation.)

Figure 23.25 shows a horizontal prism that intersects a pyramid. An auxiliary view depicts the end view of the horizontal prism with its planes as edges. Passing a series of horizontal cutting planes through the corner points of the horizontal prism and the pyramid in the auxiliary view gives the lines of intersection, which form triangular sections in the top view.

The cutting plane through corner point P in the auxiliary view is an example of a typical cutting plane. At point P the line of intersection of this plane bends around the corner of the pyramid. Other cutting planes are passed through the corner lines of the prism in the auxiliary and front views. Each corner line of the prism extends in the top view to intersect the triangular section formed by the cutting plane in the same manner P was found.

23.10 Principles of Developments

The bodies of aircraft (**Figure 23.26**) are designed to be fabricated from sheet-metal stock that is formed into shape. Although aircraft design is one of the most advanced applications of **developments,** the principles are the same as for the design of a garbage can.

Figure 23.27 illustrates some of the standard edges and joints for sheet metal. The application determines the type of seam that is used.

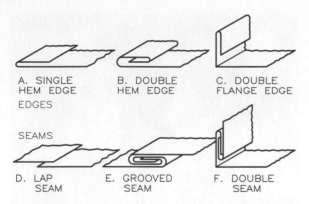

A. SINGLE HEM EDGE B. DOUBLE HEM EDGE C. DOUBLE FLANGE EDGE

EDGES

SEAMS

D. LAP SEAM E. GROOVED SEAM F. DOUBLE SEAM

23.27 These are examples of several types of edges and seams used to join sheet-metal developments. Other seams are joined by riveting and welding.

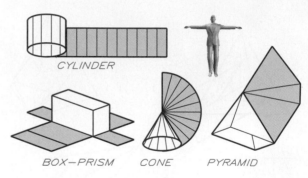

CYLINDER

BOX—PRISM CONE PYRAMID

23.28 Types of developments.
Four standard types of developments are the box, cylinder, pyramid, and cone.

The development of patterns for four typical shapes is shown in **Figure 23.28.** The sides of a box are unfolded into a common plane. The cylinder is rolled out along a stretch-out line equal in length to its circumference. The pattern of a right cone and right pyramid are developed with the length of an element serving as a radius for drawing the base arc.

The construction of patterns for geometric shapes with parallel elements, such as the prisms and cylinders shown in **Figures 23.29A and B,** begins with drawing stretch-out lines parallel to the edge views of the shapes' right sections. The distance around the right section becomes the length of the stretch-out line. The prism and cylinder in **Figures 23.29C and D** are inclined, so their right sections are perpendicular to their sides, not parallel to their bases.

In development, an inside pattern is preferable to an outside pattern for two reasons: (1) most bending machines are designed to fold metal inward, and (2) markings and scribings will be hidden. The designer labels the patterns with a series of lettered or numbered points on the layouts. **All lines on developments must**

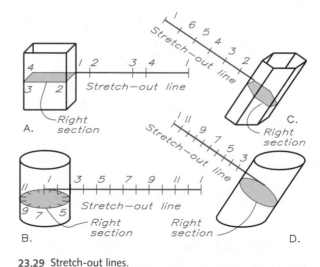

23.29 Stretch-out lines.
A–B Obtain the developments of right prisms and right cylinders by rolling out the right sections along a stretch-out line.
C–D Draw stretch-out lines parallel to the edge views of the right section of cylinders and prisms, and perpendicular to their true-length elements.

be true length. Patterns should be laid out so that the seam line (a line where the pattern is joined) is the shortest line in order to reduce the expense of riveting or welding the seams.

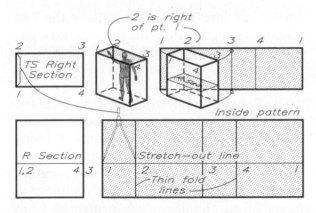

23.30 Development: Prism.
Develop the inside pattern of a rectangular prism by drawing the stretch-out line parallel to the edge view of the right section. Transfer the distances between the fold lines from the true-size right section to the stretch-out line.

23.31 The assembly of the Premier I is an example of a fuselage designed by applying principles of intersections and developments. *(Courtesy of Raytheon Aircraft.)*

23.11 Developments: Rectangular Prisms

The development of a flat pattern for a rectangular prism is illustrated in **Figure 23.30.** The edges of the prism are vertical and true length in the front view. The right section is perpendicular to these sides and the right section is true size in the top view. The stretch-out line begins with point 1 and is drawn parallel to the edge view of the right section.

If an inside pattern is to be laid out to the right, you must determine which point is to the right of the beginning point, point 1. Let's assume that you are standing inside the top view and are looking at point 1: You will see point 2 to the right of point 1.

To locate the fold lines of the pattern, transfer lines 2-3, 3-4, and 4-1 with your dividers from the right section in the top view to the stretch-out line. The length of each fold line is its projected true length from the front view. Connect the ends of the fold lines to form the boundary of the developed surface. Draw the fold lines as thin, dark lines and the outside lines as thicker, visible object lines.

The fuselage of the Premier I in **Figure 23.31** was designed using the principles of intersections and developments. Development of the prism depicted in **Figure 23.32** is similar to that shown in **Figure 23.30.** Here, though, one of its ends is beveled (truncated) rather than square. The stretch-out line is parallel to the edge view of the right section in the front view. Lay off the true-length distances around the right section along the stretch-out line (beginning with the shortest one) and locate the fold lines. Find the lengths of the fold lines by projecting from the front view of these lines.

23.12 Developments: Oblique Prisms

The prism shown in **Figure 23.33** is inclined to the horizontal plane, but its fold lines are true length in the front view. The right section is an edge perpendicular to these TL fold lines, and the stretch-out line is parallel to the edge of the right section. A true-size view of the right section is found in the auxiliary view.

Transfer the distances between the fold lines from the true-size right section to the

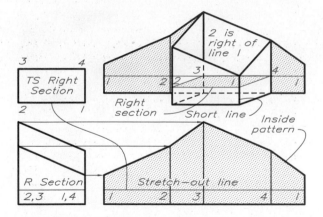

23.32 Development: Truncated prism.
Develop an inside pattern of a rectangular prism with a beveled end by drawing the stretch-out line parallel to the right section. Find the fold lines by transferring distances between the fold lines from the true-size right section to the stretch-out line.

stretch-out line. Find the lengths of the fold lines by projecting from the front view. Determine the ends of the prism and attach them to the pattern so that they can be folded into position.

In **Figure 23.34,** the fold lines of the prism are true length in the top view, and the edge view of the right section is perpendicular to them. The stretch-out line is parallel to the edge view of the right section, and the true size of the right section appears in an auxiliary view projected from the top view. Transfer the distances about the right section to the stretch-out line to locate the fold lines, beginning with the shortest line. Find the lengths of the fold lines by projecting from the top view. Attach the end portions to the pattern to complete the construction.

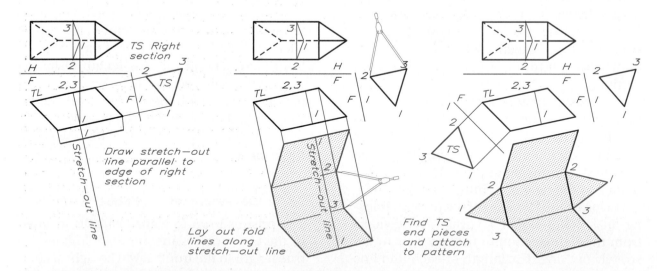

23.33 Development: Oblique prism.
Step 1 Draw the edge view of the right section perpendicular to the true-length axis in the front view. Find the true-size view of the right section in the auxiliary view. Draw the stretch-out line parallel to the edge of the right section. The line through point 1 is the first line of the development.

Step 2 Because the pattern is to be laid out to the right from line 1, the next point is line 2 (from the auxiliary view). Transfer true-length lines 1-2, 2-3, and 3-1 from the right section to the stretch-out line to locate fold lines. Determine the lengths of bend lines by projection.

Step 3 Find true-size views of the end pieces by projecting auxiliary views from the front view. Connect these ends to the development to form the completed pattern. Draw fold lines as thin, dark lines and outside lines as thicker, visible object lines.

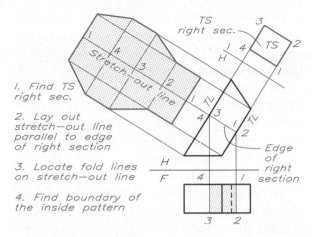

1. Find TS right sec.

2. Lay out stretch—out line parallel to edge of right section

3. Locate fold lines on stretch—out line

4. Find boundary of the inside pattern

23.34 Development: Inclined prism.
Develop this oblique chute by locating the true-size right section in the auxiliary view. Draw the stretch-out line parallel to its right section. Find fold lines by transferring their spacings from the true-size right section to the stretch-out line.

23.35 Development: Prism with a secondary auxiliary.
Develop an oblique prism by drawing a primary auxiliary view to find the fold lines true length and a secondary auxiliary view to find the true-size view of the right section. Use these views to develop the pattern the same way as in Figure 23.34.

A prism that does not project true length in either view may be developed as shown in **Figure 23.35.** The fold lines are true length in an auxiliary view projected from the front view. The right section appears as an edge perpendicular to the fold lines in the primary auxiliary view and true size in a secondary auxiliary view.

Draw the stretch-out line parallel to the edge view of the right section. Locate the fold lines on the stretch-out line by measuring around the right section in the secondary auxiliary view, beginning with the shortest one. Then project the lengths of the fold lines to the development from the primary auxiliary view.

23.13 Developments: Cylinders

Figure 23.36 illustrates how to develop a flat pattern of a right cylinder. The elements of the cylinder are true length in the front view, so

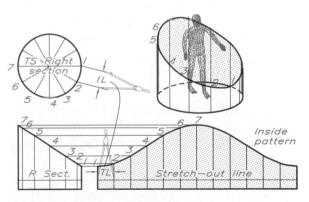

23.36 Development: Truncated cylinder.
Develop an inside pattern of a truncated right cylinder by drawing the stretch-out line parallel to the right section. Transfer points 1 thru 7 from the top view to the stretch-out line that is parallel to the right section. Point 2 is to the right of point 1 for an inside pattern.

the right section appears as an edge in this view and true size in the top view. The stretch-out line is parallel to the edge view of the right section, and point 1 is the beginning point because it lies on the shortest element.

23.37 Examples of cylindrical and rectangular developments are shown in this installation. *(Courtesy of Kirk & Blum, a CECO Environmental Company.)*

Let's assume that you are standing inside the cylinder in the top view and are looking at point 1. You will see that point 2 is to the right of point 1. Therefore, lay off point 2 to the right of point 1 for developing an inside pattern.

By drawing radial lines at 15° or 30° intervals you can equally space the elements in the top view and conveniently lay them out along the stretch-out line as equal measurements. To complete the pattern, find the lengths of the elements by projecting from the front view. An application of a large, developed cylinder joined with a curved cylinder is shown in **Figure 23.37.**

23.14 Development: Oblique Cylinders

The pattern for an oblique cylinder (**Figure 23.38**) involves the same determinations as the preceding cases, but with the additional step of finding a true-size view of the right section in an auxiliary view. First, locate a series of equally spaced elements around the right section in the auxiliary view and project them back to the true-length view. Draw the stretch-out line parallel to the edge view of the right section in the front view.

Lay out the spacing between the elements along the stretch-out line, and draw the elements through these points perpendicular to the stretch-out line. Find the lengths of the elements by projecting from the front view and complete the pattern.

A more general case is the oblique cylinder shown in **Figure 23.39,** where the elements are not true length in the given views. A primary auxiliary view gives the elements true length, and a secondary auxiliary view yields a true-size view of the right section. Draw the stretch-out line parallel to the edge view of the right section in the primary auxiliary view. Transfer the elements to the stretch-out line from the true-size right section.

Draw the elements perpendicular to the stretch-out line and find their lengths by projecting from the primary auxiliary view. Connect the endpoints with a smooth curve to complete the pattern.

23.15 Developments: Pyramids

All lines used to draw patterns must be true length, but pyramids have few lines that are true length in the given views. For this reason you must find the sloping corner lines' true length before drawing a development.

Figure 23.40 shows the method of finding the corner lines of a pyramid's true length by revolution. Revolve line O-5 into the frontal plane to line O-5′ in the top view so that it will be true length in the front view. An application of the development of a pyramid is the sheet-metal hopper shown in **Figure 23.41.**

Figure 23.42 shows the development of a right pyramid. Line O-1 is revolved into the frontal plane in the top view to find its true length in the front view. Because it is a right pyramid, all corner lines are equal in length.

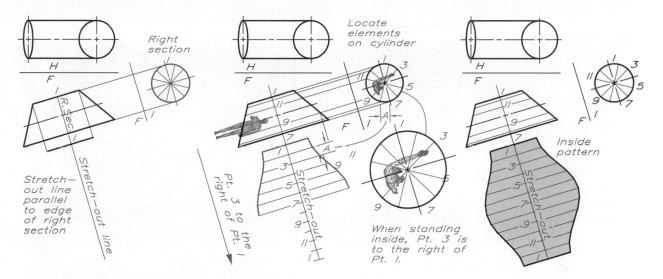

23.38 Development: Oblique cylinder:
Step 1 Draw the right section perpendicular to the true-length axis in the front view. Draw an auxiliary view to find the right section true size; divide it into equal chords. Draw a stretch-out line parallel to the edge of the right section. Locate the shortest line at 1.

Step 2 Project elements from the right section to the front view. Transfer the chordal measurements in the auxiliary view to the stretch-out line to locate cylindrical elements and determine their lengths by projection.

Step 3 Locate the remaining elements to complete the construction as begun in step 2. Connect the ends of the elements with a smooth curve. This is an inside pattern with a seam along its shortest element.

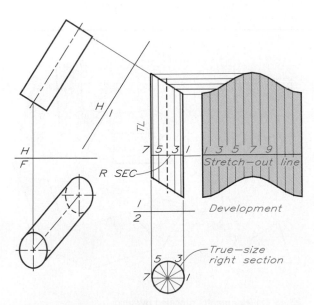

23.39 Development: Cylinder–secondary auxiliary.
Develop an oblique cylinder by drawing a primary auxiliary view showing its elements TL. Find the right section true size in a secondary auxiliary view; complete the construction as shown in Figure 23.38.

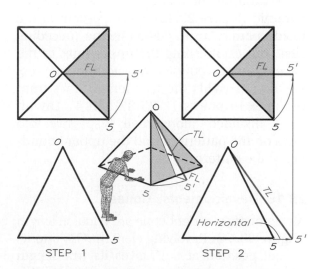

23.40 Elements true length by revolution: Pyramid.
Step 1 Find the true length of corner line 0-5 of the pyramid by revolving it into the frontal plane in the top view to 0-5'.
Step 2 Project point 5' to the front view where frontal line 0-5' is true length.

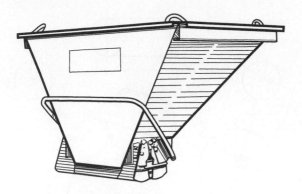

23.41 This sheet-metal hopper was designed by applying the principles of the development of a pyramid.

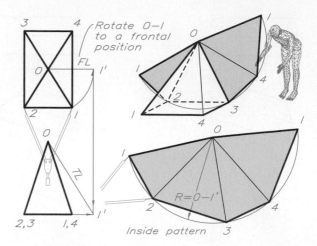

23.42 Development: Pyramid.
Develop this right pyramid by laying out an arc by using the true length of corner line O-1′ as the radius. Transfer true-length distances around the base in the top view to the arc by triangulation, and darken the lines.

Line O-1′ is the radius for the base circle of the development. When you transfer distance 1-2 from the base in the top view to the development, it forms a chord on the base circle. Find lines 2-3, 3-4, and 4-1 in the same manner and in sequence. Draw the fold lines as thin lines from the base to the apex, point O.

A variation of this case is the truncated pyramid **(Figure 23.43).** Development of the inside pattern proceeds as in the preceding case, but establishing the upper lines of the development requires an additional step. Revolution yields the true-length lines from the apex to points 1′, 2′, 3′, and 4′. Lay off these distances along their respective fold lines on the pattern to find the upper boundary of the pattern.

23.16 Developments: Cones

All elements of a right cone are equal in length **(Figure 23.44).** Revolving element O-6 into its frontal position at O-6′ gives its true length when projected to the front view. Line O-6′ is true length and is the outside element of the cone. Projecting point 7 horizontally to element O-6′ locates point 7′, and 0-7′ its true length.

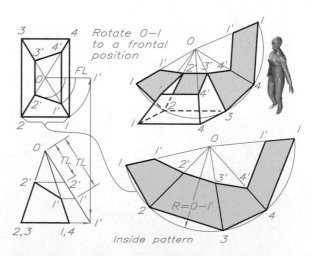

23.43 Development: Truncated pyramid.
Develop an inside pattern of a truncated right pyramid by using the method shown in Figure 23.42. Find the true lengths of elements O-1′, O-2′, O-3′ and O-4′ in the front view by revolution. Lay them off along their respective elements to find the upper boundary of the pattern.

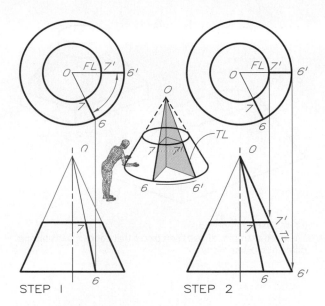

23.44 True length by revolution: Cone.
Step 1 Revolve an element of a cone, O-6, into a frontal plane in the top view, O-6'.
Step 2 Project point 6' to the front view where it is a true-length outside element of the cone. Find the true length of line O-7' by projecting point 7 to the outside element in the front view, 7'.

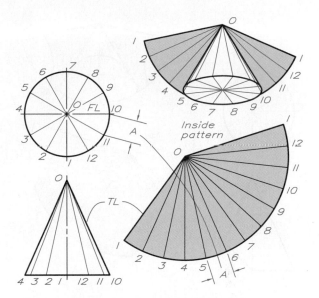

23.45 Development: Cone.
Develop an inside pattern of a right cone by using a true-length element (O-4 or O-10 in the front view) as the radius. Transfer chordal distances from the true-size base in the top view and mark them off along the arc.

To develop the right cone depicted in **Figure 23.45,** divide the base into equally spaced elements in the top view and project them to the front view, where they radiate to the apex at O. The outside elements in the front view, O-10 and O-4, are true length.

Using element O-10 as a radius, draw the base arc of the development. The spacing of the elements along the base circle is equal to the chordal distances between them on the base in the top view. Inspection of the top view from the inside, where point 2 is to the right of point 1, indicates that this is an inside pattern. The nose of the Super Hornet in **Figure 23.26** is an example of cone developed in sheet metal.

Figure 23.46 shows the development of a truncated cone. To find its pattern, lay out the entire cone by using the true-length element O-1 as the radius, ignoring the portion removed from it. Locate the hyperbolic section formed by the inclined plane through the front view of the cone in the top view by projecting points on each element of the cone to the top view of these elements. For example, determine the true length of line O-3' by projecting point 3' horizontally to the true-length element O-1 in the front view. Lay off these distances, and the others, along their respective elements to establish a smooth curve.

23.17 Developments: Transition Pieces

A transition piece changes the shape of a section at one end to a different shape at the

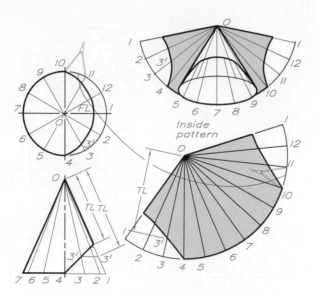

23.46 Development: Truncated cone.
To develop a conical surface with a side opening, begin by laying it out as in Figure 23.45. Find true-length elements by revolution in the front view and transfer them to their respective elements in the pattern.

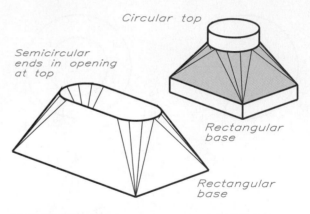

23.47 Transition pieces.
These are examples of transition pieces that connect parts having different cross sections.

23.48 These transition pieces join cylindrical ducts with rectangular hoods to keep the environment of this automotive plant mistfree.
(Courtesy of Kirk & Blum, a CECO Environmental Company.)

other end. In **Figure 23.47** you can see examples of transition pieces that convert one cross-sectional shape to another. Transition pieces in industrial applications vary from being huge to relatively small **(Figure 23.48).**

Figure 23.49 shows the steps in the development of a transition piece. Radial elements are extended from each corner to the equally spaced points on the circular end of the piece. Revolution is used to find the true length of each line. True-length lines 2-D, 3-D, and 2-3 yield the inside pattern of 2-3-D.

The true-length radial lines, used in combination with the true-length chordal distances in the top view, give a series of abutting triangles to form the pattern beginning with element D2. Adding the triangles A-1-2 and G-3-4 at each end of the pattern completes the development of a half-pattern in step 3. Only a half pattern is shown in this example.

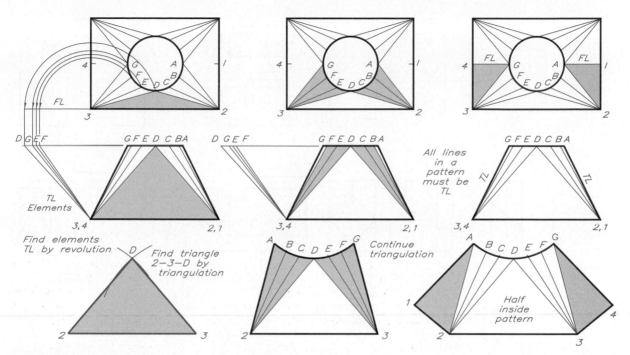

23.49 Development: Transition piece.

Step 1 Divide the circular end into equal parts in the top view and connect these points with lines to corner points 2 and 3. Find the true length of these lines by revolving and projecting them to the front view. Using the true-length lines, draw triangle 2-3-D.

Step 2 Using other true-length lines and the chordal distances on the circular end in the top view, draw a series of triangles joined at common sides. For example, draw arc 2-C from point 2. To find point C, draw arc DC from D. Chord DC is true length in the top view.

Step 3 Construct the remaining planes, A-1-2 and G-3-4, by triangulation to complete the inside half-pattern of the transition piece. Draw the fold lines, where the surface is to be bent, as thin lines. The seam line for the pattern is line A-1, the shortest line.

Problems

Two solutions can be drawn on a size A sheet when using a grid size of 0.20 inch or 5 mm. One solution can be drawn on a size A sheet when using a grid size of 0.40 inch or 10 mm.

Intersections

1–24. (Figure 23.50) Lay out the given views of the problems and find the intersections between the shapes that are necessary to complete the views.

Developments

25–48. (Figure 23.51) Lay out the problems and draw their inside developments. Orient the long side of the sheet horizontally to allow space at the right of the given views for the development.

Choose the shortest line on the pattern as the parting line, draw the inside pattern, and label all significant points with 1/8-inch letters or numerals.

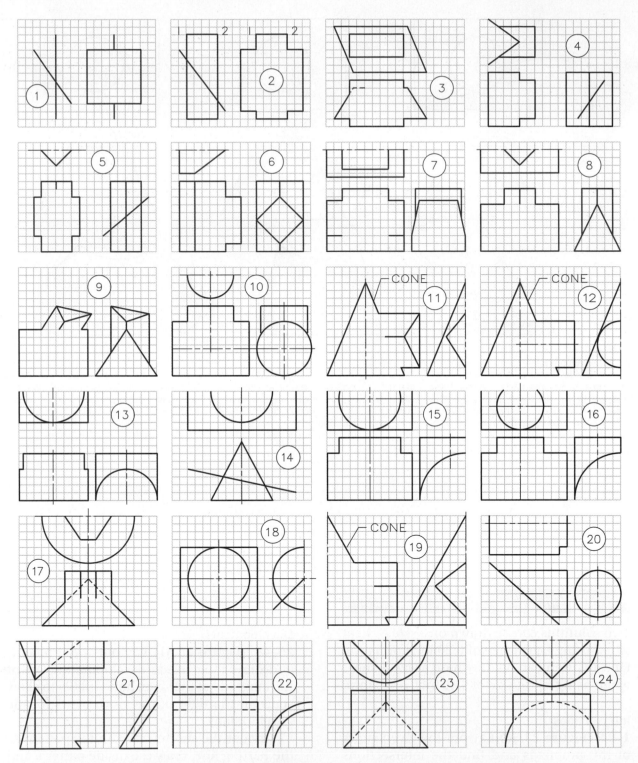

23.50 Problems 1–24. Intersections.

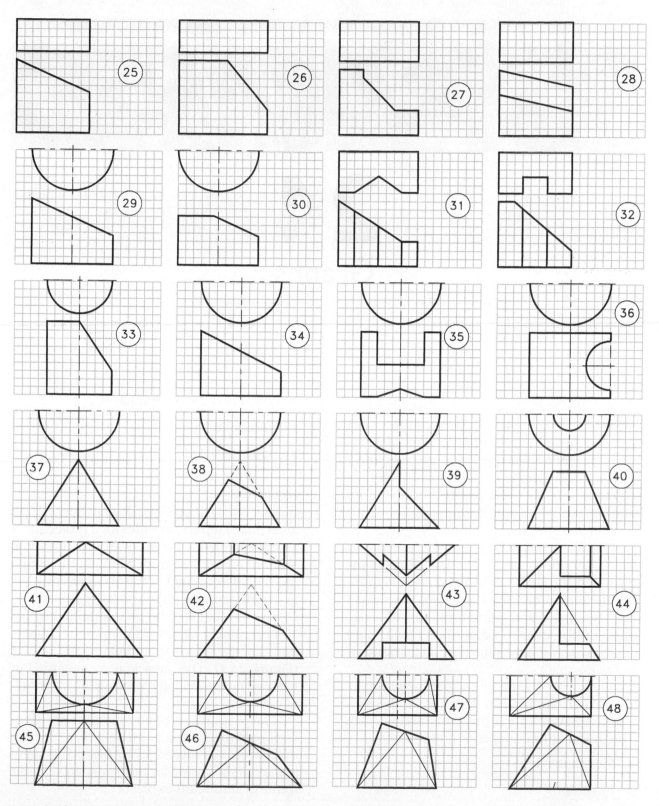

23.51 Problems 25–48. Developments.

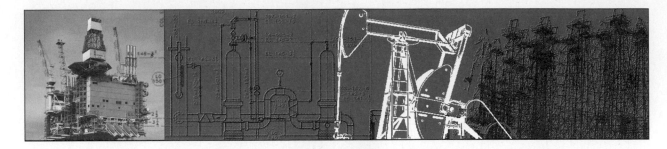

24
Graphs

24.1 Introduction

Data and information expressed as numbers and words are usually difficult to analyze or evaluate unless transcribed into graphical form, or as a **graph.** The term **chart** is an acceptable substitute for graph, but it is more appropriate when applied to maps, a specialized form of graphs.

Graphs are especially useful in presenting data at briefings where the data must be interpreted and communicated quickly to those in attendance. Graphs are convenient means of condensing and presenting data visually, allowing the data to be grasped much more easily than when presented as tables of numbers or verbally.

Several different types of graphs are widely used. Their application depends on the data and the nature of the presentation required. The most common types of graphs are:

1. Pie graphs

2. Bar graphs

3. Linear coordinate graphs

4. Logarithmic coordinate graphs

5. Semilogarithmic coordinate graphs

Proportions

Graphs are used on large display boards, and in technical reports as slides for a projector, or as transparencies for an overhead projector. Consequently, the proportion of the graph must be determined before it is constructed to match the page, slide, or transparency.

A graph that is to be photographed with a 35-mm camera must be drawn to the proportions of the film, or approximately 3×2 **(Figure 24.1).** This area may be enlarged or reduced proportionally by using the diagonal-line method.

The proportions of an overhead projector transparency are approximately 10×8. Image size should not exceed 9.5 inches $\times$ 7.5 inches to allow a mounting for the transparency on a frame of cardboard or plastic.

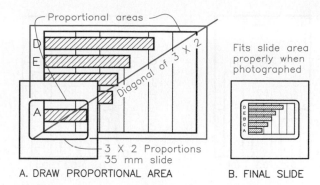

A. DRAW PROPORTIONAL AREA   B. FINAL SLIDE

24.1 This diagonal-line method may be used to lay out drawings that are proportional to the area of a 35-mm slide.

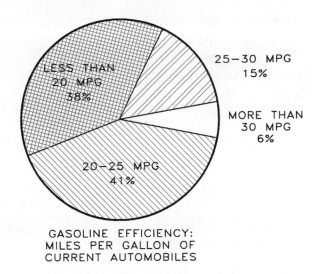

GASOLINE EFFICIENCY:
MILES PER GALLON OF
CURRENT AUTOMOBILES

24.2 A pie graph shows the relationship of parts to a whole. It is most effective when there are only a few parts.

24.2 Pie Graphs

Pie graphs compare the relationship of parts to a whole. For example, **Figure 24.2** shows a pie graph that compares the energy efficiency in miles per gallon of the current models of automobiles on the market.

Figure 24.3 illustrates the steps involved in drawing a pie graph. The data in this example, as simple as it is, are not as easily compared in numerical form as when drawn as a pie graph. Position thin sectors of a pie graph as nearly horizontal as possible to provide more space for labeling. When space is not available within the sectors, place labels outside the pie graph and, if necessary, use leaders **(Figure 24.2).** Showing the percentage represented by each sector is important and giving the actual numbers or values as part of the label is also desirable.

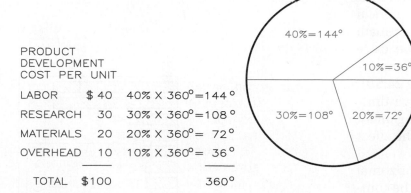

PRODUCT DEVELOPMENT COST PER UNIT		
LABOR	$ 40	40% X 360°=144°
RESEARCH	30	30% X 360°=108°
MATERIALS	20	20% X 360°= 72°
OVERHEAD	10	10% X 360°= 36°
TOTAL	$100	360°

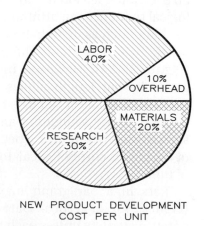

NEW PRODUCT DEVELOPMENT
COST PER UNIT

24.3 Pie graphs:
Step 1 Find the sum of the parts and the percentage that each is of the total. Multiply each percentage by 360° to obtain the angle of each sector.

Step 2 Draw the circle and construct each sector using the degrees of each from step 1. Place small sectors as nearly horizontal as possible.

Step 3 Label sectors with their proper names and percentages. Exact numbers also may be included in each sector to add more clarity.

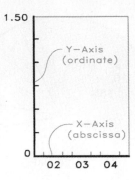

DIVIDENDS PAID BY THE APEX COMPANY	
YEAR	DIVIDEND
2002	$0.40
2003	0.60
2004	0.90

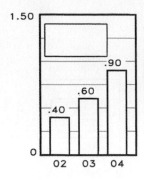

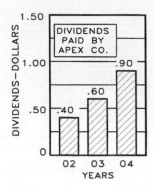

24.4 Drawing a bar graph:
Given: These numerical data are to be plotted as a bar graph for better presentation in a report.

Step 1 Scale the vertical and horizontal axes so that the data will fit on the grid. Begin the bars at zero.

Step 2 Bar widths should be greater than the space between them. Lines should not cross the bars.

Step 3 Strengthen lines, place a title in the graph, label the axes, and cross-hatch the bars.

24.3 Bar Graphs

Bar graphs are widely used for comparing values because the general public understands them. **Figure 24.4** shows how to convert data into a bar graph that can be used in a report or briefing. The axes of the graph carry labels, and its title is placed inside the graph where space is available.

The bars of a bar graph should be sorted in ascending or descending order unless there is an overriding reason not to, such as a chronological sequence. An arbitrary arrangement of the bars, such as in alphabetical or numerical order, makes a graph difficult to evaluate **(Figure 24.5A).** However, ranking the categories by bar length allows easier comparisons from smallest to largest **(Figure 24.5B).** If the data are sequential and involve time, such as sales per month, a better arrangement of the bars is chronological, to show the effect of time.

Bars in a bar graph may be horizontal **(Figure 24.6)** or vertical. Data cannot be compared accurately unless each bar is full length and originates at zero. Also, bars should not extend beyond the limits of the graph (giving the impression that the data were "too hot" to hold).

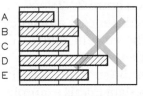

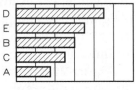

A. POOR: Not sorted   B. GOOD: Sorted

24.5 Arranging bars by length.
A When bars are arbitrarily arranged, such as alphabetically, the bar graph is difficult to interpret.
B When the bars are sorted by length, the graph is much easier to interpret.

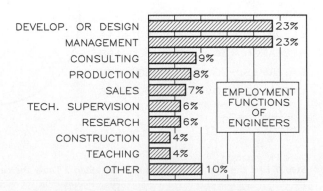

24.6 The horizontal bars of this graph are arranged in descending order to show the employment functions of engineers.

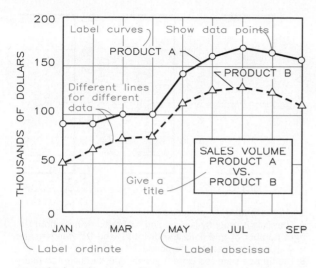

24.7 This basic linear coordinate graph illustrates the important features on a graph.

24.4 Linear Coordinate Graphs

Figure 24.7 shows a typical linear coordinate graph, with notes explaining its important features. Divided into equal divisions, the axes are referred to as linear scales. Data points are plotted on the grid by using measurements, called **coordinates,** along each axis from zero. The plotted points are marked with

symbols such as circles or squares that may be easily drawn with a template. The horizontal scale of the graph is called the **abscissa,** or *x* axis. The vertical scale is called the **ordinate,** or *y* axis.

When the points have been plotted, a curve is drawn through them to represent the data. The line drawn to represent data points is called a **curve** regardless of whether it is a straight line, smooth curve, or broken line. The curve should not extend through the plotted points; rather, the points should be left as open circles or other symbols.

The curve is the most important part of the graph, so it should be drawn as the most prominent (thickest) line. If there are two curves in a graph, they should be drawn as different line types and labeled. The title of the graph is placed in a box inside the graph, and units are given along the *x* and *y* axes with labels identifying the scales of the graph.

Broken-Line Graphs

The steps required to draw a linear coordinate graph are shown in **Figure 24.8.** Because the data points represent sales, which have no predictable pattern, the data do not give a

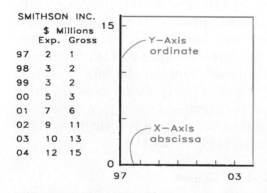
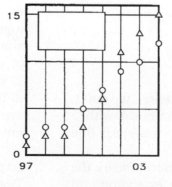
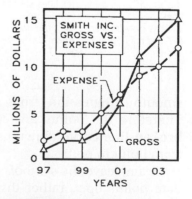

24.8 A broken-line graph:
Given: A record of the Smith Company's gross income and expenses.

Step 1 Lay off the vertical (ordinate) and horizontal (abscissa) axes to provide space for the largest values.

Step 2 Draw division lines and plot the data, using different symbols for each set of data.

Step 3 Connect points with straight lines, label the axes, title the graph, darken the lines, and label the curves.

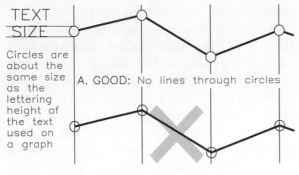

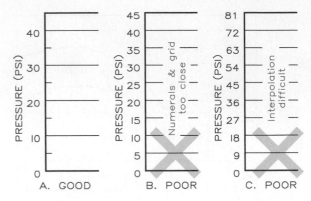

24.9 The curve of a graph drawn from point to point should not extend through the symbols used to represent data points.

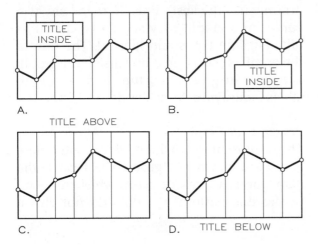

24.10 Placement of titles on a graph.
A and **B** The title of a graph may be placed inside a box within the graph. Box perimeter lines should not coincide with grid lines.
C Titles may be placed over the graph.
D Titles may be placed under the graph.

24.11 Calibrating graph scales.
A The scale is properly labeled and calibrated. It has about the right number of grid lines and divisions, and the numbers are well spaced and easy to interpolate.
B The numbers are too close together, and there are too many grid lines.
C The increments selected make interpolation difficult.

Titles The title of a graph may be located in any of the positions shown in **Figure 24.10**. A graph's title should never be as meaningless as "graph" or "coordinate graph." Instead it should identify concisely what the graph shows.

Calibration and Labeling The calibration and labeling of the axes affects the appearance and readability of a graph. **Figure 24.11A** shows a properly calibrated and labeled axis. **Figures 24.11B** and **24.11C** illustrate common mistakes: placing the grid lines too close together and labeling too many divisions along the axis. In **Figure 24.11C,** the choice of the interval between the labeled values (9 units) makes interpolation between them difficult. For example, locating the value 22 by eye is more difficult on this scale than on the one shown in **Figure 24.11A.**

Smooth-Line Graphs
The strength of concrete related to its curing time is plotted in **Figure 24.12**. The strength of concrete changes gradually and continuously in relation to curing time. Therefore the

smooth progression from point to point. Therefore the points are connected with a **broken-line curve** drawn as an angular line from point to point.

Again, leave the symbols used to mark the data points open rather than extending grid lines or data curves through them **(Figure 24.9).** Each circle or symbol used to plot points should be about 1/8 inch (3 mm) in diameter.

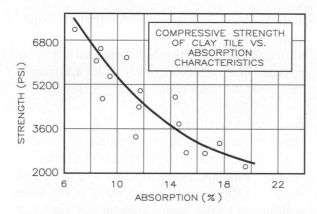

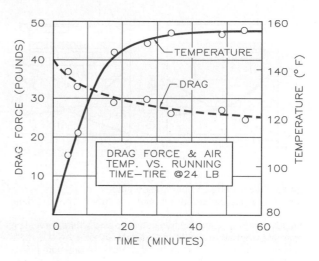

24.12 When the data being graphed involve gradual, continuous changes in relationships, the curve is drawn as a smooth curving line.

24.14 A two-scale graph has different scales along each *y* axis, and labels identify which scale applies to which curve.

interpolations between the plotted points cannot be made.

Straight-Line Graphs

Some graphs have neither broken-line curves nor smooth-line curves, but straight-line curves (**Figure 24.13**). On this graph, a third value can be determined from the two given values. For example, if you are driving 70 miles per hour and you take 5 seconds to react and apply your brakes, you will have traveled 550 feet in that time.

Two-Scale Coordinate Graphs

Graphs may contain different scales in combination, as shown in **Figure 24.14,** where the vertical scale at the left is in units of pounds and the one at the right is in degrees of temperature. Both curves are drawn with respect to their *y* axes and each curve is labeled. Two-scale graphs of this type may be confusing unless they are clearly labeled. Two-scale graphs are effective for comparing related variables, as shown here.

Optimization Graphs

Optimization graphs are effective in comparing two related variables such as an automobile's

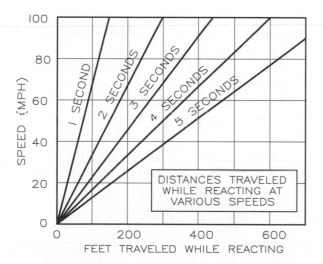

24.13 This graph may be used to determine a third value from the other two variables. For example, select a speed of 70 mph and a time of 5 seconds to find a distance traveled of 550 ft.

data points are connected with a smooth-line curve rather than a broken-line curve. These relationships are represented by the **best-fit curve,** a smooth curve that is an average representation of the points.

A smooth-line curve on a graph implies that interpolations between data points can be made to estimate other values. Data points connected by a broken-line curve imply that

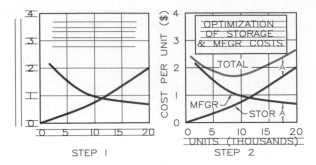

24.15 Constructing an optimization graph:
Step 1 Lay out the graph and plot the curves from the data given.
Step 2 Graphically add the two curves to find a third curve. For example, transfer distance A to locate a point on the third curve. The lowest point of the total curve is the optimum point, or 8000 units.

depreciation in comparison with increasing maintenance costs to determine the optimum time for replacing the car. The point of optimization is at the time when the cost of maintenance is equal to the value of the car.

The steps involved in drawing an optimization graph are illustrated in **Figure 24.15.** Here, the manufacturing cost per unit reduces as more units are made, causing warehousing costs to increase. Adding the two curves to get

a third (total) curve indicates that the optimum number to manufacture at a time is about 8000 units (the low point on the total curve). When more or fewer units are manufactured, the total cost per unit is greater.

Break-Even Graphs

Break-even graphs help in evaluating marketing and manufacturing costs to determine the selling price of a product. As **Figure 24.16** shows, if the desired break-even point for a product is 10,000 units, it must sell for $3.50 each to cover the costs of manufacturing and development.

24.5 Semilogarithmic–Coordinate Graphs

Semilogarithmic graphs are called ratio graphs because they graphically represent ratios. One scale, usually the vertical scale, is logarithmic, and the other is linear (divided into equal divisions).

The same data plotted on a linear grid and on a semilogarithmic grid are compared in **Figure 24.17.** The semilogarithmic graph reveals that the percentage change from 0 to 5 is greater for curve B than for curve A because

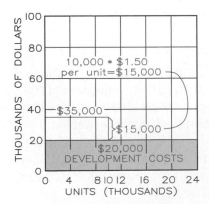

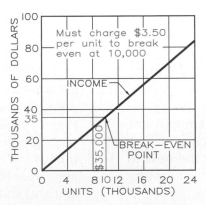

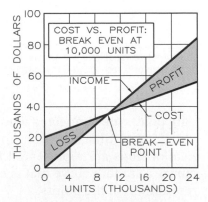

24.16 Drawing a break-even graph:
Step 1 Plot the development cost ($20,000). At $1.50 per unit to make, the total cost would be $35,000 for 10,000 units, the break-even point.

Step 2 To break even at 10,000, the manufacturer must sell each unit for $3.50. Draw a line from zero through the break-even point of $35,000 to represent income.

Step 3 There is a loss of $20,000 at zero units, but progressively less until the break-even point is reached. Profit is the difference between the curves at the right of break-even point.

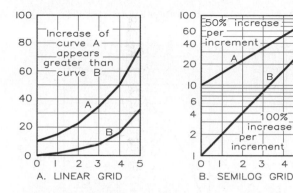

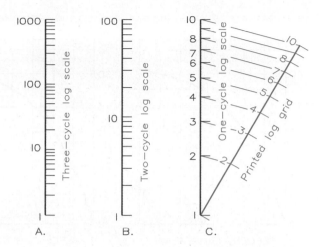

24.17 When plotted on a linear grid, curve A appears to be increasing at a greater rate than curve B. However, plotting the data on a semilogarithmic grid reveals the true rate of change.

24.19 Logarithmic scales may have several cycles: (A) three-cycle scales, (B) two-cycle scales, and (C) one-cycle scales. Calibrations may be projected to a scale of any length from a printed scale as shown here.

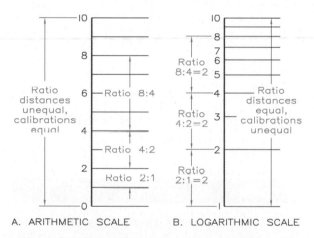

24.18 The divisions on an arithmetic scale are equal and represent unequal ratios between points. The divisions on logarithmic scales are unequal and represent equal ratios.

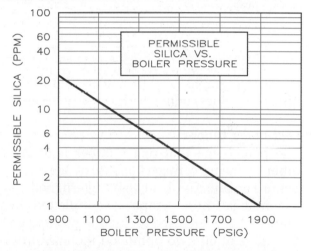

24.20 This semilogarithmic graph relates permissible silica (parts per million) to boiler pressure.

here curve B is steeper. The plot on the linear grid appears to show the opposite result.

Figure 24.18 shows the relationship between the linear scale and the logarithmic scale. Equal divisions along the linear scale have unequal ratios, but equal divisions along the log scale have equal ratios.

Log scales may have one or many cycles. Each cycle increases by a factor of 10. For example, the scale shown in **Figure 24.19A** is a

three-cycle scale, and the one shown in **Figure 24.19B** is a two-cycle scale. When scales must be drawn to a certain length, commercially printed log scales may be used to transfer graphically the calibrations to the scale being used (**Figure 24.19C**).

An application of a semilogarithmic graph for presenting industrial data is illustrated in **Figure 24.20.** People who do not realize that semilog graphs are different from linear

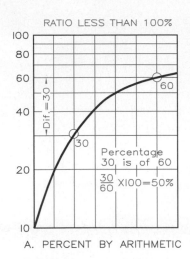

RATIO LESS THAN 100%

Percentage
30 is of 60

$\frac{30}{60}$ X100=50%

A. PERCENT BY ARITHMETIC

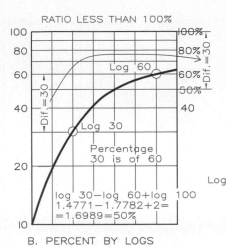

RATIO LESS THAN 100%

Log 60

Log 30

Percentage
30 is of 60

log 30−log 60+log 100
1.4771−1.7782+2=
=1.6989=50%

B. PERCENT BY LOGS

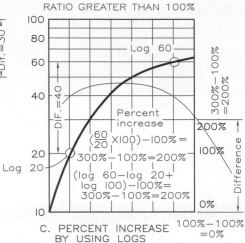

RATIO GREATER THAN 100%

Log 60

Percent
increase
$\left(\frac{60}{20}\right.$ X100)−100% =
300%−100%=200%
(log 60−log 20+
log 100)−100%=
300%−100%=200%

C. PERCENT INCREASE
BY USING LOGS

100%−100%
=0%

24.21 Percentage graphs:
A To find the percentage that one data point is of another point (the percentage that 30 is of 60, for example) you may calculate it mathematically: (30/60)(100) = 50%.

B Find the percentage that 30 is of 60 mathematically by using the logarithms of the numbers. Or find it graphically by transferring the distance between 30 and 60 to the scale at the right, which shows that 30 is 50% of 60.

C To find a percentage increase greater than 100%, divide the smaller number into the larger number. Find the difference between the logs of 60 and 20 with dividers and measure upward from 100% to find the increase of 200%.

coordinate graphs may misunderstand them. Also, zero values cannot be shown on log scales.

Percentage Graphs

The percentage that one number is of another, or the percentage increase of one number to a greater number can be determined on a semilogarithmic graph. Data plotted in **Figure 24.21A** are used to find the percentage that 30 is of 60 (two points on the curve) by arithmetic. The vertical distance between them is the difference between their logarithms, so the percentage can be found graphically in **Figure 24.21B.** The distance from 30 to 60 is transferred to the log scale at the right of the graph and subtracted from the log of 100 to find the value of 50% as a direct reading of percentage.

In **Figure 24.21C,** the percentage increase between two points is transferred from the grid to the lower end of the log scale and measured upward because the increase is greater than zero. These methods may be used

to find percentage increases or decreases for any set of points on the grid.

Problems

Draw your solutions to these problems on size A sheets. Apply the techniques and principles covered in this chapter.

Pie Graphs

1. Draw a pie graph that shows the comparative sources of retirees' income: investments, 34%; employment, 24%; social security, 21%; pensions, 19%; other, 2%.

2. Draw a pie graph that shows the number of members of the technological team: engineers, 985,000; technicians, 932,000; scientists, 410,000.

3. Construct a pie graph of the employment status of graduates of two-year technician programs a year after graduation: employed,

63%; continuing full-time study, 23%; considering job offers, 6%; military, 6%; other, 2%.

4. Draw a pie graph showing the types of degrees held by aerospace engineers: bachelor's, 65%; master's, 29%; Ph.D.'s, 6%.

Bar Graphs

5. Draw a bar graph that shows expected job growth by city: Austin, 1.3%; San Antonio, 0.8%; Houston, 1.6%; Fort Worth, 0%; Dallas, 0.4%; all of Texas, 0.8%.

6. Draw a bar graph that represents 100% of a die casting alloy. The proportional parts of the alloy are: tin, 16%; lead, 24%; zinc, 38.8%; aluminum, 16.4%; copper, 4.8%.

7. Draw a bar graph that compares the number of skilled workers employed in various occupations. Use the following data and arrange the graph for ease of comparing occupations: carpenters, 82,000; all-round machinists, 310,000; plumbers, 350,000; bricklayers, 200,000; appliance servicers, 185,000; automotive mechanics, 760,000; electricians, 380,000; and painters, 400,000.

8. Draw a bar graph that shows the characteristics of a typical U.S. family's spending: housing, 29.6%; food, 15.1%; transportation, 16.7%; clothing, 5.9%; retirement, 8.6%; entertainment, 4.9%; insurance, 5.2%; health care, 3.0%; charity, 3.1%; other, 7.9%.

9. Draw a bar graph that compares the corrosion resistance of the materials listed in the following table.

| | Loss in Weight (%) | |
	in Atmosphere	in Sea Water
Common Steel	100	100
10% nickel steel	70	80
25% nickel steel	20	55

10. Draw a bar graph of the data from Problem 1.

11. Draw a bar graph of the data from Problem 2.

12. Draw a bar graph of the data from Problem 3.

13. Construct a bar graph comparing sales and earnings of Apple Computer from 1980 through 1989. Data are by year for sales and earnings (profit) in billions of dollars: 1980, 0 and 0; 1981, 0.33 and 0.05; 1982, 0.60 and 0.07; 1983, 1.00 and 0.09; 1984, 1.51 and 0.08; 1985, 1.90 and 0.07; 1986, 1.85 and 0.12; 1987, 2.70 and 0.25; 1988, 4.15 and 0.40; 1989, 5.50 and 0.45.

Linear Coordinate Graphs

14. Draw a linear coordinate graph to show the estimated population growth in the U.S. from 1992 through 2050 in millions of people: 1992, 255; 2000, 270; 2010, 295; 2020, 325; 2030, 348; 2040, 360; 2050, 383.

15. Construct a linear coordinate graph that shows the relationship of energy costs (mills per kilowatt-hour) on the y axis to the percent capacity of a nuclear power plant and a gas- or oil-fired power plant on the x axis. Gas- or oil-fired plant data: 17 mills, 10%; 12 mills, 20%; 8 mills, 40%; 7 mills, 60%; 6 mills, 80%; 5.8 mills, 100%. Nuclear plant data: 24 mills, 10%; 14 mills, 20%; 7 mills, 40%; 5 mills, 60%; 4.2 mills, 80%; 3.7 mills, 100%.

16. Plot the data from Problem 13 as a linear coordinate graph.

17. Construct a linear coordinate graph to show the relationship between the transverse resilience in inch-pounds (ip) on the y axis and the single-blow impact in foot-pounds (fp) on the x axis of gray iron. Data: 21 fp, 375 ip; 22 fp, 350 ip; 23 fp, 380 ip; 30 fp, 400 ip; 32 fp, 420 ip; 33 fp, 410 ip; 38 fp, 510 ip; 45 fp, 615 ip; 50 fp, 585 ip; 60 fp, 785 ip; 70 fp, 900 ip; 75 fp, 920 ip.

18. Draw a linear coordinate graph to illustrate the trends in the export of U.S. services and products from 1980 through 1993: 1980, +8.5%; 1981, +2.5%; 1982, −7.5%; 1983, −4.5%; 1984, +7%; 1985, −2%; 1986, +7%; 1987, +12.5%; 1988, +17.5%; 1989, +10%; 1990, +6%; 1991, +3%; 1992, +5%; 1993, +7%.

19. Draw a linear coordinate graph for the centrifugal pump test data in the following table. The units along the x axis are to be gallons per minute. Use two curves to represent the variables given.

Gallons per Minute	Water HP	Electric HP
0	19.0	0.00
75	17.5	0.72
115	15.0	1.00
154	10.0	1.00
185	5.0	0.74
200	3.0	0.63

20. Draw a linear coordinate graph that compares two values for an alloy shown in the table below—ultimate strength and elastic limit—with degrees of temperature (x axis).

F°	Ultimate Strength	Elastic Limit
400	257,500	208,000
500	247,000	224,500
600	232,500	214,000
700	207,500	193,500
800	180,500	169,000
900	159,500	146,500
1000	142,500	128,500
1100	126,500	114,000
1200	114,500	96,500
1300	108,000	85,500

Break-Even Graphs
21. Draw a break-even graph that shows the earnings for a new product with a development cost of $12,000. The break-even point is at 8000 units and each costs $0.50 to manufacture. What would be the profit at volumes of 20,000 and 25,000?

22. Repeat Problem 21 except that the development costs are $80,000, the manufacturing cost of the first 10,000 units is $2.30 each, and the desired break-even point is 10,000 units. What is the profit at volumes of 20,000 and 30,000?

Logarithmic Graphs
23. Use the data given in **Table 24.1** to construct a logarithmic graph. Plot the vibration amplitude (A) as the ordinate and the vibration frequency (F) as the abscissa. The data for curve A1 represent the maximum limits of machinery in good condition with no danger from vibration. The data for curve A2 are the lower limits of machinery that is being vibrated excessively to the danger point. The vertical scale is three cycles, and the horizontal scale is two cycles.

24. Plot this data on a two-cycle log graph to show the current in amperes (y axis) versus the voltage in volts (x axis) of precision temperature-sensing resistors. Data: 1 volt, 1.9 amps; 2 volts, 4 amps; 4 volts, 8 amps; 8 volts, 17 amps; 10 volts, 20 amps; 20 volts, 30 amps; 40 volts, 36 amps; 80 volts, 31 amps; 100 volts, 30 amps.

25. Plot the data in Problem 14 as a logarithmic graph.

26. Plot the data in Problem 20 as a logarithmic graph.

Semilogarithmic Graphs
27. Construct a semilogarithmic graph with the y axis a two-cycle log scale from 1 to 100, and the x axis a linear scale from 1 to 7. The object of the graph is to show the survivability of a shelter at varying distances from the atmospheric detonation of a one-megaton thermonuclear bomb. Plot overpressure in psi along the y axis, and distance from ground zero in miles along the x axis. The data points represent an 80% chance of survival of the shelter. Data: 1 mile, 55 psi; 2 miles, 11 psi; 3 miles, 4.5 psi; 4 miles, 2.5 psi; 5 miles, 2.0 psi; 6 miles, 1.3 psi.

Table 24.1

F	100	200	500	1000	2000	5000	10,000
$A(1)$	0.0028	0.002	0.0015	0.001	0.0006	0.0003	0.00013
$A(2)$	0.06	0.05	0.04	0.03	0.018	0.005	0.001

28. The growth of Division A and Division B of a company is shown by the following data. Plot the data on a semilog graph with a one-cycle log scale on the y axis for sales in thousands of dollars, and a linear scale on the x axis for years. Data: first year, A = $11,700 and B = $44,000; second year, A = $19,500 and B = $50,000; third year, A = $25,000 and B = $55,000; fourth year, A = $32,000 and B = $64,000; fifth year, A = $42,000 and B = $66,000; sixth year, A = $48,000 and B = $75,000. Which division has the better growth rate?

29. Draw a semilog chart showing probable engineering progress based on the following indices: 40,000 B.C., 21; 30,000 B.C., 21.5; 20,000 B.C., 22; 16,000 B.C., 23; 10,000 B.C., 27; 6000 B.C., 34; 4000 B.C., 39; 2000 B.C., 49; 500 B.C., = 60; A.D. 1900, 100. Use a horizontal scale of 1 inch = 10,000 years, a height of about 5 inches, and two-cycle printed paper, if available.

30. Plot the data in Problem 19 as a semilogarithmic graph.

31. Plot the data in Problem 20 as a semilogarithmic graph.

Percentage Graphs

32. Plot the data below as a semilog graph to determine percentages and ratios for the data pertaining to water supply and demand. What is the percentage increase in the demand for water from 1980 to 1920? What percentage of demand is the supply for 1900, 1930, and 1970?

	Supply*	Demand*
1890	80	35
1900	90	35
1910	110	60
1920	135	80
1930	155	110
1940	240	125
1950	270	200
1960	315	320
1970	380	410
1980	450	550

*billions of gallons per day

33. Using the graph plotted in Problem 28, determine the percentage of increase of Division A and Division B growth from year 1 to year 4. What percentage of sales of Division A are the sales of Division B at the end of year 2? At the end of year 6?

34. Plot the values for water horsepower and electric horsepower from Problem 19 on semilog paper against gallons per minute along the x axis. What percentage of electric horsepower is water horsepower when 75 gallons per minute are being pumped?

25

Introduction to AutoCAD® 2005

25.1 Introduction

This chapter provides an introduction to computer graphics using AutoCAD 2005, which runs on a minimum of a Pentium III processor, 256 MB of RAM, at least a 10 Gb of hard-disk space, a mouse (or tablet), and an A-B plotter and/or printer. Windows 2000 up to Windows XP are recommended as the operating system. AutoCAD was selected as the software for presenting computer graphics because it is the most widely-used computer graphics program and it well-adapted to engineering drawings.

The coverage of AutoCAD in this book is brief and many operations could not be included because of space limitations. AutoCAD's concisely written *User's Guide* has 856 pages, and other manuals on the market have as many as 1500 pages. However, AutoCAD is covered here sufficiently well enough to guide you through an introduction to AutoCAD.

You will find that the learning of computer graphics and its successive upgrades will be a career-long experience. We recommend that you begin this self-teaching process by experimenting with the peripheral commands and options that are not covered in this book. Also, reference to *Help* should be made routinely as a means of learning new commands and refreshing your memory when necessary.

25.2 Computer Graphics Overview

The major areas of computer graphics are **CAD** (computer-aided design), **CADD** (computer-aided design and drafting), **CIM** (computer-integrated manufacturing), and **CAD/CAM** (computer-aided design and manufacturing).

CAD (computer-aided design) is used to solve design problems, analyze design data, and store design information for easy retrieval. Many CAD systems perform these functions in an integrated manner, greatly increasing the designer's productivity.

CADD (computer-aided design drafting) is the computer process of making engineering drawings and technical documents more closely related to drafting than is CAD.

25.1 This automatic assembly line uses computer-controlled robots for welding together the parts of the E-Z-GO golf car. *(Courtesy of E-Z-GO. A Textron Company.)*

CAD/CAM (computer-aided design/computer-aided manufacturing) is a system that can be used to design a part or product, devise the production steps, and electronically communicate this data to control the operation of manufacturing equipment and robots.

CIM (computer-integrated manufacturing) is a more advanced system of the CAD/CAM that coordinates and operates all stages of manufacturing from design to finished product **(Figure 25.1)**.

Advantages of CAD and CADD

Computer-graphics systems offer the designer and drafter some or all of the following advantages over hand drawings.

1. Increased accuracy. CAD systems are capable of producing drawings that are essentially 100% accurate in size, line quality, and uniformity.

2. Increased drawing speed. Engineering drawings and documents can be prepared more quickly, especially when standard details from existing libraries are incorporated into new drawings.

3. Easy to revise. Drawings can be more easily modified, changed, and revised than is possible by hand techniques.

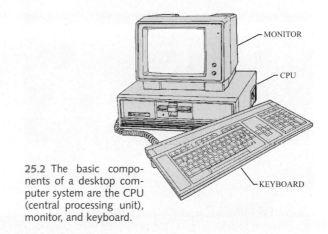

25.2 The basic components of a desktop computer system are the CPU (central processing unit), monitor, and keyboard.

4. Better design analysis. Alternative designs can be analyzed quickly and easily. Software is available to simulate a product's operation and test it under a variety of conditions, which lessens the need for models and prototypes.

5. Better presentation. Drawings can be presented in 2D or 3D and rendered as technical illustrations to better communicate designs.

6. Libraries of drawing aids. Databases of details, symbols, and figures that are used over-and-over can be archived for immediate use in making drawings.

7. Improved filing. Drawings can be conveniently filed, retrieved, and transmitted on disks and tapes.

25.3 Hardware

The hardware of a computer graphics system includes the **computer, monitor, input device** (keyboard, digitizer, mouse, or light pen), and **output device** (plotters and printers).

Computer

The computer, with an installed **program,** receives input from the user through the keyboard, executes the instructions, and produces output. The part of the computer that follows the program's instructions is the **CPU** (central processing unit). The computer graphics computer should have at least 32 MB or RAM, and

25.3 Current technology enables computers to send 3D graphics over networks for instantaneous communication. *(Courtesy of Hewlett Packard Company.)*

25.4 The DeskJet 995ck series of printers produces photoquality color images that are as close to traditional photographs as desktop printers have ever come. *(Courtesy of Hewlett Packard Company.)*

its hard-disk storage should be large, preferably 6 to 8 gigabytes and larger **(Figure 25.2).**

The monitor is a **CRT (cathode-ray tube)** and has an electron gun that emits a beam that sweeps rows of raster lines onto the screen. Each line consists of dots called **pixels.** Raster-scanned CRTs refresh the picture display many times per second. A measure of monitor quality is **resolution,** which is the number of pixels per inch that can be produced on the screen. The greater the number of pixels, the greater the clarity of the image on the screen **(Figure 25.3).**

Input Devices

Besides the standard keyboard, the **digitizer** is used to enter graphic data to the computer. The **mouse,** a hand-held device that is moved about the table top to transmit information to the computer, is the most commonly used input device. Variations of the mouse are **thumbwheels** operated by fingertips, **joysticks** that let the user "steer" about the screen by tilting a lever, and **spherical balls** that can be rotated to input 3D data to the screen.

A **tablet** and **digitizer** used in combination are an alternative to the mouse. The digitizer (stylus) can be used to select commands from the menu attached to the tablet. Also, drawings can be attached to tablets and "traced" with

the stylus to convert it to *x*- and *y*-coordinates. The **light pen** enables the user to "draw" on the screen with it to select points and lines.

Output Devices

Plotters make drawings on paper or film with a pen in the same manner a drawing is made by hand. Plotter types are **flatbed plotters, drum plotters,** and **sheet-fed plotters.** In flatbed plotters, the drawing paper is held stationary while pens are moved about its surface. Drum plotters roll the paper up-and-down over a cylinder while the pen moves left and right to make the drawing. Sheet-fed plotters hold the paper with grit wheels in a flat position as the sheet is moved forward and backward as the pen moves left and right to make the drawing.

Printers are of the impact type (much like typewriters) or nonimpact type where images are formed by sprays, laser beams, photography, or heat **(Figure 25.4).** The laser printer gives an excellent resolution of dense, accurately drawn lines in color, as well as in black and white. Inkjet technology has enabled images to be sprayed onto the drawing surface in color or in black that approaches the quality of the laser. Larger nonimpact printers (24″ × 36″ and larger) are most often inkjet printers since large lasers are much more expensive.

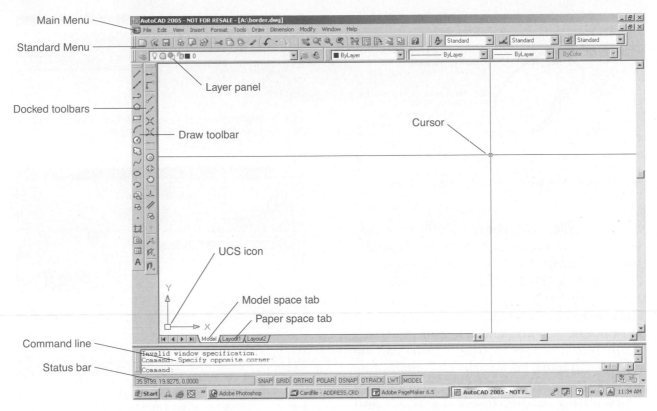

Main Menu

Standard Menu

Docked toolbars

Layer panel

Draw toolbar

Cursor

UCS icon

Model space tab

Paper space tab

Command line

Status bar

25.5 This is a typical view of the AutoCAD 2005 main screen. Additional toolbars can added to and removed from the screen by the user.

25.4 Your First Session

If this is your first session, you are anxious to turn the computer on, make a drawing on the screen, and plot it without reading the instructions. This section is what you're looking for.

Format of Presentation

In this chapter, the progression from one step of a command to the next level will be separated by an angle pointing to the right (>) in order to simplify presentation and reduce explanatory text. Since there are about three different ways of actuating most commands, these methods will be used alternatively in the following examples. The commands and prompts that appear on the screen will be given in italics to distinguish them from supplementary notes of explanation. (Enter) is the keyboard key with this name. Once a command is selected, additional prompts will be given at the *Command* line at the bottom of the screen or in dialog boxes that must be followed.

Booting   Turn on the computer and boot the system by typing AutoCAD 2005 (or the command used by your system) to activate the program **(Figure 25.5).** Experiment by moving the cursor around the screen with your mouse, select items, and try the pull-down menu.

Mouse   Most interactions with the computer will be accomplished with a mouse **(Figure 25.6),** but many commands can be entered at the keyboard (maybe more quickly, after you learn them). Press the left mouse button to click on, select, or pick a command or object; a double-click is needed in some cases. The right button has the same effect as pressing (Enter) on the keyboard.

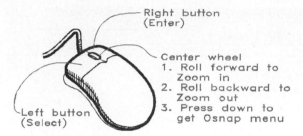

25.6 The left mouse button is used for picking points when drawing and selecting operational buttons. The right mouse button has the same effect as pressing the (Enter) button on the keyboard.

25.7 To begin a new drawing, select *Main Menu > File > New* to obtain the *Create New Drawing* box.

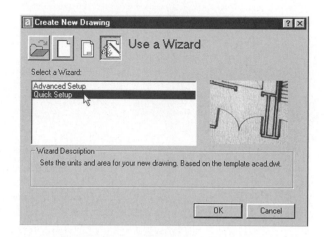

25.8 In the *Create New Drawing* box, select *Use a Wizard* icon button > *Quick Setup* > and *OK.*

Creating a File To create a new file on a disk, place your formatted disk in its slot. From the *Main Menu* bar pick *Files* and *New* from the dialogue box **(Figure 25.7).** The *Create New Drawing* dialog box will appear on the screen **(Figure 25.8).** Select the *Use a Wizard* icon button, *Quick setup,* and *OK,* and the *Quick Setup* box will appear where *Units* can be specified **(Figure 25.9);** select *Decimal* units and the *Next* button. When the *Area* box appears on the screen, insert the *Width* and *Length* (11 × 8.5) and select the *Finish* button **(Figure 25.10).** The program returns to the screen, and its command menus are ready for your drawing. (Note: The *Advanced Setup* is the same as the *Quick Setup,* but it permits the modification of angular specifications.)

Making a Drawing Since you don't have the menus and toolbars figured out, type L (for *Line*) at the *Command* line, press (Enter), and draw some lines on the screen with the mouse for the fun of it by selecting endpoints with the left button as shown in **Figure 25.11.** A line on the screen "rubberbands" from point to point. To disengage the rubberband, press the right mouse button, which is the same as pressing (Enter). Press the right button again to return to the *Line* command.

Instead of typing L to enter the *Line* command, pick the *Main Menu > Draw* and select the *Line* option to get the prompt line *Specify first:* in the *Command* line at the bottom of

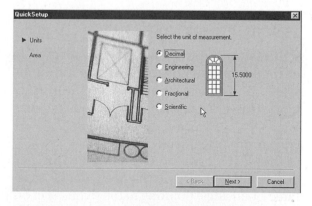

25.9 In the *Quick Setup* box specify the type of *Units* you wish to have (*Decimal* units in this example), and select *Next.*

the screen. Now, use your mouse to draw on the screen. The *Line* command can also be selected from the *Draw* toolbar that you will learn about soon. Try drawing circles and other objects on the screen by selecting icons from the *Draw* pull-down menu.

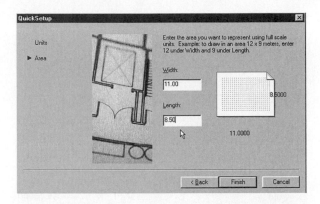

25.10 Type the dimensions for *Width* and *Length* (11 and 8.5 for an A-size sheet) and select *Finish*.

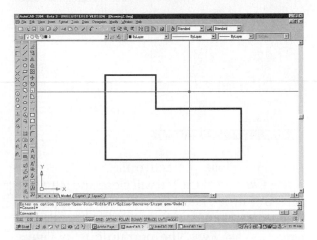

25.11 By *Command: L* (for *Line*), or selecting *Main Menu> Draw> Line*, endpoints of lines can be selected on the screen with the left mouse button to draw a figure.

Repeat Commands By pressing (Enter) on the keyboard (or the right mouse button) twice after the previous command, the command can be repeated. For example, *Line* will appear in the *Command* line at the bottom of the screen after pressing (Enter) twice, if *Line* was the previous command.

Saving Your File On the *Main Menu* click *File* and pick *Save As* from the pull-down menu to get the menu box shown in **(Figure 25.12).** Type <u>A:DRW-5</u> (if your disk is in Drive A) and select the *Save* button; the light over drive A will blink briefly and the drawing named *DRW-5* is saved to the disk in the A drive.

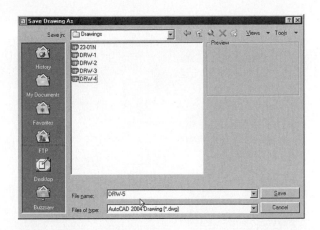

25.12 Select *Main Menu> File> Save As* to obtain the *Save Drawing As* box; type the drawing drive (A:) and the file name *DRW-5* in the *File name* box and select *Save*. File *A:DRW-5* is saved.

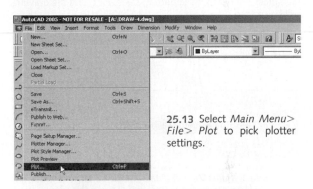

25.13 Select *Main Menu> File> Plot* to pick plotter settings.

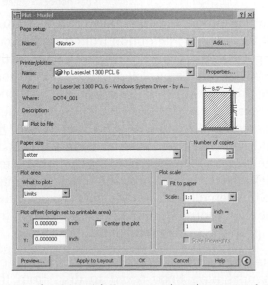

25.14 In the *Printer/plotter* area, select the *Name* of the printer or plotter that you will use, select the *Paper size*, select the *Plot area*, and pick *Center the plot* to place the drawing.

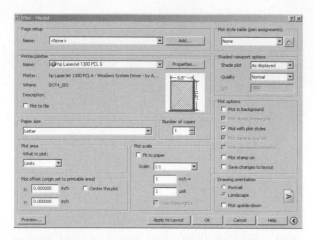

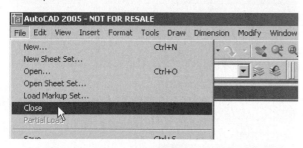

25.15 Assign the *Plot scale* (1:1) and the *Drawing orientation* (*Landscape* or *Portrait*). Click *Preview* to obtain a view of of how your drawing will appear when it is plotted. If it looks correct, press the right mouse button, select *Plot,* and the drawing will be plotted.

25.16 Select *Main Menu> File> Close* to close the current file; you will be prompted to save it if it has not been saved.

Plotting Your Drawing Select *File* **(Figure 25.13)** and the *Plot* option from its pull-down to get the *Plot* dialog box **(Figure 25.14).** In the *Printer/plotter* area select your printer from the drop-down list in the *Name* box. Set *Paper size* to <u>Letter</u>, *Plot area* to <u>Limits</u>, and select *Center the Plot.* In **Figure 25.15** set *Scale* to <u>1:1</u> and *Drawing orientation* to <u>Landscape</u>. Load the A-size paper sheet in the printer or plotter; pick *Preview* to see how the drawing will appear when plotted. If it appears correct, press the right mouse button, and select *Plot* on the pop-up menu to complete the drawing.

Quitting AutoCAD To be sure that your latest changes have been saved to disk, from the Main Menu select *File* and pick <u>Save</u> from the pull-down menu to update A:DRW-5. To quit

25.17 If the file has not been *Saved*, this dialogue prompts you to decide (*Yes, No,* or *Cancel*).

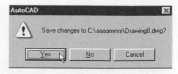

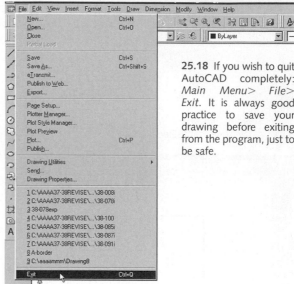

25.18 If you wish to quit AutoCAD completely: *Main Menu> File> Exit*. It is always good practice to save your drawing before exiting from the program, just to be safe.

AutoCAD, close the active file *(Main Menu> File> Close)* and the file will close and leave the screen if it has been *Saved* in its current form **(Figure 25.16)**. If changes have been made to it since its last *Save*, the pop-up box in **Figure 25.17** let's you decide to *Save* it or not (*Yes, No,* or *Cancel*). To quit your session, *Main Menu> File > Exit,* and AutoCAD closes **(Figure 25.18).** If you intend to continue your drawing session, do not *Exit* now but continue with other drawings. When working from a floppy disk, do not remove it from its drive until it has been saved with either the *Save* or *Save As* options.

That's how it works. Now, let's get into the details and learn more.

25.5 AutoCAD Windows

The recommended operating system for AutoCAD 2005 for a single user is Windows 98 or higher, which allows several programs to be open and running at the same time. For example, a word-processing program can be running in addition to AutoCAD.

25.19 The icons in the upper right of the screen can be used minimize a drawing to an icon (dash), reduce it to partial size (two boxes), or exit from the file (X).

A drawing file can be manipulated with the three buttons in the upper-right corner of the display window **(Figure 25.19).** The "overlapping-boxes" button is selected to display the file covering only a portion of the screen to allow other files or programs to share the screen. The "X" button closes the current file. The dash button minimizes a file to an icon box (that retains the three buttons of the original) located above the *Command* line **(Figure 25.20).** Maximize the minimized file to fill the screen by selecting the "box" button **(Figure 25.21)** and it will return the image to the screen when it is clicked on by the cursor.

Window areas can be resized by selecting a border, or the corner of the border, and "dragging" the window to size while holding down the left mouse button **(Figure 25.22).** When several image windows overlap on the screen, select any point on the desired window to move it to the front to make it the current program.

25.6 Command Options

AutoCAD has several ways of actuating a drawing command in almost all cases. Commands can be actuated from the *Command* line, *Main Menu,* or the *Draw* toolbar.

Command Line When a circle is drawn by typing at the *Command* line it will be presented as follows: *Command:* <u>Circle</u> (or <u>C</u>)> *Center point*> <u>P1</u>> *Radius*> <u>4</u> (Enter). Entries that are typed or picked in response to prompts are underlined. The underlined <u>P1</u> is a point selected on the screen and the underlined <u>4</u> is the radius, which is typed at the command line in response to the prompt, *Radius.*

Main Menu When the *Main Menu* is used to draw a circle, the sequence of steps is

25.20 Minimized files are displayed as an icon box above the *Command* line.

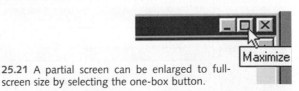

25.21 A partial screen can be enlarged to full-screen size by selecting the one-box button.

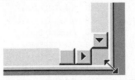

25.22 Select the lower right corner, hold down the left mouse button, and "drag" the border to size the screen as you desire.

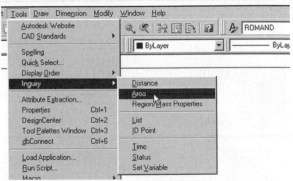

25.23 The *Main Menu* has many pull-down menus. Commands in this menu that are followed by black arrows have sub-dialog boxes. The sequence in this example is *Main Menu> Tools> Inquiry> Area.*

presented as: *Main Menu> Draw> Circle> Center, radius> Center point>* <u>P1</u>> *Radius>*. Drag to a radius of <u>4</u>.

Draw Toolbar When the *Draw* toolbar is used to draw a circle, the steps are presented as: *Draw* toolbar> *Circle* icon> *Center, radius> Center point>* <u>P1</u>> *Radius>* <u>4</u> (Enter).

25.7 Dialog Boxes

AutoCAD 2005 has many dialog boxes with names beginning with *DD* (*DDlmodes,* for example) to interact with the user. The command

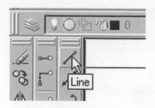

25.24 *Definition boxes* are flyout boxes that explain the functions of the various icons when the cursor is rested on them briefly.

Filedia can be used to turn off (0 = off and 1 = on) the dialog boxes if you prefer to type the commands without dialog boxes. When a command on a menu followed by three dots (**...**) or an arrow (>) is selected, supplemental dialog boxes will be displayed **(Figure 25.23)** and some of these boxes have subdialog boxes.

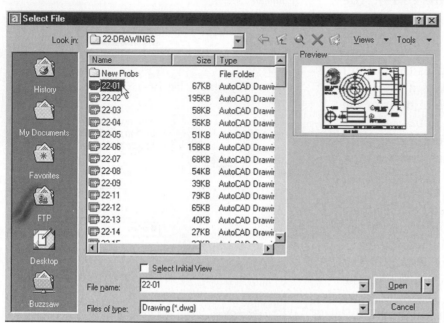

25.25 *Main Menu > File> Open* gives the *Select File* box that helps you find the file that you wish to open. A thumbnail view of the selected file is previewed in the window before it is opened.

25.26 To locate a file for which you have a name, from the *Select File* menu, select *Tools> Find> get Find box> Name & Location* tab> type file name in *Named* window> select file type in *Type* window> *Browse* to select drive or folder to search>. *Find Now* and the matching files appear in the window at the bottom of the *Find* box.

25.27 These function keys on the keyboard control the options as indicated by the notes.

Definition boxes are provided to identify the functions of each box on the screen. By resting the cursor on a box, a *flyout* will appear to define its function as shown in **Figure 25.24.**

Double-clicking the mouse (quickly pressing the left button twice) selects a file from a list and displays it. Single-clicking followed by *OK* is an alternative for activating a selection. With experimentation you will soon learn where double-clicking is best applied.

Right-clicking when the cursor (right mouse key) is placed over an icon will display a set of commands that are related to that particular command.

The *Select File* box in **Figure 25.25** has dialog boxes, lists, blanks, and buttons that can be selected by the cursor. When a file is selected it is darkened by a gray bar, and a thumbnail illustration of it is shown in the *Preview* window. To find a file for which you have a name, *Tools> Find box> Name* window (give the file name in the *Name* window, 22-03)> *Find Now* button to locate the file **(Figure 25.26).** The matching file will appear in the window at the bottom of the *Find* box.

In many cases, speed is increased if you type commands at the *Command* line instead of using dialog boxes. What could be easier than typing L and pressing (Enter) for drawing a *Line?*

25.8 Drawing Aids

Function Keys
Convenient drawing aids are available from the function keys on the keyboard, which can be used to turn settings on and off **(Figure 25.27).** *F1 (Help)* can be pressed to open the help screen for instant troubleshooting. *F2 (Flip screen)* alternates between the graphics on the screen to its corresponding text mode. *F3* controls *Snap* and *F5* controls *Isoplane.* *F4 (Tablet)* activates a digitizing tablet if one is attached to your computer. *F6 (Coordinates)* shows numerical coordinates of the cursor in the *status* bar at the bottom of the screen as the cursor is moved. *F7 (Grid)* turns the grid on or off and refreshes the screen in the process, removing any blips or erasures. *F8 (Ortho)* forces all lines to be drawn either in horizontal or vertical directions. *F9 (Snap),* when on, makes all object points lie on points defined by an invisible grid.

One of the first settings to make when beginning a drawing is that of the area size, called *Limits,* in which the drawing will be made. To set, type Limits and respond to the *Command* line prompts by typing the coordinates of the diagonal across the area. *Limits* command can also be accessed by: *Main Menu> Format> Drawing Limits.* A drawing that fills an A-size sheet (11 × 8.5 inches) has a plotting area of about 10.4 × 7.8 inches or 254 × 198 mm. The drawing area is specified with the *Limits* command as follows:

> *Command:* Limits (Enter)
> *Specify lower left corner [ON/OFF]*
> *<0.00,0.00>:* (Enter) to accept default value of 0,0.
> *Specify upper right corner <12.00,9.00>:*
> 11,8.5 (Enter)
> Press F7 *(Grid)* to see the dot pattern of the grid fill the *Limits.*
> *Command:* Zoom (Enter)> type All (Enter) and the drawing *Limits* and *Grid* will fill the screen.

Limits can be reset at any time during the drawing session by repeating these steps.

Drafting Settings
The *Drafting Settings* box is found by *Main Menu> Tools> Drafting Settings,* right-clicking on the *Osnap* icon above the *Command* line,

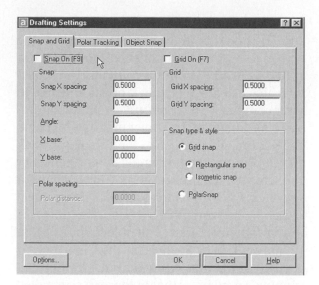

25.28 The *Drafting Settings* dialog box (or, *Command:* DDrmodes (Enter)) has three tabs from which to make settings: *Snap and Grid, Polar Tracking,* and *Object Snap.*

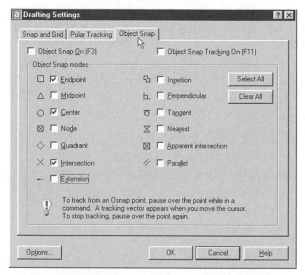

25.29 *Osnap* settings can be made from the *Object Snap* tab of the *Drafting Settings* dialog box *(DDrmodes).*

and selecting *Settings,* or by typing DDrmodes **(Figure 25.28).** The *Snap and Grid* tab gives buttons for selection and blanks for filling in to activate these settings. *Snap* forces the cursor to stop only at points on an imaginary grid of a specified spacing. The *Snap X* and *Snap Y* values can be set by typing values in the blanks. *Snap* can also be set by typing *Snap* at the *Command* line and specifying the interval

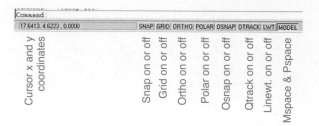

25.30 *Drawing Aides* boxes in the *Status* bar beneath the *Command* line at the bottom of the screen display the current settings. Click these buttons to turn them on or off.

desired. When *On,* the *Snap* icon in the *Status* line at the bottom of the screen is highlighted. *Snap* can be toggled on and off by clicking on this icon or by pressing *F9.*

Grid of the *Drafting Settings* box fills the *Limits* area with dots spaced apart by typing values in the *Grid X spacing* and *Grid Y spacing* boxes. When *Snap type & type> Grid snap > Rectangular snap* is selected the cursor snaps to the grid if the *Snap* and *Grid* spacings are equal.

The *Object Snap* tab **(Figure 25.29)** gives options for making lines and other geometry of a drawing snap to previously drawn objects at specified points. *Osnap* is covered in Section 25.37.

The *Status* window at the bottom of the screen displays several of the settings discussed above when they are turned *On* **(Figure 25.30).** Single-clicking on these buttons toggles them off or on.

Blips (*Command:* Blipmode) are temporary markers made on the screen when selections are made with the mouse. They are removed by refreshing the screen by typing Regen.

25.9 Help Commands

Several examples of helpful commands that can be typed at the *Command* line are shown here.

Help (Main Menu> Help> AutoCAD 2005 Help) gives menus *Contents, Index, Search, Favorites,* and *Ask me* to help you with all aspects of AutoCAD. Under *Find,* you will be able to insert key words for commands and steps that you need assistance with as shown in **Figure 25.31.**

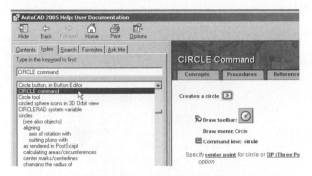

25.31 By selecting *Help* from the *Main Menu* (or pressing <u>F1</u>), the *AutoCAD 2005 Help* box with five tabs appears on the screen. Select *Index* tab, type the keyword in the blank, select the topic that matches, and pick the *Display* button to obtain instructions about your topic.

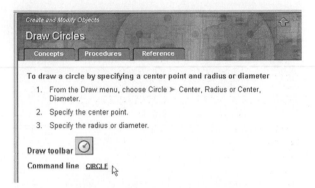

25.32 The *Help* box can be obtained by *Main Menu> Help> AutoCAD 2005 User Help>* type subject (*CIRCLE*)> Select *CIRCLE* from list (all caps for a command)> *Display* button> *Select*> *To draw a circle by specifying a center point and a radius or diameter* to get the help screen shown here.

When a topic, *Circle*, for example, is selected and *Display* is picked, a screen of instructions will appear to help you with using the *Circle* command **(Figure 25.32).** Additional options are provided under *Help*, which are self-explanatory when you experiment with them.

Purge (Main Menu> File> Drawing Utilities> Purge> All) can be used at any time to remove unused layers, blocks, and other attributes from files. *Purge* can the activated by using *Command:* <u>Purge</u> (Enter) to obtain the *Purge* box where you can select *All items* or *Blocks/ Dimension styles/ Layers/ Linetypes/ Mline styles/ Plot styles/ Shapes/ Text styles.* By purging unused attributes of a file, clutter is eliminated and disk space is saved.

25.33 Select the paper stack next to *Layer Control* panel to obtain the *Layer Properties Manager* box for setting all layer properties.

The *All* option is used to eliminate all unused references one at a time as prompted. The other purge options remove selected features of a drawing.

List (Main Menu> Tools> Inquiry> List) asks you to select any object drawn on the screen and gives information about it. For example, when a circle is selected it gives its radius, circumference, and area plus the coordinates of its center point.

Copy (Modify toolbar*> Copy* icon) is used to select objects on the screen (single objects or groups of objects) with the cursor, pick a new position, and make a duplicate of the selection. The *Multiple (M)* option can be selected for making more than a single copy. *Unlock* is used to select *Files* and *Unlock* them by picking *OK*.

25.10 Drawing Layers

An almost infinite number of layers can be created, each assigned a *Name, Color, Linetype, Lineweight,* and *Plot Style* on which to draw. For example, a yellow layer named *Hidden* for drawing dashed lines may be created.

Architects use separate copies of the same floor plan for different applications: dimensions, floor finishes, electrical details, and so forth. The same basic plan is used for all of these applications by turning on the needed layers and turning off others.

Working with Layers Layers and their settings are created and manipulated in the *Layer Properties Manager* box, which is displayed by selecting the paper-stack icon next to the *Layer Control* panel **(Figure 25.33).** For most working drawings the layers shown in the *Layer Properties Manager* box in **Figure 25.34**

25.34 *Main Menu> Format> Layer> Layer Properties Manager* box (or, *Command:* DDlmodes) lists the layers and their properties. From this box, layers can be created and deleted, linetypes and colors assigned, and other settings made.

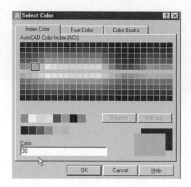

25.35 *Layer Properties Manager>* click on layer's current color> *Select Color box* to get this box from which colors can be selected and assigned to the listed layers.

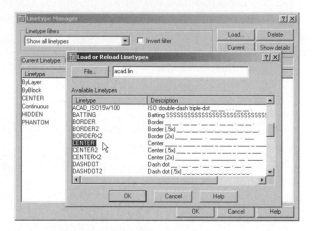

25.36 In *Main Menu> Format> Layer> Layer Properties Manager* box> Select *Linetype* (or, *Command:* DDltype) (Enter) to get the *Select Linetype* box, pick a layer, and pick *Load* to get a list of linetypes from which to select and assign to selected layers.

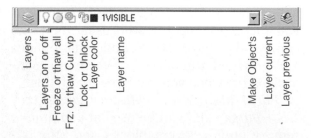

25.37 The *Layer Control* bar offers easy access to layers and their properties by selecting one of the icons shown here.

are sufficient. Layers are assigned linetypes, lineweights, and different colors so they can be easily distinguished from each other. The 0 (zero) layer is the default layer, which can be turned off or frozen but not deleted.

Layers can be created by selecting the *New* box to obtain *Layer1*—the default name that can be replaced by a new name **(Figure 25.34).** The new layer will appear in the listing of layers with a default color of white and a continuous linetype, all of which can be changed to your specifications.

Color for a new layer can be changed from its default color (white) by clicking on its current color in the *Color* column to obtain the *Select Color* menu. From this menu, pick a color for the selected layer and pick the OK button to finalize the color change **(Figure 25.35).**

Linetypes are found by clicking on the current linetype of the new layer in the *Layer Properties Manager* box and the *Select Linetype* box appears; select *Load* and the *Load or Reload Linetypes* box appears with a list of linetypes from which to select **(Figure 25.36).** To assign a hidden *Linetype* to a layer named *Hidden*, select the linetype of that layer, select *Hidden* from the *Select Linetype* box, pick *OK*, and the line is assigned to the layer. All lines drawn on the *Hidden* layer will be dashed lines.

Lineweights are found by clicking on the default lineweight of a layer in the *Layer Properties Manager* box and the *Lineweight* box appears with a list of linetypes from which to select. Select the desired lineweight, pick *OK*,

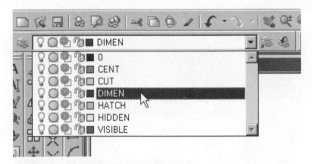

25.38 Select any point on the *Layer Control* panel where the name of the active layer's name appears, or select the down button, and this drop-down list of the layers appears. Select a layer (1OUTLINE, for example) and it become the current layer.

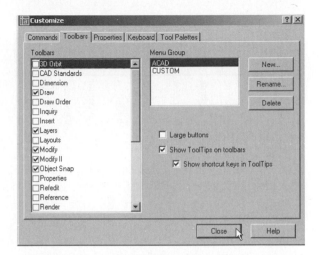

25.39 Select *Main Menu> View> Toolbars> Customize menu* to obtain this listing of toolbars that can be placed on or removed from the screen by checking a toolbar box.

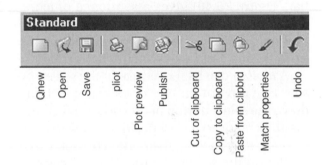

25.40 This left half of the *Standard* toolbar provides many routine operations that can be accessed with the cursor.

and the line is assigned to that layer. Lines drawn on this layer will have this line thickness.

Ltscale, when typed at the *Command* line, modifies the lengths of line segments of *all* hidden lines and other noncontinuous linetypes at one time.

Layer Control Panel

A layer must be selected as the current layer in order to draw on it using the assigned color and linetypes. The *Layer Control Panel* offers the quickest method of setting a layer (**Figure 25.37**). By picking a point anywhere within the *Layer Control Panel,* a listing of the named layers and their properties will be displayed from which a layer can be selected to make it the current layer (**Figure 25.38**). Now you can draw on this current layer.

This same portion of the *Layer Control Panel* can also be used to make other layer assignments: *On* and *Off, Freeze* and *Thaw, Lock* and *Unlock,* and others. By selecting the *Layers* icon (**Figure 25.37**), the *Layers Properties Manager* box will be displayed from which the previously covered setting can be made. This box can also be obtained by typing LA at the *Command* line and pressing (Enter).

Renaming a layer can be done from the *Layers Properties Manager* box by dragging the cursor across the existing name and typing a new one in its place.

On/Off (light bulb) is applied to a layer in the *Layer Control* panel by selecting the light-bulb icon (**Figure 25.38**). Layers can be turned on or off with buttons from the *Layers Properties Manager* box (**Figure 25.36**) (or by typing *LA* at the *Command* line) picking a layer, and *Off.* An *Off* (invisible) layer that is assigned as the current layer can be drawn on, but this is seldom done.

Freeze (snowflake icon) and *Thaw* (sun icon) under the *Layer Control Panel* are used like the *On* and *Off* options (**Figure 25.38**). *Freeze* a layer and it will (unlike an *Off* layer) be ignored by the computer until it has been *Thawed,* which makes regeneration faster than when *Off* is used.

25.41 The right half of the *Standard* toolbar has more helpful commands that you will need for most drawings.

25.11 Toolbars

Select *Main Menu> View> Toolbars> Customize* to get a menu of toolbars that can be selected to suit the current application **(Figure 25.39).** A portion of the *Standard* toolbar **(Figure 25.40)** gives a sequence of icons for commands from *QNew* to *Undo*. The remainder of the *Standard* toolbar commands from *UCS* to *Help* are given in **Figure 25.41.**

A *toolbar* can be moved about the screen by selecting a point on one of its edges, holding down the select button of the mouse, and moving the cursor to a new position. *Toolbars* can be *docked* by moving them into contact with a border on the screen **(Figure 25.42).** They can be changed from single strips, to double and triple blocks, by moving the corners of the toolbars, or into vertical or horizontal strips depending on which border of the screen they are moved to. When located in the open area of the screen, toolbars will appear as *floating menus*. When a small icon needs explanation, place the pointer on an icon and a flyout box will appear with its definition.

25.12 A New Drawing

Title Block

To create a new drawing, select *Main Menu> File> New>* **(Figure 25.43)** and the *Create New Drawing* box appears giving two options: *Advanced Setup* or *Quick Setup* as was given in **Figure 25.8.** Pick *Advanced Setup,* set the *Units* to *Decimal* with a *Precision* (decimal

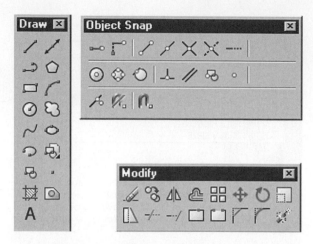

25.42 *Toolbars* can be arranged by dragging their corners to make single, double, or triple strips that are horizontal or vertical. They can be docked at the borders or left "floating" on the screen.

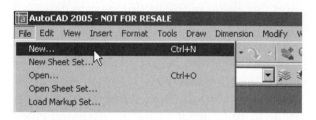

25.43 Begin the creation of a new drawing using these steps: *Main Menu> File> New.*

places) of <u>0.00</u>, and select *Next.* In the following screen set the *Angle* to *Decimal Degrees* with a *Precision* of <u>0</u> and pick *Next.*

In the *Angle Measure* box, set the angle measurement to *East* and pick *Next.* In the *Angle Direction* box, set the angle direction to *Counter-Clockwise* and pick *Next.* In the *Area* box, set the values for *Width* (<u>11</u>) and *Length* (<u>8.5</u>) and pick *Finish.*

Prototype Drawing

The drawing screen appears ready for drawing. If the grid dots do not show, select the *Grid* box below the *Command* line; the upper right dot has coordinates of about 11 and 8.5, the sheet assigned size. Draw a border using *Line>* <u>0,0</u> > <u>10.4, 0</u>> <u>10.4, 7.8</u>> <u>0, 7.8</u>> <u>0,0</u> to get a 10.4 × 7.8 border **(Figure 25.44).** Save this A-size

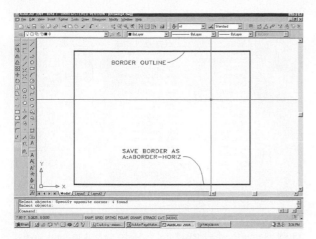

25.44 Select the *Grid* box below the *Command* line to turn the grid on and *Command:* Zoom> (Enter)> All> (Enter) to make the grid fill the drawing area. Draw a border (10.4 × 7.8) and *Save As>* A:ABORD-HORIZ and close the file.

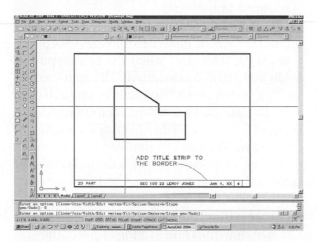

25.45 Open A:ABORD-HORIZ, save the file (*Main Menu> SaveAs>* A:DWG-5 > (Enter)) and DWG-5 becomes the current drawing. Make your drawing, add a title strip to the border, and save the file (*Main Menu> File> Save*).

border for future use as a prototype file for making drawings with these same settings *(Main Menu> File> Save As>* A:ABORD-HORIZ (Enter)). Close this file *(Main Menu> File> Close)* and the drawing leaves the screen.

Using the Prototype Drawing
So far, you have been working entirely in *Model Space*, which is adequate for two-dimensional drawing as covered in this chapter. The appli-

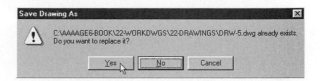

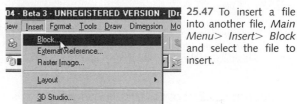

25.46 When saving a drawing file with *Save As*, you will be asked by this box if you want to update an existing file. By selecting *Yes* the named file is replaced by the current file.

25.47 To insert a file into another file, *Main Menu> Insert> Block* and select the file to insert.

cation of *Model Space* and *Paper Space* in combination will be covered later.

To make a drawing using the settings and border made in the previous sequence, open the prototype file (*Main Menu> File> Open>* A:ABORD-HORIZ), and the border and grid appear on the screen as it did at the end of the last drawing **(Figure 25.45).** Save the file with a new name (*Main Menu> File> Save As>* A:DWG-5> (Enter)) and DWG-5 becomes the current file and replaces the prototype file on the screen. (This same procedure can be used to create a new file by saving the prototype file to a new name, while preserving the prototype file in its original form.)

Add a title strip to the border for your name, date, and other information required by your instructor. You may consider making this information part of your prototype file so it will be displayed every time it is used: *File> Save As>* A:ABORD-HORIZ. Since you are saving to an existing file, the *Save Drawing As* box in **Figure 25.46** will ask you if you want to replace it; select *Yes*. Prototype file ABORD-HORIZ is updated so the title strip will be included when it is used next.

Make your drawing within the border, save it *(Main Menu> File> Save As> DWG-5),* and your drawing is ready to be plotted.

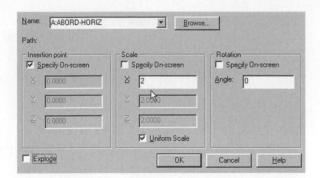

25.48 Type the name of the file to insert, <u>A:ABORD-HORI</u>,. select *Uniform Scale,* and give a *Scale* of <u>2</u>, Select *OK,* and insert the file at <u>0,0</u> of the *Command* line.

Section 25.54 covers the techniques of customizing your title block for classroom drawings, but first, you must learn more about the operation of AutoCAD. If this section seems a little advanced for you, leave it, and come back when you have learned a few more drawing principles.

25.13 Drawing Scale

It is best and easiest to work with a drawing at a full-size scale where 1 inch is equal to 1 inch. The previous examples of files and title blocks were developed as full-size layouts, which permits text size and measurements to be easily handled.

Half-size drawings can be made by creating a new drawing, DWG-6, by using the *Advanced Setup* steps covered in the last section. Insert the prototype drawing, A:ABORD-HORIZ **(Figure 25.47),** with the following steps: *(Main Menu> Insert> Block> File> A:ABORD-HORIZ).* Set parameters: check *Uniform Scale,* type <u>2</u> in *X-Scale window> OK>* type <u>0,0</u> at the *Command* line as the insertion point **(Figure 25.48).**

These commands insert the border and title strip at double size (scale = 2). For a double-size metric drawing, the scale would be 25.4 × 2 = 50.8 for millimeters. A full-size drawing is made within the double-size border. At plot time the double-size border must be reduced to half size (0.5) so it will fit

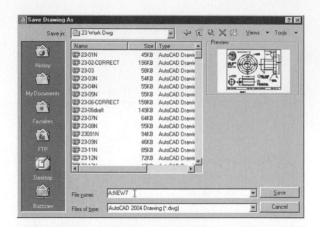

25.49 Name the file, select the drive to save it to, and save it by using this dialog box. *(Main Menu> File> Save As.)*

on its sheet, and the full-size drawing within it is reduced to half size.

Double-size drawings can be made with the same steps as half-size drawings except the *Scale* is <u>0.5</u> (half size) when the prototype file, A:ABORD-HORIZ, is inserted. (The factor would be 25.4 × 0.5 = 12.7 for millimeters.) At plot time, the half-size border must be enlarged by a factor of 2 so it will fill the sheet, and the full-size drawing within it will be doubled in size.

Using this logic, other combinations of scale factors can be determined for drawings of any scale. The most important point to remember is that you are better off working with full-size drawings and scaling at plot time.

25.14 Saving and Exiting

The pull-down file *(Main Menu> File)* gives options for saving a drawing—*Save, Save As,* and *Exit*—which can be selected from the pull-down menu, the *Standard* toolbar, or by typing one of these commands at the *Command* line.

Saving

At the *Command:* <u>Save</u> (Enter) to "quick save" to the current file's name if it has been previously named and saved. If the file is unnamed, the *Save* command prompts for a file name by displaying the *Save Drawing* As dialog box in **Figure 25.49.** Select the directory and name the file (A:NEW-7) to save it on the disk in drive A. The new drawing, *A:NEW-7*, becomes the current file on the screen.

Ending the Session

To exit the drawing session and close Auto-CAD, select *Main Menu> File> Exit*. If you have not saved immediately before selecting *Exit*, the dialog box shown in **Figure 25.46** asks if you want to save the updated drawing. Select *Yes* and the *Save Drawing As* menu box appears for assigning the drive, directory, file name, and file type. To exit without saving, respond to *Save Changes?* with *No* and the latest changes made since the last *Save* will be discarded.

Use *Command:* Close (Enter) and you will be prompted to *Save* the drawing if has been changed since the last *Save*. If the drawing has not been previously named, the *Save Drawing As* dialog box will prompt you for a file name. The previous version of the drawing is automatically saved as a backup file

with a *.bak* extension and the current drawing is saved with a *.dwg* extension. Several drawing files can be open at the same time during a session.

25.15 Plotting Parameters

To plot a drawing before ending a drawing session, select *File> Plot* to obtain the *Plot* box **(Figure 25.50)** from which a printer or a plotter can be selected. A *plotter* has pens that "draw" a drawing, whereas a *printer* has no pins but transfers the ink onto the drawing surface.

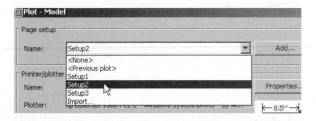

25.51 Settings made in the *Plot* menu can be saved with the *Add . . .* button for use in future drawings. The *Previous Plot* option enables you to specify settings identical to the last plot.

25.52 From the *Add . . .* button, the *User Defined Page Setups* box appears where you can name a new page setup and select *OK* to save it for future drawings.

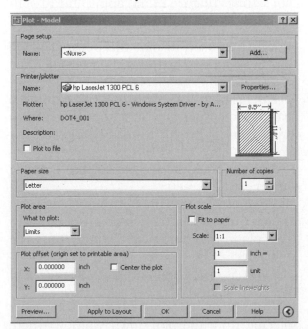

25.50 The printing options for a laser printer (without physical pens) are specified in this box *(Main Menu> Plot> Plot Device* tab). Select the printer to be used from the pull-down window across from *Name:*.

25.53 The *Plot area* section of the *Plot* box is used to specify the portion of the drawing that will be plotted: Extents, Display, Limits, or Window.

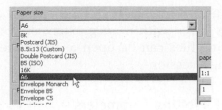

25.54 The *Paper Size* window gives a listing of the standard plot sizes that are available.

25.55 Select the desired scale (1 = 2, or half size, in this case). If *Fit to paper* is selected, the drawing fills the sheet if the placement of the origin permits.

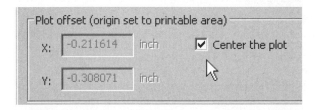

25.56 Select *Center the plot* box to position the plot at the center of the sheet; x- and y-coordinates of the drawing's origin are given.

Most size A and size B drawings are produced with a printer (laser or ink jet) whereas drawings of a larger size are rendered with a pen plotter.

Printer Settings

Use *Main Menu> Plot> Plot Device* tab to select the configured printer that you intend to use for plotting your drawing. Select the *Plot Settings* tab, select *Drawing orientation* (*Portrait* or *Landscape*), specify *Plot Area* (*Limits, Extents,* or *Display*), assign a *Scale,* set *Plot offset* to *Center the plot.* Save this *Page setup* by selecting the *Add* button (**Figure**

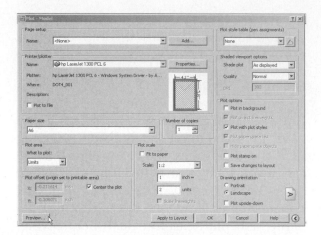

25.57 Select *Preview* to get a view of the drawing that will be plotted.

25.51) and naming it in the setup in the *User Defined Page Setups* box that appears on the screen (**Figure 25.52**) and pick <u>OK</u>. These setting will be saved with the drawing when the file is saved.

Plot Settings Definitions

Plot settings (**Figure 25.53**) applicable to both pen plotters and printers are covered below.

Limits is used to plot the portion of the drawing bounded by the grid pattern defined by its *Limits.*

Extents plots a drawing to its extents if the scale selected permits. It is good practice to apply *Zoom> Extents* to ready a drawing for plotting.

Display plots the portion of the drawing shown on the screen.

Window specifies the portion of a drawing to be plotted when the window is sized with the cursor or coordinates typed from the keyboard.

Paper Size lists standard and user-specified plot sizes (**Figure 25.54**).

Plot Scale can be typed in the edit boxes or selected from the pull-down window where 1:2 appears (**Figure 25.55**). The scale of 1 = 1 is full size, 1 = 2 is half size, and 2 = 1 is double size.

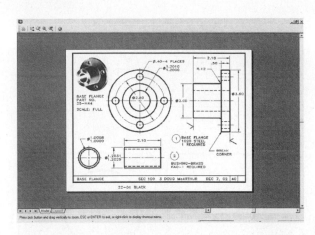

25.58 Pick the *Preview* button to get this view of the file that is to be printed when *Plot* is selected.

25.59 The Hewlett-Packard LaserJet 1300 printer has no physical pens; the ink is sprayed onto the paper. *(Courtesy of Hewlett Packard Company.)*

Scaled to Fit calculates the scale that makes the drawing's extent fill the plotting area and be as large as possible.

Center the plot is picked to specify that the drawing will be centered on the sheet (**Figure 25.56**).

Plot Offset can be set with *x* and *y* coordinates to locate a plot on a sheet (**Figure 25.56**).

Hide paperspace objects removes hidden lines from 3D drawings that are plotted.

Drawing Orientation offers *Portrait* and *Landscape* options (**Figure 25.57**).

Preview, when selected in the lower left-hand corner of **Figure 25.57,** will show a view of the plot before its execution. The *Preview* image shows the entire drawing on the screen and its relationship to the paper limits when plotted (**Figure 25.58**). Right-click to obtain *Pan, Zoom, Zoom Window, Zoom Original, Plot,* and *Exit* for inspecting the preview drawing. Press Exit to return to the *Plot* dialog box.

To add a new *Page setup,* select *Plot menu> Add> New page setup name* (Setup3, for example)> *OK* and a new setup file is created.

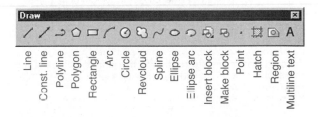

25.60 The *Draw* toolbar makes it easy for you to select commands with the cursor: *Main Menu> View> Toolbars> Draw.*

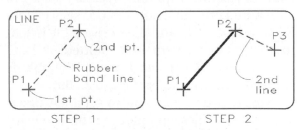

25.61 *Line.*
Step 1 *Menu Menu> Draw> Line> Specify first point:* P1. *Specify next point or [Undo]:* P2
Step 3 *Specify next point or [Close/Undo]:* P3. (Enter)

General

These approaches to making settings for plotters and printers while working from *Model Space* are the simplest to plotting drawings, yet they are sufficient for essentially all two-dimensional drawings required in a beginning graphics course.

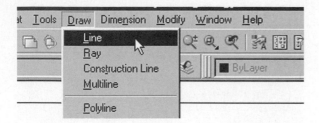

25.62 These types of lines can be selected from the *Main Menu> Draw> Line* on the drop-down menu.

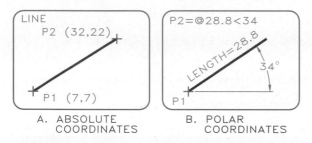

A. ABSOLUTE COORDINATES

B. POLAR COORDINATES

25.63 Lines by coordinates.

A Absolute coordinates can be typed (7,7 and 32,22) at the *Command* line to establish the ends of a line.

B Polar coordinates are relative to the current point and are specified with a length and the angle measured clockwise from the horizontal (@28.8 < 34).

25.16 Readying the Printer

Load the printer with the necessary sheets of paper, be sure the printer is on, select *Main Menu> Plot> OK,* and the drawing file is sent to the printer where it is plotted. A laser printer has no physical pens; the ink is sprayed onto the paper **(Figure 25.59).**

Now that we know how to set a few drawing aids, save files, and plot, it is time to learn how to make drawings.

25.17 Lines

Open your prototype drawing, *Main Menu> File> Open>* A:ABORD-HORIZ and use *Save As* to name the drawing as A:NO1. which becomes the current drawing with the same settings of A:ABORD-HORIZ. Load the *Draw* toolbar *(Main Menu> View> Toolbars> Draw)* to obtain the command options shown in **(Figure 25.60).**

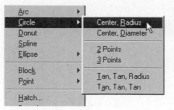

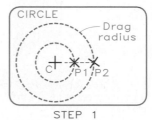

25.64 The *Circle* command *(Main Menu> Draw> Circle)* has a fly-out menu with these options for drawing circles.

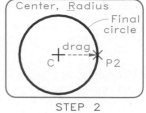

STEP 1

STEP 2

25.65 Circle.

Step 1 *Main Menu> Draw> Circle> Specify center point for circle or [3P/2P/Ttr (tan tan radius)]:* Pick C (Enter) *Specify radius of circle or [Diameter] <0.00>:* R (Enter)

Step 2 Drag radius to P1 and P2 to enlarge the circle, click the left mouse button, (Enter), and the final circle is drawn.

A *Line* (called an object) can be drawn by using the keyboard, *Draw* toolbar, or the *Main Menu (Main Menu> Draw> Line>* select endpoints). Use *Command:* Line (or L) and respond to the prompts as shown in **Figure 25.61** to draw lines by picking endpoints with the left button of the mouse. The current line will rubberband from the last point, and lines are drawn in succession until (Enter) or the right button of your mouse is pressed.

The *Draw* toolbar can be used to select: *Line, Ray,* and *Construction Line.* These types of lines (plus *Multilines*) can also be selected from *Main Menu> Draw* **(Figure 25.62).** A *Construction Line* is drawn totally across the screen, and a *Ray* is drawn from the selected point to the edge of the screen.

A comparison of absolute and polar coordinates is shown in **Figure 25.63.** *Delta coordinates* can be typed as @2,4 to specify the end of a line 2 units in the *x*-direction and 4 units in the *y*-direction from the current end.

Polar coordinates are *2D* coordinates that are typed as @ 3.6 < 56 to draw a 3.6 long line

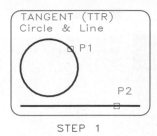
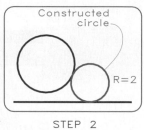

25.66 Circle: Tangent to two objects (TTR).
Step 1 *Main Menu> Draw> Circle> Specify center point for circle or [3P/2P/Ttr (tan tan radius)]:* TTR (Enter) *Specify point on object for first tangent of circle:* P1
Step 2 *Specify point on object for second tangent of circle:* P2 *Specify radius of circle <0.00>:* 2 (Enter)

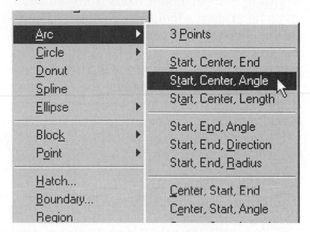

25.67 To draw an arc, select *Main Menu> Draw> Arc>* and these options are given. When using *Start, Center, Angle,* these elements must be specified in this order on the screen. This selection can be made from the *Draw* toolbar, also.

from the current (and active) end of a line at an angle of 56° with the *x*-axis. *Last coordinates* are found by typing @ Enter while in the *Line* command. This causes the cursor to move to the last point.

World coordinates locate points in the *World Coordinate System*, regardless of the *User Coordinate System* being used, by preceeding the coordinates with an asterisk (*). Examples are *4,3; *90 < 44; and @*1,3.

The *status line* at the bottom of the screen shows the length of the line and its angle from the last point as it is rubberbanded from point to point. The *Close* command will close a continuous series of lines with a line from the last to first point selected.

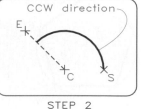
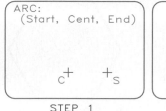

25.68 Arc: Start, Center, End option (SCE).
Step 1 *Main Menu> Draw> Arc> Start, End> Specify start point of arc or [Center]:* (Enter)
Specify center point of arc: (Enter)
Step 2 *Specify end point of arc or [Angle/chord Length]:* E Drag the arc to the end of the radial line CE, click the left mouse button (Enter) and the arc is drawn.

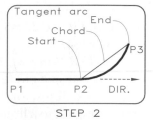
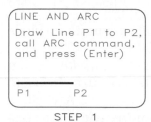

25.69 Arc: Tangent at the end of a line.
Step 1 *Draw* toolbar> *Line> Specify first point:* P1
Specify next point or [Undo]: P2 (Enter)
Step 2 *Draw* toolbar> *Arc> Specify start point of arc or [Center]:* (Enter) to grab P2 > *Specify end point of arc:* Drag the end of the tangent arc to P3.

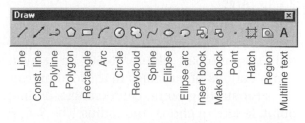

25.70 Select *Main Menu> View> Toolbars> Draw* to obtain the *Draw* toolbar. Select the *Polygon* icon to draw a polygon as shown in Figure 25.71.

25.18 Circles

The *Circle* command *(Main Menu> Draw> Circle)* draws circles when you select a center and radius, a center and diameter, or three points **(Figure 25.64).** *Command:* Circle (or C) (Enter) is the fastest means of activating the *Circle* command, but the *Circle* icon on the *Draw* toolbar **(Figure 25.60)** can also be

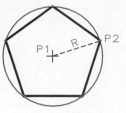
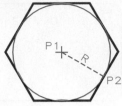

A. INSCRIBED B. CIRCUMSCRIBED

25.71 Polygons: Inscribed and circumscribed.
A *Draw toolbar> Polygon> Enter number of sides <4>: 5* (Enter)> *Specify center of polygon or [Edge]: P1*
Enter an option [Inscribed in circle/Circumscribed about circle] <1>: I (Enter)
B Select *Circumscribed* option C. *Specify radius of circle:* Drag radius to P2 to size the circumscribed circle.

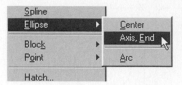

25.72 *Draw toolbar> Ellipse> Axis, End* can be used for drawing ellipses as shown in Figure 25.73.

selected. **Figure 25.65** illustrates how a circle is drawn.

The *TTR (tangent, tangent, radius)* option of *Circle* draws a circle tangent to a circle and a line, two lines, or two circles. A circle is drawn tangent to a line and a circle by selecting the circle, the line, and giving the radius **(Figure 25.66)**. The *tan, tan, tan* option calculates the radius length and draws a circle tangent to three lines.

25.19 Arcs

The *Arc* command *(Main Menu> Draw> Arc> options)* **(Figure 25.67)** or the *Arc* icon in the *Draw* toolbar **(Figure 25.60)** has eleven combinations of variables that use abbreviations for starting point, center, angle, ending point, length of chord, and radius. The *S, C, E* version requires that you locate the starting point *S*, the center *C*, and the ending point *CE* **(Figure 25.68)**. The arc begins at point *S* and is drawn counterclockwise by default to a point near *E*.

A line can be continued as an arc drawn from its last point and tangent to it for drawing runouts of fillets and rounds **(Figure 25.69)**. It can be used to draw a tangent line from an arc by applying the commands in reverse and dragging the line to its final length.

25.20 Polygons

The *Polygon* and many other objects can be drawn by selecting their icons from the *Draw* toolbar *(View> Toolbars> Draw)* **(Figure 25.70)**. A polygon is drawn from the *Draw* toolbar in **Figure 25.71**. An equal-sided polygon can be drawn by using *Main Menu> Draw> Polygon* icon and following the prompts. A polygon is drawn in a counterclockwise direction about the center point. Polygons can have a maximum of 1024 sides.

25.21 Ellipses

The *Ellipse* command in the *Main Menu> Draw* menu gives icons for three types of ellipses **(Figure 25.72)**. An ellipse is drawn by selecting the endpoints of the major axis and a third point P3, to give the length of the minor radius **(Figure 25.73)**. P3 need not lie on the ellipse; the distance from the midpoint of the axes to P3 merely gives the length of the minor axis, which is perpendicular to the major diameter, regardless of the direction in which the distance is specified.

25.22 Fillets

The corners of two lines can be rounded with the *Fillet* command *(Main Menu> Modify> Fillet)* whether or not they intersect. When the fillet is drawn, the lines are either trimmed or extended as shown in **Figure 25.74**. The assigned radius is remembered until it is changed. By setting the radius to 0, lines will be extended to a perfect intersection. Fillets

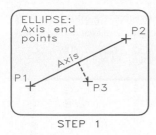 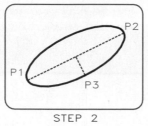

STEP 1 STEP 2

25.73 Ellipse: axis, end option.
Step 1 *Main Menu> Draw> Ellipse> Axis, End> Specify axis endpoint of ellipse or [Arc/Center]:* P1 *Specify other endpoint of axis:* P2
Step 2 *Specify distance to other axis or [Rotation]:* P3. Ellipse is drawn through P1, P2, and point established by P3.

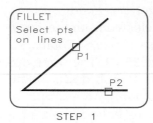 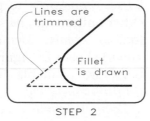

STEP 1 STEP 2

25.74 Fillet.
Step 1 *Command* line> Fillet
Current settings: Mode=TRIM, Radius= 0.0000 Select first object or [Polyline/Radius/Trim/mUltiple]: R (Enter)> *Specify fillet radius <0.0000 >;* 1.00 (Enter) (Enter)
Select first object or [Polyline/Radius/Trim/mUltiple]: P1>
Step 2 *Select second object:* P2
The tanget arc is drawn and the lines are trimmed.

of a specified radius can be drawn tangent to circles or arcs.

The *Fillet* command can also be selected by typing *Fillet* at the *Command* line. Practically all commands can be entered at the command line in this manner without referring to screen menus.

25.23 Chamfers

The *Chamfer* command *Main Menu> Modify> Chamfer)* draws angular bevels at intersections of lines or polylines. Two chamfer distances can be assigned if the bevel has unequal distances. After assigning chamfer distances (D), select two lines, which will be trimmed or extended; the *Chamfer* is then

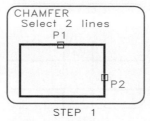 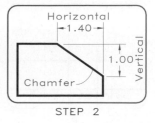

STEP 1 STEP 2

25.75 Chamfer.
Step 1 *Main Menu> Modify> Chamfer>*
(TRIM mode) Current chamfer Dist1 = 0.50, Dist2 = 0.50 Select first line or [Polyline/Distance/Angle/Trim/Method/mUltiple]: D> *Specify first chamfer distance <0.50>:* 1.40 (Enter) *Specify second chamfer distance <1.40>:* 1.00 (Enter)
Step 2 (Enter) *CHAMFER (TRIM mode) Current chamfer Dist1 = 1.4000, Dist2 = 1.0000*
Select first line or [Polyline/Distance/Angle/Trim/Method/mUltiple]: P1> *Select second line:* P2 The chamfer is drawn.

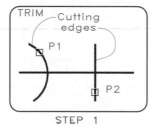 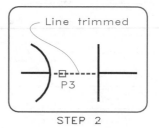

STEP 1 STEP 2

25.76 Trim: Cutting edges.
Step 1 *Modify* toolbar>Trim>*Current settings: Projection= UCS Edge=Extend Select cutting edges . . . Select objects:* P1 *1 found Select objects:* P2 *1found> Select objects:* (Enter)
Step 2 *Select object to trim or [Project/Edge/Undo]:* P3
Line between the cutting edges is removed.

drawn **(Figure 25.75).** Press (Enter) to repeat this command using the previous settings. By setting the chamfer distance to 0 (zero), this command can be used to trim nonparallel lines to perfect intersections.

25.24 Trim

The *Trim* command *(Main Menu> View> Toolbars> Modify> Trim)* selects cutting edges for trimming selected lines, arcs, or circles that cross the edges at their crossing points **(Figure 25.76).** A series of objects can be selected one at a time or selected as a group by a window. A *Crossing* window is used to select four cutting edges for trimming four lines in **Figure 25.77.**

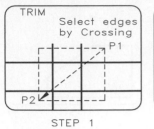 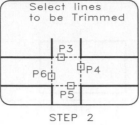

25.77 Trim: Crossing window.
Step 1 *Main Menu> Modify> Trim> Current settings: Projection=UCS, Edge=None, Select cutting edges . . . Select objects:* P1> *Specify opposite corner:* P2 *4 found.*
Step 2 *Select object to trim or shift-select or [Project/Edge/ Undo]:* P3 *Select object to trim . . . or [Project/Edge/Undo]:* P4 *Select object to trim . . . or [Project/Edge/Undo]:* P5 *Select object to trim . . . or [Project/Edge/Undo]:* P6 *(Enter)* The four lines are trimmed.

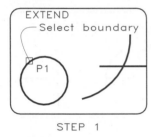 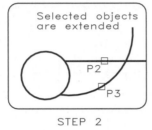

25.78 Extend.
Step 1 *Main Menu>Modify>Extend Current settings: Projection=UCS, Edge=None Select boundary edges . . . Select objects:* P1 *1 found*
Step 2 *Select objects: Select object to extend or [Project/ Edge/Undo]:* P2 *2 found Select object to extend or shift-select to trim or [Project/Edge/Undo]:* P3 *(Enter)*

25.25 Extend

The *Extend* command lengthens lines, plines, and arcs to intersect a selected boundary **(Figure 25.78).** You are prompted to select the boundary object and the object to be extended to the boundary. More than one object can be extended at a time. *Extend* will not work on "closed" polylines such as a polygon.

25.26 Zoom and Pan

Drawings can be enlarged or reduced by the *Zoom* command *(Main Menu> View> Tool-*

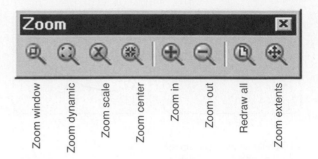

25.79 The *Zoom* toolbar contains these options for *Zooming* on the screen.

bars> Standard> to get the *Zoom* options. Other options can be selected from the *Zoom* toolbar *(Main Menu> View> Toolbars> Zoom)* shown in **Figure 25.79.** The same options are available in a pull-down menu *(Main Menu> View> Zoom).* An example of a *Zoom Window* used to enlarge part of a drawing is shown in **Figure 25.80.**

Zoom (Command:> Z) gives the following options: *All, Center, Dynamic, Extents, Previous, Scale, Window,* and *Real time.*

All expands the drawing's *Limits* (dot pattern) to fill the screen.

Center is picked to select the center of the *Zoomed* image and to specify its magnification or reduction.

Dynamic lets you *Zoom* and *Pan* by selecting points with the cursor.

Extents enlarges the drawing to its maximum size on the screen.

Previous displays the last *Zoomed* view.

Scale X/XP magnifies a drawing relative to paper space. By typing 1/4XP or .25XP the drawing will be scaled so that .25 inch equals 1 inch.

Window lets you pick the diagonal corners of a window to fill the screen.

Real time (the default) lets you drag to the left to make a crossing window, or to the right for regular window to specify the area to be enlarged.

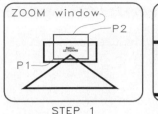
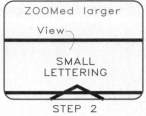

25.80 Zoom.
Step 1 *Standard* toolbar> *Zoom realtime icon*> *right-click to activate pop-up menu*> *Specify corner of window, enter a scale facor (nX or nXP), or [All/Center/Dynamic/Extents/previous/Scale/Window] <real time>* : Hold down left button and drag a window around the zoom area (P1 and P2).
Step 2 The area of the window is expanded to fill as much of the screen as possible.

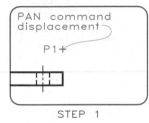
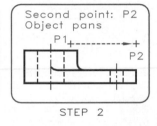

25.81 Pan.
Step 1 *Standard* toolbar> *Pan* icon
Step 2 Pick P1, hold down left button, and drag to P2 to change your view of the screen.

The *Pan (P)* command **(Figure 25.81)** is used to pan the view across the screen by selecting two points, a handle point, and its final position. The drawing is not relocated as in the *Move* command; only your viewpoint of it is changed.

25.27 Selecting Objects

A reoccurring prompt, *Select objects:*, asks you to select an object or objects that are to be *Erased, Changed,* or modified in some way. Type Select when in a current command that requires a selection (*Move,* for example) and the options will be displayed: *Window, Last, Crossing, BOX, ALL, Fence, WPolygon, CPolygon, Group, Add, Remove, Multiple, Previous, Undo, AUto,* and *Single.* **Figure 25.82** shows ways of selecting objects for applicable commands, such as *Copy,* to select objects. The

application of the options of *Single* and *Multiple, Window, Crossing,* and *BOX* are shown in parts A, B, C, and D, respectively.

Window (W) lets you select objects lying completely within it.

Crossing (C) lets you select objects lying within or crossed by the window.

Last is used to pick the most recently drawn object.

Box (B) lets you make a window by selecting a point and dragging to the right. A crossing window can be made by dragging to the left.

WPolygon (WP) forms a solid-line polygon that has the same effect as a window.

CPolygon (CP) forms a dotted-line polygon that has the same effect as a crossing window.

Fence (F) selects corner points of a polyline that will select any object it crosses the same as a crossing window.

ALL selects everything on the screen.

Remove and *Add* are used while selecting objects to remove one by typing R (remove) or to add one by typing A (*Add*). When finished, press (Enter) to end the *Select/remove* prompt.

Multiple (M) selects multiple points (without highlighting) to speed up the selection process.

Previous (P) recalls the previously selected set of objects. For example, enter *Move,* and type P, and the last selected objects are recalled.

Undo (U) removes selections in reverse order one at a time by typing U (undo) repetitively.

Auto (A) selects an object by pointing to it, and pointing to a blank area selects the first corner of a box defined by the *BOX* option.

Single (SI) causes the program to act on the object or sets of objects without pausing for a response.

A *Window (W)* or a *Crossing Window* can be obtained automatically by pressing the pick button, holding it down, and selecting the diagonal of a window **(Figure 25.83).** By dragging it to the right, a window is obtained; by dragging it to the left, a crossing window is obtained.

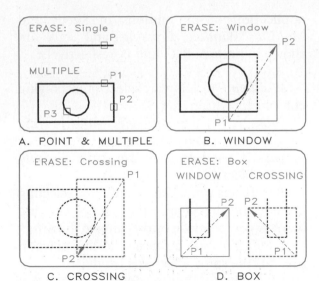

A. POINT & MULTIPLE

B. WINDOW

C. CROSSING

D. BOX

25.82 Selection options.
A *Select objects* selects the objects one at a time.
B *Window (W)* selects objects lying completely within the window; drag window left to right.
C *Crossing Window (C)* selects objects lying within or crossed by the window; drag window right to left.
D *Box* makes a *Window* or a *Crossing Window* that is determined by the direction dragging.

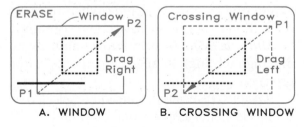

A. WINDOW

B. CROSSING WINDOW

25.83 Selection windows.
A A *Window* is formed by holding down the select button while dragging a diagonal corner to the **right.**
B A *Crossing Window* is formed in the same manner, but dragged to the **left.**

25.28 Erase and Break

The *Erase* command *(Modify* toolbar> *Erase* icon) **(Figure 25.84)** deletes specified parts of a drawing. The selection techniques described previously can be used to select objects to be erased as shown in **Figure 25.85.** The default of the *Erase* command, *Select Objects,* allows you to pick one or more objects and to delete them by pressing (Enter). Multi-

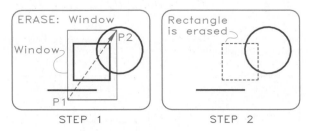

25.84 The *Modify* toolbar, a portion of which is shown here, has options for making changes in a drawing.

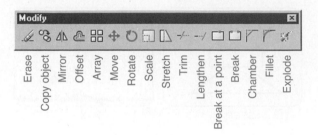

STEP 1   STEP 2

25.85 Erase: Window option.
Step 1 *Modify* toolbar> *Erase* icon *Select objects:* P1, *Other corner:* P2 (left to right)
Step 2 *Select objects:* (Enter) The box located completely within the erasing window is removed.

ple selections can be made by holding down the *Shift* key as objects are picked. Type Oops to restore the last erasure; only the last one can be removed.

The *Break* command *(Modify* toolbar> *Break* icon) removes part of a line, pline, arc, or circle **(Figure 25.86).** To specify a break at an intersection with another line as shown in **Figure 25.87,** select the line to be broken, select the *F* option, and pick the two endpoints of the line to be removed. Endpoints can be

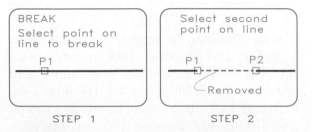

STEP 1   STEP 2

25.86 Break.
Step 1 *Modify* toolbar> *Break* icon *Select object:* P1,
Step 2 *Specify second break point or [First point]:* P2,
Line between P1 and P2 is removed.

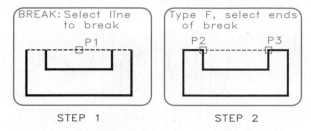

STEP 1   STEP 2

25.87 Break: F option.
Step 1 *Modify toolbar> Break icon Select object:* P1,
Specify second break point or [First point]: F (Enter)
Step 2 *Specify first break point:* P2,
Specify second break point: P3, Line P2-P3 is removed.

selected without fear of selecting the wrong lines.

25.29 Move and Copy

The *Move* command *(Modify* toolbar> *Move* icon) repositions a drawing **(Figure 25.88)** and the *Copy* command duplicates it, leaving the original in its original position. The *Copy (C)* command is applied in the same manner as the *Move* command. The *Copy* command has a *Multiple* option (type M when prompted) for locating more than one copy of the selected drawing in different positions.

25.30 Undo

The *Undo* command *(Command:* Undo, or U) can reverse the previous commands one at a time back to the beginning of a session. The *Redo* command reverses the last *Undo*; *Oops*

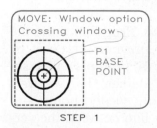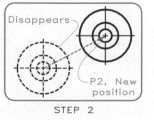

STEP 1   STEP 2

25.88 Move.
Step 1 *Modify toolbar> Move icon Select objects:* W,
Select by a Window. *Specify base point or displacement:* P1
Step 2 *Specify second point of displacement or <use first point as displacement>:* Drag to new position of P2.

will not work. The *Undo* command has options of: *Auto, Control, BEgin, End, Mark, Back,* and *Number.*

> *Command:* Undo (Enter)
> *Auto/Control/BEgin/End/Mark/Back/*
> *<Number>:* 4 (Enter)

Entering 4 has the same effect as using the *U* command four separate times.

25.31 Polyline

The *Polyline (PL)* icon of the *Draw* toolbar *(View> Toolbars> Draw)* **(Figure 25.89)** is used for drawing 2D polylines, which are lines of continuously connected segments, instead of separate segments as drawn by the *Line* command. The thickness of a *Pline* can be varied as well, which requires the pen to plot with multiple strokes when plotting rather than printing **(Figure 25.90).**

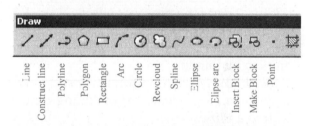

25.89 The *Polyline* icon on the *Draw* toolbar is selected for drawing *polylines*.

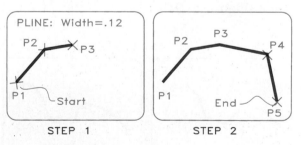

STEP 1   STEP 2

25.90 Polyline: Width option.
Step 1 *Draw toolbar> Polyline icon> Specify start point:* P1> *Current line-width is 0.00> Specify next point or [Arc/Close/ Halfwidth/Length/Undo/Width]:* W > *Specify starting width <0.00>:* .12 (Enter)> *Specify ending width <0.12>:* (Enter), *Specify next point, or [Arc/Close/Halfwidth/Length/Undo/ Width]:* P2 *Specify next point, or [Arc/Close . . ./Width]:* P3
Step 2 *Specify next point, or [Arc/Close . . . /Width]:* P4 *Specify next point, or [Arc/Close . . . /Width]:* P5 (Enter)

The *Pline* options are *Arc, Close, Halfwidth, Length, Undo,* and *Width. Close* automatically connects the last end of the polyline with its beginning point and ends the command. *Halfwidth* specifies the width of the line measured on both sides of its centerline. *Length* continues a *Pline* in the same direction by typing the length of the segment. If the first line was an arc, a line is drawn tangent to the arc. *Undo* erases the last segment of the polyline, and it can be repeated to continue erasing segments.

25.32 Pedit

The *Pedit* command (*Command:* Pedit> (Enter)) modifies *Plines* with the following options: *Close, Join, Width, Edit vertex, Fit, Spline, Decurve, Ltype gen, Undo,* and *eXit.* If the *Pline* is already closed, the *Close* command will be replaced by the *Open* option.

Join (J) gives the prompt, *Select objects,* for selecting segments to be joined into a polyline. *Segments* must have exact meeting points to be joined.

Width (W) gives the prompt, *Specifiy new width for all segments:,* for assigning a new width to a *Pline.*

Fit (F) converts a polyline into a line composed of circular arcs that pass through each point **(Figure 25.91).**

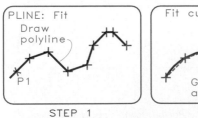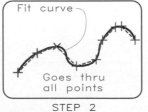

25.91 Pedit: Fit curve
Step 1 *Modify II* toolbar> *Select Pedit icon>*
Select polyline: P1 (Enter)
Step 2 [Close/Join/Width/Edit vertex/Fit/. . . Undo/eXit <X>]: Fit (Enter) The curve is drawn as a series of arcs passing through all polyline points.

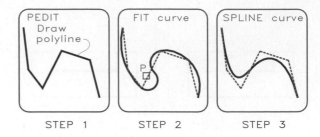

25.92 Pedit: Fit curve.
Step 1 Draw as a polyline (Pline).
Step 2 *Modify II toolbar> Select Pedit* icon> *Select polyline:* P (Select line)
Close/Join/Width/Edit vertex/Fit/ . . . Undo/eXit <X> Fit (Enter). The curve of arcs passes through all points.
Step 3 *Pedit* in the same way, but use the *Spline* option to obtain a "best curve" that may not pass through the selected points.

Spline (S) modifies the polyline as did the *Fit* curve, but it draws cubic curves passing through the first and last points, not necessarily through the other points **(Figure 25.92).**

Decurve (D) converts *Fit* or *Spline* curves to their original straight-line forms.

Ltype gen (L) applies dashed lines (such as hidden lines) in a continuous pattern on curved polylines. Without applying this option, dashed lines may omit gaps in curved lines. By setting system variable *(Setvar) Plinegen On,* linetype generation will be applied as *Plines* are drawn.

Undo (U) reverses the last *Pedit* editing step.

Edit vertex (E) selects vertexes of the *Pline* for editing by placing an X on the first vertex when the polyline is picked. The following options appear: *Next/Previous/Break/Insert/Move/ Regen/Straighten/ Tangent/Width/eXit/ <N>:* Next (Enter).

Next (N) and *Previous (P)* options move the X marker to next or previous vertexes by pressing (Enter). *Break (B)* prompts you to select a vertex with the X marker. Then use *Next* or *Previous* to move to a second point, and pick *Go* to remove the line between the vertexes. Select *eXit* to leave the *Break* command and return to *Edit Vertex.*

Insert adds a new vertex to the polyline between a selected vertex and the next vertex.

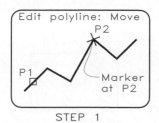
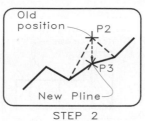

25.93 Pcdit: Edit vertex–Move.
Step 1 *Modify II toolbar> Select Pedit icon> Select polyline: P1 Enter an option [Close/Join/Width/Edit vertex . . . Undo]:* E (Enter) (Enter) (Enter) (Enter) *to place cursor on* P2*. Enter a vertex editing option [Next . . . /Move/ . . ./eXit]<N>:* Move (Enter)
Step 2 *Specify a new location for marked vertex:* P3*> Enter a vertex editing option [Next . . . /Move/ . . ./eXit] <N>:* P3 *Enter a vertex editing option* [Next/ . . . /Insert/ . . . /eXit] <N> X (Enter)

Move (M) relocates a selected vertex (**Figure 25.93**).

Straighten (S) converts the polyline into a straight line between two selected points. An X marker appears at the current vertex and the prompt, *Next/ Previous/Go/eXit/<N>*, appears. Move the X marker to a new vertex with *Next* or *Previous*, select *Go*, and the line is straightened between the vertices. Enter X to *eXit* and return to the *Edit vertex prompt*.

Tangent (T) lets a tangent direction be selected at the vertex marked by the X for curve fitting by responding to the prompt, *Direction of tangent.* Enter the angle from the keyboard or by cursor.

Width (W) sets the beginning and ending widths of an existing line segment from the X-marked vertex. Use *Next* and *Previous* to confirm in which direction the line will be drawn from the X marker. The polyline will be changed to its new thickness when the screen is regenerated with *Regen (R).* Use *eXit* to escape from the *Pedit* command.

25.33 Hatching

Hatching is a pattern of lines that fills sectioned areas, bars on graphs, and similar applications. From the *Draw* toolbar (**Figure 25.94**) the *Bhatch (boundary hatch)* dialog box is dis-

25.94 *Draw* toolbar> Select *Hatch* icon to begin to hatch a sectioned area.

played (**Figure 25.95**). By selecting the down arrow at the *Pattern* box, a listing of pattern names is given from which to select. When one is selected, a view of the pattern will appear in the *Swatch* window. Examples of some predefined patterns are shown in **Figure 25.96**.

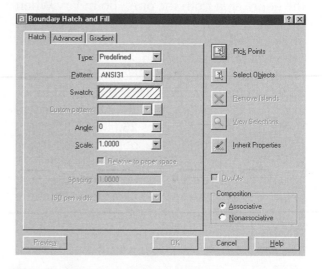

25.95 When the *Hatch* icon is selected, this *Boundary Hatch and Fill* box appears on the screen.

25.96 When the *Pattern* button in the *Boundary Hatch and Fill* box is selected, the hatch patterns and their names are displayed in this *Hatch Pattern Palette*.

The *Boundary Hatch* box **(Figure 25.95)** lets you specify *Predefined, User-defined,* or *Custom* patterns. *Predefined* patterns are those provided by AutoCAD.

Scale sets the spacing between the lines of a pattern, and *Angle* assigns their direction. The *Advanced* tab displays the options shown in **Figure 25.97** from which choices of *Normal, Outer,* or *Ignore* can be made. A square with a pentagon and a circle inside it illustrates the effect of each choice. *Normal* hatches every other nested area beginning with the outside. *Outer* hatches the outside area, and *Ignore* hatches the entire area from the outer boundary. When text within the hatching area is selected, it will appear in an opening in the hatching and hatch lines will not pass through it.

Options of *Points* and *Select Objects* **(Figure 25.95)** are used to select areas inside of boundaries and then boundaries themselves, respectively, as shown in **Figure 25.98.** When points are selected outside the boundary or if the boundary is not closed, error messages will appear. *Composition* can be specified as *Associative* or *Nonassociative. Associative* hatching is automatically updated when the size of the hatching area is changed. Select the *Inherit*

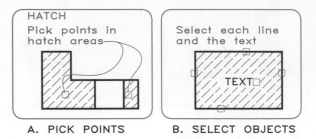

25.98 Hatching areas.
A The *Pick Points* option (Figure 25.97) of *Boundary Hatch* prompts for points inside the boundaries for hatching.
B The *Select Objects* option requires that boundary lines be selected, including text.

Properties icon, pick the symbols within a hatched area, and select the area to be hatched. Then it is filled with the same hatching symbols.

25.34 Text and Numerals

Text and numerals can be added to a drawing with the *Dtext* command. *Command:* Dtext> *Specify start point of text or [Justify/Style]:* Justify (Enter) *Enter an option [Align/Fit/Center/Middle/Right/TL/TC/TR/ML/MC/MR/BL/BC/BR]:* specify the insertion point for the text you are entering **(Figure 25.99).** For example, BC means bottom center, *RT* means right top, and so forth.

Figure 25.100 illustrates how multiple lines of *Dtext* are automatically spaced by pressing (Enter) at the end of each line and (Enter) twice to end the *Dtext* command.

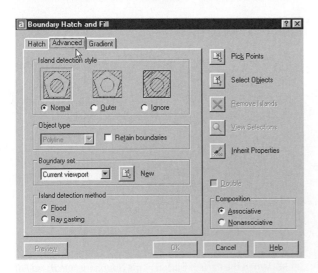

25.97 Select the *Advanced* tab in the *Boundary Hatch and Fill* box to display this dialog box.

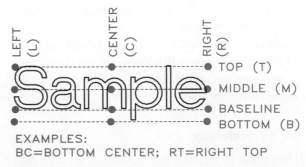

25.99 Text can be added to a drawing by using any of the insertion points above. For example, *BC* means the bottom center of a word or sentence that will be located at the cursor point.

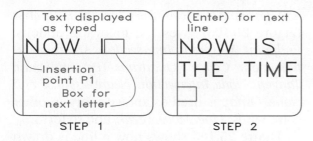

STEP 1   STEP 2

25.100 Dtext.
Step 1 *Command:* <u>Dtext</u> (Enter)
Current text style: "Standard" Text height 0.50
Specify start point of text or [Justify/Style]: <u>P1</u>
Specify height <0.50>: <u>.125</u>, *Specify rotation angle of text
<0>:* (Enter) *Enter text:* <u>NOW IS</u>
Step 2 (Enter) *Enter text:* <u>THE TIME</u> (Enter) (Enter)

25.35 Text Style

Many of AutoCAD's text fonts and their names
are shown in **Figure 25.101.** The default style,
Standard, uses the *Txt* font. From the *Format*

TXT	PRELIMINARY PLOTS
MONOTXT	FOR SPEED ONLY
	Simplex fonts
ROMANS	FOR WORKING DRAWINGS
SCRIPTS	*Handwritten Style, 1234*
GREEKS	ΓΡΕΕΚ ΣΙΜΠΛΕΞ, 12345
	Duplex fonts
ROMAND	THICK ROMAN TEXT
	Complex fonts
ROMANC	ROMAN WITH SERIFS
ITALICC	*ROMAN ITALICS TEXT*
SCRIPTC	*Thick—Stroke Script Text*
GREEKC	ΓΡΕΕΚ ΩΙΤΗ ΣΕΡΙΦΣ
	Triplex fonts
ROMANT	TRIPLE—STROKE ROMAN
ITALICT	*Triple—Stroke Italics*
	Gothic fonts
GOTHICE	English Gothic Text
GOTHICG	German Gothic Text, 12
GOTHICI	Italian Gothic Text, 12

25.101 Several examples of the many available fonts are
shown here.

menu select *Text Style,* and the *Text Style* dialog
box appears where you can assign a *New Style
Name* **(Figure 25.102).** Select the *New* button,
get the *New Style* box, type the name (<u>RD</u>, for
example) as the style name. Pick the down
arrow at the *Font Name* panel, and pick the
font (Romand, for example) that you want to
assign to the new style. Other options can be
assigned: *Height* (0 is recommended), *Width
Factor,* and *Oblique Angle.* A preview of your
preferences is shown in the *Preview* window.

The *Style names* are listed in a drop-down
menu in the window beneath the heading *Style
Name* of the *Text Style* dialog box. An example
of the text font is displayed when a *Style* is
selected **(Figure 25.102).** A defined *Style* will
retain its settings until they are changed.

If you later change a named *Style* with new
settings or fonts and select *Apply* in the *Text
Style* box, all text previously entered under
this style name will be updated with the new
properties. This technique is used to change
the *Txt* and *Monotxt* fonts to more attractive
fonts at plot time. Beforehand, time is saved
by using *Txt* and *Monotxt* fonts because they
regenerate quickly.

The *DDedit* or *(Main Menu> View>* Tool-
bars> *Modify II> Text Edit* icon*)* is used to select
a line of text to be displayed in a dialog box for

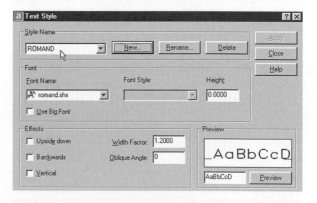

25.102 *Main Menu> Format> Text Style* (or type <u>Style</u> at the
Command line) to display this *Text Style* box. From here, a *New*
style can be named, *Fonts* assigned, *Width Factors* specified,
and other assignments made.

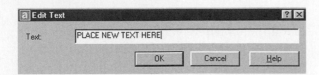

25.103 *Modify II* toolbar> Select *Edit Text* icon (or *Command:* DDedit), select a line of text on the screen and it will appear in this *Edit Text* box for modifiation.

editing **(Figure 25.103).** Correct the text, select the <u>OK</u> button, and it is revised on the screen.

25.36 Mirror

Select the *Mirror* icon (*Modify toolbar> Mirror* icon) or, (*Command:* <u>Mirror</u> (Enter)) to mirror partial figures about an axis **(Figure 25.104).** A line that coincides with the *Mirror line* (P1-P2, for example) will be drawn twice when mirrored; therefore, parting lines should not be selected, or drawn until after the drawing has been mirrored.

The system variable *Mirrtext (Command: Setvar> Mirrtext)* is used for mirroring text. By setting *Mirrtext* to <u>0,</u> it is set to *Off,* and text will not be mirrored. If *Mirrtext* is set to <u>1</u> *(On),* the text will be mirrored along with the drawing.

25.37 Osnap

By using *Osnap (Object Snap),* you can snap to objects of a drawing rather than to the *Snap*

grid. *Osnap* icons from the *Object snap* toolbar **(Figure 25.105)** gives the following options: *Endpoint, Midpoint, Intersection, Apparent Intersection, Center, Quadrant, Perpendicular, Tangent, Node, Intersection, Nearest, Quick,* and *None. Osnap* is used as an accessory to other commands: *Line, Move, Break,* and so forth.

Figure 25.106 shows how a line is drawn from an intersection to the endpoint of a line. In **Figure 25.107** a line is drawn from <u>P1</u> tangent to the circle by using the *Tangent* option

25.105 The *Object Snap* toolbar has these options for drawing to and from object features on the screen.

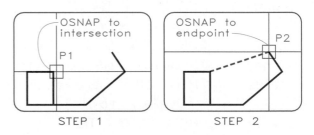

25.106 Osnap: Intersection and end.
Step 1 *Draw* toolbar> *Line* icon> *Line from point:*> *Intersection* icon on *Osnap* tool bar> *Int of:* <u>P1</u>
Step 2 *To point:*> *Endpoint* icon on *Osnap* toolbar. *To point:* <u>P2</u>. The line is drawn.

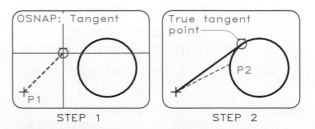

25.107 Osnap: Tangent.
Step 1 *Draw* toolbar> *Line* icon> *Line from point:* <u>P1</u>
Specify next point or [Undo]:> *Tangent Osnap* icon>
Step 2 *Tan to:* <u>P2</u>. The line is drawn from P1 tangent to the true tanget point on circle nearest P2.

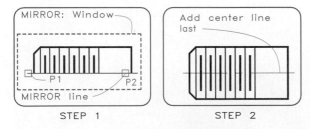

25.104 Mirror
Step 1 Draw the half to be mirrored. *Modify* toolbar> *Mirror* icon> *Select objects:* <u>W</u> (Enter) Window the drawing. *Select objects:* (Enter)
First point of mirror line: <u>P1</u>, *Second point:* <u>P2</u>
Step 2 *Delete old objects? <N>:* <u>No</u> (Enter) The drawing is mirrored.

of *Osnap*. The *Tangent* option can also be used to draw a line tangent to two arcs.

The *Node* option snaps to a *Point*, the *Quadrant* option snaps to one of the four compass points on a circle, the *Insert* option snaps to the intersection point of a *Block*, and the *None* option turns off *Osnap* for the next selection. The *Quick* option reduces searching time by selecting the first object encountered rather than searching for the one closest to the aperture's center.

Osnap settings can be temporarily retained as "running" *Osnaps* for repetitive use. One way to set running *Osnaps* is to right-click on the *Osnap* tab in the status line at the bottom of the screen **(Figure 25.30);** select *Settings* to display the *Drafting Settings* box that lists the status *Osnap* options. Settings can be added or removed, from this list of *Osnap* options.

Now the cursor has an aperture target at its intersection for picking endpoints and centers of arcs. The *Osnap* command can be turned on or off by selecting the *Osnap* button in the status line at the bottom of the screen. From the *Drafting Settings* box, select the Option button and the *Options* box will appear from which you can set *Osnap* aperture sizes from 1 to 50 pixels for snapping to objects.

25.38 Array

The *Draw* toolbar> *Array* icon is used to draw rectangular patterns (rows and columns) or polar layouts of selected drawings using the *Array* dialog box **(Figure 25.108).** A series of holes can be drawn on a bolt circle by drawing the first hole and arraying it as a polar array **(Figure 25.109).**

A rectangular array is begun by making the drawing in the lower left corner and following the steps in **Figure 25.110.** Rectangular arrays may be drawn at angles by using *Command:* Snap> Rotate to rotate the grid. The first object is drawn in the lower left corner of the array and the number of rows,

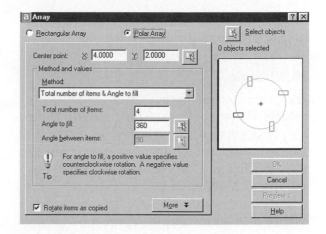

25.108 Select the *Draw* toolbar> *Array* icon to obtain the *Array* box in which to specify rectangular or polar arrays.

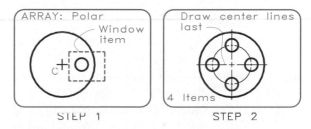

25.109 Array: Polar.
Step 1 *Modify* toolbar> *Array* icon> *Select objects:* W̲ (Enter) Window the hole> *Select objects: Enter the type of array [Rectangular/Polar] <R>:* P̲ (Enter)
Step 2 *Specify center point of array:* C̲ (Enter)
Enter the number of items in the array: 4̲ (Enter)
Specify the angle to fill(+−ccw,−−cw) <360>: 360̲ (Enter)
Rotate arrayed objects? [Yes/No] <Y>: (Enter)

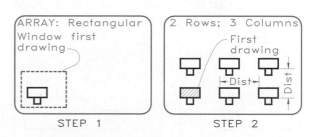

25.110 Array: Rectangle.
Step 1 *Modify* toolbar> *Array* icon> *Select objects:* W̲ (Enter) Window the desk> *Select objects: Enter the type of array [Rectangular/Polar] <R>:* R̲ (Enter)
Step 2 *Enter the number of rows (---) <1>:* 2̲ (Enter) *Enter the number of columns (|||) <1>:* 3̲ (Enter) *Enter the distance between rows or specify unit cell (---) <1>:* 4̲ (Enter) *Specify the distance between columns (|||) <1>:* 3.5̲ (Enter)

columns, and the cell distances are specified when prompted.

25.39 Scale

The *Scale* command reduces or enlarges previously drawn objects. The desk in **Figure 25.111** is enlarged by windowing it, selecting a base point, and typing a scale factor of 1.6. The drawing and its text are enlarged in both the *x*- and *y*-directions.

A second option of *Scale* lets you select a length of a given object, specify its present length, and assign a length as a ratio of the first dimension. The lengths can be given by the cursor or typed at the keyboard in numeric values.

25.40 Stretch

The *Stretch* command (*Modify* toolbar> *Stretch* icon) lengthens or shortens a portion of a drawing while one end is left stationary. The window symbol in the floor plan in **Figure 25.112** is *Stretched* to a new position, leaving the lines of the wall unchanged. A *Crossing Window* must be used to select lines that will be stretched. The *Ortho* command should be turned on in conjunction with this command to assure movement in the horizontal direction.

25.41 Rotate

A drawing can be rotated about a base point by using the *Rotate* command (*Modify* tool-

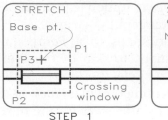 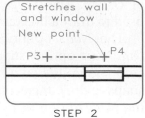

25.112 Stretch.
Step 1 *Modify* toolbar> *Stretch* icon> *Select objects:* Use P1 and P2 to form crossing window. *Specify base point or displacement:* P3
Step 2 *Specify second point of displacement:* P4 The windowed portion of the drawing is repositioned.

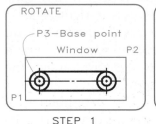 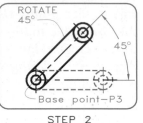

25.113 Rotate.
Step 1 *Modify* toolbar> *Rotate* icon> *Select objects:* Window part with P1 and P2.
Step 2 *Specify base point:* P3 *Specify rotation angle or [Reference]:* 45 (Enter) Object is rotated 45° counterclockwise.

bar> *Rotate* icon) as shown in **Figure 25.113**. *Window* the drawing, select a base point, and type the rotation angle or select the angle with the cursor. Drawings made on multiple layers can be rotated also.

25.42 Offset

An object (*Line* or *Polyline*) can be drawn parallel to and offset from another object, such as a polyline, by the *Offsest* command (**Figure 25.114**). *Offset* prompts for the distance or the point through which the offset line must pass, and then prompts for the side of the offset. The *Offset* command is helpful when drawing parallel lines to represent walls of a floor plan.

25.43 Blocks

One of the most productive features of computer graphics is the capability of creating

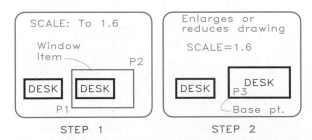

25.111 Scale: Numeric.
Step 1 *Modify* toolbar> *Scale* icon> *Select objects:* W (Enter) Window the desk with P1 and P2.
Step 2 *Specify base point:* P3 *Specify scale factor or [Reference]:* 1.6 (Enter) The desk is drawn 60% larger.

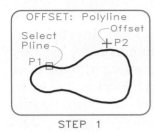
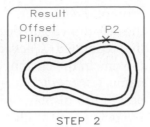

STEP 1   STEP 2

25.114 Offset.
Step 1 *Command:* <u>Offset</u> (Enter) *Specify offset distance or [Through] <1>:* <u>T</u> (Enter)
Select object to offset or <exit>: <u>P1</u> (Enter)
Step 2 *Specify through point:* <u>P2</u> (Enter).
An enlarged *Pline* is drawn that passes through <u>P2</u>.

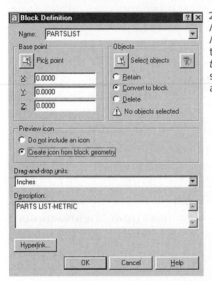

25.115 *Draw Menu> Block> Make* will open this *Block Definition* box on the screen for making a *Block*.

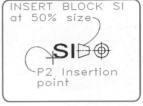

STEP 1   STEP 2

25.116 Block.
Step 1 *Draw Menu> Block> Make> Get Block Definition* box> *Name:* SI> Select *Pick point* button> <u>P1</u>
Pick *Select objects* button> Window drawing (Enter)
Step 2 *Command:*> <u>Insert</u> (Enter)> *Insert* box> *Name:* SI> *Scale:* .50> Check *Uniform Scale* box> <u>Ok</u>> *Specify insertion point:* <u>P2</u>. The block is inserted.

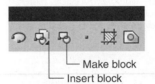

25.117 The *Make block* and *Insert block* icons can be selected from the *Main Menu> Draw* toolbar.

— Make block
— Insert block

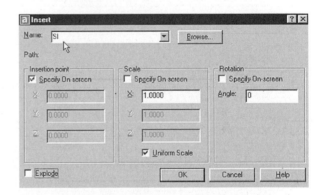

25.118 Select *Main Menu> Insert>* to get *Insert box* to specify *Blocks* or *Files (Wblocks)* for insertion in a drawing.

drawings called *Blocks* for repetitive use. *Draw Menu> Block> Make> Block definition* box is used to make a block (**Figure 25.115**). The SI symbol in **Figure 25.116** is a typical drawing that is made into a *Block* and inserted into drawings using icons from the *Draw* toolbar shown in **Figure 25.117**. The *Insert* heading on the *Main Menu (Insert> Block)* gives the *Insert* dialog box (**Figure 25.118**) for inserting *Blocks*. You can select a block from the window or *Browse* for other files to insert as *Blocks* (**Figure 25.119**).

When a *Block* is selected on the screen (to be *Moved*, for example), it is selected as a total unit. However, *Blocks* that were *Inserted* by se-

lecting the *Explode* box first, or by typing a star in front of the *Block* name (*SI*, for example), are inserted as a group of disjoint objects that can be edited one at a time. An inserted *Block* can be separated into individual entities by typing <u>Explode</u> at the *Command* line and selecting the *Block*.

25.44 Write Blocks (Wblocks)

Blocks are parts of files that can be used only in the current drawing file unless they are converted to *Wblocks (Write Blocks)*, which

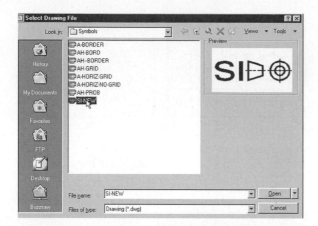

25.119 To insert a block, use *Command:* <u>Insert</u> (Enter)> *Browse*> then select a defined block from the list> *Open*> select an insertion point on the screen with the cursor (Enter).

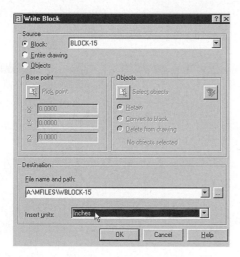

25.120 *Command:* <u>Wblock</u> (Enter) and the *Write Block* box enables you to select a *Block* as the *Source* and convert it to a *Wblock* with its *File name and path* specified in the lower window. Non-*Block* drawings can also be made into *Wblocks* by checking the *Objects* box under *Source*.

are independent files, not parts of files. This conversion is performed by typing <u>Wblock</u> at the *Command* line to get the *Write Block* dialog box **(Figure 25.120).** Select the *Block* button and type the name of the *Block* that is to be converted to a file in the window of the *Source* area. Give the *File name and path* in the *Destination* area, and pick OK.

Blocks can be redefined by selecting a previously used *Block* name to receive the prompt, *Redefine it? <N>:* <u>Y</u> *(Yes)* and select the new drawing to be blocked. After doing so, the redefined *Block* automatically replaces the one in the current drawing with the same name. *Command:* <u>Insert</u>> *Block*> *Browse* can be used to display thumbnail illustrations of *Wblocks (Files)* as illustrated in **Figure 25.119.**

25.45 Dimensioning

Figure 25.121 shows common types of dimensions that are applied to drawings. Drawings should be drawn full size on the screen to simplify their measurements and dimensions; scaling should be done at plotting time.

Dimensions can be applied as *Associative* or as *Nonassociative (Exploded)* dimensions. *Associative* dimensions (when *Dimaso* is *On*) are inserted as if the dimension line, extension lines, text, and arrows were parts of a single *Block*. Exploded dimensions (*Dimaso* set to *Off*) are applied as individual objects that can be modified independently. Except where noted, the examples that follow will be associative dimensions.

Many variables must be set before dimensioning is usable: Arrowheads and numerals must be sized, extension line offsets specified, text fonts assigned, and units adopted, to name a few.

25.46 Dimension Style Variables

To get a list of the current dimensioning variables, *Command:* <u>Dim</u>> *Dim*> <u>Status</u> (Enter). Sizes of dimensioning variables are based on

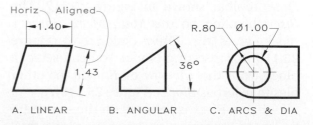

25.121 The basic types of dimensions that may appear on a drawing are shown here.

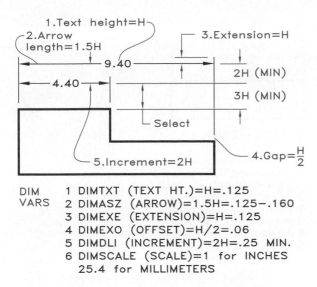

1.Text height=H
2.Arrow length=1.5H
3.Extension=H
9.40
4.40
2H (MIN)
3H (MIN)
Select
5.Increment=2H
4.Gap=$\frac{H}{2}$

DIM VARS
1 DIMTXT (TEXT HT.)=H=.125
2 DIMASZ (ARROW)=1.5H=.125–.160
3 DIMEXE (EXTENSION)=H=.125
4 DIMEXO (OFFSET)=H/2=.06
5 DIMDLI (INCREMENT)=2H=.25 MIN.
6 DIMSCALE (SCALE)=1 for INCHES
25.4 for MILLIMETERS

25.122 Dimensioning variables are based on the height of the lettering (text), which is usually about 1/8 inch high.

Dim Vars	Default	Description
DIMASO	Off	Create dimension objects
DIMSTYLE	Standard	Current dimension style
DIMADEC	-1	Angular decimal places
DIMALT	Off	Alternate units selected
DIMALTD	2	Alternate unit decimal places
DIMALTF	25.4	Alternate unit scale factor
DIMALTRND	2	Alternate units rounding value
DIMALTTD	2	Alternate tolerance dec. places
DIMALTTZ	0	Alternate tolerance zero suppress.
DIMALTU	2	Alternate units
DIMALTZ	0	Alternate unit zero suppression
DIMAPOST	-	Prefix & suffix for alternate text
DIMASZ	.125	Arrow size
DIMATFIT	3	Arrow and text fit
DIMAUNIT	0	Angular unit format
DIMAZIN	0	Angular zero suppression
DIMBLK	-	Arrow block name
DIMBLK1	-	First arrow block name
DIMBLK2	-	Second arrow block name
DIMCEN	.09	Center mark size
DIMCLRD	BYLAYER	Dimension line & leader color
DIMCLRE	BYLAYER	Extension line color
DIMCLRT	BYLAYER	Dimension text color
DIMDEC	4	Decimal places
DIMDLE	0	Dimension line extension
DIMDLI	.38	Dimension line spacing
DIMSEP	.	Decimal separator
DIMEXE	.125	Extension above dimension line
DIMEXO	.06	Extension line origin offset
DIMFRAC	0	Fraction format
DIMGAP	.06	Gap from dimension line to text
DIMJUST	0	Justification of text on dimem. line

25.123 Use *Command: Dim>* Dim> Status (Enter) to obtain a listing of the dimension variables, their settings, and definitions.

Dim Vars	Default	Description
DIMLDRBLK	ClosedFilled	Leader block name
DIMLFAC	1	Linear unit scale factor
DIMLIM	Off	Generate dimension limits
DIMLUNIT	2	Linear unit format
DIMLWD	-2	Dimension lind & leader lineweight
DIMLWE	-2	Extension line lineweight
DIMPOST	-	Prefix & suffix for dimension text
DIMRND	0	Rounding value
DIMSAH	Off	Separate arrow blocks
DIMSCALE	1	Overall scale factor
DIMSD1	Off	Suppress first dimension line
DIMSD2	Off	Suppress second dimension line
DIMSE1	Off	Suppress first extension line
DIMSE2	Off	Suppress second extension line
DIMSOXD	Off	Suppress outside dimension lines
DIMTAD	0	Place text above the dimension line
DIMTDEC	4	Tolerance decimal places
DIMTFAC	1	Tolerance text height scaling factor
DIMTIH	On	Text inside extensions is horiz.
DIMTIX	On	Place text inside extensions
DIMTM	0	Minus tolerance value
DIMTMOVE	0	Text movement
DIMTOFL	Off	Force line inside extension lines
DIMTOH	On	Text outside horizontal
DIMTOL	Off	Tolerance dimensioning
DIMTOLJ	1	Tolerance vertical justification
DIMTP	0	Plus tolerance
DIMTSZ	0	Tick size
DIMTVP	0	Text vertical position
DIMTXSTY	Standard	Text style
DIMTXT	.125	Text height
DIMTZIN	0	Tolerance zero suppression
DIMUPT	Off	User positioned text
DIMZIN	0	Zero suppression

25.124 Additional dimension variables are shown here as a continuation of Figure 25.123.

the letter height, which most often is 0.125 inch (**Figure 25.122**).

To set and save variables needed for basic applications, *Open (or Create)* the file *A:PROTO1.* Each variable is set by typing Setvar and the name of the dimensioning variable (*Dimtxt* for text height, for example) and assigning a numerical value. A list of the dimensioning variables is given in **Figures 25.123** and **25.124.** Assign the basic variable values of *Dimtxt, Dimasz, Dimexe, Dimexo, Dimtad, Dimdli, Dimaso,* and *Dimscale* shown in **Figure 25.122** to A:PROTO1 since they apply to most applications. *Command:>* Units and set decimal fractions to two decimal places for inches.

Save these settings to B:PROTO1 as an empty file with no drawings on it for use as the prototype with its dimensioning settings. Open B:PROTO1 and create a new file *Save As* A:DWG-3, for example, which becomes the current file with the same settings as A:PROTO1.

Dimensioning variables can be set from dialogue boxes also instead of being typed; these techniques are covered later. You will be more proficient by becoming familiar with both methods of assigning variables. For now, use the A:DWG-3 file to explore the fundamentals of dimensioning.

25.47 Linear Dimensions

The *Dimension* toolbar (**Figure 25.125**) provides a convenient means of selecting dimensioning commands. Select the *Linear* icon, the points as prompted, and the *Horizontal* option as shown in **Figure 25.126** to apply a horizontal dimension.

A dimension is applied semiautomatically in **Figure 25.127** by pressing (Enter) when prompted for *Endpoints*, selecting the line or circle to be dimensioned, and locating its dimension line.

Select *Main Menu> Dimension> Continue* (or type *Dimcontinue* at the *Command* line) to continue a chain of linear, angular, or ordinate dimensions from the last extension line (**Figure 25.128**). Baseline applies dimensions from a single endpoint and each

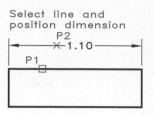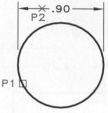

A. HORIZ. DIMEN.   B. HORIZ. DIMEN.

25.127 Dimensioning: Linear–Semiautomatic.
A *Dimensioning* toolbar> *Aligned> Specify first extension line origin or <Select object>*: (Enter)
Select object to dimension: P1
Select dimension line location or [Mtext/Text/Angle]: P2
Dimension text = 1.10: (Enter)
B Use these same steps and select a point on the circle (P1) to dimension its diameter.

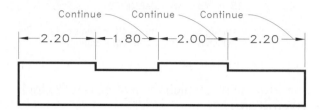

25.128 When dimensions are placed end to end, use the *Continue Dimension* option from the *Dimension* toolbar to specify the successive second extension line origins after the first dimension line has been drawn.

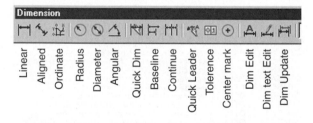

25.125 The *Dimension* toolbar has these dimensioning options from which to select.

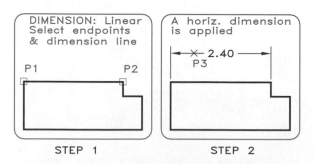

25.126 Dimensioning: Linear–Horizontal.
Step 1 *Main Menu> View> Toolbars> Dimension* toolbar> *Linear* icon> *Specify first extension line origin or <select object>*: P1 *Specify second extension line origin*: P2
Step 2 *Specify dimension line location or [Mtext/Text/Angle/Horizontal/Vertical/Rotated]*: H (Enter)
Specify dimension line location or [Mtext/Text/Angle]: P3 (Enter)> Dimension text = 2.40> (Enter) to accept the dimension.

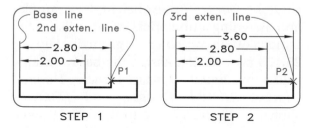

25.129 Dimensioning: Baseline option.
Step 1 *Dimension* toolbar:> *Linear*> place a 2.00 *Linear* dimension. (The first extension line becomes the baseline.) *Dimension* toolbar> *Baseline* icon> *Specify a second extension line origin or [Undo/Select] <Select>*: P1
Dimension text = 2.80 The 2.80 dimension is drawn.
Step 2 *Specify a second extension line origin or [Undo/Select] <Select>*: P2 >*Dimension text = 360* The 3.60 dimension is drawn. (Enter) to exit from last dimension.

dimension is offset incrementally by the dimension line increment variable, *Dimdli* (**Figure 25.129**).

Select the *Aligned* command from the *Dimension* toolbar and you will be prompted

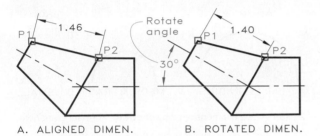

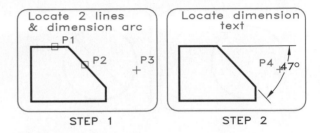

STEP 1 STEP 2

25.130 Dimensioning: Oblique lines.
A Linear dimensions can be aligned by selecting the *Align* icon and selecting endpoints P1 and P2.
B Linear dimensions can be rotated by the selecting the *Linear* icon, the *Rotate* option, assign the angle (150°), and pick endpoints P1 and P2.

25.132 Dimensioning: Angles.
Step 1 Use *Dimension* toolbar> *Angular* icon> *Select arc, circle, line, or <specify vertex>:* P1 > *Select second line:* P2 *Specify dimension arc line location or [Mtext/Text/Angle]:* P3
Step 2 *Enter dimension text <47>:* (Enter) to accept 47°. *Enter text location (or press ENTER):* (Enter)

to select the *1st* and *2nd* extension lines and the position of the dimension line **(Figure 25.130A).** The dimension line will be inserted aligned with line 1-2. Extension lines can be automatically drawn by pressing (Enter) at the first prompt and selecting the line to be dimensioned. A rotated dimension can be applied with the *Rotate* option with an assigned angle of rotation **(Figure 25.130B).**

25.48 Angular Dimensions

Figure 25.131 shows variations for dimensioning angles depending on the space available. In **Figure 25.132** an angle is dimensioned by selecting the *Angular* icon, selecting lines of the angle, and locating the dimension line arc as shown in step 1. If space permits, the dimension text will be centered in the arc between the arrows (step 2).

25.49 Diameter

Diameters of circles can be placed as shown in **Figure 25.133** depending on the available space. By changing system variables *Dimatfit*, *Dimtofl*, and *Dimtmove*, circles can be dimensioned as shown in **Figures 25.134** and **25.135.** By setting the *Dimtix* system variable *On*, the text is forced inside the extension lines regardless of the available space. The

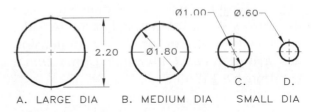

A. LARGE DIA B. MEDIUM DIA SMALL DIA

25.133 Examples of methods of dimensioning circles depending on available space.

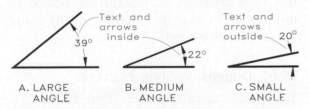

A. LARGE ANGLE B. MEDIUM ANGLE C. SMALL ANGLE

25.131 Angles will be dimensioned in any of these three formats depending on available space.

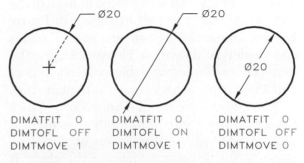

DIMATFIT 0	DIMATFIT 0	DIMATFIT 0
DIMTOFL OFF	DIMTOFL ON	DIMTOFL OFF
DIMTMOVE 1	DIMTMOVE 1	DIMTMOVE 0

25.134 Examples of circle diameters with their associated dimensioning variables. Assign variables by *Command:* Dim> Dim> type name of variable (Enter).

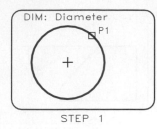
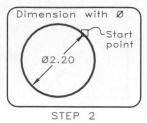

STEP 1   STEP 2

25.135 Dimensioning: Circle.
Step 1 *Dimension* toolbar> *Diameter icon*>
Select arc or circle: <u>P1</u>
Step 2 *Dimension text = <2.20>* (Enter)
Specify dimension line location or [Mtext/Text/Angle]:
Select start point (Enter)

Dimtmove variable has the following options: *0* moves the dimension line with dimension text; *1* adds a leader when dimension text is moved; *2* allows text to be moved freely without a leader.

The *Dimatfit* variable has the following options: *0* places both text and arrows outside extension lines; *1* moves arrows first, then text; *2* moves text first, then arrows; *3* moves either text or arrows, whichever fits best. The *Dimtofl (On)* dimension variable forces a dimension line to be drawn between the arrows when the text is located outside. To specify whether or not a dimension has a leader use *Command:* <u>Dim</u>> *Dim*> <u>Dimtmove</u>> <u>1</u> (Enter).

25.50 Radius

Select *Radius* from the *Dimension* toolbar to dimension arcs with an *R* placed in front of the text (R1.00, for example) as shown in **Figure 25.136.** Dimensioning an arc with a radius and leader is shown in **Figure 25.137.** The same dimensioning variable covered in Section 25.49 apply to arcs as well as to diameters.

The *Leader* command (*Dimension*> *Quick Leader* on the *Dimension* toolbar) is used to add a leader with a dimension or note to a drawing, but it cannot measure the circle; it inserts the value of the last measurement made **(Figure 25.138).** The arc's diameter or

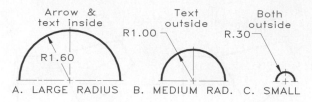

25.136 Arcs are dimensioned by one of the formats given here.

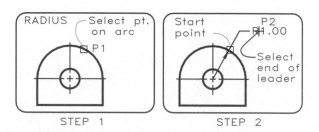

25.137 Dimensioning: Radius.
Step 1 *Command:* <u>Dim</u> (Enter)
Dim: <u>Radius</u> (Enter) *Select arc or circle:* <u>P1</u>
Step 2 *Dimension text <1.00>:* (Enter)
Specify dimension line location or [Mtext/Text/Angle]: <u>P2</u>

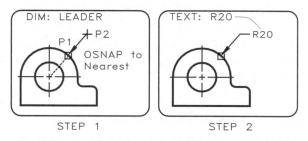

25.138 Dimensioning: Quick Leader.
Step 1 *Dimension* toolbar> *Quick Leader* icon> *Specify first leader point, or [Settings] <Settings>:* <u>P1</u>
Specify next point: <u>P2</u> (Enter)
Specify text width <1.00>: (Enter)
Step 2 *Enter first line of annotation text <Mtext>:* <u>R20</u> (Enter)
Enter next line of annotation text: (Enter) Note: The *Leader* command does not make measurements, you must insert them.

radius must be known and typed to override this measurement. Notes can be added in multiple lines at a specified length by using the *text width* option.

25.51 Dimension Style Manager

Select *Dimension Style* icon from the *Dimension* toolbar as shown in **Figure 25.139** (or *Command:* <u>DDim</u>) to display the *Dimension*

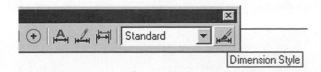

25.139 Use *Dimension* toolbar> *Dimension Style* icon to get the *Dimension Style Manager* box from which to create a new *Dimension Style* that will appear in the window with the *Standard* style.

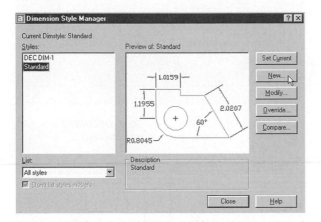

25.140 Use *Dimension* toolbar> *Dimension Style* icon> to get this *Dimension Style Manager* box for setting dimensioning variables. Pick the *New* button to create a new *Style*.

25.141 In the *New Style Name* window, type the name of the new style, <u>DEC DIM-2</u>, and pick the *Continue* button.

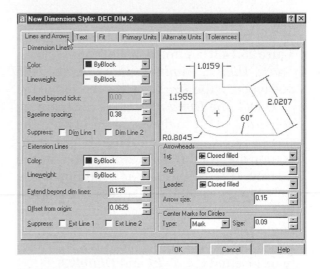

25.142 After the *Continue* button has been picked, the *New Dimension Style: DEC DIM-2* box appears. Select the *Lines and Arrows* tab for making the settings in the windows.

Style Manager dialog box shown in **Figure 25.140** from which *Dimension Styles* can be created or modified and variables can be assigned from different tabs. Each group of settings can be made and saved by style name (*DEC DIM-2,* for example) for future use to eliminate the time and effort required to make new settings for each drawing.

Lines and Arrows Tab

Click the *New* box (**Figure 25.141**) to get the *Create New Dimension Style* box in which you can make modifications on the *DEC DIM-2* style. Pick <u>OK</u> and select the *Lines and Arrows* tab shown in **Figure 25.142.** From here, settings can be made for *Dimension lines, Extension lines, Arrowheads*, and *Center marks*. When *Oblique-Stroke* arrows are used, the value placed in the *Arrow size* box specifies the distance the dimension line extends be-

yond the extension line. The *Baseline spacing* box is used to set *Dimdli*, which controls the spacing between baseline dimensions. The *Color* button displays the color menu from which to select a color for the dimension line (*Dimclrd*).

The options of the *Extension Lines* group (**Figure 25.142**) control the variables: *Lineweight, Extend beyond dim lines, Color*, and *Offset from origin*. The *Suppress 1st* and *2nd* boxes turn *On* the *Dimse1* and *Dimse2* variables to suppress the first and second extension lines. The value typed in the *Extension* box specifies the distance the extension line extends beyond the dimensioning arrowhead (*Dimexe*). The *Origin Offset* option is used to specify the size of the gap between the object and the extension line (*Dimexo*). The *Color* button lets you select the color of the extension lines (*Dimclre*).

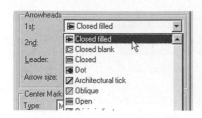

25.143 From the *Arrowheads* area, a drop-down listing of *Arrowheads* is given from which to select.

The options of the *Arrowheads* group **(Figure 25.143)** control the following variables: *Dimasz, Dimtsz, Dimblk1,* and *Dimblk2.* The value typed in the *Size* box gives the size of the arrowhead (the *Dimasz* variable). By selecting scroll arrows next to the *1st* or *2nd* boxes, the types of arrowheads for each end of the dimension are listed **(Figure 25.143).** If only the *1st* arrow type is selected, it is automatically applied to the second end unless a *2nd* arrow type is specified. *Tick marks* are given when *Oblique* is selected, the *Dimtsz* variable. When *User Arrow* is selected custom-made arrows can be inserted (*Dimblk1* and *Dimblk2*).

In the area labeled *Center Marks for Circles,* select from the the *Type* window *Mark, Line,* or *None* to draw center marks, center marks with centerlines, or neither on when arcs or diameters are dimensioned. Center marks and centerlines are applied to arcs and circles in the same manner without dimensions by *Command:* Dim> *Dim*> Center> select the object.

Text Tab

Select *Dimension* toolbar> *Dimension Style*> *Modify*> *Text* tab to get the menu shown in **Figure 25.144,** which controls the dimensioning text. *Text Appearance* includes settings for *Text color, Text style,* and *Text height.* The *Text Placement* area gives windows for specifying *Vertical* and *Horizontal* text placement and the *Offset from dim. line.* The *Text Alignment* section lets you set text as *Horizontal* (unidirectional), *Aligned* with dimension line, or *Iso,* which aligns text with the dimension line when text is inside the extension lines, but aligns it horizontally when text is outside

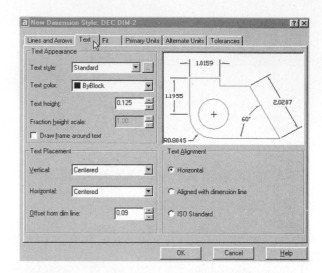

25.144 From the *New Dimension Style: DECDIM-2* box, pick the *Text* tab and make changes in text variables.

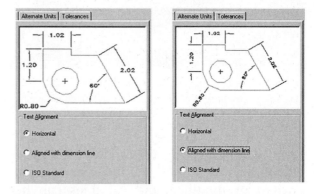

25.145 These examples show the results of having the *Unidirectional (Horizontal)* and *Aligned* text in dimension lines.

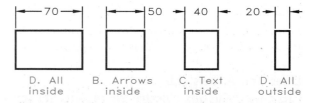

D. All inside   B. Arrows inside   C. Text inside   D. All outside

25.146 Examples of dimension applications when the *Best Fit* option is used.

(Figure 25.145). *Text Styles* can be selected from the pull-down window or created by picking the button next to the style window. *Text Placement* options let dimensioning text to be positioned as shown in **Figures 25.146** through **25.148.**

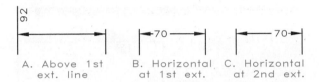

A. Above 1st ext. line   B. Horizontal at 1st ext.   C. Horizontal at 2nd ext.

25.147 Examples of *Text Placement* settings.

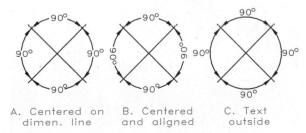

A. Centered on dimen. line   B. Centered and aligned   C. Text outside

25.148 Examples of *Text Placement* settings on arcs.

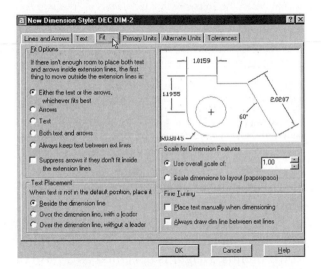

25.149 Select *Dimension Style DEC DIM-2> Fit* tab to get this menu for placing text on dimensions.

Fit Tab

Under the *Fit* tab, *Fit options, Text Placement, Scale for Dimension Features,* and *Fine Tuning* adjustments can be made **(Figure 25.149).** The *Use overall scale of:* box controls the scale of all the dimensioning variables on the screen arrow size, text height, extension-line offsets, center size, and others. It is very useful in changing the units of dimensioning from English to metric by entering a scale factor of 25.4.

Primary Units Tab

Select the *Primary Units* tab to obtain options for *Linear Dimensions, Measurement Scale, Zero Suppression,* and *Angular Dimensions* **(Figure 25.150).** Use the scroll arrow under *Precision,* and select the number of decimal places (or fractions) desired. From the *Angle* area, the *Units Format* of degrees and their *Precision* can be specified in the same manner as above.

Entries in the *Zero Suppression (Dimzin)* boxes suppress zeros that are leading or trailing decimal points. Select *Leading* to make 0.20 become .20. **Figure 25.151** shows the results of applying the four options to architectural units. From under *Measurement Scale Factor (Dimlfac)* units for measurements and be specified.

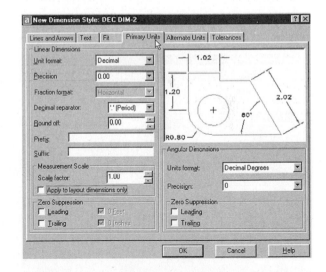

25.150 *New Dimension Style: DEC DIM-2> Primary Units* tab lets you make changes to the dimensioning text settings.

0 Ft 0 In	1/4"	4"	1'	1'-01/4"
No options	0'-01/4"	0'-6"	2'-0"	1'-01/4"
0 In	0'-01/4"	0'-4"	1'	1'-01/4"
0 Ft	1/2"	4"	1'-0"	1'-01/4"

25.151 The *Zero Suppression (Dimzin)* options control the leading and trailing zeros in dimensioning, especially when applied to architectural dimensions.

Alternate Units Tab

Select the *Display Alternate Units* button and two dimensions appear on each dimension as in the example drawing in **Figure 25.152.** *Unit Format* and *Precision* are used to define units and decimal points. If *Multiplier for all units* is set to 25.4, the millimeter equivalents for inches are given as the alternate dimension. Examples of dimensions with alternate units are shown in **Figure 25.153.**

Values can be placed in the *Prefix* and *Suffix* panels **(Figure 25.152)** so dimensions can have text (such as inches) before or after the dimensions (such as 26 mm or 42 inches). *Placement* options allow alternate dimensions to be placed over or after primary units.

Tolerances Tab

The *Tolerance Format* in **Figure 25.154** can be selected from options of *None, Symmetrical,*

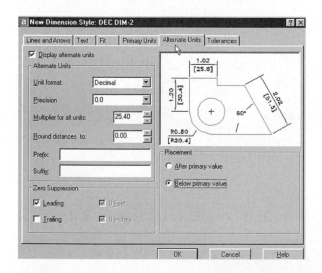

25.152 The *New Dimensioning Style: DEC DIM-2> Alternate Units* tab gives this window for assigning alternate and dual dimensions.

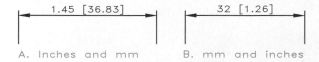

A. Inches and mm   B. mm and inches

25.153 Examples of alternate units (dual dimensions) made in inches and millimeters.

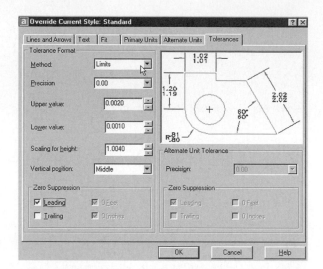

25.154 Select the *Tolerance* tab, pick *Limits* in *Tolerance Format* area, and provide the *Upper value* and *Lower value* for toleranced dimensions.

```
                              +.0030        2.0030
2.0000 ±.0020    2.0000       −.0020        1.9980

DIMTP  &  DIMTM   DIMTP  &  DIMTM           DIMLIM
     SAME             DIFFERENT
```

25.155 Examples of various formats for tolerancing using the names of the dimensioning variables.

Deviation, Limits, and *Basic.* Examples of applications of these options are shown in **Figure 25.155.** The number of decimal points is set with *Precision,* and *Upper* and *Lower* values are selected from their respective pull-down menus. *Height* is specified as a ratio of the primary text height, the basic dimension, and is recommended to be about 80%.

25.52 Saving Dimension Styles

When the settings of this new *Dimension Style* is complete, pick the *OK* button at the current tab and the first menu of the *Dimension Style Manager* will be saved and will appear on the screen showing the new style, DEC DIM-4, along with the *Standard* style **(Figure 25.139).** Options of *Modify* (change any settings), *Override* (make temporary changes in a style), and *Compare* (list the settings side-by-side of any two selected styles) are available **(Figure 25.156).** An often

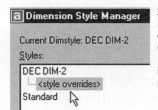

25.156 The new style, DEC DIM-2, is displayed in the list of styles. When a style is *Overridden* (values changed) a *<style overrides>* subheading is displayed beneath the primary style.

A. DIMSCALE=1   B. DIMSCALE=2

25.157 When *Dimscale* is changed from 1 to 2 and *Override* is used, the selected dimension is updated.

used override is the change of the *overall scale factor (Dimscale)* by changing a single multiplier **(Figure 25.157).** This text style can be used in the future with the assigned values.

25.53 Editing Dimensions

When the dimensioning variable *Dimaso* is set to *On*, the dimensioning entities (arrows, text, extension lines, etc.) become a single unit *(Associative)* once a dimension has been inserted into a drawing.

A related dimensioning variable, *Dimsho*, can be set *On* to show the dimensioning numerals being dynamically changed on the screen as the dimension line is *Stretched*. When *Dimaso* is *On* and *Dimsho* is *Off*, the numerals will be changed after the *Stretch* **(Figure 25.158),** but not dynamically during the *Stretch*.

When in the associative dimensioning mode, the *Dimension edit* can be selected from *Dimension* toolbar to obtain the options of *Home, New, Rotate,* and *Oblique* for changing existing dimensions. *Home* repositions text to its standard position at the center of the dimension line after being changed by the *Stretch* commands **(Figure 25.159).** *New* changes text within a dimension line **(Figure 25.160)** by pressing (Enter) when prompted and inserting the new text. *Rotate* positions dimension text

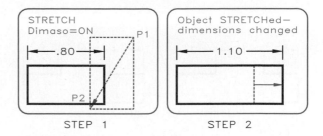

25.158 Dimensiong: Associative–Dimassoc.
Step 1 *Command:* <u>Dim</u>> *Dim*> <u>Dimassoc</u>> *Enter new value for DIMASSOC <1>:* <u>1</u> (On) and apply dimensions to the part. Use the *Stretch* command with a *Crossing* window at the end of the part.
Step 2 *Stretch* the side of the window to a new position; the object will be lengthened, a new dimension dynamically calculated, and be shown in its final position.

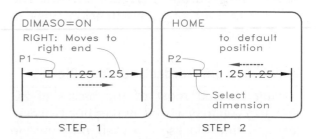

25.159 Dimensioning: Associative–Left, right, home.
Step 1 *Dimension* toolbar> *Dimension text edit* icon> *Select dimension* <u>P1</u>, *Enter text location (Left/Right/Home/Angle):* <u>Right</u> (Enter) The numeral moves to right.
Step 2 *Command:* (Enter) *Select dimension:* <u>P2</u>
Specify new location for dimension text or [Left/Right/Center/Home/Angle]: <u>Home</u> (Enter) Numeral moves to the midpoint of the dimension line.

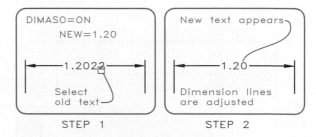

25.160 Dimensiong: Associative–New text.
Step 1 *Dimension* toolbar> *Dimension Edit* icon> *Enter type of dimension editing [Home/New/Rotate/Oblique] <Home>:* <u>New</u> (Enter)> *Select objects:* pick text
Step 2 Type new text inside the brackets of *the Type Formatting* window and pick <u>Ok</u>> *Select objects:* Pick a point on the dimension line and the text is changed.

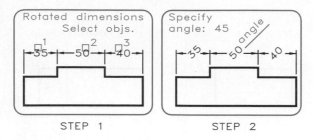

25.161 Dimensioning: Associative–Oblique extension lines.
Step 1 Dimension as usual with vertical extension lines.
Dimension toolbar> *Dimension Text Edit* icon> *Select dimension:* Pick <u>1</u>, <u>2</u>, and <u>3</u> (Enter)
Specify new location for dimension text or [Left/Right/Center/Home/Angle]: <u>Angle</u> (Enter)
Step 2 *Specify angle for dimension text:* <u>45</u> (Enter) Text is angled at 45° counterclockwise.

at any specified angle **(Figure 25.161).** The *Oblique* option converts extension lines to angular lines.

25.54 Custom Border and Title Block

Having covered most of the basics of 2D drawing, you may wish to make your own customized title blocks with their own unique parameters that can be used to start up new drawings as introduced in Section 25.12. The format shown in **Figure 25.162** is used for laying out problems given at the ends of the chapters in this book.

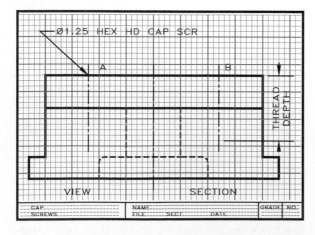

25.162 This is an example of a problem sheet border and title block that you can draw and save as a template file.

To make a border identical to the one used in this example, draw the border 10.3 inch wide × 7.6 inch high, with its lower left corner at 0,0 and a polyline that is .03 thick. Draw a title strip across the bottom that has two rows of 1/8-inch text with spacing of 1/8 inch above and below each line. Fill out the title strip using the *Romand* font (single-stroke Gothic) and fill in all blanks (name, date, etc.); they will be changed later. Set *Snap* and *Grid* to 0.2 inch, set any other dimensioning variables that you expect to use based on a letter height of 1/8 inch. Save the drawing (*Main Menu> Save As> File Name>:* <u>PROTO-1.DWG</u>).

To create a new drawing *Main Menu> File> Open>* <u>PROTO-1</u>, and the border and title block appear on the screen. *Main Menu> File> Save As>* <u>NEW-1</u> (Enter), and a new file named NEW-1 is ready for drawing. It would be saved as A:NEW-1 if it were saved to the A: drive. Use *DDedit* (*Modify II* toolbar> *Text edit* icon) to select the title strip text that needs to be edited and update each entry. Save this setup to a file *(Main Menu> File> Save>,* and the drawing is updated and saved again as NEW-1, ready for drawing.

25.55 Oblique Pictorials

An oblique pictorial can be constructed as shown in **Figure 25.163.** The front view of the oblique is drawn as an orthographic view, and

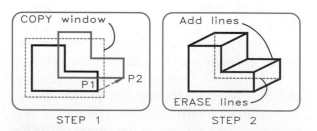

25.163 Pictorial: Oblique.
Step 1 Draw the front surface of the oblique and *Copy* the view from <u>P1</u> to <u>P2</u>.
Step 2 Connect the corner points with oblique lines from front to back and erase the invisible lines to complete the oblique. This is not a true 3D drawing but a 2D drawing with only one viewpoint.

duplicated behind the front view at the angle selected for the receding axis of the oblique with the *Copy* command. The visible endpoints are connected with the *Osnap Endpoint* option and the invisible lines are erased. Circles can be drawn as true circles on the true-size front surface, but circular features should be avoided on the receding planes since their construction is complex and their appearance is unrealistic.

This oblique is not a true, three-dimensional drawing that can be rotated and viewed from different positions. It is merely a two-dimensional drawing with a single point of view that appears three-dimensional.

25.56 Isometric Pictorials

To draw a two-dimensional isometric that appears as a true three-dimensional object, set the isometric grid *Command:* Snap> *Style>* Isometric. By using the same sequence of commands and typing *S* for *Standard,* the grid can be converted back to the rectangular grid. The *Isometric* style shows the grid dots (when the grid is turned on) in an isometric pattern—vertically and at 30° with the horizontal **(Figure 25.164).** The cursor will align with two of the isometric axes and can be made to *Snap* to the grid by activating the *Snap* comand *(F9).* The axes of the cursor are rotated 120° by pressing *Ctrl E* or by *Command:* Isoplane (Enter). When *Ortho* is *On,* lines are forced to be drawn parallel to the isometric axes. **Figure 25.165** illustrates the steps for constructing an isometric drawing.

When the *Grid* is set to the *Isometric* mode, the *Ellipse* command displays the following options: *<Axis endpoint 1>/Center/Isocircle:* Isocircle (Enter). Selection of the Isocircle option enables isometric ellipses to be positioned in each of the isometric orientations shown in **Figure 25.166** by using *Ctrl E.* An example of an isometric with partial ellipses is shown in **Figure 25.167.** The principles of drawing four-center ellipses was covered in Chapter 5 if you would like to review them.

The oblique and isometric drawings covered here are 2D drawings that appear to be 3D views, but they cannot be rotated on the screen to obtain different viewpoints of them.

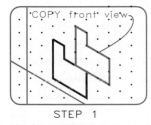
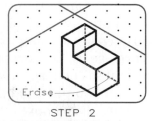

25.165 Pictorial: Isometric.
Step 1 Draw the front view of the object with the grid set to the *Isometric* style and *Copy* the view at its proper depth.
Step 2 Connect the corner points and erase the invisible lines. The cursor lines can be rotated into three positions using *Ctrl E* or by *Command:* Isoplane (Enter)

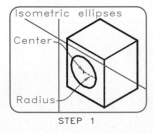
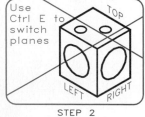

25.166 Pictorial: Isometric–isocircle.
Step 1 When in the isometric *Snap* mode:
Command: Ellipse (Enter) *<Axis endpoint 1>* /Center/Isocircle: Isocircle (Enter) *Center of circle:* Select center.
<Circle radius> Diameter: Drag or type radius size.
Step 2 Other isocircle ellipses are drawn using these same steps, and by changing the cursor to correspond to each *Isoplane* of the box by *(Ctrl E).*

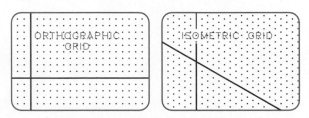

25.164 Screen Grids.
A The orthographic grid is called the *Standard* style.
(Command: Snap> *Style> Standard>* (Enter))
B The isometric grid is specified as follows:
Command: Snap> *Style>Isometric* (Enter)

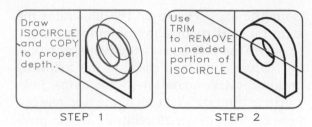

STEP 1　　　　　　STEP 2

25.167 Pictorial: Isometric–partial isocircles.
Step 1 *Command:* Ellipse (Enter) <*Axis endpoint1*>/ *Center/ Isocircle:* Isocircle (Enter) *Center of circle:* Select center. <*Circle radius*>/*Diameter:* Drag or type radius size. Draw the outside and inside *Isocircles*.
Step 2 Use the *Trim* command to remove the unneeded portions of the *Isocircles*. If the *Trim* fails to respond in some cases, use the *Break* command to select points on the *Isocircle* for line removal. Add missing lines.

25.57 Summary

The coverage of two-dimensional computer graphics and AutoCAD fundamentals has been presented in this chapter to enable you to prepare working drawings. However, this coverage is, of necessity, very brief and is meant to serve as an introduction to the basics of computer graphics. Many options of menus, dialog boxes, and toolbars could not be covered in this limited space. Therefore, you should experiment on your own with various options and commands as you progress in learning and applying the fundamentals.

The *Help* command on the *Main Menu* provides a valuable reference manual on the screen that will provide the answers to many of your questions when you wish to have them. This is one of the first menus with which you should become familiar. You will soon find that most of your knowledge of computer graphics will be self-taught.

Problems

It is suggested that problems be solved in a professional format within a border with a title block similar to the one illustrated in **Figure 25.162.**

1. A good first assignment would be the design of a prototype border and title block, as covered in Section 25.54. As this prototype is used and more variables are introduced, they can be added to the prototype file to preserve your time and effort that were required to make these settings.

2. Problems at the ends of most chapters lend themselves to solution by computer. It is always best to make rapid, freehand sketches of problem solutions before converting them to computer solutions.

3. Figures within the text of this chapter can be selected and drawn by computer as a good technique of becoming familiar with the commands and procedures of computer graphics.

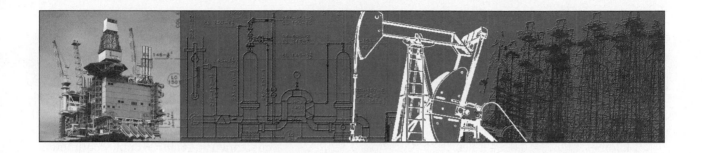

Appendix Contents

LENGTH

1 millimeter (mm) = 0.03937 inch
1 centimeter (cm) = 0.39370 inch
1 meter (m) = 39.37008 inches
1 meter = 3.2808 feet
1 meter = 1.0936 yards
1 kilometer (km) = 0.6214 miles
1 inch = 25.4 millimeters
1 inch = 2.54 centimeters
1 foot = 304.8 millimeters
1 foot = 0.3048 meters
1 yard = 0.9144 meters
1 mile = 1.609 kilometers

DRY CAPACITY

1 cubic centimeter (cm3) = 0.061 cubic inches
1 liter = 0.0353 cubic foot
1 liter = 61.023 cubic inches
1 cubic meter (m3) = 35.315 cubic feet
1 cubic meter = 1.308 cubic yards
1 cubic inch = 16.38706 cubic centimeters
1 cubic foot = 0.02832 cubic meter
1 cubic foot = 28.317 liters
1 cubic yard = 0.7646 cubic meter

AREA

1 square millimeter = 0.00155 square inch
1 square centimeter = 0.155 square inch
1 square meter = 10.764 square feet
1 square meter = 1.196 square yards
1 square kilometer = 0.3861 square mile
1 square inch = 645.2 square millimeters
1 square inch = 6.452 square centimeters
1 square foot = 929 square centimeters
1 square foot = 0.0929 square meter
1 square yard = 0.836 square meter
1 square mile = 2.5899 square kilometers

LIQUID CAPACITY

1 liter = 1.0567 U.S. quarts
1 liter = 0.2642 U.S. gallons
1 liter = 0.2200 Imperial gallons
1 cubic meter = 264.2 U.S. gallons
1 cubic meter = 219.969 Imperial gallons
1 U.S. quart = 0.946 liters
1 Imperial quart = 1.136 liters
1 U.S. gallon = 3.785 liters
1 Imperial gallon = 4.546 liters

WEIGHT

1 gram (g) = 15.432 grains
1 gram = 0.03215 ounce troy
1 gram = 0.03527 ounce avoirdupois
1 kilogram (kg) = 35.274 ounces avoirdupois
1 kilogram = 2.2046 pounds
1000 kilograms = 1 metric ton (t)
1000 kilograms = 1.1023 tons of 2000 pounds
1000 kilograms = 0.9842 tons of 2240 pounds
1 ounce avoirdupois = 28.35 grams
1 ounce troy = 31.103 grams
1 pound = 453.6 grams
1 pound = 0.4536 kilogram
1 ton of 2240 pounds = 1016 kilograms
1 ton of 2240 pounds = 1.016 metric tons
1 grain = 0.0648 gram
1 metric ton = 0.9842 tons of 2240 pounds
1 metric ton = 2204.6 pounds

APPENDIX 2 • Screw Threads: American National and Unified (inches)

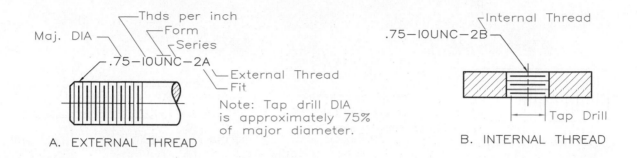

A. EXTERNAL THREAD

Maj. DIA — Thds per inch — Form — Series

.75—10UNC—2A

External Thread — Fit

Note: Tap drill DIA is approximately 75% of major diameter.

B. INTERNAL THREAD

Internal Thread

.75—10UNC—2B

Tap Drill

Nominal Diameter	Basic Diameter	Coarse NC & UNC		Fine NF & UNF		Extra Fine NEF/UNEF		Nominal Diameter	Basic Diameter	Coarse NC & UNC		Fine NF & UNF		Extra Fine NEF/UNEF	
		Thds per in.	Tap Drill DIA	Thds per in.	Tap Drill DIA	Thds per in.	Tap Drill DIA			Thds per in.	Tap Drill DIA	Thds per in.	Tap Drill DIA	Thds per in.	Tap Drill DIA
0	.060			80	.0469			1	1.000	8	.875	12	.922	20	.953
1	.073	64	No. 53	72	No. 53			1-1/16	1.063	...	...	...	...	18	1.000
2	.086	56	No. 50	64	No. 50			1-1/8	1.125	7	.904	12	1.046	18	1.070
3	.099	48	No. 47	56	No. 45			1-3/16	1.188	...	...	...	...	18	1.141
4	.112	40	No. 43	48	No. 42			1-1/4	1.250	7	1.109	12	1.172	18	1.188
5	.125	40	No. 38	44	No. 37			1-5/16	1.313	...	...	...	...	18	1.266
6	.138	32	No. 36	40	No. 33			1-3/8	1.375	6	1.219	12	1.297	18	1.313
8	.164	32	No. 29	36	No. 29			1-7/16	1.438	...	...	...	...	18	1.375
10	.190	24	No. 25	32	No. 21			1-1/2	1.500	6	1.344	12	1.422	18	1.438
12	.216	24	No. 16	28	No. 14	32	No. 13	1-9/16	1.563	...	...	...	...	18	1.500
1/4	.250	20	No. 7	28	No. 3	32	.2189	1-5/8	1.625	...	...	...	...	18	1.563
5/16	.3125	18	F	24	I	32	.2813	1-11/16	1.688	...	...	...	...	18	1.625
3/8	.375	16	.3125	24	Q	32	.3438	1-3/4	1.750	5	1.563	...	...	...	...
7/16	.4375	14	U	20	.3906	28	.4062	2	2.000	4.5	1.781	...	...	...	...
1/2	.500	13	.4219	20	.4531	28	.4688	2-1/4	2.250	4.5	2.031	...	...	...	...
9/16	.5625	12	.4844	18	.5156	24	.5156	2-1/2	2.500	4	2.250	...	...	...	...
5/8	.625	11	.5313	18	.5781	24	.5781	2-3/4	2.750	4	2.500	...	...	...	...
11/16	.6875	...	...	...	...	24	.6406	3	3.000	4	2.750	...	...	...	...
3/4	.750	10	.6563	16	.6875	20	.7031	3-1/4	3.250	4	...	...	...	...	...
13/16	.8125	...	...	...	...	20	.7656	3-1/2	3.500	4	...	...	...	...	...
7/8	.875	9	.7656	14	.8125	20	.8281	3-3/4	3.750	4	...	...	...	...	...
15/16	.9375	...	...	...	...	20	.8906	4	4.000	4	...	...	...	...	...

Source: ANSI/ASME B1.1—1989

APPENDIX 3 • Screw Threads: American National and Unified (inches) Constant-Pitch Threads

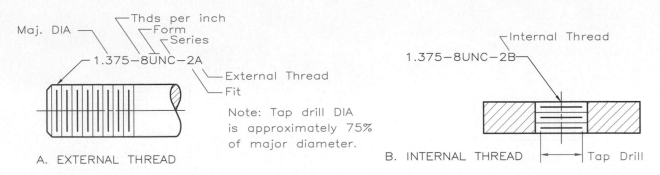

Maj. DIA

Thds per inch
Form
Series

1.375–8UNC–2A

External Thread
Fit

Note: Tap drill DIA is approximately 75% of major diameter.

A. EXTERNAL THREAD

Internal Thread

1.375–8UNC–2B

B. INTERNAL THREAD — Tap Drill

Nominal Diameter	8 Pitch 8N & 8UN		12 Pitch 12N & 12UN		16 Pitch 16N & 16UN		Nominal Diameter	8 Pitch 8N & 8UN		12 Pitch 12N & 12UN		16 Pitch 16N & 16UN	
	Thds per in.	Tap Drill DIA	Thds per in.	Tap Drill DIA	Thds per in.	Tap Drill DIA		Thds per in.	Tap Drill DIA	Thds per in.	Tap Drill DIA	Thds per in.	Tap Drill DIA
.500	...	...	12	.422	...	...	2.063	...	...	...	...	16	2.000
.563	...	...	12	.484	...	...	2.125	...	...	12	2.047	16	2.063
.625	...	...	12	.547	...	...	2.188	...	...	...	...	16	2.125
.688	...	...	12	.609	...	...	2.250	8	2.125	12	2.172	16	2.188
.750	...	...	12	.672	16	.688	2.313	...	...	...	...	16	2.250
.813	...	...	12	.734	16	.750	2.375	...	...	12	2.297	16	2.313
.875	...	...	12	.797	16	.813	2.438	...	...	...	...	16	2.375
.934	...	...	12	.859	16	.875	2.500	8	2.375	12	2.422	16	2.438
1.000	8	.875	12	.922	16	.938	2.625	...	...	12	2.547	16	2.563
1.063	...	...	12	.984	16	1.000	2.750	8	2.625	12	2.717	16	2.688
1.125	8	1.000	12	1.047	16	1.063	2.875	...	...	12	...	16	...
1.188	...	...	12	1.109	16	1.125	3.000	8	2.875	12	...	16	...
1.250	8	1.125	12	1.172	16	1.188	3.125	...	...	12	...	16	...
1.313	...	...	12	1.234	16	1.250	3.250	8	...	12	...	16	...
1.375	8	1.250	12	1.297	16	1.313	3.375	...	...	12	...	16	...
1.434	...	...	12	1.359	16	1.375	3.500	8	...	12	...	16	...
1.500	8	1.375	12	1.422	16	1.438	3.625	...	...	12	...	16	...
1.563	...	...	...	...	16	1.500	3.750	8	...	12	...	16	...
1.625	8	1.500	12	1.547	16	1.563	3.875	...	...	12	...	16	...
1.688	...	...	...	...	16	1.625	4.000	8	...	12	...	16	...
1.750	8	1.625	12	1.672	16	1.688	4.250	8	...	12	...	16	...
1.813	...	...	...	...	16	1.750	4.500	8	...	12	...	16	...
1.875	8	1.750	12	1.797	16	1.813	4.750	8	...	12	...	16	...
1.934	...	...	...	...	16	1.875	5.000	8	...	12	...	16	...
2.000	8	1.875	12	1.922	16	1.938	5.250	8	...	12	...	16	...

Source: ANSI/ASME B1.1—1989.

APPENDIX 4 • Screw Threads: American National and Unified (metric)

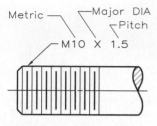

A. EXTERNAL THREAD

Note: Tap drill DIA is approximately 75% of major diameter.

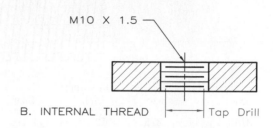

B. INTERNAL THREAD

COARSE		FINE		COARSE		FINE	
MAJ. DIA & THD PITCH	TAP DRILL	MAJ. DIA & THD PITCH	TAP DRILL	MAJ. DIA & THD PITCH	TAP DRILL	MAJ. DIA & THD PITCH	TAP DRILL
M1.6 × 0.35	1.25			M20 × 2.5	17.5	M20 × 1.5	18.5
M1.8 × 0.35	1.45			M22 × 2.5	19.5	M22 × 1.5	20.5
M2 × 0.4	1.6			M24 × 3	21.0	M24 × 2	22.0
M2.2 × 0.45	1.75			M27 × 3	24.0	M27 × 2	25.0
M2.5 × 0.45	2.05			M30 × 3.5	26.5	M30 × 2	28.0
M3 × 0.5	2.5			M33 × 3.5	29.5	M33 × 2	31.0
M3.5 × 0.6	2.9			M36 × 4	32.0	M36 × 3	33.0
M4 × 0.7	3.3			M39 × 4	35.0	M39 × 3	36.0
M4.5 × 0.75	3.75			M42 × 4.5	37.5	M42 × 3	39.0
M5 × 0.8	4.2			M45 × 4.5	40.5	M45 × 3	42.0
M6 × 1	5.0			M48 × 5	43.0	M48 × 3	45.0
M7 × 1	6.0			M52 × 5	47.0	M52 × 3	49.0
M8 × 1.25	6.8	M8 × 1	7.0	M56 × 5.5	50.5	M56 × 4	52.0
M9 × 1.25	7.75			M60 × 5.5	54.5	M60 × 4	56.0
M10 × 1.5	8.5	M10 × 1.25	8.75	M64 × 6	58.0	M64 × 4	60.0
M11 × 1.5	9.5			M68 × 6	62.0	M68 × 4	64.0
M12 × 1.75	10.3	M12 × 1.25	10.5	M72 × 6	66.0	M72 × 4	68.0
M14 × 2	12.0	M14 × 1.5	12.5	M80 × 6	74.0	M80 × 4	76.0
M16 × 2	14.0	M16 × 1.5	14.5	M90 × 6	84.0	M90 × 4	86.0
M18 × 2.5	15.5	M18 × 1.5	16.5	M100 × 6	94.0	M100 × 4	96.0

Source: ANSI/ASME B1.13

APPENDIX 5 • Square and Acme Threads

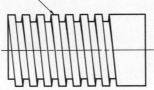

2.00—2.5 SQUARE

Typical
thread
note

Dimensions are in inches		Thds per inch	Size	Size	Thds per inch	Size	Size	Thds per inch
Size	Size							
3/8	.375	12	1-1/8	1.125	4	3	3.000	1-1/2
7/16	.438	10	1-1/4	1.250	4	3-1/4	3.125	1-1/2
1/2	.500	10	1-1/2	1.500	3	3-1/2	3.500	1-1/3
9/16	.563	8	1-3/4	1.750	2-1/2	3-3/4	3.750	1-1/3
5/8	.625	8	2	2.000	2-1/2	4	4.000	1-1/3
3/4	.75	6	2-1/4	2.250	2	4-1/4	4.250	1-1/3
7/8	.875	5	2-1/2	2.500	2	4-1/2	4.500	1
1	1.000	5	2-3/4	2.750	2	Larger		1

APPENDIX 6 • American Standard Taper Pipe Threads (NPT)

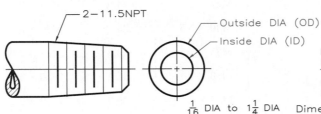

2—11.5NPT

Outside DIA (OD)

Inside DIA (ID)

PIPES THRU 12 INCHES IN
DIA ARE SPECIFIED BY
THEIR INSIDE DIAMETERS.
LARGER PIPES ARE SPECI-
FIED BY THEIR OD.

$\frac{1}{16}$ DIA to $1\frac{1}{4}$ DIA   Dimensions in inches

Nominal ID	$\frac{1}{16}$	$\frac{1}{8}$	$\frac{1}{4}$	$\frac{3}{8}$	$\frac{1}{2}$	$\frac{3}{4}$	1	1-1/4
Outside DIA	0.313	0.405	0.540	0.675	0.840	1.050	1.315	1.660
Thds/Inch	27	27	18	18	14	14	$11\frac{1}{2}$	$11\frac{1}{2}$

$1\frac{1}{2}$ DIA to 6 DIA

Nominal ID	$1\frac{1}{2}$	2	$2\frac{1}{2}$	3	$3\frac{1}{2}$	4	5	6
Outside DIA	1.900	2.375	2.875	3.500	4.000	4.500	5.563	6.625
Thds/Inch	$11\frac{1}{2}$	$11\frac{1}{2}$	8	8	8	8	8	8

8 DIA to 24 DIA

Nominal ID	8	10	12	14 OD	16 OD	18 OD	20 OD	24 OD
Outside DIA	8.625	10.750	12.750	14.000	16.000	18.000	20.000	24.000
Thds/Inch	8	8	8	8	8	8	8	8

Source: ANSI B2.1.

APPENDIX 7 • Square Bolts (inches)

DIA	E Max.	F Max.	G Avg.	H Max.	R Max.
1/4	.250	.375	.530	.188	.031
5/16	.313	.500	.707	.220	.031
3/8	.375	.563	.795	.268	.031
7/16	.438	.625	.884	.316	.031
1/2	.500	.750	1.061	.348	.031
5/8	.625	.938	1.326	.444	.062
3/4	.750	1.125	1.591	.524	.062
7/8	.875	1.313	1.856	.620	.062
1	1.000	1.500	2.121	.684	.093
1-1/8	1.125	1.688	2.386	.780	.093
1-1/4	1.250	1.875	2.652	.876	.093
1-3/8	1.375	2.625	2.917	.940	.093
1-1/2	1.500	2.250	3.182	1.036	.093

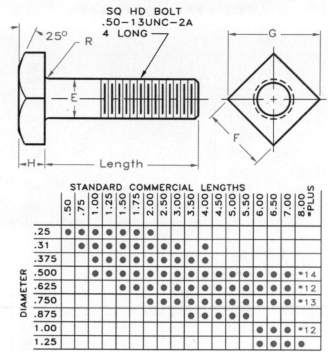

SQ HD BOLT .50-13UNC-2A 4 LONG

STANDARD COMMERCIAL LENGTHS

DIAMETER	.50	.75	1.00	1.25	1.50	1.75	2.00	2.50	3.00	3.50	4.00	4.50	5.00	5.50	6.00	6.50	7.00	8.00 *PLUS
.25	●	●	●	●	●	●	●											
.31		●	●	●	●	●	●	●	●		●							
.375			●	●	●	●	●	●	●	●	●							
.500			●	●	●	●	●	●	●	●	●	●	●	●	●	●	●	*14
.625				●	●	●	●	●	●	●	●	●	●	●	●	●	●	*12
.750						●	●	●	●	●	●	●	●	●	●	●	●	*13
.875									●	●	●	●	●					
1.00															●	●	●	*12
1.25															●	●	●	●

*14 MEANS THAT LENGTHS ARE AVAILABLE AT 1 INCH INCREMENTS UP 14 INCHES.

APPENDIX 8 • Square Nuts

Dimensions are in inches.

DIA	DIA	F Max.	G Avg.	H Max.
1/4	.250	.438	.619	.235
5/16	.313	.563	.795	.283
3/8	.375	.625	.884	.346
7/16	.438	.750	1.061	.394
1/2	.500	.813	1.149	.458
5/8	.625	1.000	1.414	.569
3/4	.750	1.125	1.591	.680
7/8	.875	1.313	1.856	.792
1	1.000	1.500	2.121	.903
1-1/8	1.125	1.688	2.386	1.030
1-1/4	1.250	1.875	2.652	1.126
1-3/8	1.375	1.063	2.917	1.237
1-1/2	1.500	2.250	3.182	1.348

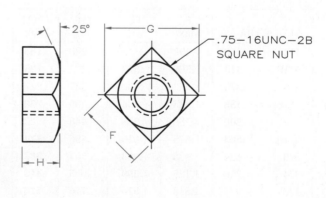

.75-16UNC-2B SQUARE NUT

APPENDIX 9 • Hexagon Head Bolts

Dimensions are in inches.

DIA	E Max.	F Max.	G Avg.	H Max.	R Max.
1/4	.250	.438	.505	.163	.025
5/16	.313	.500	.577	.211	.025
3/8	.375	.563	.650	.243	.025
7/16	.438	.625	.722	.291	.025
1/2	.500	.750	.866	.323	.025
9/16	.563	.812	.938	.371	.045
5/8	.625	.938	1.083	.403	.045
3/4	.750	1.125	1.299	.483	.045
7/8	.875	1.313	1.516	.563	.065
1	1.000	1.500	1.732	.627	.095
1-1/8	1.125	1.688	1.949	.718	.095
1-1/4	1.250	1.875	2.165	.813	.095
1-3/8	1.375	2.063	2.382	.878	.095
1-1/2	1.500	2.250	2.598	.974	.095
1-3/4	1.750	2.625	3.031	1.134	.095
2	2.000	3.000	3.464	1.263	.095
2-1/4	2.250	3.375	3.897	1.423	.095
2-1/2	2.500	3.750	4.330	1.583	.095
2-3/4	2.750	4.125	4.763	1.744	.095
3	3.000	4.500	5.196	1.935	.095

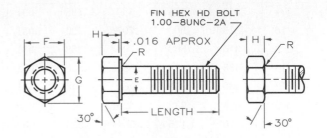

STANDARD COMMERCIAL LENGTHS

*10 MEANS THAT LENGTHS ARE AVAILABLE
AT 1 INCH INCREMENTS UP TO 10 INCHES.

APPENDIX 10 • Hex Nuts and Hex Jam Nuts

MAJOR DIA		F Max.	G Avg.	H1 Max.	H2 Max.
1/4	.250	.438	.505	.226	.163
5/16	.313	.500	.577	.273	.195
3/8	.375	.563	.650	.337	.227
7/16	.438	.688	.794	.385	.260
1/2	.500	.750	.866	.448	.323
9/16	.563	.875	1.010	.496	.324
5/8	.625	.938	1.083	.559	.387
3/4	.750	1.125	1.299	.665	.446
7/8	.875	1.313	1.516	.776	.510
1	1.000	1.500	1.732	.887	.575
1-1/8	1.125	1.688	1.949	.899	.639
1-1/4	1.250	1.875	2.165	1.094	.751
1-3/8	1.375	2.063	2.382	1.206	.815
1-1/2	1.500	2.250	2.589	1.317	.880

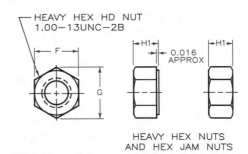

HEAVY HEX NUTS
AND HEX JAM NUTS

REGULAR HEX NUT   HEX JAM NUT

APPENDIX 11 • Round Head Cap Screws

Dimensions are in inches.

DIA	D Max.	A Max.	H Avg.	J Max.	T Max.
1/4	.250	.437	.191	.075	.117
5/16	.313	.562	.245	.084	.151
3/8	.375	.625	.273	.094	.168
7/16	.438	.750	.328	.094	.202
1/2	.500	.812	.354	.106	.218
9/16	.563	.937	.409	.118	.252
5/8	.625	1.000	.437	.133	.270
3/4	.750	1.250	.546	.149	.338

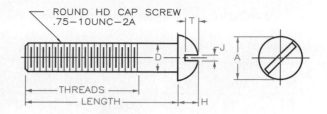

ROUND HD CAP SCREW
.75–10UNC–2A

STANDARD COMMERCIAL LENGTHS

	.50	.75	1.00	1.25	1.50	1.75	2.00	2.50	3.00	3.50	4.00
.25	●	●	●	●	●	●	●				
.31		●	●	●	●	●	●	●	●	●	●
.375			●	●	●	●	●	●	●	●	●
.500			●	●	●	●	●	●	●	●	●
.625				●	●	●	●	●	●	●	●
.750						●	●	●	●	●	●

OTHER LENGTHS AND DIAMETERS ARE AVAILABLE, BUT THESE ARE THE MORE STANDARD ONES.

APPENDIX 12 • Flat Head Cap Screws

Dimensions are in inches.

DIA	D Max.	A Max.	H Avg.	J Max.	T Max.
1/4	.250	.500	.140	.075	.068
5/16	.313	.625	.177	.084	.086
3/8	.375	.750	.210	.094	.103
7/16	.438	.813	.210	.094	.103
1/2	.500	.875	.210	.106	.103
9/16	.563	1.000	.244	.118	.120
5/8	.625	1.125	.281	.133	.137
3/4	.750	1.375	.352	.149	.171
7/8	.875	1.625	.423	.167	.206
1	1.000	1.875	.494	.188	.240
1-1/8	1.125	2.062	.529	.196	.257
1-1/4	1.250	2.312	.600	.211	.291
1-3/8	1.375	2.562	.665	.226	.326
1-1/2	1.500	2.812	.742	.258	.360

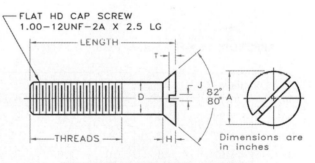

FLAT HD CAP SCREW
1.00–12UNF–2A X 2.5 LG

Dimensions are in inches

STANDARD COMMERCIAL LENGTHS

	.50	.75	1.00	1.25	1.50	1.75	2.00	2.50	3.00	3.50	4.00
.25	●	●	●	●	●	●	●				
.31		●	●	●	●	●	●	●	●	●	
.375			●	●	●	●	●	●	●	●	●
.500			●	●	●	●	●	●	●	●	●
.625				●	●	●	●	●	●	●	●
.750						●	●	●	●	●	●
.875							●	●	●	●	●
1.00								●	●	●	●
1.50								●	●	●	●

OTHER LENGTHS AND DIAMETERS ARE AVAILABLE, BUT THESE ARE THE MORE STANDARD ONES.

APPENDIX 13 • Fillister Head Cap Screws

Dimensions are in inches.

DIA	D Max.	A Max.	H Avg.	J Max.	T Max.
1/4	.250	.375	.172	.075	.097
5/16	.313	.437	.203	.084	.115
3/8	.375	.562	.250	.094	.142
7/16	.438	.625	.297	.094	.168
1/2	.500	.750	.328	.106	.193
9/16	.563	.812	.375	.118	.213
5/8	.625	.875	.422	.133	.239
3/4	.750	1.000	.500	.149	.283
7/8	.875	1.125	.594	.167	.334
1	1.000	1.312	.656	.188	.371

Source: ANSI B18.6.2.

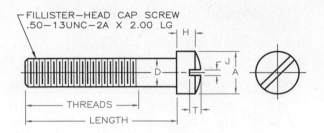

APPENDIX 14 • Flat Socket Head Cap Screws

Diameter mm	inches	Pitch	A	Ang.	W
M3	.118	.5	6	90	2
M4	.157	.7	8	90	2.5
M5	.197	.8	10	90	3
M6	.236	1	12	90	4
M8	.315	1.25	16	90	5
M10	.394	1.5	20	90	6
M12	.472	1.75	24	90	8
M14	.551	2	27	90	10
M16	.630	2	30	90	10
M20	.787	2.5	36	90	12

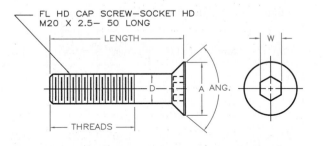

APPENDIX 15 • Socket Head Cap Screws

Diameter mm	Diameter inches	Pitch	A	H	W
M3	.118	.5	6	3	2
M4	.157	.7	8	4	3
M5	.187	.8	10	5	4
M6	.236	1	12	6	6
M8	.315	1.25	16	8	6
M10	.394	1.5	20	10	8
M12	.472	1.75	24	12	10
M14	.551	2	27	14	12
M16	.630	2	30	16	14
M20	.787	2.5	36	20	17

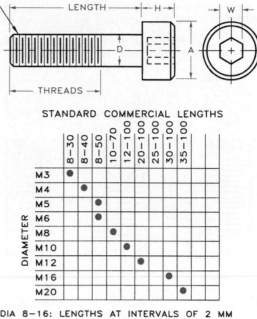

STANDARD COMMERCIAL LENGTHS

DIA 8–16: LENGTHS AT INTERVALS OF 2 MM
DIA 20–100: LENGTHS AT INTERVALS OF 5 MM

APPENDIX 16 • Round Head Machine Screws

Dimensions are in inches.

DIA	D Max.	A Max.	H Avg.	J Max.	T Max.
0	.060	.113	.053	.023	.039
1	.073	.138	.061	.026	.044
2	.086	.162	.069	.031	.048
3	.099	.187	.078	.035	.053
4	.112	.211	.086	.039	.058
5	.125	.236	.095	.043	.063
6	.138	.260	.103	.048	.068
8	.164	.309	.120	.054	.077
10	.190	.359	.137	.060	.087
12	.216	.408	.153	.067	.096
1/4	.250	.472	.175	.075	.109
5/16	.313	.590	.216	.084	.132
3/8	.375	.708	.256	.094	.155
7/16	.438	.750	.328	.094	.196
1/2	.500	.813	.355	.106	.211
9/16	.563	.938	.410	.118	.242
5/8	.625	1.000	.438	.133	.258
3/4	.750	1.250	.547	.149	.320

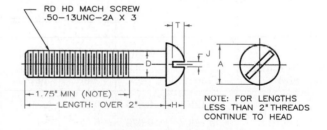

NOTE: FOR LENGTHS LESS THAN 2" THREADS CONTINUE TO HEAD

STANDARD LENGTHS

OTHER LENGTHS AND DIAMETERS ARE AVAILABLE; THESE ARE THE MORE STANDARD ONES.

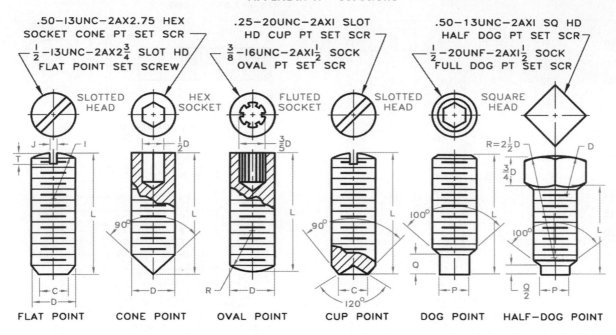

Dimensions for the set screws shown in ANSI Fig. 18.44 (dimensions in inches)										
D	I	J	T	R	C		P		Q	q
Nominal Size	Radius of Headless Crown	Width of Slot	Depth of Slot	Oval Point Radius	Diameter of Cup and Flat Points		Diameter of Dog Point		Length of Dog Point	
					Max	Min	Max	Min	Full	Half
5 0.125	0.125	0.023	0.031	0.094	0.067	0.057	0.083	0.078	0.060	0.030
6 0.138	0.138	0.025	0.035	0.109	0.047	0.064	0.092	0.087	0.070	0.035
8 0.164	0.164	0.029	0.041	0.125	0.087	0.076	0.109	0.103	0.080	0.040
10 0.190	0.190	0.032	0.048	0.141	0.102	0.088	0.127	0.120	0.090	0.045
12 0.216	0.216	0.036	0.054	0.156	0.115	0.101	0.144	0.137	0.110	0.055
$\frac{1}{4}$ 0.250	0.250	0.045	0.063	0.188	0.132	0.118	0.156	0.149	0.125	0.063
$\frac{5}{16}$ 0.3125	0.313	0.051	0.076	0.234	0.172	0.156	0.203	0.195	0.156	0.078
$\frac{3}{8}$ 0.375	0.375	0.064	0.094	0.281	0.212	0.194	0.250	0.241	0.188	0.094
$\frac{7}{16}$ 0.4375	0.438	0.072	0.190	0.328	0.252	0.232	0.297	0.287	0.219	0.109
$\frac{1}{2}$ 0.500	0.500	0.081	0.125	0.375	0.291	0.270	0.344	0.344	0.250	0.125
$\frac{9}{16}$ 0.5625	0.563	0.091	0.141	0.422	0.332	0.309	0.391	0.379	0.281	0.140
$\frac{5}{8}$ 0.625	0.625	0.102	0.156	0.469	0.371	0.347	0.469	0.456	0.313	0.156
$\frac{3}{4}$ 0.750	0.750	0.129	0.188	0.563	0.450	0.425	0.563	0.549	0.375	0.188

Source: Courtesy of ANSI; B18.6.2.

APPENDIX 18 • Cotter Pins: American National Standard

Nominal Diameter	Maximum DIA A	Minimum DIA B	Hole Size
0.031	0.032	0.063	0.047
0.047	0.048	0.094	0.063
0.062	0.060	0.125	0.078
0.078	0.076	0.156	0.094
0.094	0.090	0.188	0.109
0.109	0.104	0.219	0.125
0.125	0.120	0.250	0.141
0.141	0.176	0.281	0.156
0.156	0.207	0.313	0.172
0.188	0.176	0.375	0.203
0.219	0.207	0.438	0.234
0.250	0.225	0.500	0.266
0.312	0.280	0.625	0.313
0.375	0.335	0.750	0.375
0.438	0.406	0.875	0.438
0.500	0.473	1.000	0.500
0.625	0.598	1.250	0.625
0.750	0.723	1.500	0.750

Source: Courtesy of ANSI: B18.8.1—1983.

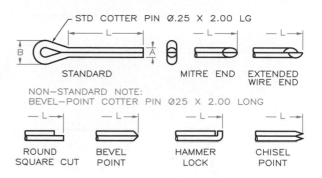

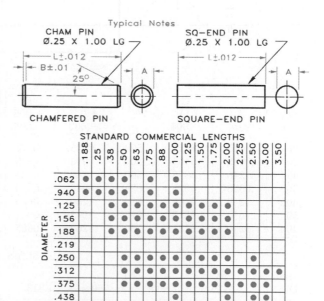

APPENDIX 19 • Straight Pins

Nominal DIA	Diameter A Max	Min	Chamfer B
0.062	0.0625	0.0605	0.015
0.094	0.0937	0.0917	0.015
0.109	0.1094	0.1074	0.015
0.125	0.1250	0.1230	0.015
0.156	0.1562	0.1542	0.015
0.188	0.1875	0.1855	0.015
0.219	0.2187	0.2167	0.015
0.250	0.2500	0.2480	0.015
0.312	0.3125	0.3095	0.015
0.375	0.3750	0.3720	0.030
0.438	0.4345	0.4345	0.030
0.500	0.4970	0.4970	0.030

Source: Courtesy of ANSI: B5.20.

APPENDIX 20 • Woodruff Keys

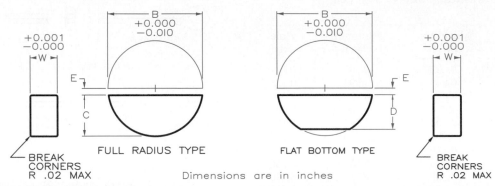

FULL RADIUS TYPE   FLAT BOTTOM TYPE

BREAK CORNERS R .02 MAX

Dimensions are in inches

Key No.	W × B	C Max.	D Max.	E		Key No.	W × B	C Max.	D Max.	E
204	1/16 × 1/2	.203	.194	.047		506	5/32 × 3/4	.313	.303	.063
304	3/32 × 1/2	.203	.194	.047		606	3/16 × 3/4	.313	.303	.063
404	1/8 × 1/2	.203	.194	.047		507	5/32 × 7/8	.375	.365	.063
305	3/32 × 5/8	.250	.240	.063		607	3/16 × 7/8	.375	.365	.063
405	1/8 × 5/8	.250	.240	.063		807	1/4 × 7/8	.375	.365	.063
505	5/32 × 5/8	.250	.240	.063		608	3/16 × 1	.438	.428	.063
406	1/8 × 3/4	.313	.303	.063		609	3/16 × 1-1/8	.484	.475	.078

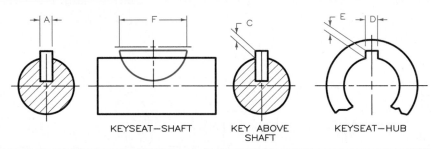

KEYSEAT–SHAFT   KEY ABOVE SHAFT   KEYSEAT–HUB

Key No.	A Min.	C +.005 −.000	F	D +.005 −.000	E +.005 −.000		Key No.	A Min.	C +.005 −.000	F	D +.005 −.000	E +.005 −.000
204	.0615	.0312	.500	.0635	.0372		506	.1553	.0781	.750	.1573	.0841
304	.0928	.0469	.500	.0948	.0529		606	.1863	.0937	.750	.1885	.0997
404	.1240	.0625	.500	.1260	.0685		507	.1553	.0781	.875	.1573	.0841
305	.0928	.0625	.625	.0948	.0529		607	.1863	.0937	.875	.1885	.0997
405	.1240	.0469	.625	.1260	.0685		807	.2487	.1250	.875	.2510	.1310
505	.1553	.0625	.625	.1573	.0841		608	.1863	.3393	1.000	.1885	.0997
406	.1240	.0781	.750	.1260	.0685		609	.1863	.3853	1.125	.1885	.0997

KEY SIZES VS. SHAFT SIZES

Shaft DIA	to .375	to .500	to .750	to 1.313	to 1.188	to 1.448	to 1.750	to 2.125	to 2.500
Key Nos.	204	304 305	404 405 406	505 506 507	606 607 608 609	807 808 809	810 811 812	1011 1012	1211 1212

APPENDIX 21 • Standard Keys and Keyways

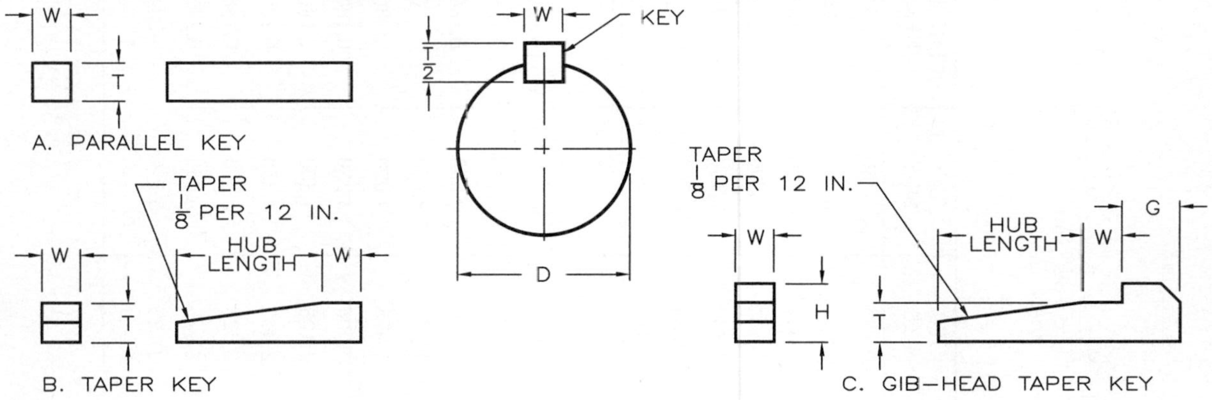

Sprocket Bore (= Shaft Diam.)	Keyway Dimensions — Inches		Key Dimensions—Inches			Gib Head Dimensions—Inches				Key Tolerances Taper and Gib Head	
	For Square Key	For Flat Key	Square	Flat	Tolerance on W and T (−)	Square Key		Flat Key			
Inches D	Width W × Depth T/2	Width W × Depth T/2	Width W × Height T	Width W × Height T		H	G	H	G	W (−)	T (−)
1/2—9/16	1/8 × 1/16	1/8 × 3/64	1/8 × 1/8	1/8 × 3/32	0.002	1/4	7/32	3/16	1/8	0.002	0.002
5/8—7/8	3/16 × 3/32	3/16 × 1/16	3/16 × 3/16	3/16 × 1/8	0.002	5/16	9/32	1/4	3/16	0.002	0.002
13/16—1 1/4	1/4 × 1/8	1/4 × 3/32	1/4 × 1/4	1/4 × 3/16	0.002	7/16	11/32	5/16	1/4	0.002	0.002
1 5/16—1 3/8	5/16 × 5/32	5/16 × 1/8	5/16 × 5/16	5/16 × 1/4	0.002	5/16	13/32	3/8	5/16	0.002	0.002
1 7/16—1 3/4	3/8 × 3/16	3/8 × 1/8	3/8 × 3/8	3/8 × 1/4	0.002	11/16	15/32	7/16	3/8	0.002	0.002
1 13/16—2 1/4	1/2 × 1/4	1/2 × 3/16	1/2 × 1/2	1/2 × 3/8	0.0025	7/8	19/32	5/8	1/2	0.0025	0.0025
2 5/16—2 3/4	5/8 × 5/16	5/8 × 7/32	5/8 × 5/8	5/8 × 7/16	0.0025	1 1/16	23/32	3/4	5/8	0.0025	0.0025
2 7/8—3 1/4	3/4 × 3/8	3/4 × 1/4	3/4 × 3/4	3/4 × 1/2	0.0025	1 1/4	7/8	7/8	3/4	0.0025	0.0025
3 3/8—3 3/4	7/8 × 7/16	7/8 × 5/16	7/8 × 7/8	7/8 × 5/8	0.003	1 1/2	1	1 1/16	7/8	0.003	0.003
3 7/8—4 1/2	1 × 1/2	1 × 3/8	1 × 1	1 × 3/4	0.003	1 3/4	1 3/16	1 1/4	1	0.003	0.003
4 3/4—5 1/2	1 1/4 × 5/8	1 1/4 × 7/16	1 1/4 × 1 1/4	1 1/4 × 7/8	0.003	2	1 7/16	1 1/2	1 1/2	0.003	0.003
5 3/4—7 3/8	1 1/2 × 3/4	1 1/2 × 1/2	1 1/2 × 1 1/2	1 1/2 × 1	0.003	2 1/2	1 3/4	1 3/4	1 1/2	0.003	0.003
7 1/2—9 7/8	1 3/4 × 7/8	.. ..	1 3/4 × 1 3/4	.. ..	0.004	3	2	..	..	0.004	0.004
10—12 1/2	2 × 1	.. ..	2 × 2	.. ..	0.004	3 1/2	2 3/8	..	..	0.004	0.004

Standard Keyway Tolerances: Straight Keyway—Width (W) +.005 / −.000; Depth (T/2) + .010 / −.000
Taper Keyway—Width (W) − .005 / −.000; Depth (T/2) + .000 / −.010

APPENDIX 22 • Plain Washers (inches)

.938 X 2.25 X .165
TYPE A PLAIN WASHER

Dimensioned
Washer

In Screw Size Column
N= Narrow washer
W= Wide washer

Narrow Washer (N)
TYPE A PLAIN WASHERS

WIDE WASHER (W)

SCREW SIZE	ID SIZE	OD SIZE	THICK-NESS	SCREW SIZE	ID SIZE	OD SIZE	THICK-NESS
0.138	0.156	0.375	0.049	0.875 N	0.938	1.750	0.134
0.164	0.188	0.438	0.049	0.875 W	0.938	2.250	0.165
0.190	0.219	0.500	0.049	1.000 N	1.062	2.000	0.134
0.188	0.250	0.562	0.049	1.000 W	1.062	2.500	0.165
0.216	0.250	0.562	0.065	1.125 N	1.250	2.250	0.134
0.250 N	0.281	0.625	0.065	1.125 W	1.250	2.750	0.165
0.250 W	0.312	0.734	0.065	1.250 N	1.375	2.500	0.165
0.312 N	0.344	0.688	0.065	1.250 W	1.375	3.000	0.165
0.312 W	0.375	0.875	0.083	1.375 N	1.500	2.750	0.165
0.375 N	0.406	0.812	0.065	1.375 W	1.500	3.250	0.180
0.375 W	0.438	1.000	0.083	1.500 N	1.625	3.000	0.165
0.438 N	0.469	0.922	0.065	1.500 W	1.625	3.500	0.180
0.438 W	0.500	1.250	0.083	1.625	1.750	3.750	0.180
0.500 N	0.531	1.062	0.095	1.750	1.875	4.000	0.180
0.500 W	0.562	1.375	0.109	1.875	2.000	4.250	0.180
0.562 N	0.594	1.156	0.095	2.000	2.125	4.500	0.180
0.562 W	0.594	1.469	0.190	2.250	2.375	4.750	0.220
0.625 N	0.625	1.312	0.095	2.500	2.625	5.000	0.238
0.625 N	0.625	1.750	0.134	2.750	2.875	5.250	0.259
0.750 W	0.812	1.469	0.134	3.000	3.125	5.500	0.284
0.750 W	0.812	2.000	0.148				

APPENDIX 23 • Metric Washers (millimeters)

SCREW SIZE	ID SIZE	OD SIZE	THICK-NESS
3	3.2	9	0.8
4	4.3	12	1
5	5.3	15	1.5
6	6.4	18	1.5
8	8.4	25	2
10	10.5	30	2.5
12	13	40	3
14	15	45	3
16	17	50	3
18	19	56	4
20	21	60	4
2.6	2.8	5.5	0.5
3	3.2	6	0.5
4	4.3	8	0.5
5	5.3	10	1.0
6	6.4	11	1.5
8	8.4	15	1.5
10	10.5	18	1.5
12	13	20	2.0
14	15	25	2.0
16	17	27	2.0
18	19	30	2.5
20	21	33	2.5

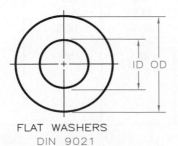

FLAT WASHERS
DIN 9021

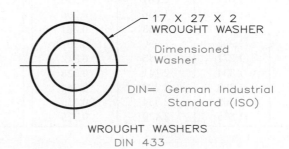

17 X 27 X 2
WROUGHT WASHER

Dimensioned Washer

DIN= German Industrial Standard (ISO)

WROUGHT WASHERS
DIN 433

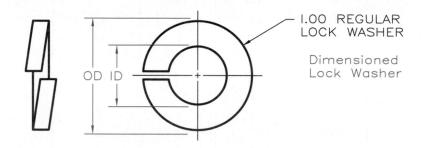

1.00 REGULAR LOCK WASHER

Dimensioned Lock Washer

LOCK WASHERS—inches			
SCREW SIZE	ID SIZE	OD SIZE	THICK-NESS
0.164	0.168	0.175	0.040
0.190	0.194	0.202	0.047
0.216	0.221	0.229	0.056
0.250	0.255	0.263	0.062
0.312	0.318	0.328	0.078
0.375	0.382	0.393	0.094
0.438	0.446	0.459	0.109
0.500	0.509	0.523	0.125
0.562	0.572	0.587	0.141
0.625	0.636	0.653	0.156
0.688	0.700	0.718	0.172
0.750	0.763	0.783	0.188
0.812	0.826	1.367	0.203
0.875	0.890	1.464	0.219
0.938	0.954	1.560	0.234
1.000	1.017	1.661	0.250
1.062	1.080	1.756	0.266
1.125	1.144	1.853	0.281
1.188	1.208	1.950	0.297
1.250	1.271	2.045	0.312
1.312	1.334	2.141	0.328
1.375	1.398	2.239	0.344
1.438	1.462	2.334	0.359
1.500	1.525	2.430	0.375

METRIC LOCK WASHERS—DIN 127 (Millimeters)			
SCREW SIZE	ID SIZE	OD SIZE	THICK-NESS
4	4.1	7.1	0.9
5	5.1	8.7	1.2
6	6.1	11.1	1.6
8	8.2	12.1	1.6
10	10.2	14.2	2
12	12.1	17.2	2.2
14	14.2	20.2	2.5
16	16.2	23.2	3
18	18.2	26.2	3.5
20	20.2	28.2	3.5
22	22.5	34.5	4
24	24.5	38.5	5
27	27.5	41.5	5
30	30.5	46.5	6
33	33.5	53.5	6
36	36.5	56.5	6
39	39.5	59.5	6
42	42.5	66.5	7
45	45.5	69.5	7
48	49	73	7

Limits are in thousandths of an inch.
Limits for hole and shaft are applied algebraically to the basic size to obtain the limits of size for the parts.
Data in bold face are in accordance with ABC agreements.
Symbols H5, g5, etc., are Hole and Shaft designations used in ABC System.

Nominal Size Range Inches		Class RC1			Class RC 2			Class RC 3			Class RC4		
		Limits of Clearance	Standard Limits		Limits of Clearance	Standard Limits		Limits of Clearance	Standard Limits		Limits of Clearance	Standard Limits	
Over	To		Hole H5	Shaft g4		Hole H6	Shaft g5		Hole H7	Shaft f6		Hole H8	Shaft f7
0	−0.12	**0.1**	**+0.2**	**−0.1**	**0.1**	**+0.25**	**−0.1**	0.3	**+0.4**	**−0.3**	0.3	**+0.6**	**−0.3**
		0.45	**0**	**−0.25**	**0.55**	**0**	**−0.3**	0.95	**0**	**−0.55**	1.3	**0**	**−0.7**
0.12	−0.24	**0.15**	**+0.2**	**−0.15**	**0.15**	**+0.3**	**−0.15**	0.4	**+0.5**	**−0.4**	0.4	**+0.7**	**−0.4**
		0.5	**0**	**−0.3**	**0.65**	**0**	**−0.35**	1.12	**0**	**−0.7**	1.6	**0**	**−0.9**
0.24	−0.40	**0.2**	**0.25**	**−0.2**	**0.2**	**+0.4**	**−0.2**	0.5	**+0.6**	**−0.5**	0.5	**+0.9**	**−0.5**
		0.6	**0**	**−0.35**	**0.85**	**0**	**−0.45**	1.5	**0**	**−0.9**	2.0	**0**	**−1.1**
0.40	−0.71	**0.25**	**+0.3**	**−0.25**	**0.25**	**+0.4**	**−0.25**	0.6	**+0.7**	**−0.6**	0.6	**+1.0**	**−0.6**
		0.75	**0**	**−0.45**	**0.95**	**0**	**−0.55**	1.7	**0**	**−1.0**	2.3	**0**	**−1.3**
0.71	−1.19	**0.3**	**+0.4**	**−0.3**	**0.3**	**+0.5**	**−0.3**	0.8	**+0.8**	**−0.8**	0.8	**+1.2**	**−0.8**
		0.95	**0**	**−0.55**	**1.2**	**0**	**−0.7**	2.1	**0**	**−1.3**	2.8	**0**	**−1.6**
1.19	−1.97	**0.4**	**+0.4**	**−0.4**	**0.4**	**+0.6**	**−0.4**	1.0	**+1.0**	**−1.0**	1.0	**+1.6**	**−1.0**
		1.1	**0**	**−0.7**	**1.4**	**0**	**−0.8**	2.6	**0**	**−1.6**	3.6	**0**	**−2.0**
1.97	−3.15	**0.4**	**+0.5**	**−0.4**	**0.4**	**+0.7**	**−0.4**	1.2	**+1.2**	**−1.2**	1.2	**+1.8**	**−1.2**
		1.2	**0**	**−0.7**	**1.6**	**0**	**−0.9**	3.1	**0**	**−1.9**	4.2	**0**	**−2.4**
3.15	−4.73	**0.5**	**+0.6**	**−0.5**	**0.5**	**+0.9**	**−0.5**	1.4	**+1.4**	**−1.4**	1.4	**+2.2**	**−1.4**
		1.5	**0**	**−0.9**	**2.0**	**0**	**−1.1**	3.7	**0**	**−2.3**	5.0	**0**	**−2.8**
4.73	−7.09	**0.6**	**+0.7**	**−0.6**	**0.6**	**+1.0**	**−0.6**	1.6	**+1.6**	**−1.6**	1.6	**+2.5**	**−1.6**
		1.8	**0**	**−1.1**	**2.3**	**0**	**−1.3**	4.2	**0**	**−2.6**	5.7	**0**	**−3.2**
7.09	−9.85	**0.6**	**+0.8**	**−0.6**	**0.6**	**+1.2**	**−0.6**	2.0	**+1.8**	**−2.0**	2.0	**+2.8**	**−2.0**
		2.0	**0**	**−1.2**	**2.6**	**0**	**−1.4**	5.0	**0**	**−3.2**	6.6	**0**	**−3.8**
9.85	−12.41	0.8	**+0.9**	−0.8	0.8	**+1.2**	−0.8	2.5	**+2.0**	−2.5	2.5	**+3.0**	−2.5
		2.3	**0**	−1.4	2.9	**0**	−1.7	5.7	**0**	−3.7	7.5	**0**	−4.5
12.41	−15.75	1.0	**+1.0**	−1.0	1.0	**+1.4**	−1.0	3.0	+	−3.0	3.0	**+3.5**	−3.0
		2.7	**0**	−1.7	3.4	**0**	−2.0	6.6	**0**	−4.4	8.7	**0**	−5.2
15.75	−19.69	1.2	**+1.0**	−1.2	1.2	**+1.6**	−1.2	4.0	**+1.6**	−4.0	4.0	**+4.0**	−4.0
		3.0	**0**	−2.0	3.8	**0**	−2.2	8.1	**0**	−5.6	10.5	**0**	−6.5
19.69	−30.09	1.6	+1.2	−1.6	1.6	+2.0	−1.6	5.0	+3.0	−5.0	5.0	+5.0	−5.0
		3.7	0	−2.5	4.8	0	−2.8	10.0	0	−7.0	13.0	0	−8.0
30.09	−41.49	2.0	+1.6	−2.0	2.0	+2.5	−2.0	6.0	+4.0	−6.0	6.0	+6.0	−6.0
		4.6	0	−3.0	6.1	0	−3.6	12.5	0	−8.5	16.0	0	−10.0
41.49	−56.19	2.5	+2.0	−2.5	2.5	+3.0	−2.5	8.0	+5.0	−8.0	8.0	+8.0	−8.0
		5.7	0	−3.7	7.5	0	−4.5	16.0	0	−11.0	21.0	0	−13.0
56.19	−76.39	3.0	+2.5	−3.0	3.0	+4.0	−3.0	10.0	+6.0	−10.0	10.0	+10.0	−10.0
		7.1	0	−4.6	9.5	0	−5.5	20.0	0	−14.0	26.0	0	−16.0
76.39	−100.9	4.0	+3.0	−4.0	4.0	+5.0	−4.0	12.0	+8.0	−12.0	12.0	+12.0	−12.0
		9.0	0	−6.0	12.0	0	−7.0	25.0	0	−17.0	32.0	0	−20.0
100.9	−131.9	5.0	+4.0	−5.0	5.0	+6.0	−5.0	16.0	+10.0	−16.0	16.0	+16.0	−16.0
		11.5	0	−7.5	15.0	0	−9.0	32.0	0	−22.0	36.0	0	−26.0
131.9	−171.9	6.0	+5.0	−6.0	6.0	+8.0	−6.0	18.0	+8.0	−18.0	18.0	+20.0	−18.0
		14.0	0	−9.0	19.0	0	−11.0	38.0	0	−26.0	50.0	0	−30.0
171.9	−200	8.0	+6.0	−8.0	8.0	+10.0	−8.0	22.0	+16.0	−22.0	22.0	+25.0	−22.0
		18.0	0	−12.0	22.0	0	−12.0	48.0	0	−32.0	63.0	0	−38.0

Source: Courtesy of USASI; B4.1—1955.

Class RC 5			Class RC 6			Class RC 7			Class RC 8			Class RC 9			Nominal Size Range Inches	
Limits of Clearance	Hole H8	Shaft e7	Limits of Clearance	Hole H9	Shaft e8	Limits of Clearance	Hole H9	Shaft d8	Limits of Clearance	Hole H10	Shaft c9	Limits of Clearance	Hole H11	Shaft	Over	To
0.6	+0.6	−0.6	0.6	+1.0	−0.6	1.0	+1.0	−1.0	2.5	+1.6	−2.5	4.0	+2.5	−4.0	0	− 0.12
1.6	−0	−1.0	2.2	−0	−1.2	2.6	0	−1.6	5.1	0	−3.5	8.1	0	−5.6		
0.8	+0.7	−0.8	0.8	+1.2	−0.8	1.2	+1.2	−1.2	2.8	+1.8	−2.8	4.5	+3.0	−4.5	0.12	− 0.24
2.0	−0	−1.3	2.7	−0	−1.5	3.1	0	−1.9	5.8	0	−4.0	9.0	0	−6.0		
1.0	+0.9	−1.0	1.0	+1.4	−1.0	1.6	+1.4	−1.6	3.0	+2.2	−3.0	5.0	+3.5	−5.0	0.24	− 0.40
2.5	−0	−1.16	3.3	−0	−1.9	3.9	0	−2.5	6.6	0	−4.4	10.7	0	−7.2		
1.2	+1.0	−1.2	1.2	+1.6	−1.2	2.0	+1.6	−2.0	3.5	+2.8	−3.5	6.0	+4.0	−6.0	0.40	− 0.71
2.9	−0	−1.9	3.8	−0	−2.2	4.6	0	−3.0	7.9	0	−5.1	12.8	−0	−8.8		
1.6	+1.2	−1.6	1.6	+2.0	−1.6	2.5	+2.0	−2.5	4.5	+3.5	−4.5	7.0	+5.0	−7.0	0.71	− 1.19
3.6	−0	−2.4	4.8	−0	−2.8	5.7	0	−3.7	10.0	0	−6.5	15.5	0	−10.5		
2.0	+1.6	−2.0	2.0	+2.5	−2.0	3.0	+2.5	−3.0	5.0	+4.0	−5.0	8.0	+6.0	−8.0	1.19	− 1.97
4.6	−0	−3.0	6.1	−0	−3.6	7.1	0	−4.6	11.5	0	−7.5	18.0	0	−12.0		
2.5	+1.8	−2.5	2.5	+3.0	−2.5	4.0	+3.0	−4.0	6.0	+4.5	−6.0	9.0	+7.0	−9.0	1.97	− 3.15
5.5	−0	−3.7	7.3	−0	−4.3	8.8	0	−5.8	13.5	0	−9.0	20.5	0	−13.5		
3.0	+2.2	−3.0	3.0	+3.5	−3.0	5.0	+3.5	−5.0	7.0	+5.0	−7.0	10.0	+9.0	−10.0	3.15	− 4.73
6.6	−0	−4.4	8.7	−0	−5.2	10.7	0	−7.2	15.5	0	−10.5	24.0	0	−15.0		
3.5	+2.5	−3.5	3.5	+4.0	−3.5	6.0	+4.0	−6.0	8.0	+6.0	−8.0	12.0	+10.0	−12.0	4.73	− 7.09
7.6	−0	−5.1	10.0	−0	−6.0	12.5	0	−8.5	18.0	0	−12.0	28.0	0	−18.0		
4.0	+2.8	−4.0	4.0	+4.5	−4.0	7.0	+4.5	−7.0	10.0	+7.0	−10.0	15.0	+12.0	−15.0	7.09	− 9.85
8.6	−0	−5.8	11.3	0	−6.8	14.3	0	−9.8	21.5	0	−14.5	34.0	0	−22.0		
5.0	+3.0	−5.0	5.0	+5.0	−5.0	8.0	+5.0	−8.0	12.0	+8.0	−12.0	18.0	+12.0	−18.0	9.85	− 12.41
10.0	0	−7.0	13.0	0	−8.0	16.0	0	−11.0	25.0	0	−17.0	38.0	0	−26.0		
6.0	+3.5	−6.0	6.0	+6.0	−6.0	10.0	+6.0	−10.0	14.0	+9.0	−14.0	22.0	+14.0	−22.0	12.41	− 15.75
11.7	0	−8.2	15.5	0	−9.5	19.5	0	13.5	29.0	0	−20.0	45.0	0	−31.0		
8.0	+4.0	−8.0	8.0	+6.0	−8.0	12.0	+6.0	−12.0	16.0	+10.0	−16.0	25.0	+16.0	−25.0	15.75	− 19.69
14.5	0	−10.5	18.0	0	−12.0	22.0	0	−16.0	32.0	0	−22.0	51.0	0	−35.0		
10.0	+5.0	−10.0	10.0	+8.0	−10.0	16.0	+8.0	−16.0	20.0	+12.0	−20.0	30.0	+20.0	−30.0	19.69	− 30.09
18.0	0	−13.0	23.0	0	−15.0	29.0	0	−21.0	40.0	0	−28.0	62.0	0	−42.0		
12.0	+6.0	−12.0	12.0	+10.0	−12.0	20.0	+10.0	−20.0	25.0	+16.0	−25.0	40.0	+25.0	−40.0	30.09	− 41.49
22.0	0	−16.0	28.0	0	−18.0	36.0	0	−26.0	51.0	0	−35.0	81.0	0	−56.0		
16.0	+8.0	−16.0	16.0	+12.0	−16.0	25.0	+12.0	−25.0	30.0	+20.0	−30.0	50.0	+30.0	−50.0	41.49	− 56.19
29.0	0	−21.0	36.0	0	−24.0	45.0	0	−33.0	62.0	0	−42.0	100	0	−70.0		
20.0	+10.0	−20.0	20.0	+16.0	−20.0	30.0	+16.0	−30.0	40.0	+25.0	−40.0	60.0	+40.0	−60.0	56.19	− 76.39
36.0	0	−26.0	46.0	0	−30.0	56.0	0	−40.0	81.0	0	−56.0	125	0	−85.0		
25.0	+12.0	−25.0	25.0	+20.0	−25.0	40.0	+20.0	−40.0	50.0	+30.0	−50.0	80.0	+50.0	−80.0	76.39	− 100.9
45.0	0	−33.0	57.0	0	−37.0	72.0	0	−52.0	100	0	−70.0	160	0	−110		
30.0	+16.0	−30.0	30.0	+35.0	−30.0	50.0	+25.0	−50.0	60.0	+40.0	−60.0	100	+60.0	−100	100.9	− 131.9
56.0	0	−40.0	71.0	0	−46.0	91.0	0	−66.0	125	0	−85.0	200	0	−140		
35.0	+20.0	−35.0	35.0	+30.0	−35.0	60.0	+30.0	−60.0	80.0	+50.0	−80.0	130	+80.0	−130	131.9	− 171.9
57.0	0	−47.0	85.0	0	−55.0	110.0	0	−80.0	160	0	−110	260	0	−180		
45.0	+25.0	−45.0	45.0	+40.0	−45.0	80.0	+40.0	−80.0	100	+60.0	−100	150	+100	−150	171.9	− 200
86.0	0	−61.0	110.0	0	−70.0	145.0	0	−105.0	200	0	−140	310	0	−210		

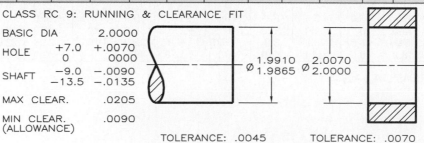

CLASS RC 9: RUNNING & CLEARANCE FIT

BASIC DIA	2.0000	
HOLE	+7.0 / 0	+.0070 / 0000
SHAFT	−9.0 / −13.5	−.0090 / −.0135
MAX CLEAR.	.0205	
MIN CLEAR. (ALLOWANCE)	.0090	

Ø 1.9910 / 1.9865   Ø 2.0070 / 2.0000

TOLERANCE: .0045   TOLERANCE: .0070

Limits are in thousandths of an inch.
Limits for hole and shaft are applied algebraically to the basic size to obtain the limits of size for the parts.
Data in bold face are in accordance with ABC agreements.
Symbols H9, f8, etc., are Hole and Shaft designations used in ABC System.

Nominal Size Range Inches Over	To	Class LC 1 Limits of Clearance	Hole H6	Shaft h5	Class LC2 Limits of Clearance	Hole H7	Shaft h6	Class LC 3 Limits of Clearance	Hole H8	Shaft h7	Class LC 4 Limits of Clearance	Hole H10	Shaft h9	Class LC 5 Limits of Clearance	Hole H7	Shaft g6
0	−0.12	0	+0.25	+0	0	+0.4	+0	0	+0.6	+0	0	+1.6	+0	0.1	+0.4	−0.1
		0.45	−0	−0.2	0.65	−0	−0.25	1	−0	−0.4	2.6	−0	−1.0	0.75	−0	−0.35
0.12	−0.24	0	+0.3	+0	0	+0.5	+0	0	+0.7	+0	0	+1.8	+0	0.15	+0.5	−0.15
		0.5	−0	−0.2	0.8	−0	−0.3	1.2	−0	−0.5	3.0	−0	−1.2	0.95	−0	−0.45
0.24	−0.40	0	+0.4	+0	0	+0.6	+0	0	+0.9	+0	0	+2.2	+0	0.2	+0.6	−0.2
		0.65	−0	−0.25	1.0	−0	−0.4	1.5	−0	−0.6	3.6	−0	−1.4	1.2	−0	−0.6
0.40	−0.71	0	+0.4	+0	0	+0.7	+0	0	+1.0	+0	0	+2.8	+0	0.25	+0.7	−0.25
		0.7	−0	−0.3	1.1	−0	−0.4	1.7	−0	−0.7	4.4	−0	−1.6	1.35	−0	−0.65
0.71	−1.19	0	+0.5	+0	0	+0.8	+0	0	+1.2	+0	0	+3.5	+0	0.3	+0.8	−0.3
		0.9	−0	−0.4	1.3	−0	−0.5	2	−0	−0.8	5.5	−0	−2.0	1.6	−0	−0.8
1.19	−1.97	0	+0.6	+0	0	+1.0	+0	0	+1.6	+0	0	+4.0	+0	0.4	+1.0	−0.4
		1.0	−0	−0.4	1.6	−0	−0.6	2.6	−0	−1	6.5	−0	−2.5	2.0	−0	−1.0
1.97	−3.15	0	+0.7	+0	0	+1.2	+0	0	+1.8	+0	0	+4.5	+0	0.4	+1.2	−0.4
		1.2	−0	−0.5	1.9	−0	−0.7	3	−0	−1.2	7.5	−0	−3	2.3	−0	−1.1
3.15	−4.73	0	+0.9	+0	0	+1.4	+0	0	+2.2	+0	0	+5.0	+0	0.5	+1.4	−0.5
		1.5	−0	−0.6	2.3	−0	−0.9	3.6	−0	−1.4	8.5	−0	−3.5	2.8	−0	−1.4
4.73	−7.09	0	+1.0	+0	0	+1.6	+0	0	+2.5	+0	0	+6.0	+0	0.6	+1.6	−0.6
		1.7	−0	−0.7	2.6	−0	−1.0	4.1	−0	−1.6	10	−0	−4	3.2	−0	1.6
7.09	−9.85	0	+1.2	+0	0	+1.8	+0	0	+2.8	+0	0	+7.0	+0	0.6	+1.8	−0.6
		2.0	−0	−0.8	3.0	−0	−1.2	4.6	−0	−1.8	11.5	−0	−4.5	3.6	−0	−1.8
9.85	12.41	0	+1.2	+0	0	+2.0	+0	0	+3.0	+0	0	8.0	+0	0.7	+2.0	−0.7
		2.1	−0	−0.9	3.2	−0	−1.2	5	−0	−2.0	13	−0	−5	3.9	−0	−1.9
12.41	−15.75	0	+1.4	+0	0	+2.2	+0	0	+3.5	+0	0	+9.0	+0	0.7	+2.2	−0.7
		2.4	−0	−1.0	3.6	−0	−1.4	5.7	−0	−2.2	15	−0	−6	4.3	−0	−2.1
15.75	−19.69	0	+1.6	+0	0	+2.5	+0	0	+4	+0	0	+10.0	+0	0.8	+2.5	−0.8
		2.6	−0	−1.0	4.1	−0	−1.6	6.5	−0	−2.5	16	−0	−6	4.9	−0	−2.4
19.69	−30.09	0	+2.0	+0	0	+3	+0	0	+5	+0	0	+12.0	+0	0.9	+3.0	−0.9
		3.2	−0	−1.2	5.0	−0	−2	8	−0	−3	20	−0	−8	5.9	−0	−2.9
30.09	−41.49	0	+2.5	+0	0	+4	+0	0	+6	+0	0	+16.0	+0	1.0	+4.0	−1.0
		4.1	−0	−1.6	6.5	−0	−2.5	10	−0	−4	26	−0	−10	7.5	−0	−3.5
41.49	−56.19	0	+3.0	+0	0	+5	+0	0	+8	+0	0	+20.0	+0	1.2	+5.0	−1.2
		5.0	−0	−2.0	8.0	−0	−3	13	−0	−5	32	−0	−12	9.2	−0	−4.2
56.19	−76.39	0	+4.0	+0	0	+6	+0	0	+10	+0	0	+25.0	+0	1.2	+6.0	−1.2
		6.5	−0	−2.5	10	−0	−4	16	−0	−6	41	−0	−16	11.2	−0	−5.2
76.39	−100.9	0	+5.0	+0	0	+8	+0	0	+12	+0	0	+30.0	+0	1.4	+8.0	−1.4
		8.0	−0	−3.0	13	−0	−5	20	−0	−8	50	−0	−20	14.4	−0	−6.4
100.9	−131.9	0	+6.0	+0	0	+10	+0	0	+16	+0	0	+40.0	+0	1.6	+10.0	−1.6
		10.0	−0	−4.0	16	−0	−6	26	−0	−10	65	−0	−25	17.6	−0	−7.6
131.9	−171.9	0	+8.0	+0	0	+12	+0	0	+20	+0	0	+50.0	+0	1.8	+12.0	−1.8
		13.0	−0	−5.0	20	−0	−8	32	−0	−12	8	−0	−30	21.8	−0	−9.8
171.9	−200	0	+10.0	+0	0	+16	+0	0	+25	+0	0	+60.0	+0	1.8	+16.0	−1.8
		16.0	−0	−6.0	26	−0	−10	41	−0	−16	100	−0	−40	27.8	−0	−11.8

Source: Courtesy of USASI; B4.1—1955.

Class LC 6			Class LC 7			Class LC 8			Class LC 9			Class LC 10			Class LC 11			Nominal Size Range Inches	
Limits of Clearance	Hole H9	Shaft f8	Limits of Clearance	Hole H10	Shaft e9	Limits of Clearance	Hole H10	Shaft d9	Limits of Clearance	Hole H11	Shaft c10	Limits of Clearance	Hole H12	Shaft	Limits of Clearance	Hole H13	Shaft	Over	To
0.3	+1.0	−0.3	0.6	+1.6	−0.6	1.0	+0.6	−1.0	2.5	+2.5	−2.5	4	+4	−4	5	+6	−5	0 − 0.12	
1.9	0	−0.9	3.2	0	−1.6	3.6	−0	−2.0	6.6	−0	−4.1	12	−0	−8	17	−0	−11		
0.4	+1.2	−0.4	0.8	+1.8	−0.8	1.2	+1.8	−1.2	2.8	+3.0	−2.8	4.5	+5	−4.5	6	+7	−6	0.12 − 0.24	
2.3	0	−1.1	3.8	0	−2.0	4.2	−0	−2.4	7.6	−0	−4.6	14.5	−0	−9.5	20	−0	−13		
0.5	+1.4	−0.5	1.0	+2.2	−1.0	1.6	+2.2	−1.6	3.0	+3.5	−3.0	5	+6	−5	7	+9	−7	0.24 − 0.40	
2.8	0	−1.4	4.6	0	−2.4	5.2	−0	−3.0	8.7	−0	−5.2	17	−0	−11	25	−0	−16		
0.6	+1.6	−0.6	1.2	+2.8	−1.2	2.0	+2.8	−2.0	3.5	+4.0	−3.5	6	+7	−6	8	+10	−8	0.40 − 0.71	
3.2	0	−1.6	5.6	0	−2.8	6.4	−0	−3.6	10.3	−0	−6.3	20	−0	−13	28	−0	−18		
0.8	+2.0	−0.8	1.6	+3.5	−1.6	2.5	+3.5	−2.5	4.5	+5.0	−4.5	7	+8	−7	10	+12	−10	0.71 − 1.19	
4.0	0	−2.0	7.1	0	−3.6	8.0	−0	−4.5	13.0	−0	−8.0	23	−0	−15	34	−0	−22		
1.0	+2.5	−1.0	2.0	+4.0	−2.0	3.0	+4.0	−3.0	5	+6	−5	8	+10	−8	12	+16	−12	1.19 − 1.97	
5.1	0	−2.6	8.5	0	−4.5	9.5	−0	−5.5	15	−0	−9	28	−0	−18	44	−0	−28		
1.2	+3.0	−1.2	2.5	+4.5	−2.5	4.0	+4.5	−4.0	6	+7	−6	10	+12	−10	14	+18	−14	1.97 − 3.15	
6.0	0	−3.0	10.0	0	−5.5	11.5	−0	−7.0	17.5	−0	−10.5	34	−0	−22	50	−0	−32		
1.4	+3.5	−1.4	3.0	+5.0	−3.0	5.0	+5.0	−5.0	7	+9	−7	11	+14	−11	16	+22	−16	3.15 − 4.73	
7.1	0	−3.6	11.5	0	−6.5	13.5	−0	−8.5	21	−0	−12	39	−0	−25	60	−0	−38		
1.6	+4.0	−1.6	3.5	+6.0	−3.5	6	+6	−6	8	+10	−8	12	+16	−12	18	+25	−18	4.73 − 7.09	
8.1	0	−4.1	13.5	0	−7.5	16	−0	−10	24	−0	−14	44	−0	−28	68	−0	−43		
2.0	+4.5	−2.0	4.0	+7.0	−4.0	7	+7	−7	10	+12	−10	16	+18	−16	22	+28	−22	7.09 − 9.85	
9.3	0	−4.8	15.5	0	−8.5	18.5	−0	−11.5	29	−0	−17	52	−0	−34	78	−0	−50		
2.2	+5.0	−2.2	4.5	+8.0	−4.5	7	+8	−7	12	+12	−12	20	+20	−20	28	+30	−28	9.85 − 12.41	
10.2	0	−5.2	17.5	0	−9.5	20	−0	−12	32	−0	−20	60	−0	−40	88	−0	−58		
2.5	+6.0	−2.5	5.0	+9.0	−5	8	+9	−8	14	+14	−14	22	+22	−22	30	+35	−30	12.41 − 15.75	
12.0	0	−6.0	20.0	0	−11	23	−0	−14	37	−0	−23	66	−0	−44	100	−0	−65		
2.8	+6.0	−2.8	5.0	+10.0	−5	9	+10	−9	16	+16	−16	25	+25	−25	35	+40	−35	15.75 − 19.69	
12.8	0	−6.8	21.0	0	−11	25	−0	−15	42	−0	−26	75	−0	−50	115	−0	−75		
3.0	+8.0	−3.0	6.0	+12.0	−6	10	+12	−10	18	+20	−18	28	+30	−28	40	+50	−40	19.69 − 30.09	
16.0	0	−8.0	26.0	−0	−14	30	−0	−18	50	−0	−30	88	−0	−58	140	−0	−90		
3.5	+10.0	−3.5	7.0	+16.0	−7	12	+16	−12	20	+25	−20	30	+40	−30	45	+60	−45	30.09 − 41.49	
19.5	0	−9.5	33.0	−0	−17	38	−0	−22	61	−0	−36	110	−0	−70	165	−0	−105		
4.0	+12.0	−4.0	8.0	+20.0	−8	14	+20	−14	25	+30	−25	40	+50	−40	60	+80	−60	41.49 − 56.19	
24.0	0	−12.0	40.0	−0	−20	46	−0	−26	75	−0	−45	140	−0	−90	220	−0	−140		
4.5	+16.0	−4.5	9.0	+25.0	−9	16	+25	−16	30	+40	−30	50	+60	−50	70	+100	−70	56.19 − 76.39	
30.5	0	−14.5	50.0	−0	−25	57	−0	−32	95	−0	−55	170	−0	−110	270	−0	−170		
5.0	+20.0	−5	10.0	+30.0	−10	18	+30	−18	35	+50	−35	50	+80	−50	80	+125	−80	76.39 − 100.9	
37.0	0	−17	60.0	−0	−30	68	−0	−38	115	−0	−65	210	−0	−130	330	−0	−205		
6.0	+25.0	−6	12.0	+40.0	−12	20	+40	−20	40	+60	−40	60	+100	−60	90	+160	−90	100.9 − 131.9	
47.0	0	−22	67.0	−0	−27	85	−0	−45	140	−0	−80	260	−0	−160	410	−0	−250		
7.0	+30.0	−7	14.0	+50.0	−14	25	+50	−25	50	+80	−50	80	+125	−80	100	+200	−100	131.9 − 171.9	
57.0	0	−27	94.0	−0	−44	105	−0	−55	180	−0	−100	330	−0	−205	500	−0	−300		
7.0	+40.0	−7	14.0	+60.0	−14	25	+60	−25	50	+100	−50	90	+160	−90	125	+250	−125	171.9 − 200	
72.0	0	−32	114.0	−0	−54	125	−0	−65	210	−0	−110	410	−0	−250	625	−0	−375		

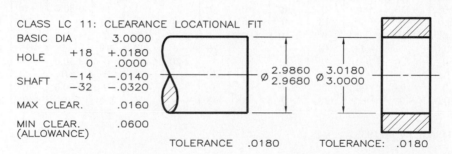

CLASS LC 11: CLEARANCE LOCATIONAL FIT

BASIC DIA		3.0000
HOLE	+18 / 0	+.0180 / .0000
SHAFT	−14 / −32	−.0140 / −.0320
MAX CLEAR.		.0160
MIN CLEAR. (ALLOWANCE)		.0600

Ø 2.9860 / 2.9680   Ø 3.0180 / 3.0000

TOLERANCE .0180   TOLERANCE: .0180

Limits are in thousandths of an inch.

Limits for hole and shaft are applied algebraically to the basic size to obtain the limits of size for the mating parts.

Data in bold face are in accordance with ABC agreements.

"Fit" represents the maximum interference (minus values) and the maximum clearance (plus values).

Symbols H7, js6, etc., are Hole and Shaft designations used in ABC System.

Nominal Size Range Inches Over	To	Class LT 1 Fit	LT 1 Hole H7	LT 1 Shaft js6	Class LT 2 Fit	LT 2 Hole H8	LT 2 Shaft js7	Class LT 3 Fit	LT 3 Hole H7	LT 3 Shaft k6	Class LT 4 Fit	LT 4 Hole H8	LT 4 Shaft k7	Class LT 5 Fit	LT 5 Hole H7	LT 5 Shaft n6	Class LT 6 Fit	LT 6 Hole H7	LT 6 Shaft n7
0	0.12	−0.10 / +0.50	+0.4 / −0	+0.10 / −0.10	−0.2 / +0.8	+0.6 / −0	+0.2 / −0.2							−0.5 / +0.15	+0.4 / −0	+0.5 / +0.25	−0.65 / +0.15	+0.4 / −0	+0.65 / +0.25
0.12	0.24	−0.15 / +0.65	+0.5 / −0	+0.15 / −0.15	−0.25 / +0.95	+0.7 / −0	+0.25 / −0.25							−0.6 / +0.2	+0.5 / −0	+0.6 / +0.3	−0.8 / +0.2	+0.5 / −0	+0.8 / +0.3
0.24	0.40	−0.2 / +0.8	+0.6 / −0	+0.2 / −0.2	−0.3 / +1.2	+0.9 / −0	+0.3 / −0.3	−0.5 / +0.5	+0.6 / −0	+0.5 / +0.1	−0.7 / +0.8	+0.9 / −0	+0.7 / +0.1	−0.8 / +0.2	+0.6 / −0	+0.8 / +0.4	−1.0 / +0.2	+0.6 / −0	+1.0 / +0.4
0.40	0.71	−0.2 / +0.9	+0.7 / −0	+0.2 / −0.2	−0.35 / +1.35	+1.0 / −0	+0.35 / −0.35	−0.5 / +0.6	+0.7 / −0	+0.5 / +0.1	−0.8 / +0.9	+1.0 / −0	+0.8 / +0.1	−0.9 / +0.2	+0.7 / −0	+0.9 / +0.5	−1.2 / +0.2	+0.7 / −0	+1.2 / +0.5
0.71	1.19	−0.25 / +1.05	+0.8 / −0	+0.25 / −0.25	−0.4 / +1.6	+1.2 / −0	+0.4 / −0.4	−0.6 / +0.7	+0.8 / −0	+0.6 / +0.1	−0.9 / +1.1	+1.2 / −0	+0.9 / +0.1	−1.1 / +0.2	+0.8 / −0	+1.1 / +0.6	−1.4 / +0.2	+0.8 / −0	+1.4 / +0.6
1.19	1.97	−0.3 / +1.3	+1.0 / −0	+0.3 / −0.3	−0.5 / +2.1	+1.6 / −0	+0.5 / −0.5	−0.7 / +0.9	+1.0 / −0	+0.7 / +0.1	−1.1 / +1.5	+1.6 / −0	+1.1 / +0.1	−1.3 / +0.3	+1.0 / −0	+1.3 / +0.7	−1.7 / +0.3	+1.0 / −0	+1.7 / +0.7
1.97	3.15	−0.3 / +1.5	+1.2 / −0	+0.3 / −0.3	−0.6 / +2.4	+1.8 / −0	+0.6 / −0.6	−0.8 / +1.1	+1.2 / −0	+0.8 / +0.1	−1.3 / +1.7	+1.8 / −0	+1.3 / +0.1	−1.5 / +0.4	+1.2 / −0	+1.5 / +0.8	−2.0 / +0.4	+1.2 / −0	+2.0 / +0.8
3.15	4.73	−0.4 / +1.8	+1.4 / −0	+0.4 / −0.4	−0.7 / +2.9	+2.2 / −0	+0.7 / −0.7	−1.0 / +1.3	+1.4 / −0	+1.0 / +0.1	−1.5 / +2.1	+2.2 / −0	+1.5 / +0.1	−1.9 / +0.4	+1.4 / −0	+1.9 / +1.0	−2.4 / +0.4	+1.4 / −0	+2.4 / +1.0
4.73	7.09	−0.5 / +2.1	+1.6 / −0	+0.5 / −0.5	−0.8 / +3.3	+2.5 / −0	+0.8 / −0.8	−1.1 / +1.5	+1.6 / −0	+1.1 / +0.1	−1.7 / +2.4	+2.5 / −0	+1.7 / +0.1	−2.2 / +0.4	+1.6 / −0	+2.2 / +1.2	−2.8 / +0.4	+1.6 / −0	+2.8 / +1.2
7.09	9.85	−0.6 / +2.4	+1.8 / −0	+0.6 / −0.6	−0.9 / +3.7	+2.8 / −0	+0.9 / −0.9	−1.4 / +1.6	+1.8 / −0	+1.4 / +0.2	−2.0 / +2.6	+2.8 / −0	+2.0 / +0.2	−2.6 / +0.4	+1.8 / −0	+2.6 / +1.4	−3.2 / +0.4	+1.8 / −0	+3.2 / +1.4
9.85	12.41	−0.6 / +2.6	+2.0 / −0	+0.6 / −0.6	−1.0 / +4.0	+3.0 / −0	+1.0 / −1.0	−1.4 / +1.8	+2.0 / −0	+1.4 / +0.2	−2.2 / +2.8	+3.0 / −0	+2.2 / +0.2	−2.6 / +0.6	+2.0 / −0	+2.6 / +1.4	−3.4 / +0.6	+2.0 / −0	+3.4 / +1.4
12.41	15.75	−0.7 / +2.9	+2.2 / −0	+0.7 / −0.7	−1.0 / +4.5	+3.5 / −0	+1.0 / −1.0	−1.6 / +2.0	+2.2 / −0	+1.6 / +0.2	−2.4 / +3.3	+3.5 / −0	+2.4 / +0.2	−3.0 / +0.6	+2.2 / −0	+3.0 / +1.6	−3.8 / +0.6	+2.2 / −0	+3.8 / +1.6
15.75	19.69	−0.8 / +3.3	+2.5 / −0	+0.8 / −0.8	−1.2 / +5.2	+4.0 / −0	+1.2 / −1.2	−1.8 / +2.3	+2.5 / −0	+1.8 / +0.2	−2.7 / +3.8	+4.0 / −0	+2.7 / +0.2	−3.4 / +0.7	+2.5 / −0	+3.4 / +1.8	−4.3 / +0.7	+2.5 / −0	+4.3 / +1.8

Source: Courtesy of ANSI; B4.1–1955.

APPENDIX 28 • American Standard Interference Locational Fits (hole basis)

Limits are in thousandths of an inch.
Limits for hole and shaft are applied algebraically to the basic size to obtain the limits of size for the parts.
Data in bold face are in accordance with ABC agreements.
Symbols H7, p6, etc., are Hole and Shaft designations used in ABC System.

Nominal Size Range Inches Over	To	Class LN 1 Limits of Interference	Class LN 1 Standard Limits Hole H6	Class LN 1 Standard Limits Shaft n5	Class LN 2 Limits of Interference	Class LN 2 Standard Limits Hole H7	Class LN 2 Standard Limits Shaft p6	Class LN 3 Limits of Interference	Class LN 3 Standard Limits Hole H7	Class LN 3 Standard Limits Shaft r6
0	−0.12	0	+0.25	+0.45	0	+0.4	+0.65	0.1	+0.4	+0.75
		0.45	−0	+0.25	0.65	−0	+0.4	0.75	−0	+0.5
0.12	−0.24	0	+0.3	+0.5	0	+0.5	+0.8	0.1	+0.5	+0.9
		0.5	−0	+0.3	0.8	−0	+0.5	0.9	0	+0.6
0.24	−0.40	0	+0.4	+0.65	0	+0.6	+1.0	0.2	+0.6	+1.2
		0.65	−0	+0.4	1.0	−0	+0.6	1.2	−0	+0.8
0.40	−0.71	0	+0.4	+0.8	0	+0.7	+1.1	0.3	+0.7	+1.4
		0.8	−0	+0.4	1.1	−0	+0.7	1.4	−0	+1.0
0.71	−1.19	0	+0.5	+1.0	0	+0.8	+1.3	0.4	+0.8	+1.7
		1.0	−0	+0.5	1.3	−0	+0.8	1.7	−0	+1.2
1.19	−1.97	0	+0.6	+1.1	0	+1.0	+1.6	0.4	+1.0	+2.0
		1.1	−0	+0.6	1.6	−0	+1.0	2.0	−0	+1.4
1.97	−3.15	0.1	+0.7	+1.3	0.2	+1.2	+2.1	0.4	+1.2	+2.3
		1.3	−0	+0.7	2.1	−0	+1.4	2.3	−0	+1.6
3.15	−4.73	0.1	+0.9	+1.6	0.2	+1.4	+2.5	0.6	+1.4	+2.9
		1.6	−0	+1.0	2.5	−0	+1.6	2.9	−0	+2.0
4.73	−7.09	0.2	+1.0	+1.9	0.2	+1.6	+2.8	0.9	+1.6	+3.5
		1.9	−0	+1.2	2.8	−0	+1.8	3.5	−0	+2.5
7.09	−9.85	0.2	+1.2	+2.2	0.2	+1.8	+3.2	1.2	+1.8	+4.2
		2.2	−0	+1.4	3.2	−0	+2.0	4.2	−0	+3.0
9.85	−12.41	0.2	+1.2	+2.3	0.2	+2.0	+3.4	1.5	+2.0	+4.7
		2.3	−0	+1.4	3.4	−0	+2.2	4.7	−0	+3.5
12.41	−15.75	0.2	+1.4	+2.6	0.3	+2.2	+3.9	2.3	+2.2	+5.9
		2.6	−0	+1.6	3.9	−0	+2.5	5.9	−0	+4.5
15.75	−19.69	0.2	+1.6	+2.8	0.3	+2.5	+4.4	2.5	+2.5	+6.6
		2.8	−0	+1.8	4.4	−0	+2.8	6.6	−0	+5.0
19.69	−30.09		+2.0		0.5	+3	+5.5	4	+3	+9
			−0		5.5	−0	+3.5	9	−0	+7
30.09	−41.49		+2.5		0.5	+4	+7.0	5	+4	+11.5
			−0		7.0	−0	+4.5	11.5	−0	+9
41.49	−56.19		+3.0		1	+5	+9	7	+5	+15
			−0		9	−0	+6	15	−0	+12
56.19	−76.39		+4.0		1	+6	+11	10	+6	+20
			−0		11	−0	+7	20	−0	+16
76.39	−100.9		+5.0		1	+8	+14	12	+8	+25
			−0		14	−0	+9	25	−0	+20
100.9	−131.9		+6.0		2	+10	+18	15	+10	+31
			−0		18	−0	+12	31	−0	+25
131.9	−171.9		+8.0		4	+12	+24	18	+12	+38
			−0		24	−0	+16	38	−0	+30
171.9	−200		+10.0		4	+16	+30	24	+16	+50
			−0		30	−0	+20	50	−0	+40

Source: Courtesy of ANSI; B4.1–1955.

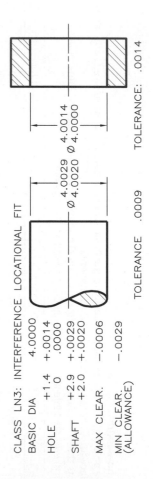

CLASS LN3: INTERFERENCE LOCATIONAL FIT

BASIC DIA.	4.0000	
HOLE	+1.4	+.0014
	0	.0000
SHAFT	+2.9	+.0029
	+2.0	+.0020
MAX CLEAR.		−.0006
MIN CLEAR. (ALLOWANCE)		−.0029

Ø 4.0029 / Ø 4.0020   TOLERANCE .0009

4.0014 / 4.0000   TOLERANCE: .0014

Limits are in thousandths of an inch.
Limits for hole and shaft are applied algebraically to the basic size to obtain the limits of size for the parts.
Data in bold face are in accordance with ABC agreements.
Symbols H7, s6, etc., are Hole and Shaft designations used in ABC System.

Nominal Size Range Inches Over	To	Class FN 1 Limits of Interference	Standard Limits Hole H6	Shaft	Class FN 2 Limits of Interference	Standard Limits Hole H7	Shaft s6	Class FN 3 Limits of Interference	Standard Limits Hole H7	Shaft t6	Class FN 4 Limits of Interference	Standard Limits Hole H7	Shaft u6	Class FN 5 Limits of Interference	Standard Limits Hole H8	Shaft x7
0	0.12	0.05	+0.25	+0.5	0.2	+0.4	+0.85				0.3	+0.4	+0.95	0.3	+0.6	+1.3
		0.5	−0	+0.3	0.85	−0	+0.6				0.95	−0	+0.7	1.3	−0	+0.9
0.12	0.24	0.1	+0.3	+0.6	0.2	+0.5	+1.0				0.4	+0.5	+1.2	0.5	+0.7	+1.7
		0.6	−0	+0.4	1.0	−0	+0.7				1.2	−0	+0.9	1.7	−0	+1.2
0.24	0.40	0.1	+0.4	+0.75	0.4	+0.6	+1.4				0.6	+0.6	+1.6	0.5	+0.9	+2.0
		0.75	−0	+0.5	1.4	−0	+1.0				1.6	−0	+1.2	2.0	−0	+1.4
0.40	0.56	0.1	+0.4	+0.8	0.5	+0.7	+1.6				0.7	+0.7	+1.8	0.6	+1.0	+2.3
		0.8	−0	+0.5	1.6	−0	+1.2				1.8	−0	+1.4	2.3	−0	+1.6
0.56	0.71	0.2	+0.4	+0.9	0.5	+0.7	+1.6				0.7	+0.7	+1.8	0.8	+1.0	+2.5
		0.9	−0	+0.6	1.6	−0	+1.2				1.8	−0	+1.4	2.5	−0	+1.8
0.71	0.95	0.2	+0.5	+1.1	0.6	+0.8	+1.9				0.8	+0.8	+2.1	1.0	+1.2	+3.0
		1.1	−0	+0.7	1.9	−0	+1.4				2.1	−0	+1.6	3.0	−0	+2.2
0.95	1.19	0.3	+0.5	+1.2	0.6	+0.8	+1.9	0.8	+0.8	+2.1	1.0	+0.8	+2.3	1.3	+1.2	+3.3
		1.2	−0	+0.8	1.9	−0	+1.4	2.1	−0	+1.6	2.3	−0	+1.8	3.3	−0	+2.5
1.19	1.58	0.3	+0.6	+1.3	0.8	+1.0	+2.4	1.0	+1.0	+2.6	1.5	+1.0	+3.1	1.4	+1.6	+4.0
		1.3	−0	+0.9	2.4	−0	+1.8	2.6	−0	+2.0	3.1	−0	+2.5	4.0	−0	+3.0
1.58	1.97	0.4	+0.6	+1.4	0.8	+1.0	+2.4	1.2	+1.0	+2.8	1.8	+1.0	+3.4	2.4	+1.6	+5.0
		1.4	−0	+1.0	2.4	−0	+1.8	2.8	−0	+2.2	3.4	−0	+2.8	5.0	−0	+4.0
1.97	2.56	0.6	+0.7	+1.8	0.8	+1.2	+2.7	1.3	+1.2	+3.2	2.3	+1.2	+4.2	3.2	+1.8	+6.2
		1.8	−0	+1.3	2.7	−0	+2.0	3.2	−0	+2.5	4.2	−0	+3.5	6.2	−0	+5.0
2.56	3.15	0.7	+0.7	+1.9	1.0	+1.2	+2.9	1.8	+1.2	+3.7	2.8	+1.2	+4.7	4.2	+1.8	+7.2
		1.9	−0	+1.4	2.9	−0	+2.2	3.7	−0	+3.0	4.7	−0	+4.0	7.2	−0	+6.0
3.15	3.94	0.9	+0.9	+2.4	1.4	+1.4	+3.7	2.1	+1.4	+4.4	3.6	+1.4	+5.9	4.8	+2.2	+8.4
		2.4	−0	+1.8	3.7	−0	+2.8	4.4	−0	+3.5	5.9	−0	+5.0	8.4	−0	+7.0
3.94	4.73	1.1	+0.9	+2.6	1.6	+1.4	+3.9	2.6	+1.4	+4.9	4.6	+1.4	+6.9	5.8	+2.2	+9.4
		2.6	−0	+2.0	3.9	−0	+3.0	4.9	−0	+4.0	6.9	−0	+6.0	9.4	−0	+8.0
4.73	5.52	1.2	+1.0	+2.9	1.9	+1.6	+4.5	3.4	+1.6	+6.0	5.4	+1.6	+8.0	7.5	+2.5	+11.6
		2.9	−0	+2.2	4.5	−0	+3.5	6.0	−0	+5.0	8.0	−0	+7.0	11.6	−0	+10.0
5.52	6.30	1.5	+1.0	+3.2	2.4	+1.6	+5.0	3.4	+1.6	+6.0	5.4	+1.6	+8.0	9.5	+2.5	+13.6
		3.2	−0	+2.5	5.0	−0	+4.0	6.0	−0	+5.0	8.0	−0	+7.0	13.6	−0	+12.0
6.30	7.09	1.8	+1.0	+3.5	2.9	+1.6	+5.5	4.4	+1.6	+7.0	6.4	+1.6	+9.0	9.5	+2.5	+13.6
		3.5	−0	+2.8	5.5	−0	+4.5	7.0	−0	+6.0	9.0	−0	+8.0	13.6	−0	+12.0
7.09	7.88	1.8	+1.2	+3.8	3.2	+1.8	+6.2	5.2	+1.8	+8.2	7.2	+1.8	+10.2	11.2	+2.8	+15.8
		3.8	−0	+3.0	6.2	−0	+5.0	8.2	−0	+7.0	10.2	−0	+9.0	15.8	−0	+14.0
7.88	8.86	2.3	+1.2	+4.3	3.2	+1.8	+6.2	5.2	+1.8	+8.2	8.2	+1.8	+11.2	13.2	+2.8	+17.8
		4.3	−0	+3.5	6.2	−0	+5.0	8.2	−0	+7.0	11.2	−0	+10.0	17.8	−0	+16.0
8.86	9.85	2.3	+1.2	+4.3	4.2	+1.8	+7.2	6.2	+1.8	+9.2	10.2	+1.8	+13.2	13.2	+2.8	+17.8
		4.3	−0	+3.5	7.2	−0	+6.0	9.2	−0	+8.0	13.2	−0	+12.0	17.8	−0	+16.0
9.85	11.03	2.8	+1.2	+4.9	4.0	+2.0	+7.2	7.0	+2.0	+10.2	10.0	+2.0	+13.2	15.0	+3.0	+20.0
		4.9	−0	+4.0	7.2	−0	+6.0	10.2	−0	+9.0	13.2	−0	+12.0	20.0	−0	+18.0
11.03	12.41	2.8	+1.2	+4.9	5.0	+2.0	+8.2	7.0	+2.0	+10.2	12.0	+2.0	+15.2	17.0	+3.0	+22.0
		4.9	−0	+4.0	8.2	−0	+7.0	10.2	−0	+9.0	15.2	−0	+14.0	22.0	−0	+20.0
12.41	13.98	3.1	+1.4	+5.5	5.8	+2.2	+9.4	7.8	+2.2	+11.4	13.8	+2.2	+17.4	18.5	+3.5	+24.2
		5.5	−0	+4.5	9.4	−0	+8.0	11.4	−0	+10.0	17.4	−0	+16.0	24.2	+0	+22.0
13.98	15.75	3.6	+1.4	+6.1	5.8	+2.2	+9.4	9.8	+2.2	+13.4	15.8	+2.2	+19.4	21.5	+3.5	+27.2
		6.1	−0	+5.0	9.4	−0	+8.0	13.4	−0	+12.0	19.4	−0	+18.0	27.2	−0	+25.0
15.75	17.72	4.4	+1.6	+7.0	6.5	+2.5	+10.6	9.5	+2.5	+13.6	17.5	+2.5	+21.6	24.0	+4.0	+30.5
		7.0	−0	+6.0	10.6	−0	+9.0	13.6	−0	+12.0	21.6	−0	+20.0	30.5	−0	+28.0
17.72	19.69	4.4	+1.6	+7.0	7.5	+2.5	+11.6	11.5	+2.5	+15.6	19.5	+2.5	+23.6	26.0	+4.0	+32.5
		7.0	−0	+6.0	11.6	−0	+10.0	15.6	−0	+14.0	23.6	−0	+22.0	32.5	−0	+30.0

Courtesy of ANSI; B4.1

APPENDIX 30 • The International Tolerance Grades (ANSI B4.2)

Dimensions are in mm.

Basic sizes		Tolerance grades[3]																	
Over	Up to and including	IT01	IT0	IT1	IT2	IT3	IT4	IT5	IT6	IT7	IT8	IT9	IT10	IT11	IT12	IT13	IT14	IT15	IT16
0	3	0.0003	0.0005	0.0008	0.0012	0.002	0.003	0.004	0.006	0.010	0.014	0.025	0.040	0.060	0.100	0.140	0.250	0.400	0.600
3	6	0.0004	0.0006	0.001	0.0015	0.0025	0.004	0.005	0.008	0.012	0.018	0.030	0.048	0.075	0.120	0.180	0.300	0.480	0.750
6	10	0.0004	0.0006	0.001	0.0015	0.0025	0.004	0.006	0.009	0.015	0.022	0.036	0.058	0.090	0.150	0.220	0.360	0.580	0.900
10	18	0.0005	0.0008	0.0012	0.002	0.003	0.005	0.008	0.011	0.018	0.027	0.043	0.070	0.110	0.180	0.270	0.430	0.700	1.100
18	30	0.0006	0.001	0.0015	0.0025	0.004	0.006	0.009	0.013	0.021	0.033	0.052	0.084	0.130	0.210	0.330	0.520	0.840	1.300
30	50	0.0006	0.001	0.0015	0.0025	0.004	0.007	0.011	0.016	0.025	0.039	0.062	0.100	0.160	0.250	0.390	0.620	1.000	1.600
50	80	0.0008	0.0012	0.002	0.003	0.005	0.008	0.013	0.019	0.030	0.046	0.074	0.120	0.190	0.300	0.460	0.740	1.200	1.900
80	120	0.001	0.0015	0.0025	0.004	0.006	0.010	0.015	0.022	0.035	0.054	0.087	0.140	0.220	0.350	0.540	0.870	1.400	2.200
120	180	0.0012	0.002	0.0036	0.005	0.008	0.012	0.018	0.025	0.040	0.063	0.100	0.160	0.250	0.400	0.630	1.000	1.600	2.500
180	250	0.002	0.003	0.0045	0.007	0.010	0.014	0.020	0.029	0.046	0.072	0.115	0.185	0.290	0.460	0.720	1.150	1.850	2.900
250	315	0.0025	0.004	0.006	0.008	0.012	0.016	0.023	0.032	0.052	0.081	0.130	0.210	0.320	0.520	0.810	1.300	2.100	3.200
315	400	0.003	0.005	0.007	0.009	0.013	0.018	0.025	0.036	0.057	0.089	0.140	0.230	0.360	0.570	0.890	1.400	2.300	3.600
400	500	0.004	0.006	0.008	0.010	0.015	0.020	0.027	0.040	0.063	0.097	0.156	0.250	0.400	0.630	0.970	1.550	2.500	4.000
500	630	0.0045	0.006	0.009	0.011	0.016	0.022	0.030	0.044	0.070	0.110	0.175	0.280	0.440	0.700	1.100	1.750	2.800	4.400
630	800	0.005	0.007	0.010	0.013	0.018	0.025	0.035	0.050	0.080	0.125	0.200	0.320	0.500	0.800	1.250	2.000	3.200	5.000
800	1000	0.0055	0.008	0.011	0.015	0.021	0.029	0.040	0.056	0.090	0.140	0.230	0.360	0.560	0.900	1.400	2.300	3.600	5.600
1000	1250	0.0065	0.009	0.013	0.018	0.024	0.034	0.046	0.066	0.105	0.165	0.260	0.420	0.660	1.050	1.650	2.600	4.200	6.600
1250	1600	0.008	0.011	0.015	0.021	0.029	0.040	0.054	0.078	0.125	0.195	0.310	0.500	0.780	1.250	1.950	3.100	5.000	7.800
1600	2000	0.009	0.013	0.018	0.025	0.035	0.048	0.065	0.092	0.150	0.230	0.370	0.600	0.920	1.500	2.300	3.700	6.000	9.200
2000	2500	0.011	0.015	0.022	0.030	0.041	0.057	0.077	0.110	0.175	0.280	0.440	0.700	1.100	1.750	2.800	4.400	7.000	11.000
2500	3150	0.013	0.018	0.026	0.036	0.050	0.069	0.093	0.135	0.210	0.330	0.540	0.860	1.350	2.100	3.300	5.400	8.600	13.500

[3]IT Values for tolerance grades larger than IT16 can be calculated by using the following formulas:
IT17 = IT12 × 10; IT18 = IT13 × 10; etc.

APPENDIX 31 • Preferred Hole Basis Clearance Fits—Cylindrical Fits (ANSI B4.2)

AMERICAN NATIONAL STANDARD PREFERRED METRIC LIMITS AND FITS ANSI B4.2—1978
Dimensions are in mm.

BASIC SIZE		LOOSE RUNNING			FREE RUNNING			CLOSE RUNNING			SLIDING			LOCATIONAL CLEARANCE		
		Hole H11	Shaft c11	Fit	Hole H9	Shaft d9	Fit	Hole H8	Shaft f7	Fit	Hole H7	Shaft g6	Fit	Hole H7	Shaft h6	Fit
1	MAX	1.060	0.940	0.180	1.025	0.980	0.070	1.014	0.994	0.030	1.010	0.998	0.018	1.010	1.000	0.016
	MIN	1.000	0.880	0.060	1.000	0.955	0.020	1.000	0.984	0.006	1.000	0.992	0.002	1.000	0.994	0.000
1.2	MAX	1.260	1.140	0.180	1.225	1.180	0.070	1.214	1.194	0.030	1.210	1.198	0.018	1.210	1.200	0.016
	MIN	1.200	1.080	0.060	1.200	1.155	0.020	1.200	1.184	0.006	1.200	1.192	0.002	1.200	1.194	0.000
1.6	MAX	1.660	1.540	0.180	1.625	1.580	0.070	1.614	1.594	0.030	1.610	1.598	0.018	1.610	1.600	0.016
	MIN	1.600	1.480	0.060	1.600	1.555	0.020	1.600	1.584	0.006	1.600	1.592	0.002	1.600	1.594	0.000
2	MAX	2.060	1.940	0.180	2.025	1.980	0.070	2.014	1.994	0.030	2.010	1.998	0.018	2.010	2.000	0.016
	MIN	2.000	1.880	0.060	2.000	1.955	0.020	2.000	1.984	0.006	2.000	1.992	0.002	2.000	1.994	0.000
2.5	MAX	2.560	2.440	0.180	2.525	2.480	0.070	2.514	2.494	0.030	2.510	2.498	0.018	2.510	2.500	0.016
	MIN	2.500	2.380	0.060	2.500	2.455	0.020	2.500	2.484	0.006	2.500	2.492	0.002	2.500	2.494	0.000
3	MAX	3.060	2.940	0.180	3.025	2.980	0.070	3.014	2.994	0.030	3.010	2.998	0.018	3.010	3.000	0.016
	MIN	3.000	2.880	0.060	3.000	2.955	0.020	3.000	2.984	0.006	3.000	2.992	0.002	3.000	2.994	0.000
4	MAX	4.075	3.930	0.220	4.030	3.970	0.090	4.018	3.990	0.040	4.012	3.996	0.024	4.012	4.000	0.020
	MIN	4.000	3.855	0.070	4.000	3.940	0.030	4.000	3.978	0.010	4.000	3.988	0.004	4.000	3.992	0.000
5	MAX	5.075	4.930	0.220	5.030	4.970	0.090	5.018	4.990	0.040	5.012	4.996	0.024	5.012	5.000	0.020
	MIN	5.000	4.855	0.070	5.000	4.940	0.030	5.000	4.978	0.010	5.000	4.988	0.004	5.000	4.992	0.000
6	MAX	6.075	5.930	0.220	6.030	5.970	0.090	6.018	5.990	0.040	6.012	5.996	0.024	6.012	6.000	0.020
	MIN	6.000	5.855	0.070	6.000	5.940	0.030	6.000	5.978	0.010	6.000	5.988	0.004	6.000	5.992	0.000
8	MAX	8.090	7.920	0.260	8.036	7.960	0.112	8.022	7.987	0.050	8.015	7.995	0.029	8.015	8.000	0.024
	MIN	8.000	7.830	0.080	8.000	7.924	0.040	8.000	7.972	0.013	8.000	7.986	0.005	8.000	7.991	0.000
10	MAX	10.090	9.920	0.260	10.036	9.960	0.112	10.022	9.987	0.050	10.015	9.995	0.029	10.015	10.000	0.024
	MIN	10.000	9.830	0.080	10.000	9.924	0.040	10.000	9.972	0.013	10.000	9.986	0.005	10.000	9.991	0.000
12	MAX	12.110	11.905	0.315	12.043	11.950	0.136	12.027	11.984	0.061	12.018	11.994	0.035	12.018	12.000	0.029
	MIN	12.000	11.795	0.095	12.000	11.907	0.050	12.000	11.966	0.016	12.000	11.983	0.006	12.000	11.989	0.000
16	MAX	16.110	15.905	0.315	16.043	15.950	0.136	16.027	15.984	0.061	16.018	15.994	0.035	16.018	16.000	0.029
	MIN	16.000	15.795	0.095	16.000	15.907	0.050	16.000	15.966	0.016	16.000	15.983	0.006	16.000	15.989	0.000
20	MAX	20.130	19.890	0.370	20.052	19.935	0.169	20.033	19.980	0.074	20.021	19.993	0.041	20.021	20.000	0.034
	MIN	20.000	19.760	0.110	20.000	19.883	0.065	20.000	19.959	0.020	20.000	19.980	0.007	20.000	19.987	0.000
25	MAX	25.130	24.890	0.370	25.052	24.935	0.169	25.033	24.980	0.074	25.021	24.993	0.041	25.021	25.000	0.034
	MIN	25.000	24.760	0.110	25.000	24.883	0.065	25.000	24.959	0.020	25.000	24.980	0.007	25.000	24.987	0.000
30	MAX	30.130	29.890	0.370	30.052	29.935	0.169	30.033	29.980	0.074	30.021	29.993	0.041	30.021	30.000	0.034
	MIN	30.000	29.760	0.110	30.000	29.883	0.065	30.000	29.959	0.020	30.000	29.980	0.007	30.000	29.987	0.000

Source: American National Standard Preferred Metric Limits and Figs, ANSI B4.2—1978.

APPENDIX 31 • Preferred Hole Basis Clearance Fits—Cylindrical Fits (ANSI B4.2)(continued)

BASIC SIZE		LOOSE RUNNING			FREE RUNNING			CLOSE RUNNING			SLIDING			LOCATIONAL CLEARANCE		
		Hole H11	Shaft c11	Fit	Hole H9	Shaft d9	Fit	Hole H8	Shaft f7	Fit	Hole H7	Shaft g6	Fit	Hole H7	Shaft h6	Fit
40	MAX	40.160	39.880	0.440	40.062	39.920	0.204	40.039	39.975	0.089	40.025	39.991	0.050	40.025	40.000	0.041
	MIN	40.000	39.720	0.120	40.000	39.858	0.080	40.000	39.950	0.025	40.000	39.975	0.009	40.000	39.984	0.000
50	MAX	50.160	49.870	0.450	50.062	49.920	0.204	50.039	49.975	0.089	50.025	49.991	0.050	50.025	50.000	0.041
	MIN	50.000	49.710	0.130	50.000	49.858	0.080	50.000	49.950	0.025	50.000	49.975	0.009	50.000	49.984	0.000
60	MAX	60.190	59.860	0.520	60.074	59.900	0.248	60.046	59.970	0.106	60.030	59.990	0.059	60.030	60.000	0.049
	MIN	60.000	59.670	0.140	60.000	59.826	0.100	60.000	59.940	0.030	60.000	59.971	0.010	60.000	59.981	0.000
80	MAX	80.190	79.850	0.530	80.074	79.900	0.248	80.046	79.970	0.106	80.030	79.990	0.059	80.030	80.000	0.049
	MIN	80.000	79.660	0.150	80.000	79.826	0.100	80.000	79.940	0.030	80.000	79.971	0.010	80.000	79.981	0.000
100	MAX	100.220	99.830	0.610	100.087	99.880	0.294	100.054	99.964	0.125	100.035	99.988	0.069	100.035	100.000	0.057
	MIN	100.000	99.610	0.170	100.000	99.793	0.120	100.000	99.929	0.036	100.000	99.966	0.012	100.000	99.978	0.000
120	MAX	120.220	119.820	0.620	120.087	119.880	0.294	120.054	119.964	0.125	120.035	119.988	0.069	120.035	120.000	0.057
	MIN	120.000	119.600	0.180	120.000	119.793	0.120	120.000	119.929	0.036	120.000	119.966	0.012	120.000	119.978	0.000
160	MAX	160.250	159.790	0.710	160.100	159.855	0.345	160.063	159.957	0.146	160.040	159.986	0.079	160.040	160.000	0.065
	MIN	160.000	159.540	0.210	160.000	159.755	0.145	160.000	159.917	0.043	160.000	159.961	0.014	160.000	159.975	0.000
200	MAX	200.290	199.760	0.820	200.115	199.830	0.400	200.072	199.950	0.168	200.046	199.985	0.090	200.046	200.000	0.075
	MIN	200.000	199.470	0.240	200.000	199.715	0.170	200.000	199.904	0.050	200.000	199.956	0.015	200.000	199.971	0.000
250	MAX	250.290	249.720	0.860	250.115	249.830	0.400	250.072	249.950	0.168	250.046	249.985	0.090	250.046	250.000	0.075
	MIN	250.000	249.430	0.280	250.000	249.715	0.170	250.000	249.904	0.050	250.000	249.956	0.015	250.000	249.971	0.000
300	MAX	300.320	299.670	0.970	300.130	299.810	0.450	300.081	299.944	0.189	300.052	299.983	0.101	300.052	300.000	0.084
	MIN	300.000	299.350	0.330	300.000	299.680	0.190	300.000	299.892	0.056	300.000	299.951	0.017	300.000	299.968	0.000
400	MAX	400.360	399.600	1.120	400.140	399.790	0.490	400.089	399.938	0.208	400.057	399.982	0.111	400.057	400.000	0.093
	MIN	400.000	399.240	0.400	400.000	399.650	0.210	400.000	399.881	0.062	400.000	399.946	0.018	400.000	399.964	0.000
500	MAX	500.400	499.520	1.280	500.155	499.770	0.540	500.097	499.932	0.228	500.063	499.980	0.123	500.063	500.000	0.103
	MIN	500.000	499.120	0.480	500.000	499.615	0.230	500.000	499.869	0.068	500.000	499.940	0.020	500.000	499.960	0.000

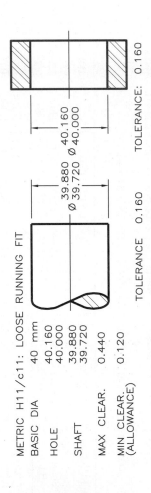

METRIC H11/c11: LOOSE RUNNING FIT

BASIC DIA	40 mm	
HOLE	40.160	
	40.000	
SHAFT	39.880	
	39.720	
MAX CLEAR.	0.440	
MIN CLEAR. (ALLOWANCE)	0.120	

Ø 39.880   Ø 40.160
Ø 39.720   Ø 40.000

TOLERANCE 0.160   TOLERANCE: 0.160

APPENDIX 32 • Preferred Hole Basis Transition and Interference Fits—Cylindrical Fits (ANSI B4.2)

Dimensions are in mm.

BASIC SIZE		LOCATIONAL TRANSN.			LOCATIONAL TRANSN.			LOCATIONAL INTERF.			MEDIUM DRIVE			FORCE		
		Hole H7	Shaft k6	Fit	Hole H7	Shaft n6	Fit	Hole H7	Shaft p6	Fit	Hole H7	Shaft s6	Fit	Hole H7	Shaft u6	Fit
1	MAX	1.010	1.006	0.010	1.010	1.010	0.006	1.010	1.012	0.004	1.010	1.020	−0.004	1.010	1.024	−0.008
	MIN	1.000	1.000	−0.006	1.000	1.004	−0.010	1.000	1.006	−0.012	1.000	1.014	−0.020	1.000	1.018	−0.024
1.2	MAX	1.210	1.206	0.010	1.210	1.210	0.006	1.210	1.212	0.004	1.210	1.220	−0.004	1.210	1.224	−0.008
	MIN	1.200	1.200	−0.006	1.200	1.204	−0.010	1.200	1.206	−0.012	1.200	1.214	−0.020	1.200	1.218	−0.024
1.6	MAX	1.610	1.606	0.010	1.610	1.610	0.006	1.610	1.612	0.004	1.610	1.620	−0.004	1.610	1.624	−0.008
	MIN	1.600	1.600	−0.006	1.600	1.604	−0.010	1.600	1.606	−0.012	1.600	1.614	−0.020	1.600	1.618	−0.024
2	MAX	2.010	2.006	0.010	2.010	2.010	0.006	2.010	2.010	0.004	2.010	2.020	−0.004	2.010	2.024	−0.008
	MIN	2.000	2.000	−0.006	2.000	2.004	−0.010	2.000	2.006	−0.012	2.000	2.014	−0.020	2.000	2.018	−0.024
2.5	MAX	2.510	2.506	0.010	2.510	2.510	0.006	2.510	2.512	0.004	2.510	2.520	−0.004	2.510	2.524	−0.008
	MIN	2.500	2.500	−0.006	2.500	2.504	−0.010	2.500	2.506	−0.012	2.500	2.514	−0.020	2.500	2.518	−0.024
3	MAX	3.010	3.006	0.010	3.010	3.010	0.006	3.010	3.012	0.004	3.010	3.020	−0.004	3.010	3.024	−0.008
	MIN	3.000	3.000	−0.006	3.000	3.004	−0.010	3.000	3.006	−0.012	3.000	3.014	−0.020	3.000	3.018	−0.024
4	MAX	4.012	4.009	0.011	4.012	4.016	0.004	4.012	4.020	0.000	4.012	4.027	−0.007	4.012	4.031	−0.011
	MIN	4.000	4.001	−0.009	4.000	4.008	−0.016	4.000	4.012	−0.020	4.000	4.019	−0.027	4.000	4.023	−0.031
5	MAX	5.012	5.009	0.011	5.012	5.016	0.004	5.012	5.020	0.000	5.012	5.027	−0.007	5.012	5.031	−0.011
	MIN	5.000	5.001	−0.009	5.000	5.008	−0.016	5.000	5.012	−0.020	5.000	5.019	−0.027	5.000	5.023	−0.031
6	MAX	6.012	6.009	0.011	6.012	6.016	0.004	6.012	6.020	0.000	6.012	6.027	−0.007	6.012	6.031	−0.011
	MIN	6.000	6.001	−0.009	6.000	6.008	−0.016	6.000	6.012	−0.020	6.000	6.019	−0.027	6.000	6.023	−0.031
8	MAX	8.015	8.010	0.014	8.015	8.019	0.005	8.015	8.024	0.000	8.015	8.032	−0.008	8.015	8.037	−0.013
	MIN	8.000	8.001	−0.010	8.000	8.010	−0.019	8.000	8.015	−0.024	8.000	8.023	−0.032	8.000	8.028	−0.037
10	MAX	10.015	10.010	0.014	10.015	10.019	0.005	10.015	10.024	0.000	10.015	10.032	−0.008	10.015	10.037	−0.013
	MIN	10.000	10.001	−0.010	10.000	10.010	−0.019	10.000	10.015	−0.024	10.000	10.023	−0.032	10.000	10.028	−0.037
12	MAX	12.018	12.012	0.017	12.018	12.023	0.006	12.018	12.029	0.000	12.018	12.039	−0.010	12.018	12.044	−0.015
	MIN	12.000	12.001	−0.012	12.000	12.012	−0.023	12.000	12.018	−0.029	12.000	12.028	−0.039	12.000	12.033	−0.044
16	MAX	16.018	16.012	0.017	16.018	16.023	0.006	16.018	16.029	0.000	16.018	16.039	−0.010	16.018	16.044	−0.015
	MIN	16.000	16.001	−0.012	16.000	16.012	−0.023	16.000	16.018	−0.029	16.000	16.028	−0.039	16.000	16.033	−0.044
20	MAX	20.021	20.015	0.019	20.021	20.028	0.006	20.021	20.035	−0.001	20.021	20.048	−0.014	20.021	20.054	−0.020
	MIN	20.000	20.002	−0.015	20.000	20.015	−0.028	20.000	20.022	−0.035	20.000	20.035	−0.048	20.000	20.041	−0.054
25	MAX	25.021	25.015	0.019	25.021	25.028	0.006	25.021	25.035	−0.001	25.021	25.048	−0.014	25.021	25.061	−0.027
	MIN	25.000	25.002	−0.015	25.000	25.015	−0.028	25.000	25.022	−0.035	25.000	25.035	−0.048	25.000	25.048	−0.061
30	MAX	30.021	30.015	0.019	30.021	30.028	0.006	30.021	30.035	−0.001	30.021	30.048	−0.014	30.021	30.061	−0.027
	MIN	30.000	30.002	−0.015	30.000	30.015	−0.028	30.000	30.022	−0.035	30.000	30.035	−0.048	30.000	30.048	−0.061

Source: American National Standard Preferred Metric Limit and Fits, ANSI B4.2—1978.

APPENDIX 32 • Preferred Hole Basis Transition and Interference Fits—Cylindrical Fits (ANSI B4.2) (continued)

Dimensions are in mm.

BASIC SIZE		LOCATIONAL TRANSN.			LOCATIONAL TRANSN.			LOCATIONAL INTERF.			MEDIUM DRIVE			FORCE		
		Hole H7	Shaft k6	Fit	Hole H7	Shaft n6	Fit	Hole H7	Shaft p6	Fit	Hole H7	Shaft s6	Fit	Hole H7	Shaft u6	Fit
40	MAX	40.025	40.018	0.023	40.025	40.033	0.008	40.025	40.042	-0.001	40.025	40.059	-0.018	40.025	40.076	-0.035
	MIN	40.000	40.002	-0.018	40.000	40.017	-0.033	40.000	40.026	-0.042	40.000	40.043	-0.059	40.000	40.060	-0.076
50	MAX	50.025	50.018	0.023	50.025	50.033	0.008	50.025	50.042	-0.001	50.025	50.059	-0.018	50.025	50.086	-0.045
	MIN	50.000	50.002	-0.018	50.000	50.017	-0.033	50.000	50.026	-0.042	50.000	50.043	-0.059	50.000	50.070	-0.086
60	MAX	60.030	60.021	0.028	60.030	60.039	0.010	60.030	60.051	-0.002	60.030	60.072	-0.023	60.030	60.106	-0.057
	MIN	60.000	60.002	-0.021	60.000	60.020	-0.039	60.000	60.032	-0.051	60.000	60.053	-0.072	60.000	60.087	-0.106
80	MAX	80.030	80.021	0.028	80.030	80.039	0.010	80.030	80.051	-0.002	80.030	80.078	-0.029	80.030	80.121	-0.072
	MIN	80.000	80.002	-0.021	80.000	80.020	-0.039	80.000	80.032	-0.051	80.000	80.059	-0.078	80.000	80.102	-0.121
100	MAX	100.035	100.025	0.032	100.035	100.045	0.012	100.035	100.059	-0.002	100.035	100.093	-0.036	100.035	100.146	-0.089
	MIN	100.000	100.003	-0.025	100.000	100.023	-0.045	100.000	100.037	-0.059	100.000	100.071	-0.093	100.000	100.124	-0.146
120	MAX	120.035	120.025	0.032	120.035	120.045	0.012	120.035	120.059	-0.002	120.035	120.101	-0.044	120.035	120.166	-0.109
	MIN	120.000	120.003	-0.025	120.000	120.023	-0.045	120.000	120.037	-0.059	120.000	120.079	-0.101	120.000	120.144	-0.166
160	MAX	160.040	160.028	0.037	160.040	160.052	0.013	160.040	160.068	-0.003	160.040	160.125	-0.060	160.040	160.215	-0.150
	MIN	160.000	160.003	-0.028	160.000	160.027	-0.052	160.000	160.043	-0.068	160.000	160.100	-0.125	160.000	160.190	-0.215
200	MAX	200.046	200.033	0.042	200.046	200.060	0.015	200.046	200.079	-0.004	200.046	200.151	-0.076	200.046	200.265	-0.190
	MIN	200.000	200.004	-0.033	200.000	200.031	-0.060	200.000	200.050	-0.079	200.000	200.122	-0.151	200.000	200.236	-0.265
250	MAX	250.046	250.033	0.042	250.046	250.060	0.015	250.046	250.079	-0.004	250.046	250.169	-0.094	250.046	250.313	-0.238
	MIN	250.000	250.004	-0.033	250.000	250.031	-0.060	250.000	250.050	-0.079	250.000	250.140	-0.169	250.000	250.284	-0.313
300	MAX	300.052	300.036	0.048	300.052	300.066	0.018	300.052	300.088	-0.004	300.052	300.202	-0.118	300.052	300.382	-0.298
	MIN	300.000	300.004	-0.036	300.000	300.034	-0.066	300.000	300.056	-0.088	300.000	300.170	-0.202	300.000	300.350	-0.382
400	MAX	400.057	400.040	0.053	400.057	400.073	0.020	400.057	400.098	-0.005	400.057	400.244	-0.151	400.057	400.471	-0.378
	MIN	400.000	400.004	-0.040	400.000	400.037	-0.073	400.000	400.062	-0.098	400.000	400.208	-0.244	400.000	400.435	-0.471
500	MAX	500.063	500.045	0.058	500.063	500.080	0.023	500.063	500.108	-0.005	500.063	500.292	-0.189	500.063	500.580	-0.477
	MIN	500.000	500.005	-0.045	500.000	500.040	-0.080	500.000	500.068	-0.108	500.000	500.252	-0.292	500.000	500.540	-0.580

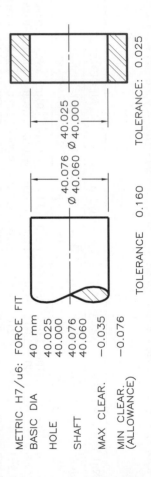

METRIC H7/u6: FORCE FIT

BASIC DIA   40 mm
HOLE   40.025
               40.000
SHAFT   40.076
               40.060
MAX CLEAR.   -0.035
MIN CLEAR.   -0.076
(ALLOWANCE)

TOLERANCE 0.160

Ø 40.076   Ø 40.060

40.025   40.000

TOLERANCE: 0.025

APPENDIX 33 • Preferred Shaft Basis Clearance Fits—Cylindrical Fits (ANSI B4.2)

Dimensions are in mm.

BASIC SIZE		LOOSE RUNNING			FREE RUNNING			CLOSE RUNNING			SLIDING			LOCATIONAL CLEARANCE		
		Hole C11	Shaft h11	Fit	Hole D9	Shaft h9	Fit	Hole F8	Shaft h7	Fit	Hole G7	Shaft h6	Fit	Hole H7	Shaft h6	Fit
1	MAX	1.120	1.000	0.180	1.045	1.000	0.070	1.020	1.000	0.030	1.012	1.000	0.018	1.010	1.000	0.016
	MIN	1.060	0.940	0.060	1.020	0.975	0.020	1.006	0.990	0.006	1.002	0.994	0.002	1.000	0.994	0.000
1.2	MAX	1.320	1.200	0.180	1.245	1.200	0.070	1.220	1.200	0.030	1.212	1.200	0.018	1.210	1.200	0.016
	MIN	1.260	1.140	0.060	1.220	1.175	0.020	1.206	1.190	0.006	1.202	1.194	0.002	1.200	1.194	0.000
1.6	MAX	1.720	1.600	0.180	1.656	1.600	0.070	1.620	1.600	0.030	1.612	1.600	0.018	1.610	1.600	0.016
	MIN	1.660	1.540	0.060	1.620	1.575	0.020	1.606	1.590	0.006	1.602	1.595	0.002	1.600	1.594	0.000
2	MAX	2.120	2.000	0.180	2.045	2.000	0.070	2.020	2.000	0.030	2.012	2.000	0.018	2.010	2.000	0.016
	MIN	2.060	1.940	0.060	2.020	1.975	0.020	2.006	1.990	0.006	2.002	1.994	0.002	2.000	1.994	0.000
2.5	MAX	2.620	2.500	0.180	2.545	2.500	0.070	2.520	2.500	0.030	2.512	2.500	0.018	2.510	2.500	0.016
	MIN	2.560	2.440	0.060	2.520	2.475	0.020	2.506	2.490	0.006	2.502	2.494	0.002	2.500	2.494	0.000
3	MAX	3.120	3.000	0.180	3.045	3.000	0.070	3.020	3.000	0.030	3.012	3.000	0.018	3.010	3.000	0.016
	MIN	3.060	2.940	0.060	3.020	2.975	0.020	3.006	2.990	0.006	3.002	2.994	0.002	3.000	2.994	0.000
4	MAX	4.145	4.000	0.220	4.060	4.000	0.090	4.028	4.000	0.040	4.016	4.000	0.024	4.012	4.000	0.020
	MIN	4.070	3.925	0.070	4.030	3.970	0.030	4.010	3.988	0.010	4.004	3.992	0.004	4.000	3.992	0.000
5	MAX	5.145	5.000	0.220	5.060	5.000	0.090	5.028	5.000	0.040	5.016	5.000	0.024	5.012	5.000	0.020
	MIN	5.070	4.925	0.070	5.030	4.970	0.030	5.010	4.988	0.010	5.004	4.992	0.004	5.000	4.992	0.000
6	MAX	6.145	6.000	0.220	6.060	6.000	0.090	6.028	6.000	0.040	6.016	6.000	0.024	6.012	6.000	0.020
	MIN	6.070	5.925	0.070	6.030	5.970	0.030	6.010	5.988	0.010	6.004	5.992	0.004	6.000	5.992	0.000
8	MAX	8.170	8.000	0.260	8.076	8.000	0.112	8.035	8.000	0.050	8.020	8.000	0.029	8.015	8.000	0.024
	MIN	8.080	7.910	0.080	8.040	7.964	0.040	8.013	7.985	0.013	8.005	7.991	0.005	8.000	7.991	0.000
10	MAX	10.170	10.000	0.260	10.076	10.000	0.112	10.035	10.000	0.050	10.020	10.000	0.029	10.015	10.000	0.024
	MIN	10.080	9.910	0.080	10.040	9.964	0.040	10.013	9.985	0.013	10.005	9.991	0.005	10.000	9.991	0.000
12	MAX	12.205	12.000	0.315	12.093	12.000	0.136	12.043	12.000	0.061	12.024	12.000	0.035	12.018	12.000	0.029
	MIN	12.095	11.890	0.095	12.050	11.957	0.050	12.016	11.982	0.016	12.006	11.989	0.006	12.000	11.989	0.000
16	MAX	16.205	16.000	0.315	16.093	16.000	0.136	16.043	16.000	0.061	16.024	16.000	0.035	16.018	16.000	0.029
	MIN	16.095	15.890	0.095	16.050	15.957	0.050	16.016	15.982	0.016	16.006	15.989	0.006	16.000	15.989	0.000
20	MAX	20.240	20.000	0.370	20.117	20.000	0.169	20.053	20.000	0.074	20.028	20.000	0.041	20.021	20.000	0.034
	MIN	20.110	19.870	0.110	20.065	19.948	0.065	20.020	19.979	0.020	20.007	19.987	0.007	20.000	19.987	0.000
25	MAX	25.240	25.000	0.370	25.117	25.000	0.169	25.053	25.000	0.074	25.028	25.000	0.041	25.021	25.000	0.034
	MIN	25.110	24.870	0.110	25.065	24.948	0.065	25.020	24.979	0.020	25.007	24.987	0.007	25.000	24.987	0.000
30	MAX	30.240	30.000	0.370	30.117	30.000	0.169	30.053	30.000	0.074	30.028	30.000	0.041	30.021	30.000	0.034
	MIN	30.110	29.870	0.110	30.065	29.948	0.065	30.020	29.979	0.020	30.007	29.987	0.007	30.000	29.987	0.000

Source: American National Standard Preferred Metric Limits and Fits, ANSI B4.2—1978.

BASIC SIZE		LOOSE RUNNING			FREE RUNNING			CLOSE RUNNING			SLIDING			LOCATIONAL CLEARANCE		
		Hole C11	Shaft h11	Fit	Hole D9	Shaft h9	Fit	Hole F8	Shaft h7	Fit	Hole G7	Shaft h6	Fit	Hole H7	Shaft h6	Fit
40	MAX	40.280	40.000	0.440	40.142	40.000	0.204	40.064	40.000	0.089	40.034	40.000	0.050	40.025	40.000	0.041
	MIN	40.120	39.840	0.120	40.080	39.938	0.080	40.025	39.975	0.025	40.009	39.984	0.009	40.000	39.984	0.000
50	MAX	50.290	50.000	0.450	50.142	50.000	0.204	50.064	50.000	0.089	50.034	50.000	0.050	50.025	50.000	0.041
	MIN	50.130	49.840	0.130	50.080	49.938	0.080	50.025	49.975	0.025	50.009	49.984	0.009	50.000	49.984	0.000
60	MAX	60.330	60.000	0.520	60.174	60.000	0.248	60.076	60.000	0.106	60.040	60.000	0.059	60.030	60.000	0.049
	MIN	60.140	59.810	0.140	60.100	59.926	0.100	60.030	59.970	0.030	60.010	59.981	0.010	60.000	59.981	0.000
80	MAX	80.340	80.000	0.530	80.174	80.000	0.248	80.076	80.000	0.106	80.040	80.000	0.059	80.030	80.000	0.049
	MIN	80.150	79.810	0.150	80.100	79.926	0.100	80.030	79.970	0.030	80.010	79.981	0.010	80.000	79.981	0.000
100	MAX	100.390	100.000	0.610	100.207	100.000	0.294	100.090	100.000	0.125	100.047	100.000	0.069	100.035	100.000	0.057
	MIN	100.170	99.780	0.170	100.120	99.913	0.120	100.036	99.965	0.036	100.012	99.979	0.012	100.000	99.979	0.000
120	MAX	120.400	120.000	0.620	120.207	120.000	0.294	120.090	120.000	0.125	120.047	120.000	0.069	120.035	120.000	0.057
	MIN	120.180	119.780	0.180	120.120	119.913	0.120	120.036	119.965	0.036	120.012	119.978	0.012	120.000	119.978	0.000
160	MAX	160.460	160.000	0.710	160.245	160.000	0.345	160.106	160.000	0.146	160.054	160.000	0.079	160.040	160.000	0.065
	MIN	160.210	159.750	0.210	160.145	159.900	0.145	160.043	159.960	0.043	160.014	159.975	0.014	160.000	159.975	0.000
200	MAX	200.530	200.000	0.820	200.285	200.000	0.400	200.122	200.000	0.168	200.061	200.000	0.090	200.046	200.000	0.075
	MIN	200.240	199.710	0.240	200.170	199.885	0.170	200.050	199.954	0.050	200.015	199.971	0.015	200.000	199.971	0.000
250	MAX	250.570	250.000	0.860	250.285	250.000	0.400	250.122	250.000	0.168	250.061	250.000	0.090	250.046	250.000	0.075
	MIN	250.280	249.710	0.280	250.170	249.885	0.170	250.050	249.954	0.050	250.015	249.971	0.015	250.000	249.971	0.000
300	MAX	300.650	300.000	0.970	300.320	300.000	0.450	300.137	300.000	0.189	300.069	300.000	0.101	300.052	300.000	0.084
	MIN	300.330	299.680	0.330	300.190	299.870	0.190	300.056	299.948	0.056	300.017	299.968	0.017	300.000	299.968	0.000
400	MAX	400.760	400.000	1.120	400.350	400.000	0.490	400.151	400.000	0.208	400.075	400.000	0.111	400.057	400.000	0.983
	MIN	400.400	399.640	0.400	400.210	399.860	0.210	400.062	399.943	0.062	400.018	399.964	0.018	400.000	399.964	0.000
500	MAX	500.880	500.000	1.280	500.385	500.000	0.540	500.165	500.000	0.228	500.083	500.000	0.123	500.063	500.000	0.103
	MIN	500.480	499.600	0.480	500.230	499.845	0.230	500.068	499.937	0.068	500.020	499.960	0.020	500.000	499.960	0.000

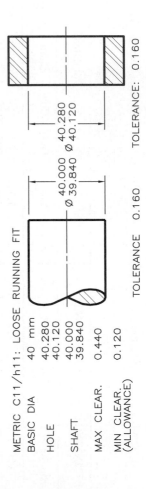

METRIC C11/h11: LOOSE RUNNING FIT

BASIC DIA	40 mm
HOLE	40.280 / 40.120
SHAFT	40.000 / 39.840
MAX CLEAR.	0.440
MIN CLEAR. (ALLOWANCE)	0.120

Ø 40.000 Ø 40.280
Ø 39.840 Ø 40.120
TOLERANCE 0.160   TOLERANCE 0.160   TOLERANCE: 0.160

APPENDIX 34 • Preferred Shaft Basis Transition and Interference Fits—Cylindrical Fits (ANSI B4.2)

Dimensions are in mm.

BASIC SIZE		LOCATIONAL TRANSN.			LOCATIONAL TRANSN.			LOCATIONAL INTERF.			MEDIUM DRIVE			FORCE		
		Hole K7	Shaft h6	Fit	Hole N7	Shaft h6	Fit	Hole P7	Shaft h6	Fit	Hole S7	Shaft h6	Fit	Hole U7	Shaft h6	Fit
1	MAX	1.000	1.000	0.006	0.996	1.000	0.002	0.994	1.000	0.000	0.986	1.000	−0.008	0.982	1.000	−0.012
	MIN	0.990	0.994	−0.010	0.986	0.994	−0.014	0.984	0.994	−0.016	0.976	0.994	−0.024	0.972	0.994	−0.028
1.2	MAX	1.200	1.200	0.006	1.196	1.200	0.002	1.194	1.200	0.000	1.186	1.200	−0.008	1.182	1.200	−0.012
	MIN	1.190	1.194	−0.010	1.186	1.194	−0.014	1.184	1.194	−0.016	1.176	1.194	−0.024	1.172	1.194	−0.028
1.6	MAX	1.600	1.600	0.006	1.596	1.600	0.002	1.594	1.600	0.000	1.586	1.600	−0.008	1.582	1.600	−0.012
	MIN	1.590	1.594	−0.010	1.586	1.594	−0.014	1.584	1.594	−0.016	1.576	1.594	−0.024	1.572	1.594	−0.028
2	MAX	2.000	2.000	0.006	1.996	2.000	0.002	1.994	2.000	0.008	1.986	2.000	−0.008	1.982	2.000	−0.012
	MIN	1.990	1.994	−0.010	1.986	1.994	−0.014	1.984	1.994	−0.016	1.976	1.994	−0.024	1.972	1.994	−0.028
2.5	MAX	2.500	2.500	0.006	2.496	2.500	0.002	2.494	2.500	0.000	2.486	2.500	−0.008	2.482	2.500	−0.012
	MIN	2.490	2.494	−0.010	2.486	2.494	−0.014	2.484	2.494	−0.016	2.476	2.494	−0.024	2.472	2.494	−0.028
3	MAX	3.000	3.000	0.006	2.996	3.000	0.002	2.994	3.000	0.000	2.986	3.000	−0.008	2.982	3.000	−0.012
	MIN	2.990	2.994	−.010	2.986	2.994	−0.014	2.984	2.994	−0.016	2.976	2.994	−0.024	2.972	2.994	−0.028
4	MAX	4.003	4.000	0.011	3.996	4.000	0.004	3.992	4.000	0.000	3.985	4.000	−0.007	3.981	4.000	−0.011
	MIN	3.991	3.992	−0.009	3.984	3.992	−0.016	3.980	3.992	−0.020	3.973	3.992	−0.027	3.969	3.992	−0.031
5	MAX	5.003	5.000	0.011	4.996	5.000	0.004	4.992	5.000	0.000	4.985	5.000	−0.007	4.981	5.000	−0.011
	MIN	4.991	4.992	−0.009	4.984	4.992	−0.016	4.980	4.992	−0.020	4.973	4.992	−0.027	4.969	4.992	−0.031
6	MAX	6.003	6.000	0.011	5.996	6.000	0.004	5.992	6.000	0.000	5.985	6.000	−0.007	5.981	6.000	−0.011
	MIN	5.991	5.992	−0.009	5.984	5.992	−0.016	5.980	5.992	−0.020	5.973	5.992	−0.027	5.969	5.992	−0.031
8	MAX	8.005	8.000	0.014	7.986	8.000	0.005	7.991	8.000	0.000	7.983	8.000	−0.008	7.978	8.000	−0.013
	MIN	7.990	7.991	−0.010	7.981	7.991	−0.019	7.976	7.991	−0.024	7.968	7.991	−0.032	7.963	7.991	−0.037
10	MAX	10.005	10.000	0.014	9.996	10.000	0.005	9.991	10.000	0.0000	9.983	10.000	−0.008	9.978	10.000	−0.013
	MIN	9.990	9.991	−0.010	9.981	9.991	−0.019	9.976	9.991	−0.024	9.968	9.991	−0.032	9.963	9.991	−0.037
12	MAX	12.006	12.000	0.017	11.995	12.000	0.006	11.989	12.000	0.000	11.979	12.000	−0.010	11.974	12.000	−0.015
	MIN	11.988	11.989	−0.012	11.977	11.989	−0.023	11.971	11.989	−0.029	11.961	11.989	−0.039	11.956	11.989	−0.044
16	MAX	16.006	16.000	0.017	15.995	16.000	0.006	15.989	16.000	0.000	15.979	16.000	−0.010	15.974	16.000	−0.015
	MIN	15.988	15.989	−0.012	15.977	15.989	−0.023	15.971	15.989	−0.029	15.961	15.989	−0.039	15.956	15.989	−0.044
20	MAX	20.006	20.000	0.019	19.993	20.000	0.006	19.986	20.000	−0.001	19.973	20.000	−0.014	19.967	20.000	−0.020
	MIN	19.985	19.987	−0.015	19.972	19.987	−0.028	19.965	19.987	−0.035	19.952	19.987	−0.048	19.946	19.987	−0.054
25	MAX	25.006	25.000	0.019	24.993	25.000	0.006	24.986	25.000	−0.001	24.973	25.000	−0.014	24.960	25.000	−0.027
	MIN	24.985	24.987	−0.015	24.972	24.987	−0.028	24.965	24.987	−0.035	24.952	24.987	−0.048	24.939	24.987	−0.061
30	MAX	30.006	30.000	0.019	29.993	30.000	0.006	29.986	30.000	−0.001	29.973	30.000	−0.014	29.960	30.000	−0.027
	MIN	29.985	29.987	−0.015	29.972	29.987	−0.028	29.965	29.987	−0.035	29.952	29.987	−0.048	29.939	29.987	−0.061

Source: American National Standard Preferred Metric Limits and Fits, ANSI B4.2—1978.

APPENDIX 34 • Preferred Shaft Basis Transition and Interference Fits—Cylindrical Fits (ANSI B4.2) (continued)

Dimensions are in mm.

BASIC SIZE		LOCATIONAL TRANSN. Hole K7	Shaft h6	Fit	LOCATIONAL TRANSN. Hole N7	Shaft h6	Fit	LOCATIONAL INTERF. Hole P7	Shaft h6	Fit	MEDIUM DRIVE Hole S7	Shaft h6	Fit	FORCE Hole U7	Shaft h6	Fit
40	MAX	40.007	40.000	0.023	39.992	40.000	0.008	39.983	40.000	-0.001	39.966	40.000	-0.018	39.949	40.000	-0.035
	MIN	39.982	39.984	-0.018	39.967	39.984	-0.033	39.958	39.984	-0.042	39.941	39.984	-0.059	39.924	39.984	-0.076
50	MAX	50.007	50.000	0.023	49.992	50.000	0.008	49.983	50.000	-0.001	49.966	50.000	-0.018	49.939	50.000	-0.045
	MIN	49.982	49.984	-0.018	49.967	49.984	-0.033	49.958	49.984	-0.042	49.941	49.984	-0.059	49.914	49.984	-0.086
60	MAX	60.009	60.000	0.028	59.991	60.000	0.010	59.979	60.000	-0.002	59.958	60.000	-0.023	59.924	60.000	-0.057
	MIN	59.979	59.981	-0.021	59.961	59.981	-0.039	59.949	59.981	-0.051	59.928	59.981	-0.072	59.894	59.981	-0.106
80	MAX	80.009	80.000	0.028	79.991	80.000	0.010	79.979	80.000	-0.002	79.952	80.000	-0.029	79.909	80.000	-0.072
	MIN	79.979	79.981	-0.021	79.961	79.981	-0.039	79.949	79.981	-0.051	79.922	79.981	-0.078	79.879	79.981	-0.121
100	MAX	100.010	100.000	0.032	99.990	100.000	0.012	99.976	100.000	-0.002	99.942	100.000	-0.036	99.889	100.000	-0.089
	MIN	99.975	99.978	-0.025	99.955	99.978	-0.045	99.941	99.978	-0.059	99.907	99.978	-0.093	99.854	99.978	-0.146
120	MAX	120.010	120.000	0.032	119.990	120.000	0.012	119.976	120.000	-0.002	119.934	120.000	-0.044	119.869	120.000	-0.109
	MIN	119.975	119.978	-0.025	119.955	119.978	-0.045	119.941	119.978	-0.059	119.899	119.978	-0.101	119.834	119.978	-0.166
160	MAX	160.012	160.000	0.037	159.988	160.000	0.013	159.972	160.000	-0.003	159.915	160.000	-0.060	159.825	160.000	-0.150
	MIN	159.972	159.975	-0.028	159.948	159.975	-0.052	159.932	159.975	-0.068	159.875	159.975	-0.125	159.785	159.975	-0.215
200	MAX	200.013	200.000	0.042	199.986	200.000	0.015	199.967	200.000	-0.004	199.895	200.000	-0.076	199.781	200.000	-0.190
	MIN	199.967	199.971	-0.033	199.940	199.971	-0.060	199.921	199.971	-0.079	199.849	199.971	-0.151	199.735	199.971	-0.265
250	MAX	250.013	250.000	0.042	249.986	250.000	0.015	249.967	250.000	-0.004	249.877	250.000	-0.094	249.733	250.000	-0.238
	MIN	249.967	249.971	-0.033	249.940	249.971	-0.060	249.921	249.971	-0.079	249.831	249.971	-0.169	249.687	249.971	-0.313
300	MAX	300.016	300.000	0.048	299.986	300.000	0.018	299.964	300.000	-0.004	299.850	300.000	-0.188	299.670	300.000	-0.298
	MIN	299.964	299.968	-0.036	299.934	299.968	-0.066	299.912	299.968	-0.088	299.798	299.968	-0.202	299.618	299.968	-0.382
400	MAX	400.017	400.000	0.053	399.984	400.000	0.020	399.959	400.000	-0.005	399.813	400.000	-0.151	399.586	400.000	-0.378
	MIN	399.960	399.964	-0.040	399.927	399.964	-0.073	399.902	399.964	-0.08	399.756	399.964	-0.244	399.529	399.964	-0.471
500	MAX	500.018	500.000	0.058	499.983	500.000	0.023	499.955	500.000	-0.005	499.771	500.000	-0.189	499.483	500.000	-0.477
	MIN	499.955	499.960	-0.045	499.920	499.960	-0.080	499.892	499.960	-0.1808	499.708	499.960	-0.292	499.420	499.960	-0.580

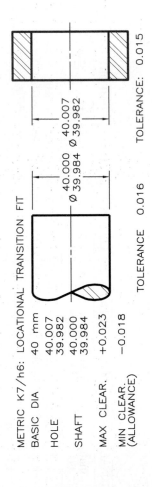

METRIC K7/h6: LOCATIONAL TRANSITION FIT

BASIC DIA	40 mm
HOLE	40.007 / 39.982
SHAFT	40.000 / 39.984
MAX CLEAR.	+0.023
MIN CLEAR. (ALLOWANCE)	-0.018

Ø40.000 Ø39.984

40.007 / 39.982   Ø39.984

TOLERANCE 0.016   TOLERANCE 0.015

Basic Size		C11	D9	F8	G7	H7	H8	H9	H11	K7	N7	P7	S7	U7
OVER	0	+0.120	+0.045	+0.020	+0.012	+0.010	+0.014	+0.025	+0.060	0.000	−0.004	−0.006	−0.014	−0.018
TO	3	+0.060	+0.020	+0.006	+0.002	0.000	0.000	0.000	0.000	−0.010	−0.014	−0.016	−0.024	−0.028
OVER	3	+0.145	+0.060	+0.028	+0.016	+0.012	+0.018	+0.030	+0.075	+0.003	−0.004	−0.008	−0.015	−0.019
TO	6	+0.070	+0.030	+0.010	+0.004	0.000	0.000	0.000	0.000	−0.009	−0.016	−0.020	−0.027	−0.031
OVER	6	+0.170	+0.076	+0.035	+0.020	+0.015	+0.022	+0.036	+0.090	+0.005	−0.004	−0.009	−0.017	−0.022
TO	10	+0.080	+0.040	+0.013	+0.005	0.000	0.000	0.000	0.000	−0.010	−0.019	−0.024	−0.032	−0.037
OVER	10	+0.205	+0.093	+0.043	+0.024	+0.018	+0.027	+0.043	+0.110	+0.006	−0.005	−0.011	−0.021	−0.026
TO	14	+0.095	+0.050	+0.016	+0.006	0.000	0.000	0.000	0.000	−0.012	−0.023	−0.029	−0.039	−0.044
OVER	14	+0.205	+0.093	+0.043	+0.024	+0.018	+0.027	+0.043	+0.110	+0.006	−0.005	−0.011	−0.021	−0.026
TO	18	+0.095	+0.050	+0.016	+0.006	0.000	0.000	0.000	0.000	−0.012	−0.023	−0.029	−0.039	−0.044
OVER	18	+0.240	+0.117	+0.053	+0.028	+0.021	+0.033	+0.052	+0.130	+0.006	−0.007	−0.014	−0.027	−0.033
TO	24	+0.110	+0.065	+0.020	+0.007	0.000	0.000	0.000	0.000	−0.015	−0.028	−0.035	−0.048	−0.054
OVER	24	+0.240	+0.117	+0.053	+0.028	+0.021	+0.033	+0.052	+0.130	+0.006	−0.007	−0.014	−0.027	−0.040
TO	30	+0.110	+0.065	+0.020	+0.007	0.000	0.000	0.000	0.000	−0.015	−0.028	−0.035	−0.048	−0.061
OVER	30	+0.280	+0.142	+0.064	+0.034	+0.025	+0.039	+0.062	+0.160	+0.007	−0.008	−0.017	−0.034	−0.051
TO	40	+0.120	+0.080	+0.025	+0.009	0.000	0.000	0.000	0.000	−0.018	−0.033	−0.042	−0.059	−0.076
OVER	40	+0.290	+0.142	+0.064	+0.034	+0.025	+0.039	+0.062	+0.160	+0.007	−0.008	−0.017	−0.034	−0.061
TO	50	+0.130	+0.080	+0.025	+0.009	0.000	0.000	0.000	0.000	−0.018	−0.033	−0.042	−0.059	−0.086
OVER	50	+0.330	+0.174	+0.076	+0.040	+0.030	+0.046	+0.074	+0.190	+0.009	−0.009	−0.021	−0.042	−0.076
TO	65	+0.140	+0.100	+0.030	+0.010	0.000	0.000	0.000	0.000	−0.021	−0.039	−0.051	−0.072	−0.106
OVER	65	+0.340	+0.174	+0.076	+0.040	+0.030	+0.046	+0.074	+0.190	+0.009	−0.009	−0.021	−0.048	−0.091
TO	80	+0.150	+0.100	+0.030	+0.010	0.000	0.000	0.000	0.000	−0.021	−0.039	−0.051	−0.078	−0.121
OVER	80	+0.390	+0.207	+0.090	+0.047	+0.035	+0.054	+0.087	+0.220	+0.010	−0.010	−0.024	−0.058	−0.111
TO	100	+0.170	+0.120	+0.036	+0.012	0.000	0.000	0.000	0.000	−0.025	−0.045	−0.059	−0.093	−0.146

Basic Size		C11	D9	F8	G7	H7	H8	H9	H11	K7	N7	P7	S7	U7
OVER	100	+0.400	+0.207	+0.090	+0.047	+0.035	+0.054	+0.087	+0.220	+0.010	−0.010	−0.024	−0.066	−0.131
TO	120	+0.180	+0.120	+0.036	+0.012	0.000	0.000	0.000	0.000	−0.025	−0.045	−0.059	−0.101	−0.166
OVER	120	+0.450	+0.245	+0.106	+0.054	+0.040	+0.063	+0.100	+0.250	+0.012	−0.012	−0.028	−0.077	−0.155
TO	140	+0.200	+0.145	+0.043	+0.014	0.000	0.000	0.000	0.000	−0.028	−0.052	−0.068	−0.117	−0.195
OVER	140	+0.460	+0.245	+0.106	+0.054	+0.040	+0.063	+0.100	+0.250	+0.012	−0.012	−0.028	−0.085	−0.175
TO	160	+0.210	+0.145	+0.043	+0.014	0.000	0.000	0.000	0.000	−0.028	−0.052	−0.068	−0.125	−0.215
OVER	160	+0.480	+0.245	+0.106	+0.054	+0.040	+0.063	+0.100	+0.250	+0.012	−0.012	−0.028	−0.093	−0.195
TO	180	+0.230	+0.145	+0.043	+0.014	0.000	0.000	0.000	0.000	−0.028	−0.052	−0.068	−0.133	−0.235
OVER	180	+0.530	+0.285	+0.122	+0.061	+0.046	+0.072	+0.115	+0.290	+0.013	−0.014	−0.033	−0.105	−0.219
TO	200	+0.240	+0.170	+0.050	+0.015	0.000	0.000	0.000	0.000	−0.033	−0.060	−0.079	−0.151	−0.265
OVER	200	+0.550	+0.285	+0.122	+0.061	+0.046	+0.072	+0.115	+0.290	+0.013	−0.014	−0.033	−0.113	−0.241
TO	225	+0.260	+0.170	+0.050	+0.015	0.000	0.000	0.000	0.000	−0.033	−0.060	−0.079	−0.159	−0.287
OVER	225	+0.570	+0.285	+0.122	+0.061	+0.046	+0.072	+0.115	+0.290	+0.013	−0.014	−0.033	−0.123	−0.267
TO	250	+0.280	+0.170	+0.050	+0.015	0.000	0.000	0.000	0.000	−0.033	−0.060	−0.079	−0.169	−0.313
OVER	250	+0.620	+0.320	+0.137	+0.069	+0.052	+0.081	+0.130	+0.320	+0.016	−0.014	−0.036	−0.138	−0.295
TO	280	+0.300	+0.190	+0.056	+0.017	0.000	0.000	0.000	0.000	−0.036	−0.066	−0.088	−0.190	−0.347
OVER	280	+0.650	+0.320	+0.137	+0.069	+0.052	+0.081	+0.130	+0.320	+0.016	−0.014	−0.036	−0.150	−0.330
TO	315	+0.330	+0.190	+0.056	0.017	0.000	0.000	0.000	0.000	−0.036	−0.066	−0.088	−0.202	−0.382
OVER	315	+0.720	+0.350	+0.151	+0.075	+0.057	+0.089	+0.140	+0.360	+0.017	−0.016	−0.041	−0.169	−0.369
TO	355	+0.360	+0.210	+0.062	+0.018	0.000	0.000	0.000	0.000	−0.040	−0.073	−0.058	−0.226	−0.426
OVER	355	+0.760	+0.350	+0.151	+0.075	+0.057	+0.089	+0.140	+0.360	+0.017	−0.016	−0.041	−0.187	−0.414
TO	400	+0.400	+0.210	+0.062	+0.018	0.000	0.000	0.000	0.000	−0.040	−0.073	−0.058	−0.244	−0.471
OVER	400	+0.840	+0.385	+0.165	+0.083	+0.063	+0.097	+0.155	+0.400	+0.018	−0.017	−0.045	−0.209	−0.467
TO	450	+0.440	+0.230	+0.068	+0.020	0.000	0.000	0.000	0.000	−0.045	−0.080	−0.108	−0.272	−0.530
OVER	450	+0.880	+0.385	+0.165	+0.083	+0.063	+0.097	+0.155	+0.400	+0.018	−0.017	−0.045	−0.229	−0.517
TO	500	+0.480	+0.230	+0.068	+0.020	0.000	0.000	0.000	0.000	−0.045	−0.080	−0.108	−0.292	−0.580

APPENDIX 36 • Shaft Sizes for Nonpreferred Diameters (millimeters)

Basic Size OVER	Basic Size TO	c11	d9	f7	g6	h6	h7	h9	h11	k6	n6	p6	s6	u6
0	3	−0.060 / −0.120	−0.020 / −0.045	−0.006 / −0.016	−0.002 / −0.008	0.000 / −0.006	0.000 / −0.010	0.000 / −0.025	0.000 / −0.060	+0.006 / 0.000	+0.010 / +0.004	+0.012 / +0.006	+0.020 / +0.014	+0.024 / +0.018
3	6	−0.070 / −0.145	−0.030 / −0.060	−0.010 / −0.022	−0.004 / −0.012	0.000 / −0.008	0.000 / −0.012	0.000 / −0.030	0.000 / −0.075	+0.009 / +0.001	+0.016 / +0.008	+0.020 / +0.012	+0.027 / +0.019	+0.031 / +0.023
6	10	−0.080 / −0.170	−0.040 / −0.076	−0.013 / −0.028	−0.005 / −0.014	0.000 / −0.009	0.000 / −0.015	0.000 / −0.036	0.000 / −0.090	+0.010 / +0.001	+0.019 / +0.010	+0.024 / +0.015	+0.032 / +0.023	+0.037 / +0.028
10	14	−0.095 / −0.205	−0.050 / −0.093	−0.016 / −0.034	−0.006 / −0.017	0.000 / −0.011	0.000 / −0.018	0.000 / −0.043	0.000 / −0.110	+0.012 / +0.001	+0.023 / +0.012	+0.029 / +0.018	+0.039 / +0.028	+0.044 / +0.033
14	18	−0.095 / −0.205	−0.050 / −0.093	−0.016 / −0.034	−0.006 / −0.017	0.000 / −0.011	0.000 / −0.018	0.000 / −0.043	0.000 / −0.110	+0.012 / +0.001	+0.023 / +0.012	+0.029 / +0.018	+0.039 / +0.028	+0.044 / +0.033
18	24	−0.110 / −0.240	−0.065 / −0.117	−0.020 / −0.041	−0.007 / −0.020	0.000 / −0.013	0.000 / −0.021	0.000 / −0.052	0.000 / −0.130	+0.015 / +0.002	+0.028 / +0.015	+0.035 / +0.022	+0.048 / +0.035	+0.054 / +0.041
24	30	−0.110 / −0.240	−0.065 / −0.117	−0.020 / −0.041	−0.007 / −0.020	0.000 / −0.013	0.000 / −0.021	0.000 / −0.052	0.000 / −0.130	+0.015 / +0.002	+0.028 / +0.015	+0.035 / +0.022	+0.048 / +0.035	+0.061 / +0.048
30	40	−0.120 / −0.280	−0.080 / −0.142	−0.025 / −0.050	−0.009 / −0.025	0.000 / −0.016	0.000 / −0.025	0.000 / −0.062	0.000 / −0.160	+0.018 / +0.002	+0.033 / +0.017	+0.042 / +0.026	+0.059 / +0.043	+0.076 / +0.060
40	50	−0.130 / −0.290	−0.080 / −0.142	−0.025 / −0.050	−0.009 / −0.025	0.000 / −0.016	0.000 / −0.025	0.000 / −0.062	0.000 / −0.160	+0.018 / +0.002	+0.033 / +0.017	+0.042 / +0.026	+0.059 / +0.043	+0.086 / +0.070
50	65	−0.140 / −0.330	−0.100 / −0.174	−0.030 / −0.060	−0.010 / −0.029	0.000 / −0.019	0.000 / −0.030	0.000 / −0.074	0.000 / −0.190	+0.021 / +0.002	+0.039 / +0.020	+0.051 / +0.032	+0.072 / +0.053	+0.106 / +0.087
65	80	−0.150 / −0.340	−0.100 / −0.174	−0.030 / −0.060	−0.010 / −0.029	0.000 / −0.019	0.000 / −0.030	0.000 / −0.074	0.000 / −0.190	+0.021 / +0.002	+0.039 / +0.020	+0.051 / +0.032	+0.078 / +0.059	+0.121 / +0.102
80	100	−0.170 / −0.390	−0.120 / −0.207	−0.036 / −0.071	−0.012 / −0.034	0.000 / −0.022	0.000 / −0.035	0.000 / −0.087	0.000 / −0.220	+0.025 / +0.003	+0.045 / +0.023	+0.059 / +0.037	+0.093 / +0.071	+0.146 / +0.124

APPENDIX 36 • Shaft Sizes for Nonpreferred Diameters (millimeters) (continued)

Basic Size		c11	d9	f7	g6	h6	h7	h9	h11	k6	n6	p6	s6	u6
OVER	100	−0.180	−0.120	−0.036	−0.012	0.000	0.000	0.000	0.000	+0.025	+0.045	+0.059	+0.101	+0.166
TO	120	−0.400	−0.207	−0.071	−0.034	−0.022	−0.035	−0.087	−0.220	+0.003	+0.023	+0.037	+0.079	+0.144
OVER	120	−0.200	−0.145	−0.043	−0.014	0.000	0.000	0.000	0.000	+0.028	+0.052	+0.068	+0.117	+0.195
TO	140	−0.450	−0.245	−0.083	−0.039	−0.025	−0.040	−0.100	−0.250	+0.003	+0.027	+0.043	+0.092	+0.170
OVER	140	−0.210	−0.145	−0.043	−0.014	0.000	0.000	0.000	0.000	+0.028	+0.052	+0.068	+0.125	+0.215
TO	160	−0.460	−0.245	−0.083	−0.039	−0.025	−0.040	−0.100	−0.250	+0.003	+0.027	+0.043	+0.100	+0.190
OVER	160	−0.230	−0.145	−0.043	−0.014	0.000	0.000	0.000	0.000	+0.028	+0.052	+0.068	+0.133	+0.235
TO	180	−0.480	−0.245	−0.083	−0.039	−0.025	−0.040	−0.100	−0.250	+0.003	+0.027	+0.043	+0.108	+0.210
OVER	180	−0.240	−0.170	−0.050	−0.015	0.000	0.000	0.000	0.000	+0.033	+0.060	+0.079	+0.151	+0.265
TO	200	−0.530	−0.285	−0.096	−0.044	−0.029	−0.046	−0.115	−0.290	+0.004	+0.031	+0.050	+0.122	+0.236
OVER	200	−0.260	−0.170	−0.050	−0.015	0.000	0.000	0.000	0.000	+0.033	+0.060	+0.079	+0.159	+0.287
TO	225	−0.550	−0.285	−0.096	−0.044	−0.029	−0.046	−0.115	−0.290	+0.004	+0.031	+0.050	+0.130	+0.258
OVER	225	−0.280	−0.170	−0.050	−0.015	0.000	0.000	0.000	0.000	+0.033	+0.060	+0.079	+0.169	+0.313
TO	250	−0.570	−0.285	−0.096	−0.044	−0.029	−0.046	−0.115	−0.290	+0.004	+0.031	+0.050	+0.140	+0.284
OVER	250	−0.300	−0.190	−0.056	−0.017	0.000	0.000	0.000	0.000	+0.036	+0.066	+0.088	+0.190	+0.347
TO	280	−0.620	−0.320	−0.108	−0.049	−0.032	−0.052	−0.130	−0.320	+0.004	+0.034	+0.056	+0.158	+0.315
OVER	280	−0.330	−0.190	−0.056	−0.017	0.000	0.000	0.000	0.000	+0.036	+0.066	+0.088	+0.202	+0.382
TO	315	−0.650	−0.320	−0.108	−0.049	−0.032	−0.052	−0.130	−0.320	+0.004	+0.034	+0.056	+0.170	+0.350
OVER	315	−0.360	−0.210	−0.062	−0.018	0.000	0.000	0.000	0.000	+0.040	+0.073	+0.098	+0.226	+0.426
TO	355	−0.720	−0.350	−0.119	−0.054	−0.036	−0.057	−0.140	−0.360	+0.004	+0.037	+0.062	+0.190	+0.390
OVER	355	−0.400	−0.210	−0.062	−0.018	0.000	0.000	0.000	0.000	+0.040	+0.073	+0.098	+0.244	+0.471
TO	400	−0.760	−0.350	−0.119	−0.054	−0.036	−0.057	−0.140	−0.360	+0.004	+0.037	+0.062	+0.208	+0.435
OVER	400	−0.440	−0.230	−0.068	−0.020	0.000	0.000	0.000	0.000	+0.045	+0.080	+0.108	+0.272	+0.530
TO	450	−0.840	−0.385	−0.131	−0.060	−0.040	−0.063	−0.155	−0.400	+0.005	+0.040	+0.068	+0.232	+0.490
OVER	450	−0.480	−0.230	−0.068	−0.020	0.000	0.000	0.000	0.000	+0.045	+0.080	+0.108	+0.292	+0.580
TO	500	−0.880	−0.385	−0.131	−0.060	−0.040	−0.063	−0.155	−0.400	+0.005	+0.040	+0.068	+0.252	+0.540

APPENDIX 37 • Descriptive Geometry LISP Programs (ACAD.LSP)

The following LISP programs were written by Professor Leendert Kersten of the University of Nebraska and are given here with his permission. These programs were introduced in Chapter 15, in which the principles of orthographic projection are covered. These very valuable programs can be duplicated and added as supplements to your AutoCAD software.

```
)
(defun C:PARALLEL ()
  (setvar "aperture" 5)
    (setq sp (getpoint "\nSelect START point of
      parallel line:"))
    (setq ep (getpoint "\nSelect END point of
      parallel line:"))
    (setvar "osmode" 1)
    (setq sl (getpoint "\nSelect 1st point on line
      for parallelism:"))
    (setq el (getpoint "\nSelect 2nd point on line
      for parallelism:"))
    (setvar "osmode" 0)
    (setq pa (angle sl el ))
    (setq la (angle sp ep))
    (setq ll (distance sp ep))
    (setq m -1)
    (setq d 0)
    (if (> pa d) (setq m 1))
    (if (> la d) (setq d 1))
    (if (/= m d) (setq pa (+ pa 3.141593)))
    (setq ep (polar sp pa ll))
    (setvar "cmdecho" 0)
    (command line sp ep "" )
    (restore)
)
(defun C:PERPLINE ()
    (setvar "aperture" 5) (setvar "cmdecho" 0)
    (setq sp (getpoint "\nSelect START point of
      perpendicular line:"))
    (setvar "osmode" 128)
    (setq cc (getpoint sp "\nSelect ANY point on
      line to which perp'lr:"))
    (setq beta (angle sp cc)) (setvar "osmode" 0)
(setq ep
(getpoint "\nSelect END point of desired
perpendicular (for length only): "))
    (setq length (distance sp ep))
    (setq ep (polar sp beta length))
    (command "line sp ep "")
    (restore)
)
(defun C:TRANSFER ()
    (setvar "aperture" 5) (setvar "cmdecho" 0)
    (setvar "osmode" 1)
    (setq aa (getpoint "\nSelect start of transfer
      distance:"))
    (setvar "osmode" 128)
    (setq bb (getpoint aa "\nSelect the reference
      plane:"))
    (setq length (distance aa bb))
    (setvar "osmode" 1)
    (setq cc (getpoint "\nSelect point to be
      projected:"))
    (setvar "osmode" 128)
    (setq dd (getpoint oc "\nSelect other reference
      plane:"))
(setvar "osmode" 0)
    (setq alpha (angle cc dd))
    (setq ep (polar dd alpha length))
    (COMMAND "CIRCLE" EP 0.05)
    (restore)
)
```

```
(defun RESTORE()
    (setvar "aperture" 10)
    (setvar "cmdecho" 1)
    (setvar "osmode" 0)
)
(DEFUN *ERROR* (MSG)
(SETVAR "OSMODE" 0)
(setvar "aperture" 10)
( setvar "cmdecho" 1)
(PRINC "error: ")
(princ msg)
(terpri)
(defun C:COPYDIST ()
    (setvar "aperture"5)
    (setvar "cmdecho" 0)
  (setvar "osmode" 1)
    (setq p1 (getpoint "\nSelect start point of
      line distance to be copied: "))
    (setq p2 (getpoint "\nEnd point?: "))
  (setvar "osmode" 0)
    (setq dist (distance p1 p2))
    (setq p1 (getpoint "\nStart point of new
      distance location:"))
    (setq ang (getangle p1 "\nWhich
      direction?: "))
    (setq p2 (polar p1 ang diet))
  (setvar "osmode" 0)
    (command "circle" p2 0.05)
    (restore)
)
(defun C:BISECT ()
    (setvar "aperture" 5)
    (setvar "osmode" 32)
    (setq sp (getpoint "\nSelect Corner of angle:"))
    (setvar "osmode" 2)
    (setq aa (getpoint "\nSelect first side
      (remember CCW):"))
    (setq alpha (angle sp aa))
    (setq bb (getpoint "\nSelect other side:"))
    (setvar "osmode" 0)
    (setq beta (angle sp bb))
    (setq m (/ (+ alpha beta) 2))
    (if (> alpha beta) (setq ang (+ pi m))
      (setq ang m))
(setq ep
  (getpoint "\nSelect endpoint of bisecting line
    (for length only): "))
    (setq length (distance sp ep))
    (setq ep (polar sp ang length))
    (setvar "cmdecho" 0)
    (command "line sp ep "")
    (restore)
}
```

This graph can be used to determine the individual grades of members of a team and to compute grade averages for those who do extra assignments.

The percent participation of each team member should be determined by the team as a whole (see Section 7.7).

Example: written or oral report grades

Overall team grade: 82

Team members N=5	Contribution C=%	F=CN	Grade (graph)
J. Doe	20%	100	82.0
H. Brown	16%	80	75.8
L. Smith	24%	120	86.0
R. Black	20%	100	82.0
T. Jones	20%	100	82.0
	100%		

Example: quiz or problem sheet grades

Number assigned: 30
Number extra: 6
Total   36

Average grade for total (36): 82

$$F = \frac{\text{No. completed} \times 100}{\text{No. assigned}} = \frac{36 \times 100}{30} = 120$$

Final grade (from graph): 86.0

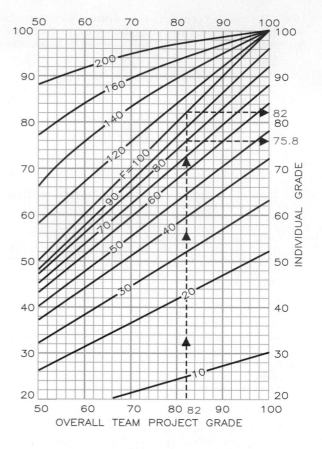

Figure A38–1   Grading graph.

SUBSTANCE	WEIGHT LB. PER CU. FT.	SPECIFIC GRAVITY	SUBSTANCE	WEIGHT LB. PER CU. FT.	SPECIFIC GRAVITY
METALS, ALLOYS, ORES			**TIMBER, U.S. SEASONED**		
Aluminum, cast, hammered	165	2.55–2.75	Moisture Content by Weight		
Brass, cast, rolled	534	8.4–8.7	Seasoned timber 15 to 20%		
Bronze, 7.9 to 14% Sn	509	7.4–8.9	Green timber up to 50%		
Bronze, aluminum	481	7.7	Ash, white, red	40	0.62–0.65
Copper, cast, rolled	556	8.8–9.0	Cedar, white, red	22	0.32–0.38
Copper ore, pyrites	262	4.1–4.3	Chestnut	41	0.66
Gold, cast, hammered	1205	19.25–19.3	Cypress	30	0.48
Iron, cast, pig	450	7.2	Fir, Douglas spruce	32	0.51
Iron, wrought	485	7.6–7.9	Fir, eastern	25	0.40
Iron, spiegel-eisen	466	7.5	Elm, white	45	0.72
Iron, ferro-silicon	437	6.7–7.3	Hemlock	29	0.42–0.52
Iron ore, hematite	325	5.2	Hickory	49	0.74–0.84
Iron ore, hematite in bank	160–180	—	Locust	46	0.73
Iron ore, hematite loose	130–160	—	Maple, hard	43	0.68
Iron ore, limonite	237	3.6–4.0	Maple, white	33	0.53
Iron ore, magnetite	315	4.9–5.2	Oak, chestnut	54	0.86
Iron slag	172	2.5–3.0	Oak, live	59	0.95
Lead	710	11.37	Oak, red, black	41	0.65
Lead ore, galena	465	7.3–7.6	Oak, white	46	0.74
Magnesium, alloys	112	1.74–1.83	Pine, Oregon	32	0.61
Manganese	475	7.2–8.0	Pine, red	30	0.48
Manganese ore, pyrolusite	259	3.7–4.6	Pine, white	26	0.41
Mercury	849	13.6	Pine, yellow, long-leaf	44	0.70
Monel Metal	556	8.8–9.0	Pine, yellow, short-leaf	38	0.61
Nickel	565	8.9–9.2	Poplar	30	0.48
Platinum, cast, hammered	1330	21.1–21.5	Redwood, California	26	0.42
Silver, cast, hammered	656	10.4–10.6	Spruce, white, black	27	0.40–0.46
Steel, rolled	490	7.85	Walnut, black	38	0.61
Tin, cast, hammered	459	7.2–7.5			
Tin ore, cassiterite	418	6.4–7.0			
Zinc, cast, rolled	440	6.9–7.2			
Zinc ore, blends	253	3.9–4.2	**VARIOUS LIQUIDS**		
			Alcohol, 100%	49	0.79
			Acids, muriatic 40%	75	1.20
VARIOUS SOLIDS			Acids, nitric 91%	94	1.50
			Acids, sulphuric 87%	112	1.80
Cereals, oats .... bulk	32	—	Lye, soda	106	1.70
Cereals, barley ...... bulk	39	—	Oils, vegetable	58	0.91–0.94
Cereals, corn, rye .. bulk	48	—	Oils, mineral, lubricants	57	0.90–0.93
Cereals, wheat ...... bulk	48	—	Water, 4°C. max, density	62.428	1.0
Hay and Straw .... bales	20	—	Water, 100°C.	59.830	0.9584
Cotton, Flax, Hemp	93	1.47–1.50	Water, ice	56	0.88–0.92
Fats	58	0.90–0.97	Water, snow, fresh fallen	8	.125
Flour, loose	28	0.40–0.50	Water, sea water	64	1.02–1.03
Flour, pressed	47	0.70–0.80			
Glass, common	156	2.40–2.60			
Glass, plate or crown	161	2.45–2.72	**GASES**		
Glass, crystal	184	2.90–3.00			
Leather	59	0.86–1.02	Air, 0°C. 760 mm.	.08071	1.0
Paper	58	0.70–1.15	Ammonia	.0478	0.5920
Potatoes, piled	42	—	Carbon dioxide	.1234	1.5291
Rubber, caostchouc	59	0.92–0.96	Carbon monoxide	.0781	0.9673
Rubber goods	94	1.0–2.0	Gas, illuminating	.028–.036	0.35–0.45
Salt, granulated, piled	48	—	Gas, natural	.038–.039	0.47–0.48
Saltpeter	67	—	Hydrogen	.00559	0.0693
Starch	96	1.53	Nitrogen	.0784	0.9714
Sulphur	125	1.93–2.07	Oxygen	.0892	1.1056
Wool	82	1.32			

The specific gravities of solids and liquids refer to water at 4°C., those of gases to air at 0°C. and 760 mm. pressure. The weights per cubic foot are derived from average specific gravities, except where stated that weights are for bulk, heaped or loose materials, etc.

(Courtesy of the American Institute of Steel Construction.)

Index